Texts in Computational Science and Engineering

24

This series contains graduate and undergraduate textbooks on topics described by the term "computational science and engineering". This includes theoretical aspects of scientific computing such as mathematical modeling, optimization methods, discretization techniques, multiscale approaches, fast solution algorithms, parallelization, and visualization methods as well as the application of these approaches throughout the disciplines of biology, chemistry, physics, engineering, earth sciences, and economics.

More information about this series at http://www.springer.com/series/5151

Francis X. Giraldo

An Introduction to Element-Based Galerkin Methods on Tensor-Product Bases

Analysis, Algorithms, and Applications

 Springer

Francis X. Giraldo
Department of Applied Mathematics
Naval Postgraduate School
Monterey, CA, USA

ISSN 1611-0994   ISSN 2197-179X   (electronic)
Texts in Computational Science and Engineering
ISBN 978-3-030-55071-4   ISBN 978-3-030-55069-1   (eBook)
https://doi.org/10.1007/978-3-030-55069-1

Mathematics Subject Classification (2010): 65M60, 65M70, 65Y20

This Springer imprint is published by the registered company Springer Nature Switzerland AG.
The registered company address is: Gewerbestrasse 11, 6330 Cham, Switzerland

This book is dedicated to my wife Jeanne and my daughter Gaby who are the pillars that I lean on. These two ladies deserve credit for allowing me to talk incessantly about Galerkin methods. No child should have to endure hearing about the finer points of Galerkin methods since she was able to talk; I am grateful nonetheless.

Preface

The focus of this book is on applying element-based Galerkin (EBG) methods to solve hyperbolic and elliptic equations with an emphasis on the resulting matrix-vector problems. Part I introduces the main topics of the book, followed by Part II which discusses one-dimensional problems. Part III treats multi-dimensional problems. In Part III, the ideas are primarily discussed for two dimensions but some concepts, such as the construction of the tensor-product basis functions, numerical integration, and metric terms, are extended to three dimensions. Part IV discusses advanced topics such as stabilization methods, adaptive mesh refinement, time-integration, and the hybridized discontinuous Galerkin (HDG) method. The contents of each part are described in more detail below and at the very end of this *preface* we include a discussion on how to use this book for teaching a course on this topic.

Because the basic building-blocks of EBG methods rely on interpolation and integration, Chs. 3 and 4 (Part II) cover these two topics in one dimension and Chs. 10 and 11 in multiple dimensions (Part III). These chapters rely on the theory of Jacobi polynomials which is covered in Appendix B.

The EBG methods discussed in this book include the continuous (CG) and discontinuous Galerkin (DG) methods. These methods are introduced for one-dimensional hyperbolic equations in Chs. 5 and 6 for explicit time-integration (Part II). We need to wait until Ch. 21 (Part IV) to discuss the hybridized discontinuous Galerkin (HDG) method because HDG only makes sense in the context of implicit time-integration, which is not introduced until Ch. 20. Chapter 7 is the heart of the manuscript where the idea of unified CG/DG methods is introduced; this chapter also presents the dissipation-dispersion analysis of both methods, and applications of unified CG/DG methods for systems of nonlinear partial differential equations including the shallow water and Euler equations. The application of CG and DG methods for one-dimensional problems is completed in Chs. 8 and 9 with a discussion on the application of these methods for elliptic equations.

Chapters 12, 13, and 14 (Part III) introduce CG, DG, and unified CG/DG methods in two dimensions for elliptic equations. Elliptic equations are handled first in order to focus on spatial discretization which, in multiple dimensions, requires a detailed discussion on the construction of metric terms (Ch. 12). After the basics of EBG methods in multiple dimensions are covered, we then move to a discussion of these methods for hyperbolic equations in Chs. 15, 16, and 17 for CG, DG, and unified CG/DG. These chapters discuss the efficient construction of CG and DG methods through complexity analysis and show how tensor-product bases allow such efficiencies. Furthermore, Ch. 17 extends further the heart of the book whereby the unified CG/DG approach presented in Ch. 7 for one-dimensional hyperbolic equations is extended to two (and multiple) dimensions.

Part IV covers advanced topics that, while important for the construction of industrial-type codes, is not strictly necessary for understanding the basics of EBG methods. Chapter 18 describes stabilization methods including filters, artificial dissipation (i.e., hyper-diffusion), Riemann solvers, limiters, and entropy-stable methods. Chapter 19 describes the three types of mesh refinement which are h-, p-, and r-refinement. Chapter 20 discusses explicit, implicit, implicit-explicit, semi-Lagrangian, and multirate time-integration methods. Part IV ends with a discussion of the hybridized discontinuous Galerkin method in one dimension in Ch. 21.

Let us now briefly discuss how one might use this book to teach a course on element-based Galerkin methods. Each of the chapters can be treated as lectures that build upon each other and so can be used to construct lectures that can be delivered sequentially. The author's book website contains a sample syllabus for a 10-week quarter-based course that can be easily extended to a 14-week semester-based course (more projects can be assigned and the advanced topics can be discussed in more detail); sample project assignments are also available on the book website. For a 10-week quarter-based course, I recommend assigning 4 projects. Project 1 treats interpolation and integration in one-dimension (Chs. 3 and 4). This then allows the students to tackle Project 2 which deals with solving a one-dimensional scalar hyperbolic equation (Chs. 5 and 6). I recommend having the students first write two different codes that treat CG and DG separately. Then they can write a unified code following the concepts presented in Ch. 7. Project 3 can then focus on solving a two-dimensional elliptic equation with CG and DG as presented in Chs. 12, 13, and 14. Project 4 would then consist of building a unified CG/DG code for solving a two-dimensional hyperbolic equation as presented in Chs. 15, 16, and 17. Assigning such a project is critical for the student to learn how to write efficient code with EBG methods since many of the optimization strategies described in the book can only be fully exploited for time-dependent problems in multi-dimensions (such as sum factorization and constructing the right-hand side vector without storing full matrices).

For a 14-week semester-based course, I recommend adding two more projects. The new Project 3 would consist of solving systems of equations as discussed in Ch. 7; this project is challenging and extremely helpful in preparing the student to tackle more interesting research problems. The new Project 4 consists of solving one-dimensional elliptic equations as described in Chs. 8 and 9; this project should be relatively simple for the student. Project 5 consists of solving two-dimensional elliptic equations and Project 6 focuses on solving two-dimensional hyperbolic equations. For ambitious students, I recommend a project on solving systems of two-dimensional equations (extension of Ch. 7); this project can be combined with evolving the equations forward in time using implicit methods as presented in Ch. 20.

Acknowledgements

Allow me to share my story of how I came upon *element-based Galerkin* methods. By telling this story, it will allow me to acknowledge many of the people that were instrumental in how the ideas in this book came about. As a PhD student at the University of Virginia, I had the remarkable opportunity to take courses in finite elements from both the Department of Applied Mathematics (under Lois Mansfield) and the Department of Mechanical and Aerospace Engineering (under Earl Thornton, my advisor). These two courses gave me the theoretical and practical foundations for both understanding the theory and giving me the insight into how to write finite element code. Because my PhD dissertation involved the construction of algorithms for handling shock waves, I decided on using the finite volume method (since I did not yet know anything about how to stabilize finite element methods). Those next four years working on finite volume methods (and adaptive mesh refinement) is what allowed me to put in my 10,000 hours on Godunov methods, dealing with issues such as Riemann solvers and limiters. This set the stage for me to understand discontinuous Galerkin methods.

Upon defending my PhD dissertation, I embarked on a postdoctoral fellowship working on finite element semi-Lagrangian methods for applications in geophysical fluid dynamics (under Beny Neta, Naval Postgraduate School). Up to this point, I had worked on both finite element and finite volume methods but only for low-order. All this changed when I visited Rutgers University sometime in 1996 where I had a chance meeting with Mohamed Iskandarani who was a researcher at the Institute for Marine and Coastal Sciences (under Dale Haidvogel). Mohamed showed me some results he had produced using something called the *spectral element method*. I was absolutely mesmerized by this idea and began to explore this method for my applications. Mohamed was very kind in answering my questions and I was able to write my first paper on spectral element methods (see [151]) [1]. For the next few years, I continued to work on spectral element methods, writing papers for both shallow water models (see [154, 161]) as well as the full hydrostatic equations (see [166, 156]). In early 2000, I began playing with the discontinuous Galerkin (DG) method looking at the papers of a student of my PhD advisor. The papers by Kim Bey allowed me to reproduce some DG results using modal basis functions. However, at this time, I was not sure how I could use high-order DG and construct an efficient model. In 1999, I was invited by George Karniadakis to visit the Division of Applied Mathematics at Brown University to give a talk on my 1998 paper on spectral element methods and semi-Lagrangian time-integration. Here, I met many young students that have since become superstars in their own fields (e.g., Dongbin Xu and Mike Kirby). During this visit, I also met Tim Warburton (a postdoc at the time) and Jan Hesthaven (an assistant professor at the time). At my talk at Brown, I made the comment that I wanted to switch my quadrilateral-based spectral element codes to triangle-based. This then led Jan to invite me the following year to work on a triangle-based spectral element method for my applications.

In the winter of 2001, I visited Brown University to work with Jan and Tim on triangle-based spectral element methods. It was during this time that we began to discuss the work on discontinuous Galerkin methods that Jan and Tim had been doing. Through discussions on a chalkboard in my temporary office at 182 George Street, they explained to me the essential ideas of the nodal discontinuous Galerkin method. Upon my return to Monterey, I began working on writing nodal DG models that were quadrilateral-based [2]. During this time, I began to see the similarities between *spectral element* and *discontinuous Galerkin* methods, especially when ideas such as *inexact* integration are used (what is now typically called the *discontinuous spectral element* method [146, 237]). Since I was working at the Naval Research Laboratory (in Monterey, California) at the time, we were mostly interested in developing atmospheric models on spherical domains. Therefore, I coded up a nodal DG shallow water model on the sphere. This resulted in the paper [159], which was one of the first papers in geophysical fluid dynamics using DG methods. This paper shows the pros and cons of using exact versus inexact integration and (similar to my paper [151]) showed that beyond a certain order, it makes little sense to use exact integration. This then opened the door to constructing efficient DG models.

Although I have benefitted greatly from many others along the way, one could say that these were my formative years in building my understanding of element-based Galerkin methods that led to this book. I have had the good fortune of meeting along the way many excellent students and postdocs. By working alongside such talented

---

[1] This paper analyzes the accuracy of the spectral element method comparing exact and inexact integration. In this paper, it is shown that beyond a certain order, there is no difference in accuracy between exact and inexact integration. This analysis is carried out for both Eulerian and Semi-Lagrangian time-integration.

[2] In years later, Tim Warburton and I worked on triangle-based spectral elements and discontinuous Galerkin methods. We published many good papers and developed code for shallow water equations on both the plane and sphere. However, since this manuscript only deals with quadrilateral-based methods, that work is not part of this manuscript.

young people, I have been motivated to improve the delivery of the lessons that I have learned in my Galerkin journey so that they can benefit. I have also been fortunate to have received funding from agencies such as the Office of Naval Research, the Air Force Office of Scientific Research, the National Science Foundation, the Department of Energy, and the Defense Advanced Research Projects Agency, which have allowed me the privilege to push my understanding of element-based Galerkin methods. Finally, we truly learn the intricacies of numerical methods when we are asked to build production-type codes. Through grants from the Office of Naval Research, I had the opportunity to test all of these methods in a real-world application (weather prediction models). In such a setting, one learns what trade-offs need to be made between accuracy and efficiency and these lessons are at the heart of this manuscript.

Finally, I would like to thank my many postdoctoral students who have motivated me to make this manuscript that much more clear because I wanted them to understand this material in order to work along side of me on our joint research projects involving element-based Galerkin methods. Some of the chapters in this book were motivated by the research of my students. For example, the chapter on adaptive mesh refinement could only have been possible due to the work of Michal Kopera. In addition, I was able to make the chapter on stabilization what it is only through the work of Simone Marras. Although I had already worked out how to combine both CG and DG within the same code-base, it was the work of Daniel Abdi that allowed us to figure out how best to combine CGc, CGd, and DG into the same piece of code which, in my opinion, is one of the unique aspects of this manuscript. Without their work, these chapters would never have been included in this manuscript. In chronological order, I would like to acknowledge my postdoctoral students who have given me feedback. They are: Jim Kelly, Shiva Gopalakrishnan, Michal Kopera, Andreas Müller, Simone Marras, Daniel Abdi, Maciek Waruszewski, Felipe Alves, Thomas Gibson, and Sohail Reddy, as well as my PhD student Patrick Mugg who has given me detailed feedback on the manuscript.

I would also like to thank Mark Taylor for discussing with me the intricacies of dispersion analysis for high-order methods. In fact, the dissipation-dispersion analysis in Ch. 7 is only possible due to our conversations during the 4-month long program on "Multiscale Numerics for the Atmosphere and Ocean" at the Newton Institute for Mathematical Sciences at Cambridge University from August through December of 2012. I think of those days with fondness because those four months allowed me the freedom to pursue many ideas that eventually went into this manuscript. During these four months, my postdoc Michal Kopera was able to figure out all of the intricacies of adaptive mesh refinement for continuous Galerkin methods. This was, in no small part, due to the discussions that Michal had with Paul Fischer and for those discussions we are both grateful to Paul. I enjoyed the circle of high-order methods *friends* that we formed at Cambridge.

Although time-integration is not central to this book, the chapter on time-integration is nonetheless an important addition. I would not have been able to write this chapter without the knowledge I have gained throughout the years working with Emil Constantinescu on these topics. Also, I would like to thank Carlos Borges for helping me find my way to linear algebra - this has reshaped my view of Galerkin methods.

I have very fond memories of writing the first few chapters of this book on a kitchen table in Phuket Thailand during the winter of 2007. You could say that Thailand was the birthplace of this text. I would like to thank my brother-in-law and his family for their hospitality in Thailand. All that excellent Thai food kept me going to write the foundation for this text.

I have always appreciated the unconditional love and support from my family (my brothers Lou and Al, my mom and dad - I wish my dad was still around to have been able to read this manuscript). My wife Jeanne has been the greatest writing coach I could have asked for and I appreciate my daughter Gaby for checking the figures for me.

I would also like to thank Martin Peters (and the staff at Springer) who encouraged me to publish this manuscript and were patient with me during the ten years that I worked on this manuscript since I showed them the first draft of this book. I remember vividly our first conversation regarding the book during a sunny day in June 2010 at the Institute for Pure and Applied Mathematics at UCLA.

Finally, I would also like to thank my colleagues who have urged me to publish this manuscript; some have used my class notes in courses at Cornell University (Pete Diamessis), the Indian Institute of Technology in Bombay India (Shiva Gopalakrishnan), and the New Jersey Institute of Technology (Simone Marras). The sections on discontinuous Galerkin methods were streamlined and used during the 2012 Gene Golub SIAM Summer School (G2S3) on "Simulation and Scientific Computing in the Geosciences" held in Monterey, California from July 29 through August 10. I was happy to see that these notes were well received and am always happy to meet one of my former G2S3 students at conferences to hear that they have begun to use element-based Galerkin methods for their research. Hearing this sort of feedback is why I continue to work in this field.

Pacific Grove, California   Frank Giraldo
June 2020

Contents

Part IV Advanced Topics

Part I consists of the two chapters 1 and 2. Chapter 1 introduces the need for numerical methods for solving complex systems of nonlinear equations. The focus of this chapter is the Lax Equivalence Theorem. In Ch. 2 we give a brief summary of existing numerical methods that treat the governing equations in differential form (i.e., finite difference methods) and in integral form (Galerkin methods).

Chapter 1
Motivation and Background

1.1 Introduction

This book deals with obtaining solutions to partial differential equations using computers. There are a number of books that also cover this topic, most notably Canuto et al. [66], Deville, Fischer, and Mund [105], Karniadakis and Sherwin [219], Kopriva [236], Hesthaven and Warburton [200], and Pozrikidis [306]. From this short list it becomes obvious that this topic is by no means easy to cover in one text. This book approaches the topic primarily from a practical viewpoint although it covers sufficient theory in order for the reader to understand how to develop rigorously designed numerical methods. The partial differential equations described in this book include hyperbolic and elliptic in one and two dimensions, and to some extent three dimensions; we do not cover parabolic equations because they can be constructed using the time-dependence of hyperbolic equations with the second order terms of the elliptic equations. Furthermore, this book also covers important implementation issues that will allow the reader to learn how to build algorithms; sample code (in Matlab and Julia) for the various projects used throughout the text can be found on the author's Github page[1]. Finally, the goal of this book is to describe a class of methods that can most aptly be described as *element-based Galerkin* (EBG) methods that include finite elements, finite volume, spectral element, and discontinuous Galerkin[2] methods and how they can be used to discretize partial differential equations.

---

[1] Most of the relevant algorithms described in this text can be found at https://github.com/fxgiraldo/Element-based-Galerkin-Methods using both Matlab and Julia. However more advanced compute-kernels for 1D, 2D, and 3D for advection, shallow water, Euler, and the compressible Navier-Stokes equations can be found at https://github.com/fxgiraldo/Canary.jl/tree/compute-kernels/examples using Julia.

[2] Finite volume and discontinuous Galerkin methods are very much related and are usually cast under the umbrella of Godunov methods [170].

F. X. Giraldo, *An Introduction to Element-Based Galerkin Methods on Tensor-Product Bases*, Texts in Computational Science and Engineering 24, https://doi.org/10.1007/978-3-030-55069-1_1

Partial differential equations are used to approximate the physical laws of nature and for this reason they are of vast importance. Examples of such physical laws include aerodynamics which is important for air travel, and ocean and atmospheric dynamics which are important in weather and climate studies, to name only a few. In order to solve any of these physical laws requires us to first define a well-posed continuous mathematical problem. This means that we need to select the proper partial differential equation that governs the physical law that we are interested in, along with proper initial and boundary data that makes the problem well-posed; we discuss *well-posedness* in Sec. 1.5.2. Once we have defined the well-posed continuous problem, we then need to select a numerical method with the following properties:

1. it must be consistent with the continuous problem;
2. it must be stable (if it is consistent and stable, then it will converge to the correct solution);
3. it must be as accurate as possible;
4. it must be efficient;
5. it should be geometrically flexible.

The first two conditions are absolutely necessary, otherwise the numerical solution will be completely useless. Due to the importance of these two conditions, we address them immediately in this chapter. However, the remaining three conditions are really more a wishlist of what we would like to see in an *optimal* numerical method and are addressed throughout this text.

Clearly, if we are given a choice between a solution of low accuracy or one of high accuracy, then we would select the higher accuracy solution but this does not make the low accuracy solution useless, just not as desirable. Depending on the application, efficiency can be an absolute necessity or not. For example, if a researcher is interested in finding the best answer to one specific question, then it makes sense to spend a lot of computational time on the problem until such a solution is obtained. For such a problem, efficiency is unnecessary. On the other hand, if one is building a model that will be run numerous times daily (as in weather prediction) then efficiency is an absolute necessity. For industrial-type models (models that are run often for specific information), efficiency more than accuracy may be one of the most important requirements. It should be mentioned that the efficiency of a particular model (and its underlying numerical methods) is very much dependent on the computer architecture being used. For example, running a model on a shared memory computer (imagine a computer with 4 processors but only one very large memory that can be accessed by all 4 processors) versus a distributed memory computer (where each of the 4 processors have their own memory that cannot be accessed directly by the other processors except through a network communication via message passing) then one would have to consider using different numerical methods. As a specific example, a global method is better suited for use on shared memory whereas local methods are perfectly made for distributed memory. We discuss efficiency continuously throughout the text; however, we do not discuss specific computer hardware since this is a fast changing topic.

Finally, geometric flexibility refers to how flexible the method is in representing various computational geometries. As an example, consider constructing a grid covering the Indian Ocean using only quadrilaterals comprising an orthogonal grid (because some numerical methods require the grid to be orthogonal). If one has a method that does not require grid orthogonality this vastly simplifies the construction of the grid. Simplifying the construction of the grid will allow a researcher much freedom in the kinds of problems that they can solve. In this text, we focus on tensor-product approaches (i.e., quadrilateral grids in two dimensions and hexahedral grids in three dimensions). To show the flexibility of unstructured grids even when using quadrilaterals we show in Fig. 1.1 a grid generated for the Indian Ocean using the freeware software GMSH[3].

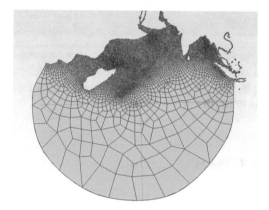

Fig. 1.1: A representation of the Indian Ocean using an unstructured quadrilateral grid.

Another reason for insisting on geometric flexibility is that this will ultimately facilitate the use of adaptive grid techniques; this is an approach by which the grid varies with the time-dependent problem. In Fig. 1.2 a discontinuous Galerkin solution of the compressible Navier-Stokes equations is illustrated using adaptive quadrilateral elements. This simulation captures the deformation of a rising thermal bubble [230]. The left part of the figure shows the adaptive mesh solution (with the corresponding grid) and the right part shows the uniform mesh solution. Note that both solutions look identical even though the adaptive solution has far fewer degrees of freedom. In fact, the adaptive solution for this problem is five times faster than the uniform mesh solution.

In essence, this text aims to address the five conditions listed above. Let us now begin to address each of these conditions in detail. We begin with the most important part of a problem, i.e., the continuous equations.

---

[3] GMSH can be found at **http://gmsh.info**.

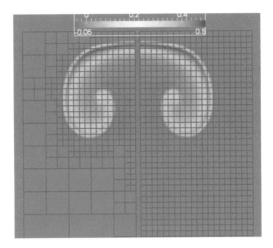

Fig. 1.2: The adaptive grid and contours of potential temperature for a rising thermal bubble using a quadrilateral-based discontinuous Galerkin compressible Navier-Stokes model. The left part of the figure shows the adaptive mesh solution and the right part shows the uniform mesh solution.

1.2 Continuous Governing Equations

The obvious question that the reader will ask is *why do we need numerical methods at all?* Let us try to answer this question by looking at the following system of nonlinear equations

$$\frac{\partial \rho}{\partial t} + \nabla \cdot (\rho \mathbf{u}) = 0$$

$$\frac{\partial \rho \mathbf{u}}{\partial t} + \nabla \cdot (\rho \mathbf{u} \otimes \mathbf{u} + P \mathcal{I}_2) = -\rho g \mathbf{k} + \nabla \cdot \mathbf{F}_{\mathbf{u}}^{\mathrm{visc}}$$

$$\frac{\partial \rho e}{\partial t} + \nabla \cdot [(\rho e + P)\mathbf{u}] = \nabla \cdot \mathbf{F}_{e}^{\mathrm{visc}} \qquad (1.1)$$

where the conserved variables are $(\rho, \rho \mathbf{u}^{T}, \rho e)^{T}$, ρ is the density, $\mathbf{u} = (u, w)^{T}$ is the velocity field, $e = c_v T + \frac{1}{2}\mathbf{u} \cdot \mathbf{u} + gz$ is the total energy, and ^{T} denotes the transpose operator. The pressure P is obtained by the equation of state which, in terms of the solution variables, is written as

$$P = \frac{R}{c_v}\rho \left(e - \frac{1}{2}\mathbf{u} \cdot \mathbf{u} - gz \right).$$

The viscous fluxes $\mathbf{F}^{\mathrm{visc}}$ are defined as follows:

$$\mathbf{F}_{\mathbf{u}}^{\mathrm{visc}} = \mu \left[\nabla \mathbf{u} + (\nabla \mathbf{u})^{T} + \lambda (\nabla \cdot \mathbf{u}) \mathcal{I}_2 \right] \qquad (1.2)$$

and

$$\mathbf{F}_e^{\text{visc}} = \mathbf{u} \cdot \mathbf{F}_{\mathbf{u}}^{\text{visc}} + \frac{\mu c_p}{\text{Pr}} \nabla T \tag{1.3}$$

where $\gamma = \frac{c_p}{c_v}$ is the specific heat ratio, R is the gas constant, c_p is the specific heat with constant pressure, c_v is the specific heat with constant volume, g is the gravitational constant, I_2 is the rank-2 identity matrix, and z is the vertical height. This set of equations is known as the compressible *Navier-Stokes equations* and they govern the motion of a viscous ($\mu \neq 0$) or inviscid ($\mu = 0$) fluid. This set of equations is important because they are the equations that govern the motion of fluids such as air (compressible) and water (incompressible) that are of great importance in aerodynamics, Newtonian fluid flow, oceanography, and meteorology, to name only a few. The importance of these equations for our daily safety and knowledge are clearly obvious.

1.3 Analytic Tools for Deriving Solutions to the Continuous Problem

In order to make the continuous equations useful one must be able to extract solutions from them. Because these equations are nonlinear and are a system (i.e., not a scalar equation such as the classical wave, heat, or Laplace equations), there is no obvious way to solve them analytically. To see this, just think back to your last methods of applied mathematics course (or your favorite partial differential equations course) and see if there is any method that you have learned that will allow you to solve these equations (e.g., why wouldn't separation of variables work? How difficult would it be to solve this system using the method of characteristics?). Let me save you some time - there are no such analytic tools, as of yet, to solve these equations analytically. In fact, the *Clay Mathematics Institute* in Cambridge MA has this equation set as one of its seven millennium prize problems with a bounty of $US 1 million for a proof on the existence of solutions to the incompressible form of this problem[4].

1.4 Basic Mathematical Concepts

Since solving most PDEs of interest cannot be done analytically, we need to understand how to solve them numerically. Before we move to discussing computational tools for obtaining solutions to the discrete problem, let us introduce some basic mathematical concepts that are used throughout the text.

---

[4] The precise definition of the problem can be found at
https://www.claymath.org/sites/default/files/navierstokes.pdf.

1.4.1 Taylor Series

The Taylor expansion of the function $f(x)$ about the point a is written as

$$f(x) = \sum_{n=0}^{\infty} \frac{f^{(n)}(a)}{n!}(x-a)^n \tag{1.4}$$

where $!$ is the factorial and $f^{(n)}$ denotes the nth derivative of the function f. The first three terms of Eq. (1.4) are

$$f(x) = f(a) + \Delta x \frac{\partial f}{\partial x}(a) + \frac{\Delta x^2}{2}\frac{\partial^2 f}{\partial x^2}(a) + O(\Delta x^3)$$

where $\Delta x = x - a$. Taylor series appear in this text when we either check for consistency or show the order of accuracy of the numerical method.

1.4.2 Product Rule of Differentiation

The *product rule* of differentiation for two functions u and v is defined as follows

$$\nabla(uv) = u\nabla v + v\nabla u. \tag{1.5}$$

The product rule arises in this text, e.g., when we discuss the weak integral form of the Galerkin method.

1.4.3 Fundamental Theorem of Calculus

The *fundamental theorem of calculus* states the following

$$\int_a^b \frac{df}{dx}dx = f(b) - f(a), \tag{1.6}$$

i.e., the integral of the derivative of a function is its definite integral. The fundamental theorem of calculus arises throughout the text when we simplify terms such as

$$\int_\Omega \frac{df}{dx}d\Omega = [f]_\Gamma$$

whereby Ω represents the domain of integration, Γ is its boundary, and $[\cdot]$ denotes a difference as defined by the right-hand side of Eq. (1.6). We use this concept to introduce boundary conditions at element interfaces and the physical boundary. In multi-dimensions, we use the *divergence theorem* which states the following

$$\int_{\Omega} \nabla \cdot \mathbf{f} \, d\Omega = \int_{\Gamma} \hat{\mathbf{n}} \cdot \mathbf{f} \, d\Gamma$$

which relates the area (in two dimensions) or volume (in three dimensions) defined by Ω to the flux (information that is entering or exiting the volume) integral through the boundary of Ω defined by Γ where the boundary has a normal vector $\hat{\mathbf{n}}$. Note that ∇ is the multi-dimensional differentiation operator (also called a gradient operator) and $\nabla\cdot$ represents the divergence operator.

1.4.4 Integration by Parts

Let us define the product rule for two functions u and $\mathbf{f}$, where u is a scalar function and $\mathbf{f}$ is a vector function, as follows

$$\nabla \cdot (u\mathbf{f}) = u\nabla \cdot \mathbf{f} + \nabla u \cdot \mathbf{f}. \tag{1.7}$$

Integrating Eq. (1.7) yields

$$\int_{\Omega} \nabla \cdot (u\mathbf{f}) \, d\Omega = \int_{\Omega} (u\nabla \cdot \mathbf{f} + \nabla u \cdot \mathbf{f}) \, d\Omega, \tag{1.8}$$

which, after using the divergence theorem, yields

$$\int_{\Gamma} \hat{\mathbf{n}} \cdot (u\mathbf{f}) \, d\Omega = \int_{\Omega} (u\nabla \cdot \mathbf{f} + \nabla u \cdot \mathbf{f}) \, d\Omega, \tag{1.9}$$

which is what we refer to as *integration by parts* in this text.

1.4.5 Green's First Identity

The idea of integration by parts can also be applied to second order operators as follows

$$\nabla \cdot (u\nabla v) = u\nabla^2 v + \nabla u \cdot \nabla v \tag{1.10}$$

for two scalar functions u and v. Integrating Eq. (1.10) and using the divergence theorem yields

$$\int_{\Gamma} \hat{\mathbf{n}} \cdot (u\nabla v) \, d\Gamma = \int_{\Omega} \left(u\nabla^2 v + \nabla u \cdot \nabla v \right) d\Omega \tag{1.11}$$

which is known as *Green's first identity* and arises when we discuss elliptic operators.

1.5 Computational Tools for Obtaining Solutions to the Discrete Problem

1.5.1 From Continuous to Discrete Formulations

At the moment, the only way of extracting useful solutions to general systems of nonlinear partial differential equations is to use numerical techniques. The idea is to replace the continuous equations into an equivalent discrete form and then solve this form numerically.

The simplest example is the one-dimensional linear wave equation

$$\frac{\partial q}{\partial t} + u\frac{\partial q}{\partial x} = 0. \tag{1.12}$$

Clearly this equation is in continuous form and has the analytic solution $q(x,t) = f(x - ut)$ (via the method of characteristics) where $f(x) = q(x,0)$ is the initial condition to the problem. This type of problem is known as a *Cauchy* or *initial value* problem because the solution for all time is determined completely from the initial conditions. Assuming that we did not know the analytic solution, then the next best way to solve this problem is via numerical methods. This means that the continuous equation must be replaced by a discrete form that can be solved numerically on a computer. We *discretize* the equation by replacing the continuous derivatives by discrete approximations in such a manner

$$\frac{q_i^{n+1} - q_i^n}{\Delta t} + u\frac{q_i^n - q_{i-1}^n}{\Delta x} = 0 \tag{1.13}$$

where n is the time-level and i is the grid point. Eq. (1.13) is in fact a first order approximation in both time and space. For $u > 0$ this discretization is called an upwind finite difference method and is one of the most stable methods for handling strong gradients (such as shock waves). It is called *upwinding* because for $u > 0$ the flow in Fig. 1.3 is moving left to right and so the advection term $u\frac{\partial q}{\partial x}$ at the point i is constructed in the direction favored by the flow; that is, the point i and $i - 1$ are upwind from $i + 1$. The point $i + 1$ should not matter in the solution of q_i since it is not felt by the flow. This is an example of using physical understanding of the problem to solve it numerically and is at the heart of many successful methods that are covered in Ch. 18.

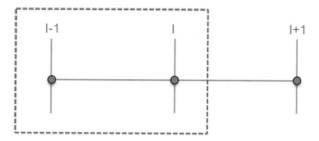

Fig. 1.3: The finite difference stencil for the first order upwind method (for $u > 0$ and moving from left to right) where the dotted (red) box denotes the differencing stencil.

1.5.2 Validity of the Numerical Solution and the Lax Equivalence Theorem

Computational scientists must know whether their numerical solution is a valid solution to a well-posed problem. Let us now introduce the following definition of *well-posedness*:

Definition 1.1 Well-Posedness: A problem is well-posed, in the Hadamard sense [188], if the partial differential equation with corresponding initial and boundary conditions satisfies the following three conditions:

1. a solution exists,
2. the solution is unique, and
3. the solution depends continuously on the initial and boundary data.

The fact that the continuous problem is well-posed and therefore, we hope, has a unique solution, tells us that we ought to be able to derive a unique numerical solution as long as the following theorem is satisfied:

Theorem 1.1 Lax Equivalence Theorem *[248]: A numerical method used to solve a well-posed problem converges to the correct solution if the method is consistent with the partial differential equation and if the method is stable. Hence, consistency + stability = convergence.*

Consistency is generally the easier condition to prove whereas stability is more difficult. Let us first discuss consistency.

1.5.3 Consistency of a Numerical Method

Looking at Eq. (1.13) we can clearly see that we have replaced the continuous equation, Eq. (1.12), with a discrete approximation. However, such approximations used must be based on rigorous theoretical arguments. In this case, we have used the finite difference method that is based on the mathematics of Taylor series which is discussed in detail in Ch. 3. The condition of *consistency* requires that in the limit as the discrete time-step, Δt, and the grid spacing, Δx, both go to zero that we recover the continuous equations; this implies that the numerical problem we are solving is in fact the same as the original continuous problem[5]. For the specific example given in Eq. (1.13) we can show that this condition is satisfied by using a Taylor series expansion about the state $q(x, t)$ or, in discrete form, q_i^n. Applying such an expansion to the terms in Eq. (1.13) yields

$$q_i^n \equiv q(x, t)$$

$$q_i^{n+1} \equiv q(x, t + \Delta t) = q(x, t) + \Delta t \frac{\partial q(x, t)}{\partial t} + O(\Delta t^2)$$

$$q_{i-1}^n \equiv q(x - \Delta x, t) = q(x, t) - \Delta x \frac{\partial q(x, t)}{\partial x} + O(\Delta x^2),$$

where, for example, $O(\Delta x^2)$ represents the fact that we are lumping all terms of order Δx^2 and greater in this *big O* notation (O). Substituting these expressions into Eq. (1.13) yields

$$\frac{q(x, t) + \Delta t \frac{\partial q(x,t)}{\partial t} + O(\Delta t^2) - q(x, t)}{\Delta t} + u \frac{q(x, t) - q(x, t) + \Delta x \frac{\partial q(x,t)}{\partial x} + O(\Delta x^2)}{\Delta x} = 0.$$

Simplifying this expression yields

$$\frac{\partial q(x, t)}{\partial t} + u \frac{\partial q(x, t)}{\partial x} + O(\Delta t, \Delta x) = 0$$

which tells us two things:

1. the method is first order in both time and space, and
2. as $(\Delta t, \Delta x) \rightarrow 0$ we recover the continuous partial differential equation and therefore the method is consistent.

Let us now look at a generalized discretization in order to see what conditions must be met to ensure that a method is indeed consistent with the underlying partial differential equations (PDEs). Let us write this generalized discretization as follows

$$\sum_{k=-\frac{K}{2}}^{\frac{K}{2}} \frac{\alpha_k q_i^{n+k}}{\Delta t} + u \sum_{k=-\frac{N}{2}}^{\frac{N}{2}} \frac{\beta_k q_{i+k}^n}{\Delta x} = 0 \qquad (1.14)$$

where K and N denote the orders of accuracy in time and space, respectively. Making this general form slightly simpler, let us only look at the following form (for $K = N = 2$)

---

[5] If consistency is not satisfied then it means that the numerical approximation of the continuous problem has changed the original continuous problem.

$$\frac{\alpha_1 q_i^{n+1} + \alpha_0 q_i^n + \alpha_{-1} q_i^{n-1}}{\Delta t} + u \frac{\beta_1 q_{i+1}^n + \beta_0 q_i^n + \beta_{-1} q_{i-1}^n}{\Delta x} = 0. \quad (1.15)$$

Expanding each term using Taylor series yields

$$q_i^{n+1} \equiv q(x, t + \Delta t) = q(x,t) + \Delta t \frac{\partial q(x,t)}{\partial t} + O(\Delta t^2)$$

$$q_i^{n-1} \equiv q(x, t - \Delta t) = q(x,t) - \Delta t \frac{\partial q(x,t)}{\partial t} + O(\Delta t^2)$$

$$q_{i+1}^n \equiv q(x + \Delta x, t) = q(x,t) + \Delta x \frac{\partial q(x,t)}{\partial x} + \frac{\Delta x^2}{2} \frac{\partial^2 q(x,t)}{\partial x^2} + O(\Delta x^3)$$

$$q_i^n \equiv q(x,t)$$

$$q_{i-1}^n \equiv q(x - \Delta x, t) = q(x,t) - \Delta x \frac{\partial q(x,t)}{\partial x} + \frac{\Delta x^2}{2} \frac{\partial^2 q(x,t)}{\partial x^2} - O(\Delta x^3)$$

that, after substituting into Eq. (1.15) yields

$$\frac{\alpha_1 \left(q(x,t) + \Delta t \frac{\partial q(x,t)}{\partial t} + O(\Delta t^2) \right) + \alpha_0 q(x,t) + \alpha_{-1} \left(q(x,t) - \Delta t \frac{\partial q(x,t)}{\partial t} + O(\Delta t^2) \right)}{\Delta t}$$

$$+ u \frac{\beta_1 \left(q(x,t) + \Delta x \frac{\partial q(x,t)}{\partial x} + \frac{\Delta x^2}{2} \frac{\partial^2 q(x,t)}{\partial x^2} + O(\Delta x^3) \right)}{\Delta x} \quad (1.16)$$

$$+ u \frac{\beta_0 q(x,t) + \beta_{-1} \left(q(x,t) - \Delta x \frac{\partial q(x,t)}{\partial x} + \frac{\Delta x^2}{2} \frac{\partial^2 q(x,t)}{\partial x^2} - O(\Delta x^3) \right)}{\Delta x} = 0.$$

Looking at the time-derivative term in Eq. (1.16) we can see that in order to recover the term $\frac{\partial q(x,t)}{\partial t}$ requires that

$$\alpha_1 + \alpha_0 + \alpha_{-1} = 0 \quad \text{and} \quad \alpha_1 - \alpha_{-1} = 1$$

which means that we have a free parameter since there are only two equations but three unknowns. If we let $\alpha_{-1} = 0$ then we determine that $\alpha_1 = 1$ and $\alpha_0 = -1$. Imposing these conditions allows us to eliminate the terms $q(x,t)$ from the time-derivative term leaving the following expression

$$\frac{\partial q(x,t)}{\partial t} + O(\Delta t)$$

$$+ u \frac{\beta_1 \left(q(x,t) + \Delta x \frac{\partial q(x,t)}{\partial x} + \frac{\Delta x^2}{2} \frac{\partial^2 q(x,t)}{\partial x^2} + O(\Delta x^3) \right)}{\Delta x} \quad (1.17)$$

$$+ u \frac{\beta_0 q(x,t) + \beta_{-1} \left(q(x,t) - \Delta x \frac{\partial q(x,t)}{\partial x} + \frac{\Delta x^2}{2} \frac{\partial^2 q(x,t)}{\partial x^2} - O(\Delta x^3) \right)}{\Delta x} = 0.$$

We could have used other combinations of α_k but we chose the values for simplicity since this text primarily deals with the spatial derivatives.

Next, we note that to eliminate the term $q(x, t)$ from the spatial derivative requires that $\beta_1 + \beta_0 + \beta_{-1} = 0$ which then yields

$$
\begin{aligned}
& \frac{\partial q(x,t)}{\partial t} + O(\Delta t) \\
& + u\beta_1 \left(\frac{\partial q(x,t)}{\partial x} + \frac{\Delta x}{2} \frac{\partial^2 q(x,t)}{\partial x^2} + O(\Delta x^2) \right) \\
& + u\beta_{-1} \left(-\frac{\partial q(x,t)}{\partial x} + \frac{\Delta x}{2} \frac{\partial^2 q(x,t)}{\partial x^2} - O(\Delta x^2) \right) = 0
\end{aligned} \tag{1.18}
$$

where we require $\beta_1 - \beta_{-1} = 1$ in order to retain the first spatial derivative. Using this condition allows us to simplify Eq. (1.18) to the following expression

$$
\begin{aligned}
& \frac{\partial q(x,t)}{\partial t} + O(\Delta t) \\
& + u \frac{\partial q(x,t)}{\partial x} + u\,(\beta_1 + \beta_{-1})\, \frac{\Delta x}{2} \frac{\partial^2 q(x,t)}{\partial x^2} + O(\Delta x^2) = 0.
\end{aligned} \tag{1.19}
$$

We now have two conditions for the spatial derivative coefficients: $\beta_1 + \beta_0 + \beta_{-1} = 0$ and $\beta_1 - \beta_{-1} = 1$ that, when combined, gives $2\beta_1 + \beta_0 = 1$. Let us take a closer look at what these two conditions tell us.

Case I

Let us pick $\beta_0 = 1$ that yields $\beta_1 = 0$ and $\beta_{-1} = -1$ allowing us to write Eq. (1.19) as follows

$$
\begin{aligned}
& \frac{\partial q(x,t)}{\partial t} + O(\Delta t) \\
& + u \frac{\partial q(x,t)}{\partial x} - O(\Delta x) = 0.
\end{aligned} \tag{1.20}
$$

Substituting $\alpha_1 = 1$, $\alpha_0 = -1$, $\alpha_{-1} = 0$ and $\beta_1 = 0$, $\beta_0 = 1$, $\beta_{-1} = -1$ into Eq. (1.15) yields

$$
\frac{q_i^{n+1} - q_i^n}{\Delta t} + u \frac{q_i^n - q_{i-1}^n}{\Delta x} = 0 \tag{1.21}
$$

which is in fact the first-order upwind scheme given in Eq. (1.13).

Case II

Let us pick $\beta_0 = 0$ that yields $\beta_1 = \frac{1}{2}$ and $\beta_{-1} = -\frac{1}{2}$ allowing us to write Eq. (1.19) as follows

$$\frac{\partial q(x,t)}{\partial t} + O(\Delta t)$$
$$+ u\frac{\partial q(x,t)}{\partial x} + O(\Delta x^2) = 0. \tag{1.22}$$

This method is first order in time but second order in space. Substituting $\alpha_1 = 1$, $\alpha_0 = -1$, $\alpha_{-1} = 0$ and $\beta_1 = \frac{1}{2}$, $\beta_0 = 0$, $\beta_{-1} = -\frac{1}{2}$ into Eq. (1.15) yields

$$\frac{q_i^{n+1} - q_i^n}{\Delta t} + u\frac{q_{i+1}^n - q_{i-1}^n}{2\Delta x} = 0. \tag{1.23}$$

Case III

Finally, let us choose $\beta_0 = -1$ that gives $\beta_1 = 1$ and $\beta_{-1} = 0$ allowing us to write Eq. (1.19) as follows

$$\frac{\partial q(x,t)}{\partial t} + O(\Delta t)$$
$$+ u\frac{\partial q(x,t)}{\partial x} + O(\Delta x) = 0. \tag{1.24}$$

This method is first order in time and space but is a downwind scheme rather than an upwind scheme (Eq. (1.13)). Substituting $\alpha_1 = 1$, $\alpha_0 = -1$, $\alpha_{-1} = 0$ and $\beta_1 = 1$, $\beta_0 = -1$, $\beta_{-1} = 0$ into Eq. (1.15) yields

$$\frac{q_i^{n+1} - q_i^n}{\Delta t} + u\frac{q_{i+1}^n - q_i^n}{\Delta x} = 0 \tag{1.25}$$

and shows that the spatial derivative is still first order but now the points $i+1$ and i are used which, if the flow is traveling from left to right, then the discrete approximation of the spatial derivative does not match the physical situation and is thereby not optimal. This idea is discussed in more depth when we deal with Riemann solvers in Ch. 18.

In summary, if we are not careful in selecting the coefficients α_k and β_k then we would not be guaranteed to enforce the numerical method to be consistent with the underlying continuous problem. However, if one always selects approximations to derivatives based on rigorous mathematics (such as Taylor series) then one is always guaranteed to construct consistent numerical methods.

Remark 1.1 The method we have just used to check for consistency (and order of accuracy) for the finite difference method is also valid for Galerkin methods. When applying this analysis to Galerkin methods, we first obtain the associated difference equation and then apply this analysis.

Throughout the text, we always derive (finite) difference representations of the element-based Galerkin formulation in order to glean insight into the discretization.

1.5.4 Stability of a Numerical Method: Von Neumann Stability Analysis

Before discussing the stability properties of a numerical method we must first introduce the concept of Fourier components: let

$$q_j^n = \sum_{l=-N}^{N} \tilde{q}_l^{(n)} e^{ik_l j \Delta x} \tag{1.26}$$

be the *Fourier series* representation of the variable q at the grid point j at time level n; on the right-hand side of the equations, the superscript (n) denotes an exponential. In Eq. (1.26) the quantities $\tilde{q}_l$ are the amplitudes of the wave decomposition of the physical variable q, $i = \sqrt{-1}$, k_l are the wave numbers, the product $k_l \Delta x$ is often times written as ϕ_l and denotes the phase angle which for $\phi = 0$ denotes the low frequency (large amplitude waves) and $\phi = \pi$ are the high frequency (small amplitude waves). The variable N denotes the number of points in the global domain where we are solving the PDE.

To simplify the problem let us take just one Fourier mode

$$q_j^n = \tilde{q}^{(n)} e^{ij\phi}. \tag{1.27}$$

To maintain *stability*, we must ensure that the amplitude of this Fourier mode remains bounded, that is, we cannot have it growing for all time. This means that the magnitude of these amplitudes must be less than or equal to one, i.e., $|\tilde{q}| \leq 1$. With this in mind, let us now revisit the three cases we discussed briefly for discretizing Eq. (1.12).

Case I

Let us now substitute this single *Fourier mode* into Eq. (1.21) which yields

$$\tilde{q}^{(n+1)} e^{ij\phi} = \tilde{q}^{(n)} e^{ij\phi} - \sigma \left(\tilde{q}^{(n)} e^{ij\phi} - \tilde{q}^{(n)} e^{i(j-1)\phi} \right) \tag{1.28}$$

where $\sigma = \frac{u \Delta t}{\Delta x}$ is a non-dimensional parameter that measures how fast information propagates across grid points and is called the *Courant* number. Dividing Eq. (1.28) by $\tilde{q}^{(n)} e^{ij\phi}$ yields

$$\tilde{q} = 1 - \sigma(1 - e^{-i\phi})$$

which, after substituting Euler's formula, $e^{\pm i\phi} = \cos \phi \pm i \sin \phi$, yields

$$\tilde{q} = 1 - \sigma(1 - \cos \phi) - i\sigma \sin \phi.$$

Note that $Re(\tilde{q}) = (1 - \sigma) + \sigma \cos \phi$ and $Im(\tilde{q}) = -\sigma \sin \phi$ which, in the Re-Im (real-imaginary, or complex) plane, represents a circle centered at $(1 - \sigma, 0)$ with radius σ (recall polar coordinates). Figure 1.4 shows the stability plot in the Re-Im plane for σ values $\{0.25, 0.5, 0.75, 1.0\}$. Note that all the dashed blue circles remain

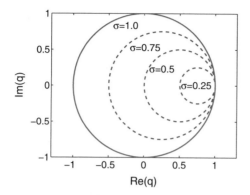

Fig. 1.4: Case I. The unit circle (solid red line) and the amplitude q for several values of the Courant number (dashed blue lines), σ, for the first order upwind method (with explicit first order time-integration).

inside the unit circle (solid red line) if and only if $0 \leq \sigma \leq 1$.

Before continuing, let us say something about the physical significance of $\tilde{q}$; this is nothing more than the amplitude of the solution. Clearly, we want this amplitude to be less than or equal to 1 because if it is greater than one, then as we increase its exponent by the value n, it will make this amplitude grow without bound. As $n \rightarrow \infty$, the amplitude will also go to infinity meaning that the solution is unbounded as time increases (i.e., it is unstable); therefore we want $\tilde{q}$ to be less than or equal to one. The condition that

$$\sigma \equiv \frac{u \Delta t}{\Delta x} \leq 1$$

is known as the *Courants-Friedrichs-Lewy* (CFL) condition [96] and is the statement of conditional stability for this particular numerical method.

Case II

Let us now apply the stability analysis to Eq. (1.23)

$$\frac{q_i^{n+1} - q_i^n}{\Delta t} + u \frac{q_{i+1}^n - q_{i-1}^n}{2 \Delta x} + O(\Delta t, \Delta x^2) = 0 \qquad (1.29)$$

which we write in the following form

$$q_i^{n+1} = q_i^n - \frac{\sigma}{2}(q_{i+1}^n - q_{i-1}^n).$$

Next, substituting the Fourier mode in Eq. (1.27) yields

$$\tilde{q} = 1 - \frac{\sigma}{2}(e^{i\phi} - e^{-i\phi}).$$

Introducing *Euler's formula* yields

$$\tilde{q} = 1 - \frac{\sigma}{2}(\cos\phi + i\sin\phi - \cos\phi + i\sin\phi)$$

which, after simplifying, yields

$$\tilde{q} = 1 - i\sigma\sin\phi$$

and results in $Re(\tilde{q}) = 1$ and $Im(\tilde{q}) = -\sigma\sin\phi$. In the Re-Im plane, this represents a line centered at $(1, 0)$ and extending vertically from $(1, -\sigma)$ to $(1, \sigma)$; this result is shown in Fig. 1.5. Note that stability is only achieved for $\sigma = 0$, thus this method is unconditionally unstable since the Fourier modes will grow unbounded with time for any nonzero Courant number[6].

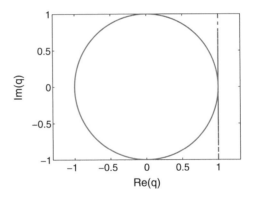

Fig. 1.5: Case II. The unit circle (solid red line) and the amplitude q for several values of the Courant number (dashed blue lines), σ, for the second order centered method (with explicit first order time-integration).

Case III

Let us now apply the stability analysis to Eq. (1.25)

---

[6] Even though the solution is stable for $\sigma = 0$ we say that the method is unstable because $\sigma = 0$ implies no flow, i.e., the solution is perfectly still. If we have any motion at all then $\sigma > 0$ and the solution of this method is unstable.

$$\frac{q_i^{n+1} - q_i^n}{\Delta t} + u\frac{q_{i+1}^n - q_i^n}{\Delta x} + O(\Delta t, \Delta x) = 0 \tag{1.30}$$

which we write in the following form

$$q_i^{n+1} = q_i^n - \sigma(q_{i+1}^n - q_i^n).$$

Next, substituting the Fourier mode in Eq. (1.27) yields

$$\tilde{q} = 1 - \sigma(e^{i\phi} - 1).$$

Introducing Euler's formula yields

$$\tilde{q} = 1 - \sigma(\cos\phi + i\sin\phi - 1)$$

which, after simplifying, yields

$$\tilde{q} = 1 + \sigma - \sigma\cos\phi - i\sigma\sin\phi$$

and results in $Re(\tilde{q}) = (1 + \sigma) - \sigma\cos\phi$ and $Im(\tilde{q}) = -\sigma\sin\phi$. In the Re-Im plane, this represents a circle with radius σ centered at $(1 + \sigma, 0)$; this result is shown in Fig. 1.6. Stability is only achieved for $\sigma = 0$, thus this method is unconditionally unstable since the Fourier modes will grow unbounded with time as shown in Fig. 1.6.

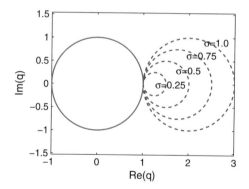

Fig. 1.6: Case III. The unit circle (solid red line) and the amplitude q for several values of the Courant number (dashed blue lines), σ, for the first order downwind method (with explicit first order time-integration).

Remark 1.2 The stability analysis just described is valid not just for the finite difference method but also for all other methods. Once again, we must first obtain the differencing stencil before applying the stability analysis. This is true for all Galerkin methods presented in this text.

1.5.5 Convergence

Recall that the *Lax Equivalence Theorem* states that if a method is consistent and stable, then it must be convergent. But what exactly does convergence mean? Let us consider the following types of series

$$\sum_{i=1}^{N} a_i = s_N \tag{1.31}$$

where a are the terms in the series and s_N is its Nth partial sum. If s_N is finite, then we say that the series a_i is convergent. Similarly, one should expect that if one is solving a PDE that has a unique solution, then a discrete approximation to the continuous PDE ought to be *close* to the unique solution; by close we mean that it is close within the discretization error, that is, the error due to the order of accuracy of the method. Furthermore, one would expect that as we let $(\Delta t, \Delta x) \to 0$ that the error decreases at the (*convergence*) rate $O(\Delta t^K, \Delta x^N)$ where K and N denote the order of accuracy in the time and space approximations. A method that behaves in this fashion is said to *converge* or to be *convergent*.

However, note that to speak of convergence requires us to also talk about errors and to quantify what we mean by error we must introduce error norms and normed vector spaces. For our purposes let us introduce the following error norms,

$$||q^{(num)} - q^{(exact)}||_{L^P} = \left(\sum_{i=1}^{N} |q_i^{(num)} - q_i^{(exact)}|^P \right)^{\frac{1}{P}} \tag{1.32}$$

which is the general p-norm error between the numerical (*num*) and exact (*exact*) solutions. The most common error norms discussed in numerical solutions of PDEs are the L^1

$$||q^{(num)} - q^{(exact)}||_{L^1} = \sum_{i=1}^{N} |q_i^{(num)} - q_i^{(exact)}|, \tag{1.33}$$

L^2

$$||q^{(num)} - q^{(exact)}||_{L^2} = \sqrt{\sum_{i=1}^{N} \left(q_i^{(num)} - q_i^{(exact)} \right)^2}, \tag{1.34}$$

and L^∞

$$||q^{(num)} - q^{(exact)}||_{L^\infty} = \max \left(|q_i^{(num)} - q_i^{(exact)}| \right) \; \forall i = 1, ..., N \tag{1.35}$$

norms. Clearly, if we know the *exact* solution then computing these norms is rather straightforward. However, often times we are interested in deriving numerical approximations to continuous problems with no known analytic solutions. So how then do we define the *exact* solution? In this case, we must replace the word *exact* with *reference* where we now use a sufficiently high-order method (high-order in space, time, boundary conditions, etc.) and with very high resolution (small time-steps and many gridpoints) and then use this as our reference solution to see if our method is indeed convergent. In other words, as we increase the resolution of our newly derived method, do we see the error (compared to our reference solution) decrease as $O\left(\Delta t^K, \Delta x^N\right)$ as predicted by theory? If so, then we can feel at ease that our method is indeed convergent.

1.6 Efficiency of Galerkin Methods on Distributed-Memory Computers

In this text, we focus on Galerkin methods for the discretization of partial differential equations. We have not yet discussed what we mean by a *Galerkin* method (this is done in Ch. 2) but for now let us simply state that a Galerkin method uses basis functions to approximate the solution. These basis functions can be either global or local as illustrated in Fig. 1.7. In this figure, we can assume that the one-dimensional domain is defined as $x \in [-1, 1]$. Figure 1.7a shows the basis functions spanning four waves while 1.7b shows an equivalent resolution using four elements using linear polynomials. Figure 1.7b requires some explanation. The basis functions ψ_1 and ψ_2 have compact support meaning that they are only defined within a specific element, e.g., for the first element which is defined by $x \in [-1, -0.5]$ the basis functions $\psi_i^{(e)}$ are defined for the specific element (e) (the first set of ψ_i, $i = 1, 2$ illustrated). Similarly, for the second element defined by $x \in [-0.5, 0]$ we have the next two basis functions ψ_i, $i = 1, 2$, and so on. The power of local Galerkin methods (i.e.,

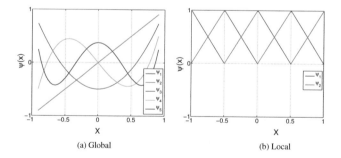

(a) Global   (b) Local

Fig. 1.7: Two Galerkin methods in their (a) global form (using 4th order polynomials) and (b) local forms (using linear polynomials but with four elements).

element-based Galerkin methods) is that they can use the partitioning of the domain into elements as in the local form shown in Fig. 1.7b yet they can use the same high-order basis functions used in the global form (Fig. 1.7a) inside each element.

In Fig. 1.8 we show the basis functions of the element-based Galerkin (EBG) method using four elements each comprised of fourth order polynomials. The order

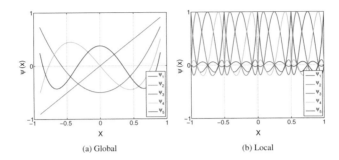

(a) Global   (b) Local

Fig. 1.8: Two Galerkin methods in their (a) global form (using 4th order polynomials) and (b) local forms (using 4th order polynomials with four elements).

of accuracy of both the local and global forms are the same but the cost for this particular example is different. For the moment let us state, without proof, that the complexity of a one-dimensional EBG model increases as $O(N_e N^2)$ where N_e denotes the number of elements (partitions) of the grid and N is the order of the polynomial of the basis functions[7].

In general, the resolution (number of degrees of freedom) of an EBG model increases as $O(N_e N^{d+c})$ where d is the dimension of the problem. Clearly, to construct a more efficient model requires increasing the elements (N_e) rather than the polynomial order (N) since the cost of the model increases exponentially (as $d + c$) with N. The study of the number of operations of a numerical model is known as *complexity analysis* and is an invaluable tool for all computational scientists.

To develop efficient computer models requires knowledge not only of complexity analysis but also of the computer architecture used to run the models. For example, if one is using a shared-memory computer (think of a large computer, such as the old Cray vector computers) then one can store the entire grid on the main chunk of memory and then use various processors (we now call them *cores* or, even better, *nodes* since this way we can lump x86 CPUs with accelerators such as GPUs)[8] to compute the equations of the model on various locations of the computational grid. For this type of computer architecture, global and local Galerkin methods can be implemented efficiently. However, if one is using a distributed-memory computer (think of a commodity computer with very cheap cores connected by fast network switches) then the model can be made efficient if and only if the problem *domain* can be *decomposed* into numerous partitions. To take full advantage of distributed-

---

[7] In general, the complexity of EBG methods is $O(N_e N^{d+c})$ for d-dimensional problems where $c = 1$ for tensor-product methods and as high as $c = d$ for non-tensor-product methods.

[8] CPU is a central processing units and GPU is a graphics processing unit.

memory computers requires the use of a numerical method that allows for the discrete problem to be represented as a collection of partitions (or elements) that requires heavy on-processor computations with little inter-processor communication. The reason for this is simple: arithmetic operations (computations) cost very little whereas moving data (communication) is the main bottleneck. Figure 1.9 shows a

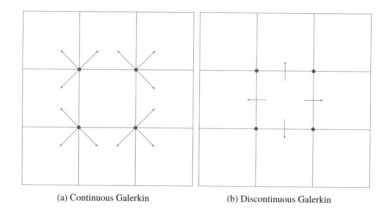

(a) Continuous Galerkin   (b) Discontinuous Galerkin

Fig. 1.9: Communication stencil of Galerkin methods on distributed-memory computers for (a) continuous Galerkin methods and (b) discontinuous Galerkin methods.

square domain partitioned (called domain decomposition) into numerous processors of a distributed-memory computer. If a global Galerkin method is used then the basis functions are defined throughout the entire domain. What this means is that if one processor would like to know the solution of the values that it claims, it will require knowing the information from all the other processors; this process is known as an *all-to-all* communication[9] and is a trademark of global methods with the most well-known of these being the *spectral transform* method. However, if one is able to decompose the domain into small chunks that fit inside each processor as shown in Fig. 1.9 for element-based Galerkin methods, then one no longer needs to perform an all-to-all communication but rather only a small information exchange.

For example, Fig. 1.9a shows the communication stencil for a version of the element-based Galerkin (EBG) method called the *continuous* Galerkin method; examples of this class of method include the finite element and spectral element methods which we introduce in Ch. 2 and describe at length in Ch. 5. Figure 1.9b shows the communication stencil of another version of EBG methods called the *discontinuous* Galerkin method; examples of this class of method include the finite volume, discontinuous finite element, and discontinuous spectral element methods which are all typically referred to simply as discontinuous Galerkin methods and are covered in Ch. 6. Discontinuous methods have small communication stencils

---

[9] On most current computers, executing an *all-to-all* communication is quite expensive. An exception was the IBM Blue Gene series.

and, for this reason, are tailor-made for both distributed-memory computers and accelerator-based processors (e.g., graphics processing units or GPUs).

In Fig. 1.10 we show the strong scaling performance of the continuous Galerkin (CG) method on a distributed-memory computer[10] using up to 3 million message-passing interface (MPI) ranks (see [280]) which are denoted as *compute-cores* in the figure. The dotted line in Fig. 1.10 represents the linear (perfect) scaling based on

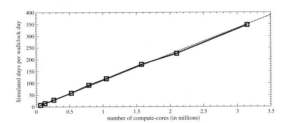

Fig. 1.10: Strong scaling of the CG method for the compressible Euler equations (line with squares). The domain is partitioned into 62 million elements with 3rd order polynomials (totaling 1.8 billion degrees of freedom). The dotted line represents the linear (perfect) strong scaling curve.

the reported time using 65,000 MPI ranks; this was the minimum number of ranks required to fit this particular problem. The linear perfect scaling curve is computed based on the following definition of speedup

$$\text{Speedup} = \frac{T_{\text{baseline}}}{T_{N_{\text{cores}}}} \tag{1.36}$$

where T_{baseline} is the time required by the baseline simulation (using 65,000 ranks in this example) and $T_{N_{\text{cores}}}$ is the time required by a simulation using N_{cores} compute-cores. In strong scaling, we maintain the same problem size and seek to increase the speedup by using more compute-cores. Figure 1.10 shows that the CG method strong scales almost perfectly; although the DG method is not shown the strong scaling results would be similar.

In Fig. 1.11 we show the weak scaling performance of the discontinuous Galerkin method on a GPU computer[11] using up to 16,000 GPU cards (see [2]). The dotted line denotes the perfect weak scaling efficiency. In weak scaling, we maintain the workload constant across the hardware and see if there is a degradation of the performance. In other words in perfect weak scaling, the time-to-solution remains constant for increased problem size by a factor of X provided that we also increase

---

[10] The Mira supercomputer comprised of IBM Blue Gene Q hardware formerly at Argonne National Laboratory.

[11] The Titan supercomputer comprised of Nvidia K20x GPUs with AMD Opteron CPUs formerly at Oak Ridge National Laboratory.

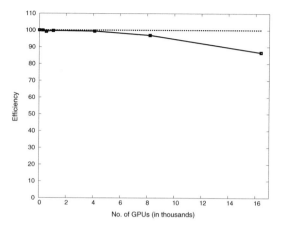

Fig. 1.11: Weak scaling of the DG method (solid line with squares) for the compressible Euler equations. Each GPU card gets 900 elements of order 7.

the number of compute-cores by the same factor. We can define the weak scaling efficiency as

$$\text{Efficiency} = \frac{T_1}{T_{N_{\text{cards}}}} \times 100 \tag{1.37}$$

where T_1 is the time required by the model on one card whereas $T_{N_{\text{cards}}}$ is the time required on N_{cards} cards where the workload per card is maintained constant. Achieving perfect weak scaling efficiency is generally not possible because there is overhead in communication (data transfer, in particular). Figure 1.11 shows that the DG method is able to achieve close to 90% weak scaling efficiency if we overlap communication with computation[12]. The CG method (without this overlap strategy) would yield approximately 80% weak scaling efficiency (these results are discussed in detail in [2, 3, 4]).

---

[12] This a useful strategy, that is, communicate what is non-local and compute what is local with the aim to mitigate the latency associated with the communication.

Chapter 2
Overview of Galerkin Methods

2.1 Introduction

Before we can understand the nuances of the various methods for discretizing non-linear partial differential equations in space, we must first realize the choices that we have at our disposal. We can categorize the possible methods as follows:

1. methods that use the differential form of the equations and
2. methods that use the integral form of the equations.

Generally speaking, the most widely used differential form method is the finite difference method while the most widely used integral form method is the Galerkin method (e.g., finite elements).

Let us take as an example the 1D conservation law

$$\frac{\partial q}{\partial t} + \frac{\partial f}{\partial x} = 0$$

where $f = qu$ and describe how to discretize this equation with both types of methods (differential and integral forms).

2.2 Differential Form: Finite Differences

The continuous conservation law can be discretized in differential form as follows

$$\frac{\partial q_i}{\partial t} + \frac{f_{i+1}^n - f_{i-1}^n}{2\Delta x} = 0$$

F. X. Giraldo, *An Introduction to Element-Based Galerkin Methods on Tensor-Product Bases*, Texts in Computational Science and Engineering 24,
https://doi.org/10.1007/978-3-030-55069-1_2

where the differencing stencil is constructed via Taylor series expansions. With this in mind, the derivative $\frac{\partial f}{\partial x}$ is approximated as follows

$$f_{i+1} = f_i + \Delta x \frac{\partial f_i}{\partial x} + \frac{\Delta x^2}{2} \frac{\partial^2 f_i}{\partial x^2} + O(\Delta x^3)$$

and

$$f_{i-1} = f_i - \Delta x \frac{\partial f_i}{\partial x} + \frac{\Delta x^2}{2} \frac{\partial^2 f_i}{\partial x^2} - O(\Delta x^3)$$

which, subtracting, gives

$$f_{i+1} - f_{i-1} = 2\Delta x \frac{\partial f_i}{\partial x} + O(\Delta x^3)$$

that then yields the following approximation to the spatial derivative

$$\frac{\partial f_i}{\partial x} = \frac{f_{i+1} - f_{i-1}}{2\Delta x} - O(\Delta x^2).$$

If a higher order approximation is desired we then require information from additional grid points. For example, the following three gridpoint information

$$f_{i+1} = f_i + \Delta x \frac{\partial f_i}{\partial x} + \frac{\Delta x^2}{2} \frac{\partial^2 f_i}{\partial x^2} + \frac{\Delta x^3}{6} \frac{\partial^3 f_i}{\partial x^3} + O(\Delta x^4),$$

$$f_{i-1} = f_i - \Delta x \frac{\partial f_i}{\partial x} + \frac{\Delta x^2}{2} \frac{\partial^2 f_i}{\partial x^2} - \frac{\Delta x^3}{6} \frac{\partial^3 f_i}{\partial x^3} + O(\Delta x^4),$$

and

$$f_{i+2} = f_i + 2\Delta x \frac{\partial f_i}{\partial x} + \frac{(2\Delta x)^2}{2} \frac{\partial^2 f_i}{\partial x^2} + \frac{(2\Delta x)^3}{6} \frac{\partial^3 f_i}{\partial x^3} + O(\Delta x^4)$$

along with f_i can be used to construct a third order accurate derivative. It should be understood that this simple finite difference method is not the only method possible when the differential form of the equations are desired. There exists a large class of high-order finite difference methods [69, 288], others known as the staggered-grid spectral methods [238, 239], and the spectral multi-domain penalty methods [232,233,234,198,196,197,173,106,121]. Recently, a bridge has been found between the differential and integral forms of the equations. We refer you to papers on the flux reconstruction method [210, 65, 8, 390, 99].

Exercise Using information at the 4 gridpoints $i - 1, i, i + 1$, and $i + 2$, construct a third order accurate first derivative of the function $f(x)$.   □

2.3 Integral Form: Galerkin Methods

The other possibility for discretizing the equations is to use the integral form of the equations. The first step is to represent the solution via a polynomial representation as follows

$$q_N(x, t) = \sum_{i=0}^{N} \tilde{q}_i(t) \Psi_i(x)$$

where $\tilde{q}$ are the expansion coefficients, Ψ are the basis functions that are represented as polynomials, and N is the order of the basis functions. Once we approximate the solution as $q_N(x, t)$, we can then approximate the flux function using similar means as follows

$$f_N(x, t) = f(q_N(x, t)).$$

Introducing these approximations into the original continuous PDE yields

$$\frac{\partial q_N}{\partial t} + \frac{\partial f_N}{\partial x} = \epsilon \tag{2.1}$$

where $\epsilon \neq 0$, which states that because we are using a finite dimensional polynomial representation, that we cannot *exactly* satisfy the original homogeneous PDE (i.e., we incur some error using a finite dimensional representation of the solution). For $N \to \infty$ we would expect to satisfy the original PDE. We know from linear algebra (e.g., see [380]) that, in a least-squares sense, the error incurred by using a lower-dimensional space (q_N) to represent a higher-dimensional space (q) which we call ϵ (i.e., the residual) is minimized when it is orthogonal to the lower-dimensional space that we represent by the basis functions Ψ; this is depicted in Fig. 2.1. Therefore,

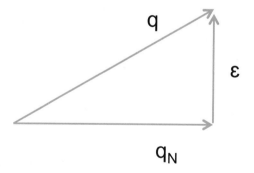

Fig. 2.1: The error ϵ between the "truth" solution q and the Galerkin approximation q_N is minimized, in a least-square sense, when it is orthogonal to the approximation space (which is defined by the set of all basis functions ψ).

imposing the orthogonality of Ψ and ϵ via the following inner product

$$\int_{\Omega} \Psi \epsilon \, d\Omega = 0$$

allows us to write the integral form of Eq. (2.1) as follows

$$\int_\Omega \Psi_i \frac{\partial q_N}{\partial t}\, d\Omega + \int_\Omega \Psi_i \frac{\partial f_N}{\partial x}\, d\Omega = 0. \tag{2.2}$$

Using the product rule of differentiation for the second term we rewrite Eq. (2.2) as follows

$$\int_\Omega \Psi_i \frac{\partial q_N}{\partial t}\, d\Omega + \int_\Omega \frac{\partial}{\partial x}\left(\Psi_i f_N\right) d\Omega - \int_\Omega \frac{d\Psi_i}{dx} f_N\, d\Omega = 0. \tag{2.3}$$

Integrating the second term (using the fundamental theorem of calculus) gives

$$\int_\Omega \Psi_i \frac{\partial q_N}{\partial t}\, d\Omega + [\Psi_i f_N]_\Gamma - \int_\Omega \frac{d\Psi_i}{dx} f_N\, d\Omega = 0$$

where the term in the square brackets is evaluated at the boundary, Γ, of the domain Ω. At this point we can construct either continuous or discontinuous Galerkin methods provided that we replace the global functions Ψ with locally defined functions ψ and integrate within each subdomain Ω_e instead of the global domain Ω and define the boundaries of Ω_e as Γ_e.

2.3.1 Continuous Galerkin Method

In continuous Galerkin methods the basis functions ψ are generally assumed to be C^0 continuous across element interfaces. Therefore the coefficients and basis functions are chosen such that the values at the interfaces are continuous. Thus, for the continuous Galerkin method the flux term (jump condition evaluated at Γ) will vanish at all element interfaces except at the physical boundary Γ. To simplify the following discussion, let us assume homogeneous Dirichlet boundary conditions of the type $f(0,t) = f(L,t) = 0$ where the domain in x is defined as $x \in [0, L]$. We can now write the continuous Galerkin approximation as follows

$$\int_\Omega \Psi_i \frac{\partial q_N}{\partial t}\, d\Omega - \int_\Omega \frac{d\Psi_i}{dx} f_N\, d\Omega = 0$$

which is quite convenient since we do not have to take the derivative of f but instead the derivative of Ψ. Figure 2.2 shows a graphical representation of the interface between two elements. Note how the left and right elements share the values at the interface. This is because in continuous Galerkin methods the gridpoints are typically defined globally.

Although in the continuous Galerkin method we speak of the global problem as that presented in Eq. (2.3), in practice we never define the global basis functions Ψ because they have compact support meaning that they are only defined within an element and are zero everywhere else. What this means is that we can construct the global problem by simply summing up the smaller local problems defined by the elements themselves. The concept of summing the local problem to construct the global problem is known as *global assembly* or *direct stiffness summation* (DSS) and is discussed in more detail later in this chapter.

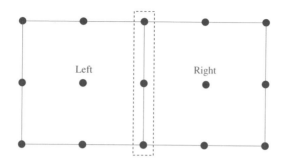

Fig. 2.2: In continuous Galerkin methods the interface values are shared by neighboring elements.

The continuous basis functions selected in this method are typically Lagrange polynomials and are the types of basis functions used in classical finite element and spectral element methods. We discuss interpolating polynomials in more detail in Ch. 3. Examples of continuous Galerkin methods include the following methods:

1. spectral (transform),
2. pseudo-spectral,
3. finite element, and
4. spectral element methods.

However, in order to distinguish between global (e.g., spectral and pseudo-spectral) and local continuous Galerkin (e.g., finite and spectral element) methods we refer to the local methods simply as element-based Galerkin (EBG) methods. Since, as you will see shortly, discontinuous Galerkin methods are also built around *elements*, we also include discontinuous Galerkin and finite volume methods within the newly defined class of element-based Galerkin methods. Let us now discuss the discontinuous Galerkin method.

2.3.2 Discontinuous Galerkin Method

In the discontinuous Galerkin method each element has its own polynomial expansion and thus no continuity is assumed in the choice of coefficients and basis functions. In fact, the discontinuous Galerkin method further differs from the continuous Galerkin method in that the problem statement is, by construction, always a local one. That is, since there is no global problem, then there is no need for *direct stiffness summation*.

Multiplying the original PDE by a local test function ψ and integrating within each local element yields

$$\int_{\Omega_e} \psi_i \left(\frac{\partial q_N^{(e)}}{\partial t} + \frac{\partial f_N^{(e)}}{\partial x} \right) d\Omega_e = 0 \qquad (2.4)$$

where $\Omega = \bigcup_{e=1}^{N_e} \Omega_e$ defines the total domain, and $\Omega_e \; \forall \, e = 1, ..., N_e$ are the elements, thus Eq. (2.4) is imposed on each element Ω_e. As in the continuous Galerkin method, we also approximate the solution variable for the discontinuous Galerkin method as follows

$$q_N^{(e)}(x, t) = \sum_{i=0}^{N} \tilde{q}_i^{(e)}(t) \psi_i(x)$$

with $f_N^{(e)} = f\left((q_N^{(e)}(x, t) \right)$, $\tilde{q}$ are the expansion coefficients, ψ the local polynomial functions, and N the order of the polynomial. Using the same calculus identities we used in the continuous Galerkin method, we arrive at the following integral form

$$\int_{\Omega_e} \psi_i \frac{\partial q_N^{(e)}}{\partial t} d\Omega_e + \left[\psi_i f_N^{(e)} \right]_{\Gamma_e} - \int_{\Omega_e} \frac{\partial \psi_i}{\partial x} f_N^{(e)} d\Omega_e = 0 \qquad (2.5)$$

where the term in the square brackets is evaluated at the element boundary Γ_e for each element Ω_e. For purposes of illustration and extension to multi-dimensions, let us write this expression in the following form

$$\int_{\Omega_e} \psi_i \frac{\partial q_N^{(e)}}{\partial t} d\Omega_e + \sum_{k=1}^{N_{FN}} \hat{\mathbf{n}}^{(e,k)} \psi_i(x_{(e,k)}) f_N(x_{(e,k)}) - \int_{\Omega_e} \frac{\partial \psi_i}{\partial x} f_N^{(e)} d\Omega_e = 0 \quad (2.6)$$

where $\hat{\mathbf{n}}^{(e,k)}$ is the unit normal vector of the element e in the direction of its neighbor k, and $x_{(e,k)}$ is the position at the interface between the elements e and k, and $N_{FN} = 2d$ is the number of face neighbors (where d is the spatial dimension). There is a dilemma, however, in Eqs. (2.5) and (2.6). The issue is that by definition we have let q (and consequently f) be discontinuous across element edges Γ_e. Therefore, if we sampled the function q at the interface between elements e and k then we would get different values for q depending on whether we used element e or element k. This inconsistency is easily overcome by taking some weighted average (the simplest choice would be the mean value). To denote that the interface value is a weighted average, we write it as $f_N^{(*)}$ that then gives for Eq. (2.6)

$$\int_{\Omega_e} \psi_i \frac{\partial q_N^{(e)}}{\partial t} d\Omega_e + \sum_{k=1}^{2} \hat{\mathbf{n}}^{(e,k)} \psi_i(x_{(e,k)}) f_N^{(*,e,k)} - \int_{\Omega_e} \frac{\partial \psi_i}{\partial x} f_N^{(e)} d\Omega_e = 0 \qquad (2.7)$$

which appears like additional work as compared to the continuous Galerkin method since we have to deal with the flux (jump) terms for each element Ω_e. The advantage of the DG method is that the method is now formally both locally and globally conservative. We discuss the numerical flux function $f^{(*,e,k)}$ in more detail later in this chapter but for now let us just state that one of the most commonly used numerical fluxes is the Rusanov[1] flux that is defined as follows

$$f_N^{(*,e,k)} = \frac{1}{2}\left[f_N^{(k)} + f_N^{(e)} - \lambda_{max}\hat{\mathbf{n}}^{(e,k)}\left(q_N^{(k)} - q_N^{(e)} \right)\right] \tag{2.8}$$

where $\lambda_{max} = |\frac{\partial f^{(e)}}{\partial q}, \frac{\partial f^{(k)}}{\partial q}|$ is the maximum wave propagation speed of the system; in other words, it is the maximum eigenvalue of the Jacobian matrix $A = \frac{\partial f}{\partial q}$. The reason why the eigenvalues of A represent the wave speed of the system can be clearly seen by rewriting the original PDE

$$\frac{\partial q}{\partial t} + \frac{\partial f}{\partial x} = 0$$

as follows

$$\frac{\partial q}{\partial t} + \frac{\partial f}{\partial q}\frac{\partial q}{\partial x} = 0$$

where we have used the chain rule since $f = f(q)$ and $q = q(x,t)$. This equation is (using the method of characteristics) often written as

$$\frac{\partial q}{\partial t} + A\frac{\partial q}{\partial x} = 0$$

where we can now see that the eigenvalues of A are similar to the wave speed of a scalar hyperbolic equation.

Figure 2.3 shows a graphical representation of the discontinuous Galerkin method. Since in this method there are no global gridpoints (only local ones) then the flux terms at the interface of neighboring elements do not cancel. This is obviously so because the values at the left and right elements are not the same (as denoted by solid and open circles). Therefore the solution values and basis functions need not be continuous across element interfaces.

Because the basis functions need not be continuous across element interfaces we can choose any type of basis function including modal basis functions such as Legendre or Fourier polynomials (i.e., not necessarily the Lagrange polynomials typically used in continuous Galerkin methods). However, a discrete integration-by-parts can only be satisfied exactly for Lagrange polynomials based on Lobatto points[2]. This is an important property in order to derive stable numerical discretizations (see, e.g., [241, 242, 142, 406]).

If we choose constant basis functions, say $\psi = 1$, Eq. (2.7) becomes

$$\int_{\Omega_e} \frac{\partial q_N^{(e)}}{\partial t}\, d\Omega_e + \sum_{k=1}^{2} \hat{\mathbf{n}}^{(e,k)} f_N^{(*,e,k)} = 0 \tag{2.9}$$

---

[1] The Rusanov flux is also known as the local Lax-Friedrichs flux.
[2] This is discussed in Sec. 6.8.

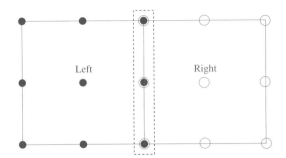

Fig. 2.3: In discontinuous Galerkin methods each element has its own set of grid-points and therefore element boundary values are not shared by neighbors. For example, at the interface between the left (blue closed circles) and right (red open circles) elements, there are two sets of values for each physical point.

which, upon integrating, yields

$$\Delta x \frac{\partial q_N^{(e)}}{\partial t} + \left(\hat{\mathbf{n}}^{(e,e+1)} f_N^{(*,e,e+1)} + \hat{\mathbf{n}}^{(e,e-1)} f_N^{(*,e,e-1)} \right) = 0$$

where we have let $k = 1$ be the element $e - 1$ and $k = 2$ the element $e + 1$. Note that this gives the following normal vectors: $\hat{\mathbf{n}}^{(e,e-1)} = -1$ and $\hat{\mathbf{n}}^{(e,e+1)} = +1$. Dividing by Δx gives the final form

$$\frac{\partial q_N^{(e)}}{\partial t} + \frac{f_N^{(*,e,e+1)} - f_N^{(*,e,e-1)}}{\Delta x} = 0. \tag{2.10}$$

Let us assume that $f = qu$ so that $\frac{\partial f}{\partial q} = u$ and so $\lambda_{max} = |u|$. If we further assume u is constant and moves from left to right then we can write $\lambda_{max} = u$ which we now use to rewrite Eq. (2.8) as follows

$$f_N^{(*,e,e+1)} = \frac{1}{2} \left[f_N^{(e+1)} + f_N^{(e)} - u \left(q_N^{(e+1)} - q_N^{(e)} \right) \right] \equiv f^{(e)}$$

$$f_N^{(*,e,e-1)} = \frac{1}{2} \left[f_N^{(e-1)} + f_N^{(e)} + u \left(q_N^{(e-1)} - q_N^{(e)} \right) \right] \equiv f^{(e-1)} \tag{2.11}$$

where we have used the fact that u is constant and so $f^{(k)} = q^{(k)} u$ for all elements k. Substituting the resulting flux of Eq. (2.11) into Eq. (2.10) yields

$$\frac{\partial q_N^{(e)}}{\partial t} + \frac{f^{(e)} - f^{(e-1)}}{\Delta x} = 0 \tag{2.12}$$

which is in fact the upwind finite volume discretization of the wave equation. Note that this discretization is identical to the upwind finite difference discretization we obtained previously (Case I in Ch. 1). In one dimension, there is no difference between a cell-centered upwind finite volume method and an upwind finite difference method.

This simple example also shows that we can, in fact, interpret the discontinuous Galerkin (DG) method as the high-order generalization of the finite volume (FV) method; this viewpoint of the DG method is true not just for one dimension but for all dimensions. The benefit of this view of DG methods is that much machinery has been developed for FV methods since the 1970s that can be recycled (albeit in a modified form) for use with DG methods. We will see this to be the case when we discuss Riemann solvers (a fancy word for numerical fluxes) and slope limiters which are used to enforce the solution to remain monotonic near sharp gradients. In this text, when we refer to discontinuous methods we are typically referring to the discontinuous Galerkin method although finite volumes certainly are a subclass of the DG method.

2.4 A Formal Introduction to Galerkin Methods

2.4.1 Problem Statement

For all Galerkin methods the idea is to solve any differential operator

$$L(q) = 0$$

in integral form by first approximating q as a polynomial expansion

$$q_N(x) = \sum_{i=0}^{N} \tilde{q}_i \psi_i(x) \tag{2.13}$$

where ψ are the polynomials (basis functions) and $\tilde{q}$ are the expansion coefficients.

The statement of the problem is: find $q \in \mathcal{V}_N$ such that

$$\int_{\Omega} L(q)\psi \, d\Omega = 0 \qquad \forall \psi \in \mathcal{V}_N$$

where $\mathcal{V}_N$ is the finite-dimensional vector space where the solution q and the basis functions ψ live. For the continuous Galerkin method, the finite-dimensional vector space is defined as follows

$$\mathcal{V}_N^{CG} = \left\{ \psi \in H^1(\Omega) : \psi \in \mathcal{P}_N \right\} \tag{2.14}$$

whereas for the discontinuous Galerkin method it is

$$\mathcal{V}_N^{DG} = \left\{ \psi \in L^2(\Omega) : \psi \in \mathcal{P}_N \right\} \tag{2.15}$$

where $\mathcal{P}_N$ is an Nth order polynomial that we define in Chs. 3 (for one-dimensional problems) and 10 (for two-dimensional problems). The space $\mathcal{V}_N$ is called a finite-dimensional vector space because it only contains a subset of the infinite-dimensional function spaces H^1 and L^2.

2.4.2 Function Spaces

The difference between H^1 and L^2, at least for our purposes, is that H^1 consists of both q and its first derivatives both living in L^2. This along with the basis functions also living in this space implies that $\mathcal{V}_N^{CG} \subset C^0$; i.e., the solution (and the basis functions) are continuous everywhere in the domain. In the DG method, this condition is relaxed so that $\mathcal{V}_N^{DG} \in L^2$ and $\mathcal{V}_N^{DG} \not\subset C^0$.

In the Galerkin method, we seek a set of basis functions ψ and coefficients $\tilde{q}$ such that the inner product of the basis functions to the differential operator is orthogonal

$$(L(q), \psi) = 0$$

where the inner product is defined in the *Hilbert space* (see, e.g., [101] for further details) of square integrable functions

$$\int_\Omega L(q)\psi \, d\Omega = 0.$$

Depending on what we pick for $\tilde{q}$ and ψ will determine which spatial discretization method we will use. Let us now define the Hilbert space that is of enormous importance in the construction of spatial discretization methods for partial differential equations. Before we do, let us first define a convergent sequence (see [213]).

Definition 2.1 A sequence $u_1, u_2, \ldots, u_i \in S$ is called convergent if

$$\lim_{i \to \infty} ||u - u_i|| = 0.$$

for $u \in S$. Furthermore, if

$$\lim_{i,j \to \infty} ||u_i - u_j|| = 0$$

then we call it a *Cauchy sequence*. The space is *complete* if every Cauchy sequence in S converges.

Definition 2.2 A Hilbert space is a complete inner product space, with the inner product defined as

$$(u, v) = \int_\Omega uv \, d\Omega$$

for two functions u and v. In other words, there exists a set of linearly independent vectors (v) in which any point in the space can be expanded in terms of these linearly independent vectors. A complete set of linearly independent vectors forms a basis. The integral is defined in the L^2 space.

Let us now define the L^2 space that is pivotal to the Hilbert space.

Definition 2.3 The L^2 space is defined as the space with square-integrable functions u such that

$$\int_\Omega |u|^2 d\Omega < \infty$$

where the integral is not the usual Riemann sum, but is instead defined in terms of the Lebesgue integral that allows for discontinuities within the domain Ω.

Let us now introduce the final space that plays a crucial role in partial differential equations and is in fact constructed on the principles of Hilbert spaces.

Definition 2.4 A Sobolev space is formed from a hierarchy of Hilbert spaces such that for two functions u and v we can write the Sobolev inner product

$$(u, v)_{H^k} = \int_\Omega \sum_{i=0}^k u^{(i)} v^{(i)} d\Omega$$

where the superscript (i) denotes the ith derivative.

The typical Sobolev space we encounter in this text is H^1 (where, in general $H^p = W_2^p$) defined as

$$(u, v)_{H^1} = \int_\Omega (\nabla u \cdot \nabla v + uv) \, d\Omega$$

which states that the functions and their first derivatives live in Hilbert spaces, i.e., they are both square-integrable functions; this is the function space required by the continuous Galerkin (CG) method. Note that the following spaces are equivalent: $H^0 \equiv L^2$, which explains why we are interested in the functions being square-integrable. In fact, we will see that when u and v are both equal to ψ (the basis function), we recover a square-integrability condition for the function ψ and this resulting entity is nothing more than the Galerkin mass matrix and this matrix must exist (be finitely-bounded) for us to construct a meaningful numerical solution to the continuous PDEs. CG methods require the derivatives to also be square-integrable because the fact that the solution vector is constrained to be continuous imposes that the derivatives also exist (this will become clear when we discuss elliptic operators where the square-integrability of the derivatives becomes obvious).

For discontinuous Galerkin methods, we are only interested in solutions in L^2 because the functions must only be square-integrable in order to satisfy a well-conditioned mass matrix. No constraint on the continuity of the solution is imposed which means that the derivatives need not be bounded. This means that the DG method admits a larger set of function spaces for both its solution and basis functions than the CG method and is one of the major advantages of the DG method.

With these definitions in place, we are now prepared to understand why these spaces play such a vital role in the construction of numerical methods. The connection between these spaces and PDEs is seen through the *Sturm-Liouville operator* [47].

2.4.3 Sturm-Liouville Operator

Definition 2.5 The Sturm-Liouville operator can be written as

$$\frac{d}{dx}\left(p(x)\frac{d\phi(x)}{dx}\right) + \lambda w(x)\phi(x) = 0 \tag{2.16}$$

which contains the largest class of operators contained in most differential equations encountered; the differential equation is uniquely defined for specific choices of boundary conditions. In this operator, $p(x)$ and $w(x)$ are real-valued functions and λ are the eigenvalues of the orthogonal eigenfunctions $\phi(x)$ with respect to the weighting function $w(x)$. The *singular* Sturm-Liouville operator is obtained when $p(x)$ vanishes at either one or both boundaries; otherwise it is called *regular*.

The eigenfunctions $\phi(x)$ (and corresponding eigenvectors, in the discrete case) of this operator form a complete basis in the space where the operator is defined. This means that as $N \to \infty$ where N are the number of eigenfunctions used in the expansion, the error of the solution goes to zero converging exponentially. This can be proven for the *singular Sturm-Liouville* operator and thus is the specific operator that is of interest to us; this cannot be guaranteed for the *regular Sturm-Liouville* operator. The importance that the eigenfunctions of the singular Sturm-Liouville operator have is that if we were to solve a differential equation analytically we would expand the approximation of this solution in terms of such functions $\phi(x)$. In fact, the classical way to derive the Sturm-Liouville operator (Eq. (2.16)) is to begin with the system of nonlinear partial differential equations that you wish to solve numerically. Next, we linearize the system of equations and collapse the system to one variable. We then solve this resulting high-order (with respect to derivatives) linear scalar equation using separation of variables; for first order systems (such as for hyperbolic equations) we obtain a second order eigenvalue problem. The solution of this eigenvalue problem are the orthogonal (eigenfunctions) polynomials that form the natural basis for representing the solution. The advantages of using such orthogonal functions derived from the specific singular Sturm-Liouville operator resulting from the specific system of equations is that semi-analytical methods are incorporated into the numerical solution of the continuous problem. For this reason, it is said that Galerkin methods are optimal for certain classes of differential operators (most notably, elliptic operators). Let us now discuss certain types of boundary conditions for the singular Sturm-Liouville operator that result in very specific well-known eigenfunctions.

Solving the Sturm-Liouville operator with periodic boundary conditions in the domain $x \in [0, 2\pi]$ with specific choices for $p(x)$ and $w(x)$ results in the Fourier series. Thus the Fourier series is the natural basis for the approximation on periodic domains. In the case of non-periodic boundary conditions (say $x \in [-1, +1]$) with specific choices for $p(x)$ and $w(x)$ results in Legendre polynomials. Another choice for $p(x)$ and $w(x)$ results in Chebyshev polynomials. These choices are described in more detail in Ch. 3.

Remark 2.1 Global Galerkin methods (such as the spectral transform or spherical harmonics method) on the sphere use Fourier series in the latitude circles ($\lambda \in [0, 2\pi]$) (i.e., the domain is periodic in this direction) and Legendre polynomials in the longitude semi-circles ($\theta \in [-\frac{\pi}{2}, +\frac{\pi}{2}]$) (this is mapped to the line $[-1, +1]$)

because these orthogonal functions are the solutions to the Sturm-Liouville operator on the sphere.

Remark 2.2 Continuous element-based Galerkin (EBG) methods (such as finite element and spectral element methods) remap the global domain into pieces with domains $x \in [-1, +1]$ where the Legendre polynomials are the natural basis. One can also use Chebyshev (or other Jacobi) polynomials but the advantage of Legendre polynomials is that their weighting function is $w(x) = 1$.

Exercise Solve the singular Sturm-Liouville operator in one-dimension with homogeneous boundary conditions in the domain $x \in [-1, +1]$.   □

2.4.4 Basis Functions

Once we have constructed the *optimal* basis function for our application we then need to decide whether we are going to use it as a modal or nodal function. A few more definitions are required.

Definition 2.6 In an Nth degree modal expansion, the $\tilde{q}$ and ψ in Eq. (2.13) are generally non-zero everywhere. Therefore if you know $\tilde{q}$ and ψ and want to know the value of q at some grid point x_j then we need to sum all N terms. Therefore we need to carry out the full summation given by

$$q_N(x_j) = \sum_{i=0}^{N} \tilde{q}_i \psi_i(x_j)$$

because the value of $q_N(x_j)$ is dependent on all the terms in the series. Therefore the $\tilde{q}_i$ represent spectral coefficients; the term *spectral* derives from the analogy with the full *spectrum* of waves represented. In other words $\tilde{q}_i, \psi_i$ represent the amplitude and frequency of the mode i.

Definition 2.7 In a nodal expansion, ψ are cardinal functions (Lagrange polynomials) meaning that they have the property

$$\psi_i(x_j) = \begin{cases} 1 \text{ for } i = j \\ 0 \text{ for } i \neq j \end{cases}$$

thus the expansion coefficients $\tilde{q}_i$ correspond to the value of $q(x_j)$ at the point x_j. In this case the series can be written as

$$q_N(x) = \sum_{i=0}^{N} q_i \psi_i(x)$$

where q_i represent the physical values at the gridpoints. Thus the value at x_j is $q_N(x_j) = q_j$. In other words q_i, ψ_i represent the value and cardinal function at the gridpoint i.

Remark 2.3 Modal expansions are typically more costly to implement because they require forward and backward transformations to get the solution values from spectral (modal) to physical (nodal) space. However, a fast transform for Fourier series (the FFT) revolutionized spectral methods because it meant that this transformation could be carried out quickly (from $O(N^2)$ to $O(N \log N)$. Unfortunately there is no such fast Legendre transform (this operation costs $O(N^2)$). For this reason it has been popular to find stable double Fourier series (e.g., [77]) for solving PDEs on the sphere using spectral methods.

2.5 Global Galerkin Methods

Global Galerkin methods, such as the spectral (transform) method (i.e., spherical harmonics on the sphere) only use modal expansions. The basis functions ψ are defined globally throughout the domain and the expansion coefficients $\tilde{q}$ have no direct one-to-one correspondence to the solution q at specific gridpoints, rather, the expansion coefficients represent the amplitudes of various frequencies that need to be summed in order to represent the solution at a physical coordinate in space. An example of a global modal expansion is illustrated in Fig. 2.4 where the first five modes for a modal Legendre polynomial expansion are shown. This method is well grounded mathematically as it comes directly from semi-analytic methods and approximation theory. Thus for a smooth function q the method is guaranteed to converge exponentially provided that the natural basis for the space is used. However, because it is a semi-analytic method it may not necessarily exploit the power of the computer, and certainly not of distributed-memory computers because the approximating functions are defined everywhere throughout the domain. This method is challenging to implement efficiently on a distributed-memory machine because all of the spectral coefficients $(\tilde{q})$ must be on-processor[3] in order to evaluate the solution at a given gridpoint. In Fig. 2.4 we can see that the approximating functions (basis functions) are truly defined everywhere on the domain. Thus if the horizontal axis represents the global domain, we would require all of the spectral coefficients and basis functions in order to construct an approximation at a given position in space. There are no gridpoints or elements in this method, only quadrature points used for the numerical integration (we cover numerical integration in Ch. 4).

2.6 Element-based Galerkin Methods

In contrast to global Galerkin methods, local Galerkin methods use elements as the basic building-blocks for the discrete representation of the solution. Inside each

---

[3] We use the term "on-processor" to mean the compute-kernel, i.e., whatever is required to perform the computation. This way we can include not just a compute-core of a parallel computer but also accelerators, e.g., graphics processing units, or any other type of computing device.

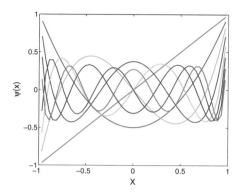

Fig. 2.4: A spectral basis for $N = 8$.

element, a basis functions expansion (as in the global Galerkin method) is used; the difference between element-based Galerkin (EBG) methods and global Galerkin methods is that the basis functions are now defined to only have compact support, i.e., they are only defined within an element. Examples of EBG methods are the finite element [289, 361, 208, 423, 349, 424], spectral element [105, 219, 306, 236], finite volume [295], and discontinuous Galerkin methods [200, 108]. Other methods belonging to the general class of EBG methods, although not covered in this text, are spectral multi-domain penalty methods [232, 233, 234, 196, 197, 173], the closely related spectral difference method [254, 395], spectral finite-volume method [393, 394], as well as the flux-reconstruction methods introduced in [210]. EBG methods can use either modal or nodal expansions defined within each element. In each of these elements, the differential equation is solved. Because the equations are enforced to hold locally, this method not only allows the solution to converge in a global sense but also in a local sense as well. In fact if you compare this family of methods with the global Galerkin method you will find that although the global method converges faster in the L^∞ sense (because it sees the solution throughout the domain as being resolved by smooth polynomials) it does not necessarily converge in the L^2 sense; whereas EBG methods (as any local method) converges in both senses.

For continuous EBG methods it makes sense to use basis functions that are continuous across element boundaries. In other words, using the nodal expansion makes sense because at the boundaries we need a matching condition and this matching condition is the value of the function q at that point. If we chose to use modal expansions we would still need to come up with matching condition at the boundaries. Therefore we would have to replace the $i = 0$ and $i = N$ modal basis functions by nodal ones (see, e.g., [361, 349, 219]). However, if we are using a discontinuous Galerkin approach, then it does not matter whether we use modal or nodal expansions.

Nodal expansions are defined as follows

$$q_N(\xi) = \sum_{i=0}^{N} q_i \psi_i(\xi)$$

where q_i, ψ_i are the values and cardinal functions at the ith node, respectively.

Remark 2.4 A cardinal function is a Lagrange polynomial with the characteristic of a Kronecker delta function δ_{ij}. Simply put, a cardinal function is 1 at its gridpoint and zero at all other gridpoints. However, between grid points the function will oscillate and will be an Nth degree polynomial. Figure 2.5 shows an example of 8th order cardinal functions.

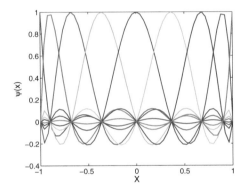

Fig. 2.5: Nodal Basis: The Lagrange cardinal functions for $N = 8$.

A modal expansion

$$q_N(\xi) = \sum_{m=0}^{N} \widetilde{q}_m \phi_m(\xi)$$

can be mapped to the nodal expansion

$$q_N(\xi) = \sum_{n=0}^{N} q_n \psi_n(\xi)$$

via a bijective (i.e., one-to-one) map, thus these two expansions are equivalent and we exploit this equivalence in later chapters (particularly in Ch. 18 when building low-pass filters). Regardless of which expansion we choose, we always have the following choices regarding EBG methods:

1. h-type (finite element method): use low order basis functions (N) but many elements.
2. p-type (finite/spectral element method): use high order basis functions (N) but few elements.

3. $N \rightarrow \infty$ yields the spectral transform method.

In addition, we show in Ch. 7 that selecting nodal expansions allows us to derive unified continuous and discontinuous Galerkin methods.

In this brief introduction to Galerkin methods we have seen that there are two things that we must do well in order to construct a successful Galerkin model:

1. interpolate the function q_N and
2. evaluate integrals of the form $\int_\Omega \psi q_N \, d\Omega$, where ψ is either a global or local basis function.

In the next two chapters we discuss interpolation and integration in detail and in so doing, we select the best methods available in order to construct numerical models.

Part II
One-Dimensional Problems

Part II focuses on solving one-dimensional problems. We begin with Chs. 3 and 4 that discuss interpolation and integration. These are then followed by Chs. 5 and 6 that treat the solution of scalar hyperbolic equations with the continuous (CG) and discontinuous Galerkin (DG) methods, respectively. Chapter 7 is the heart of this part and deals with constructing the unified CG/DG method. This chapter also shows the dissipation-dispersion analyses of the methods and discusses how to extend the methods to solve systems of nonlinear partial differential equations (e.g., shallow water and Euler). We close Part II with Chs. 8 and 9 which discuss CG and DG methods for the solution of elliptic scalar equations.

Chapter 3
Interpolation in One Dimension

3.1 Introduction

Interpolation is the act of approximating a function $f(x)$ by an Nth degree interpolant I_N such that

$$I_N(f(x_i)) = f(x_i) \tag{3.1}$$

where x_i are $i = 0, ..., N$ specific points where the function is evaluated.

Let us approximate the function $f(x)$ by the finite expansion

$$f(x) = \sum_{i=0}^{N} \phi_i(x)\tilde{f}_i + e_N(x) \tag{3.2}$$

where ϕ is the interpolating (approximating) function and is also called a basis function, $\tilde{f}$ is the expansion coefficient, and e is the error incurred by using N terms to approximate the function. The interpolation can be modal (spectral space) or nodal (physical space) which we now review.

3.2 Modal Interpolation

When modal functions are used this means that we are representing the function as the sum of multiple waves having $N + 1$ distinct frequencies and amplitudes. The question is how do we pick which functions to use? One could use a simple monomial expansion but, as we shall see, this is a bad idea and only useful when considering low order expansions. The best answer to this question is the following: we must find the *natural* basis for the function space. This means that we should first solve the singular Sturm-Liouville operator on the domain we wish to create basis

© The Editor(s) (if applicable) and The Author(s), under exclusive license to Springer Nature Switzerland AG 2020
F. X. Giraldo, *An Introduction to Element-Based Galerkin Methods on Tensor-Product Bases*, Texts in Computational Science and Engineering 24,
https://doi.org/10.1007/978-3-030-55069-1_3

functions for and then use the eigenfunctions that we obtain (see, e.g., [47]). Let us first explain what we mean by monomials and move to the Sturm-Liouville problem.

3.2.1 Monomial Expansion

Let us call $\phi(x)$ the basis function that we wish to construct. One could construct Nth degree polynomials by the following expansion

$$\phi_i(x) = x^i \ \forall \ i = 0, ..., N.$$

The first three monomials are

$$\phi_0(x) = 1, \ \phi_1(x) = x, \ \phi_2(x) = x^2.$$

This polynomial expansion has one critical deficiency - it is comprised of non-orthogonal polynomials. This is a big concern because it means that as the order increases it is possible (due to truncation error) to not be able to distinguish one basis function from another. When this happens, a discrete linear independence can no longer be guaranteed. If a set of polynomials are not linearly independent, then they cannot form a proper basis. While this issue will only arise for very large order it is best from the outset to do things in a mathematically rigorous way. Therefore, the best approach is to use orthogonal polynomials derived from the eigenvalue problem of the Sturm-Liouville operator on the domain of interest (e.g., $x \in [-1, +1]$).

3.2.2 Sturm-Liouville Operator

In one space dimension, the singular Sturm-Liouville eigenvalue problem is written as

$$\frac{d}{dx}\left[p(x)\frac{d}{dx}\phi(x)\right] + \lambda_N w(x)\phi(x) = 0 \tag{3.3}$$

where $p(x)$ and $w(x)$ are functions with $w(x)$ being the weighting function, λ are the eigenvalues and $\phi(x)$ are the eigenfunctions which satisfy this differential equation for specific domain bounds (a, b) and with specific boundary conditions. Examples of functions that can be derived from this operator include: 1) Fourier, 2) Bessel, 3) Laguerre, 4) Legendre, and 5) Chebyshev, to name only a few. From this list the only functions that are defined for finite bounds $(a, b) < \infty$ are: 1) Fourier, 2) Legendre, and 3) Chebyshev.

3.2.3 Fourier Functions

Selecting $a = 0$ and $b = 2\pi$ and imposing periodic boundary conditions results in the following simplifications for Eq. (3.3): $p(x) = w(x) = 1$ which results in

$$\frac{d^2}{dx^2}\phi(x) + \lambda\phi(x) = 0.$$

This differential equation admits solutions of the type $\phi(x) = \exp\left(i\sqrt{\lambda}x\right)$ where $i = \sqrt{-1}$. Using Euler's formula and absorbing the imaginary terms into the coefficients allow us to write

$$\phi(x) = \sum_{n=0}^{\infty} a_n \cos(nx) + b_n \sin(nx)$$

where we now call $F_n = a_n \cos(nx) + b_n \sin(nx)$ the nth degree Fourier function and $n = \sqrt{\lambda_n}$. These functions satisfy the orthogonality condition

$$\int_0^{2\pi} w(x)F_i(x)F_j(x)dx = \delta_{ij} \ \ \forall \ (i, j) = 0, ..., N$$

where N is the degree of the expansion, $w(x) = 1$ is the weighting function of this inner product, and δ_{ij} is the *Kronecker delta function* which is only non-zero for $i = j$.

3.2.4 Jacobi Polynomials

Legendre and Chebyshev polynomials fall under the general class of Jacobi polynomials for which the Jacobi differential equation is defined as

$$\frac{d}{dx}\left[\left(1 - x^2\right) w(x)\frac{d}{dx}P_N^{(\alpha,\beta)}(x)\right] + \lambda_N w(x)P_N^{(\alpha,\beta)}(x) = 0 \tag{3.4}$$

where $w(x) = (1 - x)^\alpha(1 + x)^\beta$ and $\lambda_N = N(N + \alpha + \beta + 1)$ with $x \in [-1, +1]$. The importance of this equation is that its solution is the set of orthogonal polynomials with real eigenvalues; these orthogonal polynomials, $P_N^{(\alpha,\beta)}(x)$, are the Jacobi polynomials and are orthonormal with respect to the weighting function $w(x)$. That is, the following integral is satisfied

$$\int_{-1}^{+1} w(x)P_i^{(\alpha,\beta)}(x)P_j^{(\alpha,\beta)}(x)dx = \delta_{ij} \ \ \forall \ (i, j) = 0, ..., N.$$

These orthonormal functions form a complete basis in the function space defined by the boundary conditions, i.e., the domain in which the Sturm-Liouville operator is solved. The terms α and β in the Jacobi polynomials should be assumed to be integers for our purposes in constructing basis functions. The Jacobi polynomials are described in Appendix B. From this entire family of orthogonal polynomials, Legendre polynomials (and other related polynomials) are the most widely used with EBG methods.

3.2.5 Legendre Polynomials

Legendre polynomials are a special case of Jacobi polynomials and are defined as $L_N(x) = P_N^{(\alpha,\beta)}(x)$, with $\alpha = \beta = 0$ and $a = -1$ and $b = +1$. The Legendre polynomials are the solution to the Sturm-Liouville operator

$$\frac{d}{dx}\left[\left(1 - x^2\right)\frac{d}{dx}\phi(x)\right] + \lambda_N \phi(x) = 0 \qquad (3.5)$$

where $w(x) = 1$. From this differential equation, we can see that it supports power series solutions of the type

$$\sum_{k=0}^{N} a_k x^k. \qquad (3.6)$$

with $\lambda_N = N(N+1)$. These polynomials are in fact the simplest Jacobi polynomials to consider and for this reason are typically the preferred choice for constructing orthogonal polynomials in the space $x \in [-1, +1]$ [1].

The recurrence relations for the coefficients a_k can be obtained by substituting the power series solution (3.6) into the Sturm-Liouville operator (3.5), yielding the following relations

$$\phi_0(x) = 1$$
$$\phi_1(x) = x,$$
$$\phi_N(x) = \frac{2N-1}{N}x\phi_{N-1} - \frac{N-1}{N}\phi_{N-2}(x) \quad \forall N \geq 2.$$

If we let $P_N(x) = \phi_N(x)$ then we can state the Legendre polynomials orthogonality condition as

$$\int_{-1}^{+1} w(x)P_i(x)P_j(x)dx = \delta_{ij} \quad \forall \ (i, j) = 0, ..., N$$

where the weighting function in the integral is $w(x) = 1$, if and only if $P_i(x)$ have been orthonormalized.

Exercise From the recurrence relations above, confirm that the first three Legendre polynomials are 1, x, and $\frac{1}{2}\left(3x^2 - 1\right)$. □

Exercise Starting with the monomials $1, x, x^2$, use Gram-Schmidt orthogonalization to derive the first three Legendre polynomials. □

---

[1] If we used the monomial expansion and then applied Gram-Schmidt orthogonalization we would recover the Legendre polynomials. Using Gram-Schmidt, we use the orthogonality condition for all the previous Legendre polynomials in order to derive the $N - 1$ coefficients. The N coefficient is obtained using the constraint on orthonormality which is different from the recurrence relation given above which does not assume that the polynomials have norm of unity.

3.2.6 Chebyshev Polynomials

Chebyshev polynomials are a special case of Jacobi polynomials and are defined as $T_N(x) = P_N^{(\alpha,\beta)}(x)$, with $\alpha = \beta = -\frac{1}{2}$ and $a = -1$ and $b = +1$. The Chebyshev polynomials are the solution to the Sturm-Liouville operator

$$\frac{d}{dx}\left[\left(1 - x^2\right)\frac{d}{dx}\phi(x)\right] + \frac{\lambda_N}{\sqrt{1 - x^2}}\phi(x) = 0$$

where $w(x) = \frac{1}{\sqrt{1-x^2}}$. From this differential equation, we can see that it also supports power series solutions of the type

$$\sum_{k=0}^{N} a_k x^k.$$

The solution to this differential equation results in the following set of polynomials

$$\phi_0(x) = 1$$
$$\phi_1(x) = x,$$
$$\phi_N(x) = 2x\phi_{N-1} - \phi_{N-2}(x) \quad \forall N \geq 2.$$

If we let $T_N(x) = \phi_N(x)$ then we can state that the Chebyshev polynomials satisfy the orthogonality condition

$$\int_{-1}^{+1} w(x)T_i(x)T_j(x)dx = \delta_{ij} \quad \forall \, (i, j) = 0, ..., N$$

where the weighting function is $w(x) = \frac{1}{\sqrt{1-x^2}}$.

3.2.7 More on Legendre Polynomials

Let us take a closer look at the first three Legendre polynomials

$$\phi_0(x) = 1, \ \phi_1(x) = x, \ \phi_2(x) = \frac{3}{2}x^2 - \frac{1}{2}.$$

If the function that we wish to represent is $f(x) = a$, where a is a constant, then the approximation of this function is achieved by setting

$$\tilde{f}_0 = a, \ \tilde{f}_1 = 0, \ \text{and} \ \tilde{f}_2 = 0.$$

Similarly, if $f(x) = bx$ then we can represent this function exactly with

$$\tilde{f}_0 = 0, \ \tilde{f}_1 = b, \ \text{and} \ \tilde{f}_2 = 0.$$

For a general polynomial $f(x) = a + bx + cx^2$ we can represent this function exactly
with

$$\tilde{f}_0 = a + \frac{1}{3}c, \ \tilde{f}_1 = b, \text{ and } \tilde{f}_2 = \frac{2}{3}c.$$

From this very simple example, we can see that if $f(x)$ is in fact a polynomial of
degree N we can represent the function exactly by using an Nth order Legendre
basis.

In the *modal* interpolation approach, the interpolating (basis) functions ϕ represent the various frequencies whereas the expansion coefficients $\tilde{f}$ are the amplitudes.
The amplitude-frequency space is often called *spectral* space because we are decomposing the function $f(x)$ into its wave *spectrum* which, when doing so, means that we
are evaluating wave *modes* and for this reason we refer to this type of approximation
as a *modal* approach [2]. Before discussing the *nodal* approach, it should be mentioned
that any set of polynomials will work but the best choice (as we shall see in later
sections) is to use orthogonal polynomials such as Legendre, Lobatto, Chebyshev,
Fourier, or Jacobi.

3.2.8 Vandermonde Matrix

We have described interpolation using orthogonal polynomials which is nothing more
than a finite series approximation of the function that we would like to interpolate.
For example, if we take the monomial (power series) approximation of a function,
we can write

$$f(x) = \sum_{j=0}^{N} \phi_j(x) \tilde{f}_j + e_N(x) \equiv \sum_{j=0}^{N} (x - x_0)^j \tilde{f}_j + e_N(x) \tag{3.7}$$

where x_0 is the point at which we construct our expansion about. Expanding Eq.
(3.7) up to a few terms reveals that

$$f(x) = \tilde{f}_0 + \tilde{f}_1(x - x_0) + \tilde{f}_2(x - x_0)^2 + \dots + \tilde{f}_N(x - x_0)^N + e_N(x). \tag{3.8}$$

Sampling the function at $x = x_0$ shows that in fact we have the following conditions
on the expansion coefficients

$$f(x_0) = \tilde{f}_0$$
$$f^{(1)}(x_0) = \tilde{f}_1$$
$$f^{(2)}(x_0) = 2\tilde{f}_2$$
$$f^{(3)}(x_0) = 6\tilde{f}_3$$
$$\dots = \dots$$

---

[2] In the next section we discuss the connection between modes and high-order moments (i.e.,
derivatives) and expand on this topic further in Ch. 18.

$$f^{(i)}(x_0) = i!\tilde{f}_i$$

where the superscripts on the left, e.g., $f^{(1)}$, denote derivatives. These terms in fact give the Taylor series approximation of $f(x)$ about x_0 and can be written compactly as

$$f(x) = \sum_{j=0}^{N} \frac{f^{(j)}(x_0)(x-x_0)^j}{j!} + e_N(x). \qquad (3.9)$$

Sampling the function $f(x)$ at $i = 0, ..., N$ unique x points yields

$$f(x_i) \equiv f_i = \sum_{j=0}^{N} \phi_j(x_i)\tilde{f}_j$$

which, if we expand it, yields

$$\begin{pmatrix} f_0 \\ f_1 \\ \cdots \\ f_N \end{pmatrix} = \begin{pmatrix} 1 & x_0 & x_0^2 & \cdots & x_0^N \\ 1 & x_1 & x_1^2 & \cdots & x_1^N \\ \cdots & \cdots & \cdots & \cdots & \cdots \\ 1 & x_N & x_N^2 & \cdots & x_N^N \end{pmatrix} \begin{pmatrix} \tilde{f}_0 \\ \tilde{f}_1 \\ \cdots \\ \tilde{f}_N \end{pmatrix}. \qquad (3.10)$$

This relation can be written in the following matrix-vector form

$$\mathbf{f} = \mathbf{V}\tilde{\mathbf{f}}$$

where the matrix V is known as the standard *Vandermonde* matrix . Note that if we would like to map back and forth between modal ($\tilde{f}$) and nodal space (f) requires the matrix V to be invertible and therefore non-singular (i.e., $\det(V) \neq 0$). Therefore, if V is a non-singular matrix then we can write

$$\tilde{\mathbf{f}} = \mathbf{V}^{-1}\mathbf{f}.$$

We will see later in this chapter and in Ch. 10 that the fact that $\mathbf{V}$ must be invertible imposes some restrictions not only on the polynomial functions used to construct the Vandermonde matrix (i.e., monomials versus Jacobi polynomials) but also on the interpolation points used to sample the polynomial functions. These interpolation points (or nodal points) are the building blocks of the Lagrange polynomials that we use in a nodal interpolation approach.

3.3 Nodal Interpolation

The modal interpolation approach and its corresponding Vandermonde matrix implies that there is another set of functions that can be used to approximate $f(x)$ which we write as follows

$$f(x) = \sum_{j=0}^{N} L_j(x) f_j \tag{3.11}$$

where L are the Lagrange polynomials that have the cardinal property

$$L_{ij} = \begin{cases} 1 \text{ if } i = j \\ 0 \text{ if } i \neq j \end{cases}$$

which is immediately obvious from Eq. (3.11) by substituting $x = x_i$. The question is how do we construct these functions? One way of doing this is to use the natural functions of the space, that is, the modal functions. The cardinality property of the Lagrange polynomials implies that the following condition must always be satisfied

$$\phi_i(x) = \sum_{j=0}^{N} \phi_i(x_j) L_j(x) \tag{3.12}$$

where $x_j \; \forall \; j = 0, ..., N$ are any set of interpolation points at which the modal functions, $\phi(x)$, have been sampled. Taking $x = x_k$ for some $k = 0, ..., N$ we see that the cardinality of the Lagrange polynomials yields the desired identity

$$\phi_i(x_k) = \phi_i(x_k)$$

because

$$L_j(x_k) = \begin{cases} 1 \text{ if } j = k \\ 0 \text{ if } j \neq k. \end{cases}$$

Taking advantage of the fact that $V_{ij} = \phi_i(x_j)$ is the generalized *Vandermonde* matrix [3] allows us to write Eq. (3.12) in matrix-vector form as follows

$$V_{ij} L_j(x) = \phi_i(x)$$

where left-multiplying by the inverse of V yields

$$L_i(x) = V_{ij}^{-1} \phi_j(x)$$

---

[3] The generalized Vandermonde matrix is different from the standard Vandermonde in that any modal function is used in the general but only the monomials are used in the standard.

that defines the Lagrange polynomials as functions of the orthogonal modal polynomials. Let us define the Lagrange polynomials as follows

$$L_i(x) = \sum_{j=0}^{N} V_{i,j}^{-1} \phi_j(x) \qquad (3.13)$$

because this form is the one that we shall use to construct the Lebesgue function, which is a measure of the quality of an interpolant.

3.3.1 Lebesgue Function and Lebesgue Constant

Note that the definition of the Lagrange polynomials that we use in Eq. (3.13) only ensures that the functions are cardinal but it says nothing about how good these functions will be for interpolation. Let us now introduce a measure of interpolation that we will use frequently. For a Lagrange interpolation function (note that we will also call these functions Lagrange polynomials or Lagrangian basis functions) a good measure of its interpolation quality is the Lebesgue function

$$\Lambda_N(x) = \sum_{i=0}^{N} |L_i(x)| \qquad (3.14)$$

and its associated Lebesgue constant defined as

$$\Lambda_N = max \left(\sum_{i=0}^{N} | L_i(x) | \right) \qquad (3.15)$$

where the maximum is obtained for the entire domain $x \in D$ where D is the domain of interest (e.g., $x \in [-1, +1]$).

3.3.2 Lagrange Polynomials

In one dimension, the general definition of Lagrange polynomials is

$$L_i(x) = \prod_{\substack{j=0 \\ j \neq i}}^{N} \frac{(x - x_j)}{(x_i - x_j)}. \qquad (3.16)$$

Let ϕ_N be the following *generating* polynomial [4]

$$\phi_N(x) = \prod_{j=0}^{N} (x - x_j) \qquad (3.17)$$

---

[4] To be more precise, it should be called the Lagrange generating polynomial because it is the form of the polynomial that we shall use to generate the Lagrange polynomials.

that we will use to construct the Lagrange basis functions where the points x_j are in fact its roots. The derivative of ϕ_N is

$$\phi'_N(x) = \sum_{k=0}^{N} \frac{1}{(x - x_k)} \left(\prod_{j=0}^{N}(x - x_j) \right) \tag{3.18}$$

which, for a specific value of x, that is $x = x_i \in [x_0, x_1, ..., x_N]$, yields

$$\phi'_N(x_i) = \prod_{\substack{j=0 \\ j \neq i}}^{N}(x_i - x_j). \tag{3.19}$$

Combining Eqs. (3.17) and (3.19) yields

$$L_i(x) = \frac{\phi_N(x)}{\phi'_N(x_i)(x - x_i)} \equiv \prod_{\substack{j=0 \\ j \neq i}}^{N} \frac{(x - x_j)}{(x_i - x_j)} \tag{3.20}$$

which now allows us to see the exact relationship between the generating polynomials (which are of modal form) and the Lagrange basis functions (which are of nodal form).

Exercise Use Eq. (3.20) to derive (using the generating polynomials) Lagrange polynomials for linear and quadratic polynomials for equi-spaced points.   □

Let us simplify Eq. (3.20) by only writing down the form on the far right as

$$L_i(x) = \prod_{\substack{j=0 \\ j \neq i}}^{N} \frac{(x - x_j)}{(x_i - x_j)} \tag{3.21}$$

which now defines the form of the Lagrange polynomials that we shall use to construct our Lagrange basis functions. Before discussing the algorithm for constructing Lagrange polynomials, we should mention that these polynomials are endowed with a special property called the *Partition of Unity* . The partition of unity means that for any point in which the Lagrange polynomials are defined we are guaranteed the following:

$$\sum_{i=0}^{N} L_i(x_j) = 1 \;\; \forall x_j \in [-1, +1];$$

this relation says that for any point in the space, we are guaranteed that the Lagrange polynomials will sum to unity. This is an important property because it can be easily exploited to prove, e.g., conservation of the direct-stiffness-summation operator (see, e.g., [370, 230, 231]). Let us now turn to the algorithm for constructing Lagrange polynomials. In Algorithm 3.1 we present pseudocode for building the Lagrange polynomials using the form given in Eq. (3.21).

Algorithm 3.1 Recipe for constructing Lagrange polynomials.

```
function LAGRANGE_BASIS
    for l = 0 : Q do
        x_l = x_sample(l)
        for i = 0 : N do
            x_i = x_roots(i)
            L(i, l) = 1
            for j = 0 : N do
                x_j = x_roots(j)
                if (j ≠ i) then
                    L(i, l) *= (x_l − x_j)/(x_i − x_j)
                end if
            end for
        end for
    end for
end function
```

In Alg. 3.1 $x_{sample}(l) \; \forall \; l = 0, \ldots, Q$ represents the locations where one wishes to sample the basis functions. The variable $x_{roots}(i) \; \forall \; i = 0, \ldots, N$ refers to the points (typically the roots of the generating polynomials) that are used to construct the associated Lagrange polynomials. The operator $*=$ means that we take the current value of $L(i, l)$, multiply it by the right-hand side and then update it.

To use Lagrange polynomials for interpolating the solution of a differential equation will also require knowing how to approximate the derivative of a function. Therefore, we must know how to define the derivative of Lagrange polynomials. The definition of the derivative of a Lagrange polynomial can be constructed directly from Eq. (3.21) to give

$$\frac{dL_i}{dx}(x) = \sum_{\substack{k=0 \\ k \neq i}}^{N} \left(\frac{1}{x_i - x_k} \right) \prod_{\substack{j=0 \\ j \neq i \\ j \neq k}}^{N} \frac{(x - x_j)}{(x_i - x_j)}. \tag{3.22}$$

The algorithm for constructing the derivative of the Lagrange polynomials is given in Algorithm 3.2.

In Alg. 3.2, the operator $+=$ means that we take the current value of $\frac{dL}{dx}(i, l)$ and then add to it what is on the right-hand side. Note that Algs. 3.1 and 3.2 can be combined into one function.

Exercise Combine Algs. 3.1 and 3.2 within one set of loops.   □

Algorithm 3.2 Recipe for constructing Lagrange polynomials derivatives.

```
function LAGRANGE_BASIS_DERIVATIVE
    for l = 0 : Q do
        x_l = x_sample(l)
        for i = 0 : N do
            x_i = x_roots(i)
            for j = 0 : N do
                x_j = x_roots(j)
                prod = 1
                if (j ≠ i) then
                    for k = 0 : N do
                        x_k = x_roots(k)
                        if (k ≠ i & k ≠ j) then
                            prod *= (x_l − x_k)/(x_i − x_k)
                        end if
                    end for
                    dL/dx(i, l) += prod/(x_i − x_j)
                end if
            end for
        end for
    end for
end function
```

3.3.3 Quality of Lagrange Polynomial Interpolation

Now that we know how to construct Lagrange polynomials, let us see the quality of interpolation that can be achieved with different sets of interpolation points. When we say *interpolation points* we mean the points used as the *basis* of our interpolation functions. In Eq. (3.16) these are the points x_i and x_j. In the next sections, we shall explore various sets of points on the interval $x \in [-1, 1]$. To test these various point sets, we shall use the Runge function (also known as the Agnesi mountain)

$$f(x) = \frac{1}{1 + 50x^2} \tag{3.23}$$

defined on the line $x \in [-1, +1]$.

3.3.3.1 Equi-Spaced Interpolation Points

Figure 3.1 shows the 8th order ($N = 8$) Lagrange polynomials based on equi-spaced (uniform spacing) points. The left panel shows the basis functions themselves (various colors) as well as their associated Lebesgue function (dashed line) and the right panel shows the quality of the interpolation for the Runge function. What is important about Fig. 3.1 is the following: note how the Lagrange polynomials oscillate with large amplitudes for all values of x that are not the interpolation points (the interpolation points are marked by the dots). These large oscillations (known as

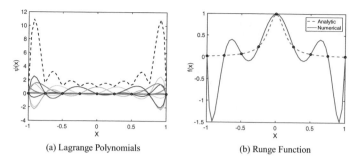

(a) Lagrange Polynomials   (b) Runge Function

Fig. 3.1: Equi-Spaced Points: The (a) Lagrange polynomials and associated Lebesgue function (dashed line) and (b) interpolation of the Runge function for 8th order equispaced points.

the Runge effect) can be seen especially near the end points (near $x = -1$ and $x = 1$) where the basis functions (labeled as ψ) reach values above $\psi(x) = 1$. The points at which these oscillations are largest can be discerned by looking at the single (dashed black) curve above the Lagrange polynomials in Fig. 3.1a which is the Lebesgue function. Ideal values of the Lebesgue function are near 1 but clearly this will only be true at the interpolation points.

These wild oscillations have a severe effect on the quality of the interpolant. Specifically, for the Runge function we show in Fig. 3.1b that the interpolation based on equi-spaced points oscillates wildly near the end points precisely because the Lagrange polynomials are poorly behaved in this region.

3.3.3.2 Chebyshev Points

The Chebyshev polynomials are defined by the following recurrence relation

$$
\begin{aligned}
\phi_0^{Cheb}(x) &= 1 \\
\phi_1^{Cheb}(x) &= x, \\
\phi_N^{Cheb}(x) &= 2x\phi_{N-1}^{Cheb} - \phi_{N-2}^{Cheb}(x) \;\; \forall N \geq 2
\end{aligned}
\tag{3.24}
$$

where the Nth order interpolation points are obtained from the roots of the (N+1)th order Chebyshev polynomials. The roots of the Chebyshev polynomials can be obtained in closed form and are given by

$$
x_i = \cos\left(\left[\frac{2i+1}{2N+2}\right]\pi\right) \;\; i = 0, \ldots, N.
$$

Figure 3.2 shows the 8th order ($N = 8$) Lagrange polynomials based on the roots of the Chebyshev points. The left panel in Fig. 3.2 shows the basis functions and their associated Lebesgue function (dashed line) and the right panel shows the quality of the interpolation for the Runge function. Note how uniform the Lebesgue function (Fig.

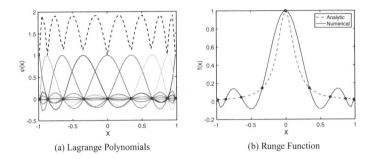

(a) Lagrange Polynomials   (b) Runge Function

Fig. 3.2: Chebyshev Points: The (a) Lagrange polynomials and associated Lebesgue function (dashed line) and (b) interpolation of the Runge function for 8th order Chebyshev points.

3.2a) is compared to the equi-spaced points. This behavior in the Lebesgue function clearly has a positive effect on the interpolation quality of the Lagrange polynomials based on these points. Figure 3.2b shows that these interpolation functions do very well interpolating the Runge function.

3.3.3.3 Legendre Points

The Legendre polynomials are constructed via the recurrence relations

$$
\begin{aligned}
\phi_0^{Leg}(x) &= 1 \\
\phi_1^{Leg}(x) &= x, \\
\phi_N^{Leg}(x) &= \frac{2N-1}{N} x \phi_{N-1}^{Leg} - \frac{N-1}{N} \phi_{N-2}^{Leg}(x) \quad \forall N \geq 2
\end{aligned}
\tag{3.25}
$$

and the Nth order interpolation points are obtained from the roots of the (N+1)th order Legendre polynomial. Algorithm 3.3 shows the pseudocode for constructing the Legendre points, where ϕ refers to the Legendre polynomials.

Figure 3.3 shows the 8th order ($N = 8$) Lagrange polynomials based on the roots of Legendre polynomials. The left panel shows the basis functions as well as their associated Lebesgue function (dashed line) and the right panel shows the quality of the interpolation for the Runge function. The Lagrange polynomials based on the roots of the Legendre polynomials are better behaved than the equi-spaced points as is evident by the well-behaved Lebesgue function which shows the spikes near the end points much less pronounced than those for the equi-spaced points. The Lebesgue

Algorithm 3.3 Algorithm to construct Legendre and Lobatto points in the interval $x \in [-1, +1]$.

for $i = 0 : N$ **do**
    pick $x_i^0 = x_i^{(Cheb)}$
    for $k = 0 : K$ **do**
        $x_i^{k+1} = x_i^k - \phi_N(x_i^k)/\phi_N'(x_i^k)$
        if $|x_i^{k+1} - x_i^k| < \epsilon$ **then**
            exit k-loop
        end if
    end for
end for

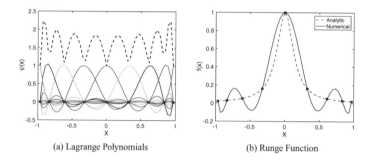

(a) Lagrange Polynomials   (b) Runge Function

Fig. 3.3: Legendre Points: The (a) Lagrange polynomials and associated Lebesgue function (dashed line) and (b) interpolation of the Runge function for 8th order polynomials.

function in Fig. 3.3a shows that the interpolation functions are well-behaved indeed and Fig. 3.3b confirms this fact by not exhibiting the oscillations near the endpoints exhibited by the equi-spaced points.

Both roots of the Legendre and Chebyshev polynomials do extremely well in interpolating the Runge function; however, these roots do not contain the endpoints of the 1-simplex and therefore are of little use in constructing continuous EBG methods such as finite element and spectral element methods. For continuous EBG methods, we need to look for a good set of interpolation points that also contain the endpoints.

3.3.3.4 Lobatto Points

Another good set of polynomials are known as the Lobatto polynomials and we shall next use the roots of these polynomials. The Lobatto polynomials are derived from the Legendre polynomials as follows

$$\phi_N^{Lob}(x) = (1 + x)(1 - x)\frac{d}{dx}\phi_{N-1}^{Leg}(x) \equiv (1 - x^2)\frac{d}{dx}\phi_{N-1}^{Leg}(x) \quad \forall N \geq 2$$

that is, the Nth order Lobatto polynomials are nothing more than the derivative of the (N-1)th Legendre polynomials with the end points $x = \pm 1$ included. The Nth order Lagrange polynomial is constructed using the roots of the (N+1)th order Lobatto polynomials; e.g., for linear Lagrange polynomial (N=1), we need to find the roots of the 2nd order Lobatto polynomials, $\phi_2^{Lob}(x)$, etc. The roots of the Lobatto polynomials are given by Algorithm 3.3 where ϕ now must be the Lobatto polynomials.

Figure 3.4 shows the 8th order ($N = 8$) Lagrange polynomials based on Lobatto points. The left panel shows the basis functions themselves as well as their associated Lebesgue function (dashed line) and the right panel shows the quality of the interpolation for the Runge function.

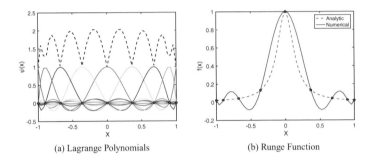

(a) Lagrange Polynomials   (b) Runge Function

Fig. 3.4: Lobatto Points: The (a) Lagrange polynomials and associated Lebesgue function (dashed line) and (b) interpolation of the Runge function for 8th order Lobatto points.

Figure 3.4 shows that the Lagrange polynomials based on the Lobatto points are well-behaved. Figure 3.4a shows this via the corresponding Lebesgue function while Fig. 3.4b shows that no wild oscillations are exhibited when interpolating the Runge function.

3.3.3.5 Comparison of Lebesgue Constants

Figure 3.5 shows the Lebesgue constants for various Lagrange polynomial orders for (a) equi-spaced points, (b) Chebyshev, (c) Legendre, and (d) Lobatto points. The interpolation based on equi-spaced points yields very large Lebesgue constants for modest degrees of N; for example, for $N = 5$ the Lebesgue constant approaches a value of 4. For the equi-spaced points, the Lebesgue constant increases exponentially with N while for all the other points it is bounded ($\Lambda_N < 3$).

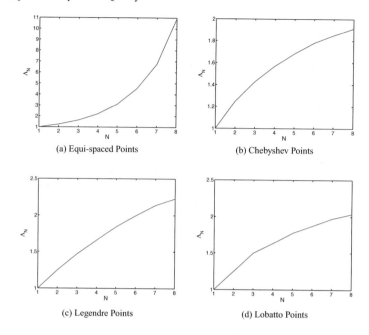

Fig. 3.5: Lebesgue constant for various Lagrange polynomial orders N for (a) equi-spaced, (b) Chebyshev, (c) Legendre, and (d) Lobatto points.

The Chebyshev points perform the best with respect to the Lebesgue constant, with the Lobatto and Legendre points following closely. However, Chebyshev points tend not to be used in EBG codes because they are only orthogonal (in an L^2 sense, i.e., inner product) with respect to a weighting function $w(x) \neq 1$ which makes them unwieldy to use in Gauss quadrature formulas. For this reason, Lobatto and Legendre points tend to be the preferred choice.

3.4 Example of Interpolation Quality

Let us assume that we wish to interpolate the function

$$q(x) = \cos\left(\frac{\pi}{2}x\right) \quad x \in [-1, +1]$$

and its derivative

$$\frac{dq(x)}{dx} = -\frac{\pi}{2}\sin\left(\frac{\pi}{2}x\right) \quad x \in [-1, +1]$$

using Lagrange polynomials as follows

$$q(x) = \sum_{j=0}^{N} L_j(x) q_j$$

and

$$\frac{dq(x)}{dx} = \sum_{j=0}^{N} \frac{dL_j(x)}{dx} q_j$$

where $L(x)$ are based on various point sets (e.g., equi-spaced, Chebyshev, Legendre, and Lobatto points).

Let us compute the error based on sampling the known function, $q(x)$, at 101 equi-spaced points by varying the polynomial order of $L(x)$ from N=1 to N=64. We use the following normalized error norm definitions

$$L^1 = \frac{\sum_{i=1}^{N_P} |q_i^{(num)} - q_i^{(exact)}|}{\sum_{i=1}^{N_P} |q_i^{(exact)}|}, \tag{3.26}$$

$$L^2 = \frac{\sqrt{\sum_{i=1}^{N_P} \left(q_i^{(num)} - q_i^{(exact)}\right)^2}}{\sqrt{\sum_{i=1}^{N_P} \left(q_i^{(exact)}\right)^2}}, \tag{3.27}$$

and

$$L^\infty = \frac{\max_{1 \leq i \leq N_P} \left(|q_i^{(num)} - q_i^{(exact)}|\right)}{\max_{1 \leq i \leq N_P} \left(|q_i^{(exact)}|\right)} \tag{3.28}$$

where $N_P = 101$ and q represents the function values as well as its derivative.

3.4.1 Equi-Spaced Points

Figure 3.6 shows the results using equi-spaced points. Figure 3.6a and 3.6b show the approximation of the function and its derivative using $N = 64$ order polynomials. Clearly, the interpolation of the function and its derivative breaks down meaning that the Lagrange interpolation based on equi-spaced points is not able to represent the function and its derivative. Figure 3.6c and 3.6d show the errors for the approximation of the function and its derivative for various orders of N. For low orders of N, the Lagrange polynomials based on equi-spaced points does an adequate job, but as N increases, the interpolation breaks down. The problem is that the Lebesgue constant for these points increases exponentially meaning that the polynomials are oscillating wildly and, consequently, the approximation breaks down.

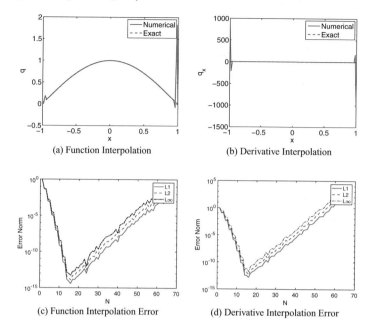

(a) Function Interpolation   (b) Derivative Interpolation

(c) Function Interpolation Error   (d) Derivative Interpolation Error

Fig. 3.6: Equi-Spaced Points. The (a) interpolation of the function using N=64, (b) the interpolation of the derivative of the function using N=64, (c) the interpolation error for various N, and (d) the derivative error for various N.

3.4.2 Chebyshev Points

Figure 3.7 shows the results using Chebyshev points. Note that these are defined in closed form as follows

$$x_i^{Cheb} = \cos\left(\frac{2i+1}{2N+2}\pi\right) \tag{3.29}$$

where $i = 0, \ldots, N$ and N is the order of the Chebyshev polynomial. Figure 3.7a and 3.7b show the approximation of the function and its derivative using $N = 64$ order polynomials. The results for the Chebyshev points are much better than those for the equi-spaced points since they decrease the error exponentially with increasing order N until the error levels off because it has reached machine zero.

3.4.3 Legendre Points

Figure 3.8 shows the results using Legendre points. Figure 3.8a and 3.8b show the approximation of the function and its derivative using $N = 64$ order polynomials.

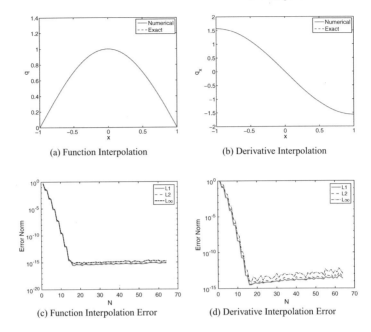

(a) Function Interpolation   (b) Derivative Interpolation

(c) Function Interpolation Error   (d) Derivative Interpolation Error

Fig. 3.7: Chebyshev Points. The (a) interpolation of the function using N=64, (b) the interpolation of the derivative of the function using N=64, (c) the interpolation errors for various N, and (d) the derivative errors for various N.

The function and its derivative are well-represented. Figure 3.8c and 3.8d show that the errors incurred in approximating the function and its derivative decrease exponentially where, at $N = 16$, the error reaches machine double precision. The errors incurred by the Legendre points are similar to those given by the Chebyshev points.

3.4.4 Lobatto Points

Figure 3.9 shows the results using Lobatto points. Figure 3.9a and 3.9b show the approximation of the function and its derivative using $N = 64$ order polynomials. The results for the Lobatto points are similar to those for the Chebyshev and Legendre points, i.e., that the error decreases exponentially with increasing order N (see Figure 3.9c and 3.9d).

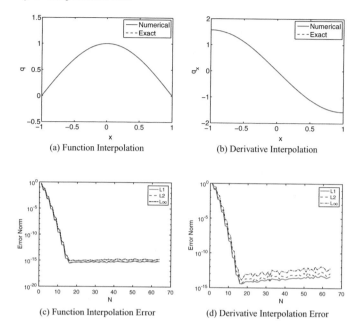

Fig. 3.8: Legendre Points. The (a) interpolation of the function using N=64, (b) the interpolation of the derivative of the function using N=64, (c) the interpolation errors for various N, and (d) the derivative errors for various N.

3.5 Summary of Interpolation Points

In this chapter, we have described the various forms of interpolation, including both modal and nodal approximations. We have addressed how to derive the generating (modal) polynomials via the solution of the Sturm-Liouville operator. Therefore, we can always derive the natural basis for a specific domain by solving for the eigenfunctions of the Sturm-Liouville operator with the proper boundary conditions. Once this is done, we then need to decide if we will use the basis functions in their modal or nodal form. Although either may be used, we will mainly focus on nodal basis functions in this text. If nodal basis functions are to be used then we need to decide which interpolation points to use. We have seen that we have many choices including Chebyshev, Legendre, and Lobatto points. If discontinuous EBG methods are to be used then any of these points are possible. However, if continuous EBG methods are to be used then Lobatto points are the only viable option because they contain the endpoints. We will see in the next chapter that the Legendre and Lobatto points are also endowed with good integration properties and, for this reason, are primarily used throughout this text.

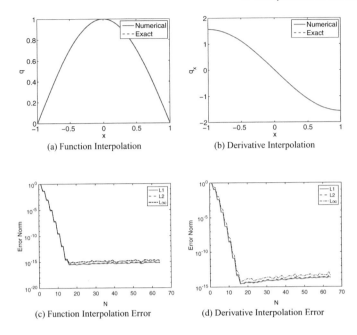

(a) Function Interpolation   (b) Derivative Interpolation

(c) Function Interpolation Error   (d) Derivative Interpolation Error

Fig. 3.9: Lobatto Points. The (a) interpolation of the function using N=64, (b) the interpolation of the derivative of the function using N=64, (c) the interpolation errors for various N, and (d) the derivative errors for various N.

Chapter 4
Numerical Integration in One Dimension

4.1 Introduction

Galerkin methods require the evaluation of integrals of the type

$$A = \int_{\Omega} f(\mathbf{x})d\Omega \ \text{ and } \ B = \int_{\Gamma} f(\mathbf{x})d\Gamma \tag{4.1}$$

where Ω is the domain that we wish to integrate within and Γ are its boundaries. Thus A and B are element and trace integrals, respectively. By element integrals we mean either area or volume integrals in 2D and 3D, respectively. By trace integrals we mean integrals along the boundary of the element which could be line or surface area integrals in 2D and 3D, respectively.

We cannot integrate $f(\mathbf{x})$ analytically because f is an unknown function that we, unfortunately, do not know in closed form. Instead, we only have f sampled at certain selected spatial positions. This then motives the need for numerical integration which in one dimension is referred to as *quadrature* or in more dimensions as *cubature*. The idea behind numerical integration is very much tied to the ideas of interpolating functions by using polynomial representations. Thus if we assume that a function takes the form of a well-known polynomial then we can figure out how to integrate this polynomial within a given *a priori* error bound. In order to simplify the discussion of integration we need to first introduce the concept of Newton divided difference formulas for representing polynomials.

F. X. Giraldo, *An Introduction to Element-Based Galerkin Methods on Tensor-Product Bases*, Texts in Computational Science and Engineering 24, https://doi.org/10.1007/978-3-030-55069-1_4

4.2 Divided Differences

Recall that we can approximate a function $f(x)$, that is the map $\mathcal{R} \to \mathcal{R}$, by the polynomial series

$$f(x) = \sum_{i=0}^{N} a_i x^i + e_N(x)$$

where a_i are the expansion coefficients and $e_N(x)$ the error remaining after using only N terms in the truncated series. However, for the purposes of discussing integration we will use the following slightly different form of this approximation

$$f(x) = \sum_{i=0}^{N} a_i (x - x_0)...(x - x_{i-1}) + e_N(x) \tag{4.2}$$

where it should be understand that the product $(x - x_0)...(x - x_{i-1})$ excludes x_i such that for $i = 0$ we only get a_0 on the right-hand side. Let us write Eq. (4.2) in the following form

$$f(x) = p_N(x) + e_N(x) \tag{4.3}$$

where $p_N(x) = \sum_{i=0}^{N} a_i (x - x_0)...(x - x_{i-1})$ is the Nth degree polynomial representation of $f(x)$. Let us now concentrate on $p_N(x)$.

If we want to match the function f at specific points $x_0, x_1, ..., x_N$ then we can figure out unique values of $a_0, a_1, ..., a_N$. If we let $x = x_0$ in Eq. (4.2) then we see that

$$f(x_0) = a_0$$

and for $x = x_1$ we get

$$f(x_1) = a_0 + a_1(x_1 - x_0)$$

that yields

$$a_1 = \frac{f(x_1) - f(x_0)}{x_1 - x_0}.$$

Introducing the divided difference notation:

$$\begin{aligned} f[x_i] &= f(x_i) \\ f[x_i, x_{i+1}] &= \frac{f[x_{i+1}] - f[x_i]}{x_{i+1} - x_i} \\ f[x_i, x_{i+1}, x_{i+2}] &= \frac{f[x_{i+1}, x_{i+2}] - f[x_i, x_{i+1}]}{x_{i+2} - x_i} \end{aligned} \tag{4.4}$$

allows us to write for $x = x_2$

$$f(x_2) = a_0 + a_1(x_2 - x_0) + a_2(x_2 - x_0)(x_2 - x_1)$$

that, when substituting for a_0 and a_1 yields

$$f(x_2) = f(x_0) + \frac{f(x_1) - f(x_0)}{x_1 - x_0}(x_2 - x_0) + a_2(x_2 - x_0)(x_2 - x_1).$$

Rearranging gives

$$a_2 = \frac{f(x_2) - f(x_0) - \frac{f(x_1)-f(x_0)}{x_1-x_0}(x_2 - x_0)}{(x_2 - x_0)(x_2 - x_1)}$$

and, adding $f(x_1) - f(x_1)$ in the numerator yields

$$a_2 = \frac{f(x_2) - f(x_1) + f(x_1) - f(x_0) - \frac{f(x_1)-f(x_0)}{x_1-x_0}(x_2 - x_0)}{(x_2 - x_0)(x_2 - x_1)}. \tag{4.5}$$

Upon simplifying this expression [1] we get

$$a_2 = \frac{f[x_1, x_2] - f[x_0, x_1]}{x_2 - x_0} \equiv f[x_0, x_1, x_2].$$

In fact, it turns out that if we continued with this approach we would get the following expression for the polynomial representation

$$p_N(x) = a_0 + a_1(x - x_0) + a_2(x - x_0)(x - x_1) + \dots + a_N(x - x_0)\dots(x - x_{N-1})$$

which can be written more compactly as

$$p_N(x) = \sum_{i=0}^{N} f[x_0, \dots, x_i] \prod_{j=0}^{i-1}(x - x_j);$$

this now leads us to the following theorem that we state without proof.

Theorem 4.1 *For $f \in C^N[a, b]$ and a set of distinct points $(x_0, \dots, x_N) \in [a, b]$ then there exists a value $\xi \in (a, b)$ such that*

$$f[x_0, x_1, \dots, x_N] = \frac{f^{(N)}(\xi)}{N!}.$$

The importance of this theorem is that it tells us that the polynomial approximation takes the following form

$$f(x) = p_N(x) + \frac{f^{(N+1)}(\xi)}{(N+1)!} \prod_{i=0}^{N}(x - x_i) \tag{4.6}$$

where the last term on the right is in fact the error term that clearly vanishes when x is one of the points x_i for $i = 0, \dots, N$ since we imposed that the approximation satisfy the true function at these points. Let us rewrite Eq. (4.2) in the following form

$$f(x) = p_N(x) + \phi_N(x)f[x_0, x_1, \dots, x_N, x] \tag{4.7}$$

that allows us to see that if $f(x)$ is of degree M and $p_N(x)$ is of degree N, then $\phi_N(x)$ is of degree N and $f[x_0, \dots, x_N, x]$ is of degree $M - N$; note that $\phi(x) = \prod_{i=0}^{N}(x - x_i)$ is the generating polynomial.

---

[1] The last two terms in Eq. (4.5) can be combined to give $-\frac{f[x_0, x_1]}{(x_2-x_0)}$.

4.3 Numerical Integration

4.3.1 Introduction

Let us suppose that we wish to integrate Eq. (4.7) in the interval $[a, b]$. We can then write

$$I = \int_a^b f(x)\, dx = \int_a^b p_N(x)\, dx + \int_a^b e_N(x)\, dx \tag{4.8}$$

where

$$e_N(x) = \phi_N(x) f[x_0, x_1, ..., x_N, x]$$

is the error term and if we wish to satisfy the integral exactly requires that

$$\int_a^b e_N(x)\, dx = 0.$$

This condition is satisfied provided that ϕ_N is of degree N and $f[x_0, x_1, ..., x_N, x]$ is of degree less than or equal to N-1 since, for families of orthogonal polynomials, all Nth degree polynomials are orthogonal to all polynomials of degree less than or equal to N-1 (you can see this by recalling how Gram-Schmidt orthogonalization builds families of orthogonal polynomials). Note that the product $\phi_N f[x_0, x_1, ..., x_N, x]$ is a polynomial of degree $2N - 1$.

 This brief analysis shows that one can achieve order $2N - 1$ integration accuracy by choosing orthogonal polynomials of degree N for the same family for ϕ_N and $f[x_0, ..., x_N, x]$, where the sampling points for the numerical integration are the roots of orthogonal polynomials. We have not yet mentioned how to compute the quadrature weights; we discuss how to compute these weights in the sections below.

4.3.2 Quadrature Roots

Since we know that the sampling points for numerical integration must be the roots of orthogonal polynomials, let us now describe how to compute them.

 Let $\phi_N(x)$ be an Nth degree orthogonal polynomial. We can write the solution to this nonlinear problem using Newton's method as follows

$$\phi_N\left(x^{k+1}\right) = \phi_N\left(x^k\right) + \left(x^{k+1} - x^k\right)\phi_N'\left(x^k\right) \equiv 0 \tag{4.9}$$

where x^k is the kth approximation to the roots. Eq. (4.9) becomes zero when x^k is equal to the root of ϕ_N. Therefore, we iterate to find the roots as follows

$$x^{k+1} = x^k - \frac{\phi_N\left(x^k\right)}{\phi_N'\left(x^k\right)}.$$

Recall that for Newton's method, we need a good initial guess in order to converge to the proper solution. We use the Chebyshev roots as the initial guess that are quite near to the Legendre and Lobatto roots and are obtained in closed form as follows

$$x_i^0 = \cos\left(\pi\frac{2i+1}{2N+2}\right) \forall i = 0, \ldots, N.$$

4.3.3 Quadrature Weights

Let us write the numerical integration (quadrature) problem as follows

$$I = \int_{-1}^{+1} f(x)\,dx \tag{4.10}$$

where, without loss of generality, we have assumed that the interval of integration is defined on the one-dimensional reference element $x \in [-1, +1]$.

Next, let us approximate the function $f(x)$ by an Nth degree Lagrange polynomial interpolant

$$f_N(x) = \sum_{j=0}^{N} L_j(x)f_j \tag{4.11}$$

where L_j are the nodal interpolation functions (i.e., Lagrange polynomials) and $f_j = f(x_j)$ are the values of $f(x)$ at specific points $x_j \in [-1, +1]$. Substituting Eq. (4.11) into (4.10) yields

$$I \approx \int_{-1}^{+1} \left(\sum_{j=0}^{N} L_j(x)f_j\right) dx. \tag{4.12}$$

Using a quadrature formula, we can represent Eq. (4.12) as follows

$$\int_{-1}^{+1} \left(\sum_{j=0}^{N} L_j(x)f_j\right) dx = \sum_{i=0}^{N} w_i \left(\sum_{j=0}^{N} L_j(x_i)f_j\right) \tag{4.13}$$

where w_i are the weights corresponding to the quadrature points $x_i \forall i = 0, \ldots, N$ which are the roots of the Nth degree generating polynomials ϕ_N (i.e., the modal functions). Reordering the summation and integral signs in Eq. (4.13) yields

$$\sum_{j=0}^{N} \left(\int_{-1}^{+1} L_j(x)\,dx\right) f_j = \sum_{j=0}^{N} \left(\sum_{i=0}^{N} w_i L_j(x_i)\right) f_j. \tag{4.14}$$

From Eq. (4.14), it becomes evident that we can solve for the quadrature weights from the linear matrix problem

$$w_i = \int_{-1}^{+1} L_i(x)\, dx \tag{4.15}$$

where we have used the cardinality property of Lagrange polynomials to cancel the term $L_j(x_i)$ from the right-hand side of Eq. (4.14).

For certain orthogonal polynomials, the quadrature weights can be shown to have very simple closed form representations.

4.3.3.1 Legendre-Gauss Weights

Legendre-Gauss (LG) quadrature formulas correspond to the orthogonal generating polynomial ϕ_N being the Legendre polynomials with the weights given as follows

$$w_i = \frac{2}{(1 - x^2)\left(\phi_N'(x_i)\right)^2}$$

where x_i are the roots of $\phi_N(x)$. These quadrature formulas can be used to evaluate integrals of the type

$$\int_{-1}^{+1} f(x)\,dx = \sum_{i=0}^{N} w_i f(x_i)$$

where x_i are the zeros of the Legendre polynomials given in Ch. 3 and are accurate for up to $2N + 1$ degree polynomials. This is the case because Legendre-Gauss has N+1 roots plus N+1 weights as the total degrees of freedom (2N+2) which allows a maximum of 2N+1 accuracy (order=DOF-1).

4.3.3.2 Legendre-Gauss-Lobatto Weights

Legendre-Gauss-Lobatto (LGL) quadrature formulas correspond to the orthogonal generating polynomial ϕ_N being the Lobatto polynomials. However, we express the weights below in terms of the Legendre polynomials since the Lobatto polynomials are related to the derivatives of the Legendre polynomials. We express the LGL weights as

$$w_i = \frac{2}{N(N + 1)\left(P_N(x_i)\right)^2}$$

where $P_N(x)$ are the Nth degree Legendre polynomials and x_i are the roots of the Lobatto polynomials ϕ_N. These quadrature formulas can be used to evaulate integrals of the type

$$\int_{-1}^{+1} f(x)\,dx = \sum_{i=0}^{N} w_i f(x_i)$$

where x_i are the zeros of the Lobatto polynomials given in Ch. 3 and are accurate up to $2N - 1$ degree polynomials. The reason why these integration formulas are less accurate than Legendre-Gauss is due to the fact that with Legendre-Gauss-Lobatto we have two fewer degrees of freedom to choose from since we insist that the end points ($x = \pm 1$) are included in the set of quadrature points (N-1 roots plus N+1 weights = 2N DOFS which allows a maximum of 2N-1 accuracy).

4.3.3.3 Chebyshev-Gauss Weights

Chebyshev-Gauss (CG) quadrature formulas correspond to the orthogonal generating polynomial ϕ_N being the Chebyshev polynomials with the weighting functions defined as $w(x) = \frac{1}{\sqrt{1-x^2}}$. The Chebyshev-Gauss weights are quite simple and are all constant and defined as

$$w_i = \frac{\pi}{N+1}$$

where N is the order of the Chebyshev polynomials having $N + 1$ points. These quadrature formulas can be used to evaulate integrals of the type

$$\int_{-1}^{+1} \frac{1}{\sqrt{1-x^2}} f(x)dx = \sum_{i=0}^{N} w_i f(x_i)$$

where x_i are the zeros of the Chebyshev polynomials given in Ch. 3 and are accurate up to $2N + 1$ degree polynomials. The fact that we must contend with a weighting function means that Chebyshev-Gauss integration is not as convenient for integrating functions that do not have weighting functions of unity.

For this reason, we shall restrict ourselves to Legendre-Gauss and Legendre-Gauss-Lobatto in most of the text. It vastly simplifies the construction of numerical algorithms if we are able to satisfy orthogonality in the L^2 sense (L^2 integral) as follows

$$\int_{-1}^{+1} \phi_i(x)\phi_j(x)w(x) = \delta_{ij}$$

especially when $w(x) = 1$, which is the case for both Legendre and Lobatto polynomials whereas $w(x) = \frac{1}{\sqrt{1-x^2}}$ for Chebyshev polynomials. The unity of the weighting function $w(x)$ simplifies the use of both Legendre and Lobatto polynomials and is the reason why they have become the *de facto* orthogonal polynomials used in element-based Galerkin methods.

4.4 Example of Integration Quality

Let us suppose that we wish to integrate the function

$$q(x) = \cos\left(\frac{\pi}{2}x\right) \quad x \in [-1, +1]$$

using Gauss quadrature that replaces the integral by a summation as follows

$$\int_{-1}^{+1} q(x)dx \approx \sum_{k=0}^{Q} w_k q(x_k).$$

We need to approximate the function $q(x)$ at the quadrature points x_k and we shall do this by using the Lagrange polynomial expansion

$$q(x_k) = \sum_{j=0}^{N} L_j(x_k)q_j$$

where $L(x)$ are constructed on specific point sets (e.g., equi-spaced, Chebyshev, Legendre, and Lobatto points).

To compute the error, we use the analytic integral which is

$$\int_{-1}^{+1} \cos\left(\frac{\pi}{2}x\right) dx = \frac{4}{\pi}.$$

Let us see how some of these point sets perform as we vary the polynomial order of the basis functions, $L(x)$, from N=1 to N=64. To measure the integration error, we use the following normalized error norm definitions

$$L^1 = \frac{|\ I^{(num)} - I^{(exact)}\ |}{|\ I^{(exact)}\ |}, \tag{4.16}$$

and

$$L^2 = \frac{\sqrt{\left(I^{(num)} - I^{(exact)}\right)^2}}{\sqrt{\left(I^{(exact)}\right)^2}} \tag{4.17}$$

where I represents the integral .

Figure 4.1 shows the integration error for Legendre, Lobatto, Chebyshev, and equi-spaced points. Figure 4.1a shows the results for both the interpolation and integration points based on Legendre points. That is, Legendre points are used to construct the Lagrange polynomials and these exact same points are used as the quadrature roots with their corresponding quadrature weights. The results using Legendre points are very good and in fact reach machine double precision very quickly.

Figure 4.1b shows the results for both the interpolation and integration points based on Lobatto points. Once again, we see that the error incurred by these points reaches machine double precision for small values of N.

Figure 4.1c shows the results using Chebyshev points for interpolation and Legendre points for integration. Although Chebyshev points integrate well, their associated quadrature weights are only conveniently derived in closed form if the weighting function $w(x) = \frac{1}{\sqrt{1-x^2}}$ is included in the integral. Since we do not need this weighting function then we cannot use the standard Chebyshev-Gauss weights and would

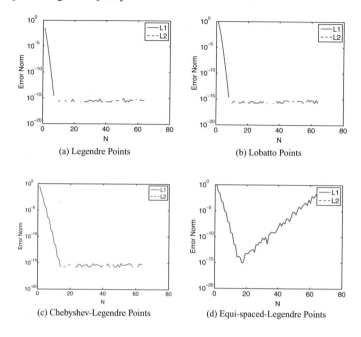

Fig. 4.1: The integration error for (a) Legendre points, (b) Lobatto points, (c) Chebyshev-Legendre, and (d) equi-spaced-Legendre points for various N.

need to derive them from the following definition of quadrature weights

$$w_i = \int_{-1}^{+1} L_i(x)$$

which would have to be derived by some high-order quadrature. Rather than doing this, we simply use the Legendre points as our quadrature points which we show in Fig. 4.1c. Since the Chebyshev points have excellent interpolation properties and the Legendre points are accurate up to $2N + 1$ degree polynomials, then the results are expected to be quite good which they indeed are. Note that this combination of interpolation and integration points achieve machine precision quite quickly.

Figure 4.1d shows the results for the interpolation points based on equi-spaced points. Since this set of points do not integrate well, we have chosen to use the Legendre points for integration. Note that even though the Legendre points are known to integrate very accurately, $O(2N + 1)$, the results for the equi-spaced points are not very good. This is because the interpolation step using equi-spaced points breaks down; we saw this already in Ch. 3.

4.5 Summary of Integration Points

In this chapter, we have described the construction of quadrature formulas. Specifically, we have described where the quadrature points come from and how to compute the corresponding quadrature weights. In addition, we have analyzed (via Newton's divided differences) the error incurred by using quadrature and have discussed how one can achieve exact integration for polynomial functions. Recall, that for polynomial functions, one can compute exact integrals given a sufficient number of points and these points are derived from the zeros of the generating polynomials which, again, are derived from the Sturm-Liouville operator.

Chapter 5
1D Continuous Galerkin Method for Hyperbolic Equations

5.1 Introduction

In this chapter we describe the implementation of the continuous Galerkin (CG) method. We show how this method can be used for the 1D wave equation which is representative of hyperbolic equations. It will be shown why we needed to spend time in Ch. 3 on interpolation and in Ch. 4 on numerical integration. Although we only focus on scalar equations in this chapter, it should be understood that the same approach can be used to discretize systems of hyperbolic equations. We tackle systems of nonlinear hyperbolic equations in Ch. 7. In particular we discuss the discretization of both the CG and DG methods for 1) the shallow water equations and 2) the Euler equations. However, in order to understand how to use the CG method for such systems we must first understand how to apply CG to a scalar equation.

This chapter is organized as follows. Section 5.2 describes the partial differential equation that we focus on for the construction of the CG method. To show how to build the CG discretization of the 1D wave equation, we need to speak specifically about basis functions. In Sec. 5.3 we introduce linear basis functions and derive the map between physical and computational space that we use to map to the reference element where all the computations are performed. Sections 5.4 and 5.5 describe the construction of the mass and differentiation matrices which are then used to construct the elemental equations described in Sec. 5.6. Since the CG method is essentially a global discretization method, the elemental solutions need to be gathered to the global gridpoints. The solution at the global gridpoints is described in Sec. 5.7. The process to obtain the global gridpoint solutions from the local element-wise solutions is known as *direct stiffness summation* and is introduced in Sec. 5.8. After we build all the matrices in the discretization we then analyze the sparsity of the global mass and differentiation matrices and explore their eigenvalues in Sec. 5.9.

© The Editor(s) (if applicable) and The Author(s), under exclusive license
to Springer Nature Switzerland AG 2020
F. X. Giraldo, *An Introduction to Element-Based Galerkin Methods on
Tensor-Product Bases*, Texts in Computational Science and Engineering 24,
https://doi.org/10.1007/978-3-030-55069-1_5

Finally, in Sec. 5.10 we close the chapter with an example CG solution for the 1D (scalar) wave equation.

5.2 Continuous Galerkin Representation of the 1D Wave Equation

The 1D wave equation in conservation (flux) form is

$$\frac{\partial q}{\partial t} + \frac{\partial f}{\partial x} = 0 \quad \text{where} \quad f = qu \tag{5.1}$$

and u is the velocity; this equation represents a conservation law since, as we show in Ch. 6, the quantity q is conserved. Taking the integral form [1] inside a single element Ω_e we get

$$\int_{\Omega_e} \psi_i \left(\frac{\partial q_N^{(e)}}{\partial t} + \frac{\partial f_N^{(e)}}{\partial x} \right) d\Omega_e = \mathbf{0} \tag{5.2}$$

where ψ_i is the basis function with $i = 0, \ldots, N$, and $\mathbf{0}$ is the zero vector with indices $i = 0, \ldots, N$; however, it should be clear from the remaining terms what the correct dimensionality of $\mathbf{0}$ needs to be. In Ch. 3 we used L to refer to the Lagrange polynomial basis functions but in this chapter (and hereafter) we use ψ to denote any basis function.

Expanding the variables inside a single element and using a linear nodal approximation ($N = 1$)

$$q_N^{(e)}(x, t) = \sum_{j=0}^{1} q_j^{(e)}(t)\psi_j(x) \quad \text{and} \quad f_N^{(e)}(x, t) = f\left(q_N^{(e)} \right) = u q_N^{(e)}$$

where $f_N^{(e)} = \sum_{j=0}^{1} f_j^{(e)}(t)\psi_j(x)$ for a linear function [2] f yields

$$\int_{\Omega_e} \psi_i \sum_{j=0}^{1} \psi_j \frac{dq_j^{(e)}}{dt} d\Omega_e + \int_{\Omega_e} \psi_i \sum_{j=0}^{1} \frac{d\psi_j}{dx} f_j^{(e)} d\Omega_e = \mathbf{0}, \quad i = 0, \ldots, 1. \tag{5.3}$$

The matrix form of Eq. (5.3) is written

$$M_{ij}^{(e)} \frac{dq_j^{(e)}}{dt} + D_{ij}^{(e)} f_j^{(e)} = \mathbf{0}, \quad i, j = 0, \ldots, N \tag{5.4}$$

---

[1] Traditionally, the differential form given in Eq. (5.1) is referred to as the *strong* form whereas the integal form given by Eq. (5.2) is known as the *weak* form. Since this text mainly deals with integral forms of the equations, we reserve the use of *strong* and *weak* to refer to two types of integral forms. This will be made clear in the next section.

[2] By stating that f is linear we are assuming that u is constant. We only do this for simplification purposes and if u is not constant then we would either define f_N simply as $f(q_N, u_N) = q_N u_N$ (which is a nonlinear function) or we could group $f(q_N, u_N) = (qu)_N$.

where the superscript e is written to remind the reader that this is defined inside an element, and we have used total derivatives (e.g., $\frac{d}{dt}$ and $\frac{d}{dx}$) to emphasize that the quantities being differentiated are only functions of that variable. The matrix

$$M_{ij}^{(e)} = \int_{\Omega_e} \psi_i \psi_j \, d\Omega_e$$

is the mass matrix and

$$D_{ij}^{(e)} = \int_{\Omega_e} \psi_i \frac{d\psi_j}{dx} \, d\Omega_e$$

is the differentiation matrix, where the indices are defined as $i, j = 0, ..., 1$. We shall refer to Eq. (5.3) as the *strong* integral form (or strong form for short) because the strong differential operator is still apparent even within this integral form. The *weak* form, on the other hand, is obtained by using integration by parts in order to move the derivative of f onto the basis function ψ. We introduce the *weak* form in Ch. 6.

Let us now discuss the mass and differentiation matrices in detail. Before doing so, however, requires us to explicitly define the basis functions that are defined within the reference element.

5.3 Basis Functions and the Reference Element

For two gridpoints per element (linear) let

$$q = \begin{cases} q_0 \text{ for } x = x_0 \\ q_1 \text{ for } x = x_1 \end{cases}.$$

Mapping from physical space $x \in [x_0, x_1]$ to computational space $\xi \in [\xi_0, \xi_1]$ yields

$$q = \begin{cases} q_0 \text{ for } \xi = \xi_0 = -1 \\ q_1 \text{ for } \xi = \xi_1 = +1 \end{cases}$$

with the basis functions

$$\psi_0 = \frac{1}{2}(1 - \xi)$$

$$\psi_1 = \frac{1}{2}(1 + \xi) \tag{5.5}$$

obtained from

$$\psi_i(\xi) = \prod_{\substack{j=0, \\ j \neq i}}^{1} \left(\frac{\xi - \xi_j}{\xi_i - \xi_j} \right)$$

which is, in fact, the general definition of Lagrange polynomials that we defined in Ch. 3. We can now approximate the coordinates of the element by the expansion

$$x(\xi) = \sum_{j=0}^{1} x_j \psi_j(\xi) = \frac{1}{2}(1 - \xi)x_0 + \frac{1}{2}(1 + \xi)x_1$$

which immediately yields $dx = \frac{\Delta x}{2} d\xi$ and, conversely

$$\xi = \frac{2(x - x_0)}{x_1 - x_0} - 1$$

which then yields $\frac{d\xi}{dx} = \frac{2}{\Delta x}$.

The reason why we map from physical space (x) to the computational space (ξ) is that this change of variable simplifies the construction of local (element-based) Galerkin methods because we then do not have to solve matrices for every single element in our grid but rather only do it for the reference element and then use metric terms to scale the reference element to the true size. Let us review the reference element in one-dimension which is simply the line $\xi \in [-1, +1]$. Note that we can now construct all the relevant matrices required by our partial differential equations in terms of this reference element. The construction of these matrices can be done somewhat independently [3] of the size of the physical domain, say $x \in [-L, +L]$, and also quite independently of how small or how large each of our elements are.

5.4 Mass Matrix

Mapping from the physical space $x \in [x_0, x_1]$ to computational space $\xi \in [-1, +1]$ we get for the mass matrix

$$M_{ij}^{(e)} = \int_{x_0}^{x_1} \psi_i(x)\psi_j(x)\, dx = \int_{-1}^{+1} \psi_i(\xi)\psi_j(\xi)\frac{\Delta x^{(e)}}{2} d\xi$$

where the the superscript e in the element length $\Delta x^{(e)}$ represents the fact that each element can be of different length. In matrix form the mass matrix becomes

$$M_{ij}^{(e)} = \frac{\Delta x^{(e)}}{2} \int_{-1}^{+1} \begin{pmatrix} \psi_0\psi_0 & \psi_0\psi_1 \\ \psi_1\psi_0 & \psi_1\psi_1 \end{pmatrix} d\xi.$$

Substituting ψ from Eq. (5.5) yields

$$M_{ij}^{(e)} = \frac{\Delta x^{(e)}}{2} \int_{-1}^{+1} \begin{pmatrix} \frac{1}{2}(1 - \xi)\frac{1}{2}(1 - \xi) & \frac{1}{2}(1 - \xi)\frac{1}{2}(1 + \xi) \\ \frac{1}{2}(1 + \xi)\frac{1}{2}(1 - \xi) & \frac{1}{2}(1 + \xi)\frac{1}{2}(1 + \xi) \end{pmatrix} d\xi. \qquad (5.6)$$

Integrating analytically

$$M_{ij}^{(e)} = \frac{\Delta x^{(e)}}{8} \begin{pmatrix} \xi - \xi^2 + \frac{1}{3}\xi^3 & \xi - \frac{1}{3}\xi^3 \\ \xi - \frac{1}{3}\xi^3 & \xi + \xi^2 + \frac{1}{3}\xi^3 \end{pmatrix}\Bigg|_{-1}^{+1}$$

---

[3] This is true in one-dimension as well as in multiple dimensions provided that the elements are essentially linear, i.e., that the edges are straight-sided which then means that all we have to do is augment each matrix by the area (in 2D) or volume (in 3D) of the element. We discuss this matter in Ch. 15.

and evaluating at the limits of integration gives

$$M_{ij}^{(e)} = \frac{\Delta x^{(e)}}{8} \begin{pmatrix} \frac{8}{3} & \frac{4}{3} \\ \frac{4}{3} & \frac{8}{3} \end{pmatrix}.$$

Factoring out the term $\frac{4}{3}$ yields

$$M_{ij}^{(e)} = \frac{\Delta x^{(e)}}{6} \begin{pmatrix} 2 & 1 \\ 1 & 2 \end{pmatrix}.$$

Remark 5.1 This matrix is clearly full. However, if we decided to use numerical integration we can use one of two options. We can mimic the analytic integration and use $Q = N + 1$. This is exact for the mass matrix because the mass matrix is a $2N$ polynomial and using $Q = N + 1$ Lobatto points will integrate exactly $2N + 1$ polynomials. Using $Q = N + 1$, which is $Q = 2$ in this case, yields

$$M_{ij}^{(e)} = \int_{x_0}^{x_1} \psi_i(x)\psi_j(x)\,dx = \int_{-1}^{+1} \psi_i(\xi)\psi_j(\xi)\frac{\Delta x^{(e)}}{2}\,d\xi$$

$$= \frac{\Delta x^{(e)}}{2} \sum_{k=0}^{2} w_k \psi_i(\xi_k)\psi_j(\xi_k)$$

where the integration weights are

$$w_{0,1,2} = \frac{1}{3}, \frac{4}{3}, \frac{1}{3}$$

and the integration points are

$$\xi_{0,1,2} = -1, 0, +1.$$

Substituting these values yields

$$M_{ij}^{(e)} = \frac{\Delta x^{(e)}}{2} \sum_{k=0}^{2} w_k \psi_i(\xi_k)\psi_j(\xi_k) = \frac{\Delta x^{(e)}}{6} \begin{pmatrix} 2 & 1 \\ 1 & 2 \end{pmatrix}.$$

Alternatively we could take $Q = N$, which is inexact in this case since this will only integrate $2N - 1$ polynomials. Using this quadrature rule yields

$$M_{ij}^{(e)} = \int_{-1}^{+1} \psi_i(\xi)\psi_j(\xi)\frac{\Delta x^{(e)}}{2}\,d\xi = \frac{\Delta x^{(e)}}{2} \sum_{k=0}^{1} w_k \psi_i(\xi_k)\psi_j(\xi_k)$$

where

$$w_{0,1} = 1, 1$$

and

$$\xi_{0,1} = -1, +1.$$

Substituting these values yields

$$M_{ij}^{(e)} = \frac{\Delta x^{(e)}}{2} \sum_{k=0}^{1} w_k \psi_i(\xi_k)\psi_j(\xi_k) = \frac{\Delta x^{(e)}}{2} \begin{pmatrix} 1 & 0 \\ 0 & 1 \end{pmatrix}$$

which is a diagonal matrix. However, we would not want to do this for $N < 4$ because the accuracy is seriously compromised. This issue will be revisited later in this chapter.

Exercise For $Q = 1$ use the Legendre roots and weights $(\xi_0, \xi_1) = \left(-\frac{1}{\sqrt{3}}, \frac{1}{\sqrt{3}}\right)$, $(w_0, w_1) = (1, 1)$ as the quadrature formula for evaluating the mass matrix. Is this quadrature formula exact?   □

5.4.1 General Form of the Mass Matrix

Let us write the mass matrix for the general case of polynomial order N with integration order Q as follows

$$M_{ij}^{(e)} = \frac{\Delta x^{(e)}}{2} \sum_{k=0}^{Q} w_k \psi_i(\xi_k)\psi_j(\xi_k) \tag{5.7}$$

where $i, j = 0, \ldots, N$ and $\psi_i(\xi_k) \equiv \psi_{ik}$ denotes the basis function at the gridpoint i evaluated at the quadrature point k. Algorithm 5.1 describes the recipe for constructing a general mass matrix. The only additional information required by this algorithm are the quadrature points and weights, which can be found in Ch. 4, and the element size $\Delta x^{(e)}$ which can be determined from the grid information as such: $\Delta x^{(e)} = x_N^{(e)} - x_0^{(e)}$ where $x_i^{(e)}$ for $i = 0, \ldots, N$ are the coordinates of the element. The operator += denotes that the value on the left side is augmented by the value on the right. Algorithm 5.1 requires storing a mass matrix for each element. Another approach, that would require less storage, is to construct the element mass matrix on the reference element as shown in Alg. 5.2.

5.5 Differentiation Matrix

The differentiation matrix in computational space becomes

$$D_{ij}^{(e)} = \int_{x_0}^{x_1} \psi_i(x)\frac{d\psi_j}{dx}(x)\,dx = \int_{-1}^{+1} \psi_i(\xi)\left(\frac{d\psi_j}{d\xi}(\xi)\frac{d\xi}{dx}\right)\frac{dx}{d\xi}\,d\xi \tag{5.8}$$

Algorithm 5.1 Construction of the 1D element mass matrix.

function ELEMENT_MASS_MATRIX
  $M^{(e)} = \mathbf{0}$
  for $e = 1 : N_e$ **do**   ▷ loop over elements
    for $k = 0 : Q$ **do**   ▷ loop over integration points
      for $j = 0 : N$ **do**   ▷ loop over columns
        for $i = 0 : N$ **do**   ▷ loop over rows
          $M_{ij}^{(e)} \mathrel{+}= \frac{\Delta x^{(e)}}{2} w_k \psi_{ik} \psi_{jk}$
        end for
      end for
    end for
  end for
end function

Algorithm 5.2 Construction of the 1D element mass matrix on the reference element.

function ELEMENT_MASS_MATRIX
  $M^{(e)} = \mathbf{0}$
  for $k = 0 : Q$ **do**   ▷ loop over integration points
    for $j = 0 : N$ **do**   ▷ loop over columns
      for $i = 0 : N$ **do**   ▷ loop over rows
        $M_{ij}^{(e)} \mathrel{+}= w_k \psi_{ik} \psi_{jk}$
      end for
    end for
  end for
end function

---

since

$$\frac{d\psi_j}{dx}(x) = \frac{d\psi_j}{d\xi}\frac{d\xi}{dx}.$$

Differentiating Eq. (5.5) yields

$$\frac{d\psi_0}{d\xi} = -\frac{1}{2}$$
$$\frac{d\psi_1}{d\xi} = +\frac{1}{2} \tag{5.9}$$

which can be written compactly as $\frac{d\psi_j}{d\xi} = \frac{1}{2}\xi_j$. This now allows us to write the derivative in physical space as

$$\frac{d\psi_j}{dx}(x) = \frac{d\psi_j}{d\xi}\frac{d\xi}{dx} = \frac{1}{2}\xi_j\frac{2}{\Delta x^{(e)}}. \tag{5.10}$$

Substituting ψ from Eq. (5.5) and $\frac{d\psi}{dx}$ from (5.10) into (5.8) yields the matrix form

$$D_{ij}^{(e)} = \frac{2}{\Delta x^{(e)}}\frac{\Delta x^{(e)}}{2}\int_{-1}^{+1}\begin{pmatrix} \frac{1}{2}(1-\xi)\frac{1}{2}(-1) & \frac{1}{2}(1-\xi)\frac{1}{2}(+1) \\ \frac{1}{2}(1+\xi)\frac{1}{2}(-1) & \frac{1}{2}(1+\xi)\frac{1}{2}(+1) \end{pmatrix} d\xi,$$

since $\frac{dx}{d\xi} = \frac{\Delta x^{(e)}}{2}$. Integrating yields

$$D_{ij}^{(e)} = \frac{1}{4}\left.\begin{pmatrix} -\xi + \frac{1}{2}\xi^2 & \xi - \frac{1}{2}\xi^2 \\ -\xi - \frac{1}{2}\xi^2 & \xi + \frac{1}{2}\xi^2 \end{pmatrix}\right|_{-1}^{+1}$$

and evaluating at the integration limits gives

$$D_{ij}^{(e)} = \frac{1}{2}\begin{pmatrix} -1 & +1 \\ -1 & +1 \end{pmatrix}.$$

Remark 5.2 The polynomial in the differentiation matrix that must be integrated is of order $2N - 1$ and in this case $Q = N$ Lobatto points will integrate it exactly. Let us show that indeed we get the integral exactly. Beginning with the integral

$$D_{ij}^{(e)} = \frac{1}{4}\int_{-1}^{+1}\begin{pmatrix} -1 + \xi & 1 - \xi \\ -1 - \xi & 1 + \xi \end{pmatrix} d\xi$$

we now evaluate it using $Q = N$ Lobatto points with weights

$$w_{0,1} = 1, 1$$

and integration points

$$\xi_{0,1} = -1, +1.$$

Substituting these values

$$D_{ij}^{(e)} = \frac{1}{4}\sum_{k=0}^{Q} w_k \begin{pmatrix} -1 + \xi_k & 1 - \xi_k \\ -1 - \xi_k & 1 + \xi_k \end{pmatrix}$$

yields the final matrix

$$D_{ij}^{(e)} = \frac{1}{2}\begin{pmatrix} -1 & +1 \\ -1 & +1 \end{pmatrix}$$

which is clearly exact.

Remark 5.3 What happens if we use $Q = N + 1$ as we did previously for the mass matrix? Obviously, since $Q = N$ is exact, then we better get the same integral when we use $Q = N + 1$. The weights and integration points for this case are

$$w_{0,1,2} = \frac{1}{3}, \frac{4}{3}, \frac{1}{3}$$

and

$$\xi_{0,1,2} = -1, 0, +1.$$

Substituting these values yields

$$D_{ij}^{(e)} = \frac{1}{2}\begin{pmatrix} -1 & +1 \\ -1 & +1 \end{pmatrix}. \tag{5.11}$$

Exercise Compute the differentiation matrix using the Legendre quadrature formula $Q = 1$ with roots and weights $(\xi_0, \xi_1) = \left(-\frac{1}{\sqrt{3}}, \frac{1}{\sqrt{3}}\right)$, $(w_0, w_1) = (1, 1)$. Is this quadrature formula exact?   □

5.5.1 General Form of the Differentiation Matrix

In a similar fashion to the general mass matrix definition, we can write the differentiation matrix for the general case of polynomial order N with integration order Q as follows

$$D_{ij}^{(e)} = \frac{\Delta x^{(e)}}{2} \sum_{k=0}^{Q} w_k \psi_i(\xi_k) \frac{d\psi_j}{dx}(\xi_k) \tag{5.12}$$

where $i, j = 0, \dots, N$ and $\frac{d\psi_j}{dx}(\xi_k) \equiv \frac{d\psi_{jk}}{d\xi} \frac{2}{\Delta x^{(e)}}$ denotes the derivative of the basis function at the gridpoint j evaluated at the quadrature point k. Algorithm 5.3 describes the recipe for constructing a general differentiation matrix; note that constructing the differentiation matrix on the reference element yields the same algorithm.

Exercise Why is the algorithm for constructing the general differentiation matrix and for the differentiation matrix on the reference element the same? One stores the elements on each physical element while the other only stores one matrix on the reference element. Note that this is only the case in 1D.   □

Algorithm 5.3 Construction of the 1D element differentiation matrix.

---

function ELEMENT_DIFFERENTIATION_MATRIX
    $D^{(e)} = 0$
    for $e = 1 : N_e$ **do**   ▷ loop over elements
        for $k = 0 : Q$ **do**   ▷ loop over integration points
            for $j = 0 : N$ **do**   ▷ loop over columns
                for $i = 0 : N$ **do**   ▷ loop over rows
                    $D_{ij}^{(e)} \mathrel{+}= w_k \psi_{ik} \frac{d\psi_{jk}}{d\xi}$
                end for
            end for
        end for
    end for
end function

5.6 Resulting Element Equations

Substituting the mass and differentiation matrices from Secs. 5.4 and 5.5 into Eq. (5.4) yields the resulting equation that must be satisfied within each element

$$\frac{\Delta x^{(e)}}{6} \begin{pmatrix} 2 & 1 \\ 1 & 2 \end{pmatrix} \frac{d}{dt} \begin{pmatrix} q_0^{(e)} \\ q_1^{(e)} \end{pmatrix} + \frac{1}{2} \begin{pmatrix} -1 & +1 \\ -1 & +1 \end{pmatrix} \begin{pmatrix} f_0^{(e)} \\ f_1^{(e)} \end{pmatrix} = \mathbf{0} \qquad (5.13)$$

where the subscripts 0 and 1 denote the element's left and right gridpoints; this local gridpoint numbering convention is shown in Figure 5.1.

Remark 5.4 The coefficients of the mass matrix should add up to the area of the element (excluding Δx, it should add up to one). The differentiation matrix should add up to zero. This can be shown by Taylor series expansion via the consistency arguments we described in Ch. 1. To satisfy consistency, the first order time-derivative terms need to add up to one (mass matrix), while the function terms in the spatial derivative have to sum to zero (differentiation matrix).

Exercise Using consistency arguments, prove that the mass matrix should add to one and the differentiation matrix should sum to zero. To make it simpler, start with lumped forms of the matrices.   □

5.7 Element Contribution to Gridpoint I

Thus far we have only described the equations on a single element. However, looking at Fig. 5.1 we see that in one-dimension there are two elements which claim the global gridpoint I. Since for continuous Galerkin methods we require the solution to be continuous at all global gridpoints, this means that in Fig. 5.1, the left element $(I-1, I)$ and right element $(I, I+1)$ both will contribute a portion of their solutions to the global gridpoint I; let us now describe this procedure. In Fig. 5.1 the blue points denote the global gridpoints whereas the red points the local gridpoints. Note that there are two red points per element which means that we are using linear polynomials (i.e., $N = 1$). Although neighboring red points (near an element interface) are not touching, in reality they will be since they denote the same global degree of freedom. For example, in Fig. 5.1 the (local) red point directly to the left and right of the global gridpoint I denote the same degree of freedom. On the left, it is the local point 1 defined in element $e = 1$ (left element) and on the right it is the local point 0 defined in element $e = 2$.

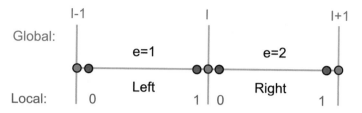

Fig. 5.1: Contribution to the gridpoint I from the left $(I - 1, I)$ and right $(I, I + 1)$ elements.

5.7.1 Left Element

Starting from Eq. (5.13) we may write the equations for the left element as follows

$$\frac{\Delta x^{(L)}}{6} \begin{pmatrix} 2 & 1 \\ 1 & 2 \end{pmatrix} \frac{d}{dt} \begin{pmatrix} q_{I-1} \\ q_I \end{pmatrix} + \frac{1}{2} \begin{pmatrix} -1 & +1 \\ -1 & +1 \end{pmatrix} \begin{pmatrix} f_{I-1} \\ f_I \end{pmatrix} = 0$$

where the first row

$$\frac{\Delta x^{(L)}}{6} \left(2 \frac{dq_{I-1}}{dt} + \frac{dq_I}{dt} \right) + \frac{1}{2} (-f_{I-1} + f_I) = 0$$

gives the contribution of the left element to the global gridpoint $I - 1$ while the second row

$$\frac{\Delta x^{(L)}}{6} \left(\frac{dq_{I-1}}{dt} + 2 \frac{dq_I}{dt} \right) + \frac{1}{2} (-f_{I-1} + f_I) = 0 \qquad (5.14)$$

gives the contribution of the left element to the global gridpoint I.

5.7.2 Right Element

From Eq. (5.13) we may write the equations for the right element as follows

$$\frac{\Delta x^{(R)}}{6} \begin{pmatrix} 2 & 1 \\ 1 & 2 \end{pmatrix} \frac{d}{dt} \begin{pmatrix} q_I \\ q_{I+1} \end{pmatrix} + \frac{1}{2} \begin{pmatrix} -1 & +1 \\ -1 & +1 \end{pmatrix} \begin{pmatrix} f_I \\ f_{I+1} \end{pmatrix} = 0$$

where the contribution to the global gridpoint I is given by the first row

$$\frac{\Delta x^{(R)}}{6} \left(2 \frac{dq_I}{dt} + \frac{dq_{I+1}}{dt} \right) + \frac{1}{2} (-f_I + f_{I+1}) = 0 \qquad (5.15)$$

and the second row

$$\frac{\Delta x^{(R)}}{6} \left(\frac{dq_I}{dt} + 2 \frac{dq_{I+1}}{dt} \right) + \frac{1}{2} (-f_I + f_{I+1}) = 0$$

gives the contribution of the right element to the global gridpoint $I + 1$.

5.7.3 Total Contribution

Assuming that $\Delta x = \Delta x^{(L)} = \Delta x^{(R)}$ and summing up the contributions from the left and right elements results in the global matrix-vector problem:

$$\frac{\Delta x}{6}\begin{pmatrix} 2 & 1 & 0 \\ 1 & 4 & 1 \\ 0 & 1 & 2 \end{pmatrix}\frac{d}{dt}\begin{pmatrix} q_{I-1} \\ q_I \\ q_{I+1} \end{pmatrix} + \frac{1}{2}\begin{pmatrix} -1 & 1 & 0 \\ -1 & 0 & 1 \\ 0 & -1 & 1 \end{pmatrix}\begin{pmatrix} f_{I-1} \\ f_I \\ f_{I+1} \end{pmatrix} = \mathbf{0} \tag{5.16}$$

where we see that the first row of the global matrix-vector system corresponds to the solution for q_{I-1}, the second row to q_I, and the third row to q_{I+1}.

To understand the underlying stencil for CG for $N = 1$ let us compute the inverse mass matrix, which is defined as follows

$$M^{-1} = \frac{1}{2\Delta x}\begin{pmatrix} 7 & -2 & 1 \\ -2 & 4 & -2 \\ 1 & -2 & 7 \end{pmatrix}. \tag{5.17}$$

If we now left-multiply Eq. (5.16) by M^{-1} we get

$$\frac{d}{dt}\begin{pmatrix} q_{I-1} \\ q_I \\ q_{I+1} \end{pmatrix} + \frac{1}{4\Delta x}\begin{pmatrix} -5 & 6 & -1 \\ -2 & 0 & 2 \\ 1 & -6 & 5 \end{pmatrix}\begin{pmatrix} f_{I-1} \\ f_I \\ f_{I+1} \end{pmatrix} = \mathbf{0}. \tag{5.18}$$

If we consider only the solution at I (second row in Eq. (5.18)) we see that the resulting equation is

$$\frac{dq_I}{dt} + \frac{f_{I+1} - f_{I-1}}{2\Delta x} = 0 \tag{5.19}$$

which is the same stencil obtained for a 2nd order centered finite difference method.

Remark 5.5 While the CG method is very much local in nature (since the solution is constructed in an element-by-element procedure) each element then contributes a portion of its solution to global gridpoints. This is the only communication required by the CG method. Here we define *communication* as the information required by one element from its neighbors in order to complete the computation.

Exercise Check to see if the flux derivative term in Eq. (5.19) is indeed consistent. To do this expand via Taylor series about the global gridpoint I and show the order of accuracy of this approximation.   ☐

Before moving on to the higher order representations, it is instructional to show the global matrix-vector problem for the three element configuration shown in Fig. 5.2 which we now write

$$\frac{\Delta x}{6}\begin{pmatrix} 2 & 1 & 0 & 0 \\ 1 & 4 & 1 & 0 \\ 0 & 1 & 4 & 1 \\ 0 & 0 & 1 & 2 \end{pmatrix}\frac{d}{dt}\begin{pmatrix} q_{I-2} \\ q_{I-1} \\ q_I \\ q_{I+1} \end{pmatrix} + \frac{1}{2}\begin{pmatrix} -1 & 1 & 0 & 0 \\ -1 & 0 & 1 & 0 \\ 0 & -1 & 0 & 1 \\ 0 & 0 & -1 & 1 \end{pmatrix}\begin{pmatrix} f_{I-2} \\ f_{I-1} \\ f_I \\ f_{I+1} \end{pmatrix} = \mathbf{0} \tag{5.20}$$

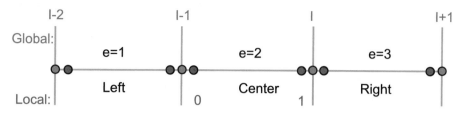

Fig. 5.2: Global matrix-vector problem for a three element configuration.

where we have assumed that the length of each element is constant $\Delta x = \Delta x^{(L)} = \Delta x^{(C)} = \Delta x^{(R)}$ and have not imposed any boundary conditions. After imposing periodic boundary conditions, the global matrix-vector problem becomes

$$\frac{\Delta x}{6}\begin{pmatrix} 4 & 1 & 1 & 0 \\ 1 & 4 & 1 & 0 \\ 0 & 1 & 4 & 1 \\ 0 & 0 & 0 & 1 \end{pmatrix}\frac{d}{dt}\begin{pmatrix} q_{I-2} \\ q_{I-1} \\ q_I \\ q_{I+1} \end{pmatrix} + \frac{1}{2}\begin{pmatrix} 0 & 1 & -1 & 0 \\ -1 & 0 & 1 & 0 \\ 0 & -1 & 0 & 1 \\ 0 & 0 & 0 & 0 \end{pmatrix}\begin{pmatrix} f_{I-2} \\ f_{I-1} \\ f_I \\ f_{I+1} \end{pmatrix} = \mathbf{0} \qquad (5.21)$$

where we impose *a posteriori* that $q_{I+1} = q_{I-2}$. In principle, we do not need the fourth row but leave it for completeness and modify the fourth row of the mass matrix in order to avoid a singular mass matrix (only the diagonal term has a value).

5.7.4 High-Order Approximation

Note that the differencing stencil for linear polynomials given in Eq. (5.19) can be written in the general form

$$\frac{dq_I}{dt} + \sum_{k=I-N}^{I+N} \frac{\alpha_k f_k}{\Delta x} = 0$$

where $N = 1$ and $\alpha_{I-1} = -\frac{1}{2}$, $\alpha_I = 0$, and $\alpha_{I+1} = \frac{1}{2}$. Therefore, if we increase the polynomial order to, say, $N = 2$ then we obtain the following differencing stencil

$$\frac{dq_I}{dt} + \sum_{k=I-2}^{I+2} \frac{\alpha_k f_k}{\Delta x} = 0.$$

Letting $Q = N$ yields the configuration illustrated in Fig. 5.3. In practice, however, we always build the discrete solutions using mass and differentiation matrices. Figure 5.3 shows two different polynomial configurations for a two-element domain. Figure 5.3a shows the linear polynomial ($N = 1$) case as in Fig. 5.1 except now we also include the basis functions (dashed color lines). Figure 5.3b shows the same case but for quadratic ($N = 2$) polynomials. We can continue to increase the order of N to arbitrarily high-order.

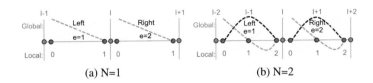

$$(a)\ N=1 \qquad\qquad\qquad (b)\ N=2$$

Fig. 5.3: Continuous Galerkin elements for polynomial order (a) $N = 1$ and (b) $N = 2$.

In one-dimension, the bandwidth of the global mass matrix is $2N + 1$. This means that increasing N increases the complexity of inverting this matrix. However, we can get around this issue by using inexact integration which results in a diagonal mass matrix but one that is only accurate up to $O(2N - 1)$ polynomials (assuming we use a Lobatto quadrature formula of order N).

Let us show an example of the resulting matrices when using $N = 2$. For $N = 2$ we can derive all of the element matrices (mass and differentiation) by going through the same exercise we followed for $N = 1$. For $N = 2$ we use the quadratic basis functions

$$\psi_0(\xi) = -\frac{1}{2}\xi(1 - \xi),$$
$$\psi_1(\xi) = (1 - \xi)(1 + \xi),$$
$$\psi_2(\xi) = +\frac{1}{2}\xi(1 + \xi),$$

where the subscripts 0, 1, 2 refer to the left, middle, and right points for each element. Going through all the steps for the element numbering given in Fig. 5.3b we find that the left element equations are

$$\frac{\Delta x^{(L)}}{30}\begin{pmatrix} 4 & 2 & -1 \\ 2 & 16 & 2 \\ -1 & 2 & 4 \end{pmatrix}\frac{d}{dt}\begin{pmatrix} q_{I-2} \\ q_{I-1} \\ q_I \end{pmatrix} + \frac{1}{6}\begin{pmatrix} -3 & 4 & -1 \\ -4 & 0 & 4 \\ 1 & -4 & 3 \end{pmatrix}\begin{pmatrix} f_{I-2} \\ f_{I-1} \\ f_I \end{pmatrix} = 0 \qquad (5.22)$$

and the right element equations are

$$\frac{\Delta x^{(R)}}{30}\begin{pmatrix} 4 & 2 & -1 \\ 2 & 16 & 2 \\ -1 & 2 & 4 \end{pmatrix}\frac{d}{dt}\begin{pmatrix} q_I \\ q_{I+1} \\ q_{I+2} \end{pmatrix} + \frac{1}{6}\begin{pmatrix} -3 & 4 & -1 \\ -4 & 0 & 4 \\ 1 & -4 & 3 \end{pmatrix}\begin{pmatrix} f_I \\ f_{I+1} \\ f_{I+2} \end{pmatrix} = 0. \qquad (5.23)$$

Note that the sum of the mass matrix equals Δx and that the sum of the differentiation matrix is zero. Assuming $\Delta x = \Delta x^{(L)} = \Delta x^{(R)}$ and summing up the contributions (via DSS) for both the left and right element equations (Eqs. (5.22) and (5.23)) results in the global matrix-vector problem (with no boundary conditions imposed)

$$
\frac{\Delta x}{30}
\begin{pmatrix}
4 & 2 & -1 & 0 & 0 \\
2 & 16 & 2 & 0 & 0 \\
-1 & 2 & 8 & 2 & -1 \\
0 & 0 & 2 & 16 & 2 \\
0 & 0 & -1 & 2 & 4
\end{pmatrix}
\frac{d}{dt}
\begin{pmatrix}
q_{I-2} \\
q_{I-1} \\
q_{I} \\
q_{I+1} \\
q_{I+2}
\end{pmatrix}
+ \frac{1}{6}
\begin{pmatrix}
-3 & 4 & -1 & 0 & 0 \\
-4 & 0 & 4 & 0 & 0 \\
1 & -4 & 0 & 4 & -1 \\
0 & 0 & -4 & 0 & 4 \\
0 & 0 & 1 & -4 & 3
\end{pmatrix}
\begin{pmatrix}
f_{I-2} \\
f_{I-1} \\
f_{I} \\
f_{I+1} \\
f_{I+2}
\end{pmatrix}
= \mathbf{0} \quad (5.24)
$$

which clearly shows the $2N + 1$ bandwidth structure for both the global mass and differentiation matrices. Let us now isolate the equation for the global gridpoint I where we lump[4] the mass matrix for convenience which results in the equation

$$
\frac{\Delta x}{3} \frac{dq_I}{dt} + \frac{1}{6}(f_{I-2} - 4f_{I-1} + 4f_{I+1} - f_{I+2}) = 0
$$

where, upon simplifying, yields the difference equation

$$
\frac{dq_I}{dt} + \frac{f_{I-2} - 4f_{I-1} + 4f_{I+1} - f_{I+2}}{2\Delta x} = 0. \quad (5.25)
$$

Exercise Compute the lumped version of the mass matrix given in Eq. (5.24). □

Exercise Derive the elemental equations for just one element to arrive at Eqs. (5.22) and (5.23). □

Exercise Using the elemental equations from the previous exercise, derive the total contribution as shown in Eq. (5.24). □

Exercise Check to see if the flux derivative term in Eq. (5.25) is indeed consistent. To do this expand via Taylor series about the global gridpoint I and show the order of accuracy of this approximation. □

Exercise Check consistency and order of accuracy for the flux derivative at the global gridpoint $I - 1$ in Eq. (5.24). Use a lumped mass matrix for convenience. □

Exercise Starting from Eq. (5.24), derive the differencing stencil at the global gridpoint I without lumping. This will require inverting the mass matrix. Note that you will get the same answer given in Eq. (5.25). □

5.8 Direct Stiffness Summation

Let us revisit the element contribution to the global gridpoint I in Fig. 5.1. Recall that we began with the elemental equation

$$
M_{ij}^{(e)} \frac{dq_j^{(e)}}{dt} + D_{ij}^{(e)} f_j^{(e)} = \mathbf{0}, \quad i, j = 0, \dots, N
$$

---

[4] Lumping means we sum all the columns of a row and position the value on the diagonal. This concept is akin to using inexact integration.

which we wrote for both the left and right elements. Then we summed the contributions of each element to the gridpoint I in order to obtain the total (global) equation that must be satisfied by the solution at the gridpoint I. This process of summing elements contributing to a gridpoint is known in the literature as either *global assembly* (because the procedure aims to assemble the global problem) or *direct stiffness summation* . The term *direct stiffness summation* arises from the origins of the finite element method which was applied to self-adjoint operators (such as the Poisson equation) where the resulting discrete form of the Laplacian is known as the stiffness matrix (from solid mechanics and structural analysis).

Note that in our elemental integral form, the indices i, j are defined as $(i, j) = 0, ..., N$ where N is the order of the polynomial used for the local basis function ψ. A mathematical way of writing the global matrix-vector problem is to use a summation operator, call it $\bigwedge_{e=1}^{N_e}$, which sums the contributions of each element e and local gridpoint i to the global gridpoint I. Using the *direct stiffness summation* (DSS) operator [5] we can write the global matrix-vector problem as follows

$$M_{IJ} \frac{dq_J}{dt} + D_{IJ} f_J = 0, \quad I, J = 1, \ldots, N_p \tag{5.26}$$

where the global matrices M and D are defined as follows:

$$M_{IJ} = \bigwedge_{e=1}^{N_e} M_{ij}^{(e)}, \qquad D_{IJ} = \bigwedge_{e=1}^{N_e} D_{ij}^{(e)},$$

where the DSS operator performs the summation via the mapping $(i, e) \rightarrow (I)$ where $(i, j) = 0, ..., N$, $e = 1, ..., N_e$, $(I, J) = 1, ..., N_p$, and N_e, N_p are the total number of elements and gridpoints spanning the domain, respectively. Note that in one-dimension $N_p = N_e N + 1$ are the total (global) degrees of freedom. To see this, let us look at Fig. 5.3a which shows $N_e = 2$ elements (left and right elements) with linear functions ($N = 1$) which gives $N_p = 3$ global gridpoints, and Fig. 5.3b which shows $N_e = 2$ elements (left and right elements) with quadratic functions ($N = 2$) which gives $N_p = 5$ global gridpoints. Algorithm 5.4 shows pseudocode for constructing the global mass matrix (M) once the local element-wise matrices ($M^{(e)}$) have been created. The map $(i, e) \rightarrow (I)$ is defined by the integer matrix *intma*[6] in Alg. 5.4. Let us now describe how *intma* is defined for the specific example in Fig. 5.3. Table 5.1 shows that for $N = 1$ the pointers to *intma* are as follows: for the left element $intma(0 : 1, e = 1) = (I - 1, I)$ and for the right $intma(0 : 1, e = 2) = (I, I + 1)$. Table 5.2 shows that for $N = 2$ the pointers to *intma* are: for the left element $intma(0 : 2, e = 1) = (I - 2, I - 1, I)$ and for the right $intma(0 : 2, e = 2) = (I, I + 1, I + 2)$.

---

[5] Deville et al. [105] describe the DSS operator as a rectangular matrix Q (gather) with the scatter operation represented by its transpose Q^T but we choose to use a different symbol because 1) we wish to emphasize that no DSS matrix is ever constructed and 2) because at the moment we do not require the scatter operation Q^T. In Ch. 7 we will require this operation.

[6] *intma* stands for the *INTer-connectivity MAtrix*.

Table 5.1: intma(i,e) array for $N = 1$ associated with Fig. 5.3a.

i	e=1 (Left)	e=2 (Right)
0	I-1	I
1	I	I+1

Table 5.2: intma(i,e) array for $N = 2$ associated with Fig. 5.3b.

i	e=1 (Left)	e=2 (Right)
0	I-2	I
1	I-1	I+1
2	I	I+2

Algorithm 5.4 shows that the goal of the DSS operator is to take the local element-wise gridpoint locator (i, e) and map it to a global gridpoint identifier (I). In Alg. 5.4, the element loop goes through all the elements N_e and then through all the interpolation points within the element $(0 : N)$. Note that Alg. 5.4 assumes that the element matrices have Jacobian information already. If this is not the case then we need to include the Jacobian as described in the line in italic font.

Algorithm 5.4 DSS operation for the mass matrix.

```
function GLOBAL_MASS_MATRIX
   M = 0
   for e = 1 : N_e do                          ▷ loop over elements
      for j = 0 : N do                         ▷ loop over columns of M^(e)
         J = intma(j, e)                       ▷ point to global gridpoint number
         for i = 0 : N do                      ▷ loop over rows of M^(e)
            I = intma(i, e)                    ▷ point to global gridpoint number
            M_IJ += M_ij^(e)                   ▷ Include Δx^(e)/2 factor if M^(e) is on reference element
         end for
      end for
   end for
end function
```

From Eq. (5.26) it can be seen that the discrete version of the continuous PDE using CG can be written as

$$\frac{dq_I}{dt} + M_{IK}^{-1}D_{KJ}f_J = 0, \quad I, J, K = 1, \ldots, N_p$$

which shows that whenever we want to replace any continuous PDE with the CG discrete form, all we need to do is to replace the continuous derivatives by the modified differentiation matrix

$$\hat{D} = M^{-1}D \tag{5.27}$$

which then yields

$$\frac{dq_I}{dt} + \hat{D}_{IJ} f_J = \mathbf{0}, \quad I, J = 1, \ldots, N_p. \tag{5.28}$$

Viewing CG methods this way allows for very easy construction of numerical models for PDEs because all we need to do is to know how to construct the differentiation matrix and the rest of the model components are trivial to assemble, this is true if and only if M is diagonal. If M is not diagonal then we must leave the problem as

$$M_{IJ}\frac{dq_J}{dt} + D_{IJ} f_J = \mathbf{0}, \quad I, J = 1, \ldots, N_p \tag{5.29}$$

and invert a matrix even when using explicit time-integrators.

Exercise Compute the global matrix-vector problem for $N = 1$ with exact integration for $N_e = 4$ elements. Assume all elements have the same length Δx. ☐

Exercise Compute the global matrix-vector problem for $N = 2$ with exact integration for $N_e = 1$ elements. ☐

5.9 Analysis of the Matrix Properties of the Spatial Operators

Now that we know that the gridpoint representation of our equation is given by Eq. (5.29) we would like to see the sparsity patterns of both M and D, i.e., the mass and differentiation matrices. In addition, we would also like to compute the eigenvalues of the right-hand side system because this will tell us whether our numerical discretization is indeed stable and can also give us an idea on the class of time-integrators that we can use to advance our solution forward in time.

For solving the 1D wave equation, we assume periodic boundary conditions and use a total of $N_p = 17$ gridpoints to completely cover the domain; to achieve this we use $N_e = 4$ elements each of order $N = 4$ which results in $N_p = N_e N + 1$ total gridpoints. We shall define the initial value problem explicitly below. Let us first look at the sparsity patterns of the matrices and the eigenvalues of the fully discretized system.

5.9.1 Sparsity Pattern of the Mass Matrix

Figure 5.4 shows the sparsity pattern for the mass matrix M using both exact ($Q = N + 1$) and inexact ($Q = N$) integration. Figure 5.4 shows that the mass matrix is not full but rather sparse; one can see the overlap where the four elements are

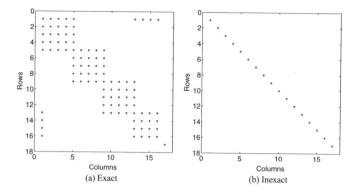

Fig. 5.4: Sparsity patterns for the mass matrix M with $N_e = 4$, $N = 4$ for (a) exact and (b) inexact integration.

touching. This figure shows that the mass matrix is tightly banded, i.e., it has a small bandwidth $(2N + 1)$ and that the matrix becomes diagonal when inexact integration is used. In Fig. 5.4a the block structure form arises from the values for each of the four elements. The top right and bottom left values of the matrix arise due to the imposition of periodic boundary conditions. The bottom right single point is in fact the definition of the periodic boundary condition since we assume that anytime we point to the global gridpoint $I = N_p$ we are actually pointing to the global gridpoint $I = 1$ [7].

5.9.2 Sparsity Pattern of the Differentiation Matrix

Figure 5.5 shows the sparsity pattern for the differentiation matrix D for exact and inexact integration. Similarly to the mass matrix, Fig. 5.5 shows that the differen-tiation matrix is not completely full, i.e., it has a small bandwidth $(2N + 1$, as in the case of the mass matrix). The most important information to glean from these figures is that they reveal the element-wise construction of both M and D.

5.9.3 Eigenvalue Analysis of the Spatial Operator

To analyze the stability of Eq. (5.29) first requires using the form given in Eq. (5.28) with $\hat{D}$ given by Eq. (5.27). Let us now rewrite Eq. (5.28) as follows

---

[7] A value is required at this location to avoid having a singular mass matrix.

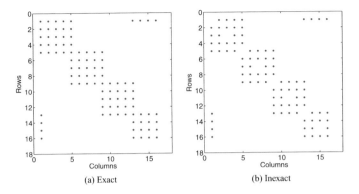

(a) Exact   (b) Inexact

Fig. 5.5: Sparsity patterns for the differentiation matrix D with $N_e = 4$, $N = 4$ for (a) exact and (b) inexact integration.

$$\frac{dq_I}{dt} = R_{IJ}q_J, \quad I, J = 1, \ldots, N_p \tag{5.30}$$

where $R = -u\hat{D}$. We can now replace the matrix R by its eigenvalues, i.e.,

$$Rx = \lambda x$$

to arrive at

$$\frac{dq_I}{dt} = \lambda q_I. \tag{5.31}$$

This equation has the exact solution

$$q(x, t) = q_0 \exp(\lambda(x)t)$$

that will be bounded (less than infinity) for $Re(\lambda) \leq 0$.

Let us now look at the eigenvalues of R for the example 1D wave equation problem that we defined earlier ($N_p = 17$ gridpoints). Figure 5.6 shows the eigenvalues of the right-hand side matrix.

This matrix is a representation of the complete discretization of the spatial operators. Note that the real part of the eigenvalues is very near zero. In fact, it is zero up to machine double precision (2.09×10^{-15}); this is true for both exact and inexact integration. However, to guarantee stability we must insist that the real part of the eigenvalues be less than zero. The concern of the CG method in its pure form is that it is non-dissipative (we analyze the dissipation properties of the method in Ch. 7). This is desirable for certain wave propagation problems but one is also interested in maintaining numerical stability. Therefore, the CG method is often used with an additional dissipative mechanism. In Fig. 5.6c we show the eigenvalues using inexact integration but with an additional Laplacian diffusion operator with an artificial

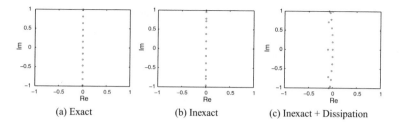

Fig. 5.6: Eigenvalues of the right-hand side matrix R with $N_e = 4$, $N = 4$ for (a) exact integration, (b) inexact integration, and (c) inexact integration with $\mu = 0.0052$ dissipation.

viscosity of $\mu = 0.0052$. This small amount of viscosity forces all the eigenvalues to be negative (maximum real eigenvalue is now -1.93×10^{-15}). We address how to construct such Laplacian operators in Chs. 8 and 18.

Before closing this section, let us look at the structure of the matrix R. Figure 5.7a

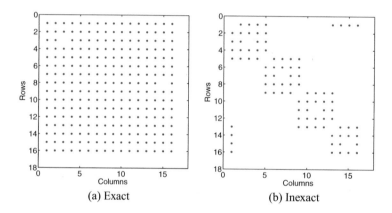

Fig. 5.7: Sparsity patterns for the right-hand side matrix R with $N_e = 4$, $N = 4$ for (a) exact and (b) inexact integration.

shows that for exact integration, the matrix R is, in fact, full. This figure illustrates that although the continuous Galerkin method is constructed in an element-wise manner, the final solution of the problem is very much a global problem. However, note that using inexact integration makes the problem far more local where the outline of the elements can be seen in the structure of R.

5.10 Example of 1D Wave Equation Problem

Let us now put everything together that we have learned in this chapter to solve a complete partial differential equation (PDE). Suppose we wish to solve the continuous PDE

$$\frac{\partial q}{\partial t} + \frac{\partial f}{\partial x} = 0 \qquad \forall x \in [-1, +1]$$

for the solution variable $q(x, t)$ such that $f(x, t) = q(x, t)u$, $u = 2$, with periodic boundary conditions $q(-1, t) = q(+1, t)$. Using the method of characteristics, we find the exact solution to this problem to be $q(x, t) = Q(x + ut)$ where $Q(x) = q(x, 0)$ is the initial condition. Furthermore, since $u = 2$ is a constant and the domain size is of length 2, this tells us that the initial condition should end at exactly the same place it started for all integer values of t. That is, it will take exactly $t = 1$ time units for the wave to complete one full revolution of the domain.

5.10.1 Initial Condition

Since the governing PDE is a hyperbolic system, then this problem represents an initial value problem (IVP or Cauchy problem). We, therefore, need an initial condition. Let it be the following Gaussian

$$q(x, 0) = \exp(-64x^2). \tag{5.32}$$

5.10.2 Boundary Condition

To reiterate, we have assumed periodic boundary conditions meaning that the domain at $x = +1$ should wrap around and back to $x = -1$. The solution variable $q(x, t)$ must have the same solution at $x = -1$ and $x = +1$.

5.10.3 Error Norm

We define the normalized L^2 error norm as follows

$$L^2 = \sqrt{\frac{\sum_{k=1}^{N_P} \left(q_k^{(num)} - q_k^{(exact)} \right)^2}{\sum_{k=1}^{N_P} \left(q_k^{(exact)} \right)^2}} \tag{5.33}$$

where $k = 1, ..., N_p$ are $N_p = N_e N + 1$ global gridpoints and $q^{(num)}$ and $q^{(exact)}$ are the numerical and exact solutions after one full revolution of the wave. Note that the wave should stop exactly where it began without changing shape for all $t \in I$ where I denotes integer values[8].

To be more precise in this definition, and when elements of different sizes are used, it is more accurate to write the norm as follows

$$L^2 = \sqrt{\frac{\int_{\Omega} \left(q^{(num)} - q^{(exact)}\right)^2 d\Omega}{\int_{\Omega} \left(q^{(exact)}\right)^2 d\Omega}} \tag{5.34}$$

where we use quadrature for the integrals as follows: let

$$\int_{\Omega} f \, d\Omega = \sum_{e=1}^{N_e} \sum_{k=0}^{Q} \frac{\Delta x^{(e)}}{2} w_k f_k$$

for any function f with quadrature weights w.

5.10.4 Time-Integrator

To solve the time-dependent portion of the problem we use the 3rd order strong stability-preserving (SSP) Runge-Kutta (RK) method by Shu [341] : for $\frac{dq}{dt} = R(q)$ let

$$q^{(1)} = q^n + \Delta t R(q^n)$$

$$q^{(2)} = \frac{3}{4}q^n + \frac{1}{4}q^{(1)} + \frac{1}{4}\Delta t R(q^{(1)})$$

$$q^{n+1} = \frac{1}{3}q^n + \frac{2}{3}q^{(2)} + \frac{2}{3}\Delta t R(q^{(2)}).$$

Definition 5.1 A strong stability-preserving (SSP) method is one that satisfies the following property:

$$||q^{n+1}|| \le ||q^n||.$$

Remark 5.6 Strong stability-preserving methods can be shown to be a linear convex combination of forward Euler methods.

Recall that the Courant number (see, e.g., [53])

$$C = u\frac{\Delta t}{\Delta x}$$

must be within a certain value for stability. For the 3rd order RK method one should use time-steps such that $C \le \frac{1}{3}$. For Δx we take the difference of the first point in the domain $x_1 = -1$ and the next point x_2 since this will be the tightest clustering

---

[8] Due to the dissipation errors of the particular numerical method, the wave may in fact diminish in size. Also, due to dispersion errors, the wave may not be exactly at the same point it began.

of points in your model. Another, more general, way is to take the minimum value of $x_{I+1} - x_I$ for all points $I = 1, ..., N_p - 1$, which is the best approach for general unstructured grids.

5.10.5 Construction of the CG Solution

Algorithms 5.5 and 5.6 highlight the main steps in constructing the CG solution. First we construct the local element-wise mass and differentiation matrices for each element and then we use DSS to construct the global matrices. In recipe 1 (Alg. 5.5) we construct the global mass matrices and then multiply $\hat{D}$ with the global solution vector q at each time-step of the time-integration loop. This approach follows the strategy of constructing and solving an entirely global matrix problem and is perhaps conceptually the easiest approach to understand, although not the most efficient. In recipe 2 (Alg. 5.6) we only construct the global mass matrix. Then for each time-step $(n = 1, ..., N_t$, where $N_t = \frac{\text{time}_{\text{final}} - \text{time}_{\text{initial}}}{\Delta t})$ in the time loop we loop through all the elements and construct the element-wise integrals, which we call the volume integrals, and store the information in the right-hand side (RHS) vector $R_i^{(e)}$. Note that $f^{(e)}$ is a function of q^n, i.e., the solution vector at time loop n. At this point we apply DSS to the RHS solution vector $R_i^{(e)}$ to form R_I using Alg. 5.7. Next we multiply by the inverse of the global mass matrix to construct a global gridpoint-wise solution. At this point we can enforce the boundary conditions in a global gridpoint approach. Now we are ready to evolve the equations forward one time-step and continue until we reach the desired number of time-steps required.

Algorithm 5.5 Recipe 1: CG solution algorithm for the 1D wave equation - global view

---

> **function** CG(q)
>> Construct $M_{ij}^{(e)}$ and $D_{ij}^{(e)}$ ▷ use Algs. 5.1, 5.3
>> Construct M_{IJ} and D_{IJ} via DSS   ▷ use Alg. 5.4
>> Construct $\hat{D} = M^{-1}D$
>> **for** $n = 1 : N_t$ **do**   ▷ perform time-integration loop
>>> $R_I = -\hat{D}_{IJ}q_J$ ▷ apply boundary conditions here
>>> $\frac{d}{dt}q_I = R_I$ ▷ evolve equations in time to get q^{n+1}
>> **end for**
> **end function**

5.10.6 Solution Accuracy for a Smooth Problem

Figure 5.8 shows the snapshot of the exact and CG solutions after one revolution $(t = 1)$ using $N = 8$ order polynomials and $N_e = 8$ elements for a total of $N_p = 65$

Algorithm 5.6 Recipe 2: CG solution algorithm for the 1D wave equation - local view

---

function CG(q)
    Construct $M_{ij}^{(e)}$ and $D_{ij}^{(e)}$ ▷ Algs. 5.1, 5.3
    Construct M_{IJ} via DSS   ▷ use Alg. 5.4
    for $n = 1 : N_t$ **do**   ▷ perform time-integration loop
        for $e = 1 : N_e$ **do**   ▷ loop over elements
            for $i = 0 : N$ **do**   ▷ loop over interpolation points
               $R_i^{(e)} = -D_{ij}^{(e)} f_j^{(e)}(q)$ ▷ construct element-wise solution
           end for
        end for
        Construct R_I via DSS   ▷ use Alg. 5.7
        $R_I = (M_{IJ})^{-1} R_J$ ▷ apply boundary conditions here
        $\frac{d}{dt} q_I = R_I$ ▷ evolve equations in time to get q^{n+1}
    end for
    end function

Algorithm 5.7 DSS operation for the vector R.

---

function GLOBAL_R_VECTOR($R^{(e)}$)
    $R = \mathbf{0}$
    for $e = 1 : N_e$ **do**   ▷ loop over elements
        for $i = 0 : N$ **do**   ▷ loop over rows of $R^{(e)}$
           $I = intma(i, e)$ ▷ point to global gridpoint number
           $R_I += R_i^{(e)}$ ▷ Gather data to CG storage
        end for
    end for
    end function

---

gridpoints. Figure 5.8 shows that there is no noticeable difference between the exact

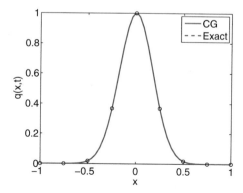

Fig. 5.8: Snapshot of the exact and CG numerical solution for the 1D wave equation using $N = 8$ order polynomials with $N_e = 8$ elements with exact integration ($Q = 9$) after one revolution ($t = 1$ time unit).

and numerical solutions. Let us now see how well our model behaves by varying the number of elements and the polynomial order of the basis functions. We use a Courant number of $C = 0.25$ for all the simulations in the convergence study.

Figure 5.9 shows the convergence rates for various polynomial orders, N, for a total number of gridpoints N_p. Figure 5.9 shows that there is little difference between

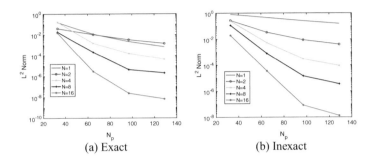

(a) Exact   (b) Inexact

Fig. 5.9: Convergence rates of CG for the 1D wave equation for polynomial orders $N = 1$, $N = 2$, $N = 4$, $N = 8$, and $N = 16$ using a total number of gridpoints N_p for (a) exact integration ($Q = N + 1$) and (b) inexact integration ($Q = N$).

exact and inexact integration at least in the rates of convergence. The main differences occur for low values of N (≤ 4), however, as N increases the differences between exact and inexact integration all but disappear. This was shown to be the case for $N \geq 4$ in [151]. One comment needs to be made regarding the unusual result in Fig. 5.9a for $N = 2$; the results for $N = 2$ are not as good as those for $N = 1$. This is due to aliasing errors - ways to combat aliasing are covered in Ch. 18. Note, however, that this does not happen for inexact integration. This is merely fortuitous but inexact integration will also suffer aliasing errors and must be treated in the same manner as in exact integration.

Upon gazing at the results of Fig. 5.9, where it is evident that using high-order polynomials give better results than low-order polynomials, the question to ask is what is the cost of using high-order polynomials. To answer this question, let us look at Fig. 5.10 where the L^2 error norm is plotted as a function of wallclock time in seconds. Figure 5.10 shows that the high-order methods are, in fact, more efficient to reach a certain level of accuracy than the low-order methods. For example, to achieve an accuracy of 10^{-4} is most efficiently achieved with $N = 8$; in comparison, $N = 1$, $N = 2$, or even $N = 4$ would require prohibitively large computational times to achieve these levels of accuracy. This result holds for both exact and inexact integration. Plots such as the one shown in Fig. 5.10 are known as *work-precision diagrams* and are the most useful types of plots for discerning the cost to solution of a given numerical method because they measure the actual time required to reach a certain level of accuracy. Because measuring time on a computer can be

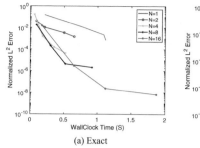
(a) Exact

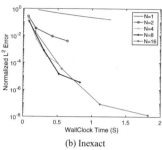
(b) Inexact

Fig. 5.10: Work-Precision diagram for CG for the 1D wave equation for polynomial orders $N = 1$, $N = 2$, $N = 4$, $N = 8$, and $N = 16$ as a function of wallclock time (in seconds) for (a) exact integration ($Q = N + 1$) and (b) inexact integration ($Q = N$).

problematic[9], it is sometimes best to use floating point operations (FLOPS) as a proxy for time because we know approximately[10] how many compute-cycles are required for a given arithmetic operation [11].

5.10.7 Solution Accuracy for a Non-Smooth Problem

In Fig. 5.8 we showed the solution after one revolution for a smooth problem (i.e. infinitely differentiable solution). For this class of problems, any high-order method will do very well. However, for purposes of comparison with the discontinuous Galerkin (DG) in the next chapter (Ch. 6), let us now show a snapshot of the CG solution after one revolution for a non-smooth initial condition. On the domain $x \in [-1, 1]$, let the initial condition be defined as

$$q(x, 0) = \begin{cases} 1 \text{ for } |x| \le R_c \\ 0 \text{ for } |x| > R_c \end{cases}$$

where $R_c = \frac{1}{8}$.

---

[9] It is difficult to measure time because, depending on the computer, other operations may run through the compute-device that you are timing on yielding different compute-times for each measure of the same computer code. For this reason, if time is measured, it should be done a few times and then an average should be taken.

[10] Here we say "approximately" because we can count all of the arithmetic operations and assume an average cost for the operations. In truth, operations such as addition/subtraction, multiplication, division, and exponents have different costs. The science of counting operations/instructions is called *complexity analysis* and is an essential ingredient in the study of scientific computing.

[11] Using FLOPS as a proxy for time is valid if we are only considering serial computing. In the case of manycore computers, the main bottleneck is in data movement through memory, which we ignore here.

Figure 5.11 shows the exact and CG solutions after one revolution ($t = 1$) using $N = 8$ order polynomials and $N_e = 16$ elements for a total of $N_p = 129$ gridpoints for a non-smooth initial condition (a square wave). Figure 5.11a shows that for

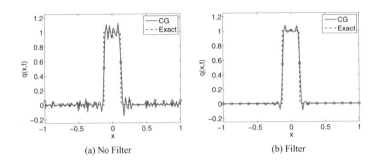

(a) No Filter   (b) Filter

Fig. 5.11: Snapshot of the exact and CG numerical solution for the 1D wave equation witn a non-smooth initial condition using $N = 8$ order polynomials with $N_e = 16$ elements with exact integration ($Q = 9$) after one revolution ($t = 1$ time unit) with (a) no filtering and (b) filtering.

discontinuous solutions the CG method is able to track the propagation of the discontinuity but the results are very noisy. In a real application, we would not use the unadulterated CG method as we have done here but would introduce some form of artificial dissipation in order to allow the method to handle such problems with steep gradients. Figure 5.11b shows the same simulation but with filtering included; clearly, the filtered solution has removed much of the noise from the unadulterated CG method. In Ch. 18 we discuss filters and other stabilization mechanisms in order to eliminate this kind of numerical noise. In Ch. 7 we discuss the dissipation and dispersion errors of numerical methods, specifically those pertaining to CG and DG methods.

Chapter 6
1D Discontinuous Galerkin Methods for Hyperbolic Equations

In Ch. 5 we described the implementation of the continuous Galerkin (CG) method for scalar hyperbolic equations. In this chapter we describe the implementation of the discontinuous Galerkin (DG) method. We show how this method can be used for the 1D wave equation which is representative of hyperbolic equations. We rely heavily on the theory that we have already presented in Ch. 3 on interpolation and in Ch. 4 on numerical integration. We only consider scalar equations in this chapter and extend the implementation of the DG method to systems of nonlinear partial differential equations in Ch. 7.

This chapter is organized as follows. Section 6.1 describes the partial differential equation that we focus on for the construction of the DG method, namely, the one-dimensional scalar wave equation. Sections 6.2, 6.3, and 6.4 describe the construction of the mass, differentiation, and flux matrices for the specific example when using linear basis functions. The reason for only considering linear basis functions is to be able to describe the process by which one would obtain these matrices analytically to ensure that the algorithms that you write are indeed correct. Once these matrices are constructed by hand, they can then be generalized to higher order using algorithms as described in Sec. 6.6. In Sec. 6.8 we discuss the conservation property of the DG method and show the specific example for linear basis functions. For our purposes in this chapter, we concentrate on the linear polynomial case, use these matrices to show the conservation property in Sec. 6.8, and describe the construction of the element equations in Sec. 6.5. After we build all the matrices in the discretization we then analyze the sparsity of the matrices and explore their eigenvalues in Sec. 6.7. Finally, in Sec. 6.9 we close the chapter with an example DG solution for the one-dimensional wave equation.

© The Editor(s) (if applicable) and The Author(s), under exclusive license to Springer Nature Switzerland AG 2020
F. X. Giraldo, *An Introduction to Element-Based Galerkin Methods on Tensor-Product Bases*, Texts in Computational Science and Engineering 24, https://doi.org/10.1007/978-3-030-55069-1_6

6.1 Discontinuous Galerkin Representation of the 1D Wave Equation

The one-dimensional (1D) wave equation in conservation (flux) form is

$$\frac{\partial q}{\partial t} + \frac{\partial f}{\partial x} = 0 \quad \text{where} \quad f = qu$$

and u is the velocity; we discuss the necessary initial and boundary conditions in Sec. 6.9. Using the integral form within a single element Ω_e, yields

$$\int_{\Omega_e} \psi_i \left(\frac{\partial q_N^{(e)}}{\partial t} + \frac{\partial f_N^{(e)}}{\partial x} \right) d\Omega_e = \mathbf{0} \tag{6.1}$$

where $q_N^{(e)}$ is the Nth degree approximation of q and, similarly, $f_N^{(e)}$ is the Nth degree representation of f and $\mathbf{0}$ is the zero vector with $N + 1$ rows. Integrating the second term by parts gives

$$\int_{\Omega_e} \psi_i \frac{\partial q_N^{(e)}}{\partial t} d\Omega_e - \int_{\Omega_e} \frac{d\psi_i}{dx} f_N^{(e)} d\Omega_e + \int_{\Omega_e} \frac{\partial}{\partial x} \left(\psi_i f_N^{(e)} \right) d\Omega_e = \mathbf{0}.$$

Using the Fundamental Theorem of Calculus on the third term yields

$$\int_{\Omega_e} \psi_i \frac{\partial q_N^{(e)}}{\partial t} d\Omega_e - \int_{\Omega_e} \frac{d\psi_i}{dx} f_N^{(e)} d\Omega_e + \left[\psi_i f_N^{(e)} \right]_{\Gamma_e} = \mathbf{0}. \tag{6.2}$$

Remark 6.1 We can also use Eq. (6.2) for the CG method but we would then have to apply the DSS operator to arrive at the global system of equations. However, the third term in Eq. (6.2) would be zero for the CG method everywhere except at the domain boundaries; this is because the basis functions and solution values are continuous across element interfaces. Recall that in the CG method we are constructing the solution at global gridpoints shared by adjacent elements. However, for DG this is not the case because the solution variables and basis functions need not be continuous across element interfaces. For the DG method the third term represents the jump condition across the element interfaces. In Ch. 7, we describe a unified CG and DG approach using the DG formulation given by Eq. (6.2).

In order to simplify the description of the DG method, let us consider linear basis functions. Letting

$$q_N^{(e)}(x, t) = \sum_{j=0}^{1} q_j^{(e)}(t) \psi_j(x) \quad \text{and} \quad f_N^{(e)}(x, t) = \sum_{j=0}^{1} f_j^{(e)}(t) \psi_j(x)$$

gives

$$\int_{\Omega_e} \psi_i \psi_j \, d\Omega_e \frac{dq_j^{(e)}}{dt} - \int_{\Omega_e} \frac{d\psi_i}{dx} \psi_j \, d\Omega_e \, f_j^{(e)} + \left[\psi_i \psi_j \right]_{\Gamma_e} f_j^{(e)} = \mathbf{0}, \quad i, j = 0, ..., 1. \tag{6.3}$$

We can use the expansion $f_N^{(e)}(x, t) = \sum_{j=0}^{1} f_j^{(e)}(t)\psi_j(x)$ here because $f_N = uq_N$ in this example, where u is constant. For nonlinear equations, we could use this approach or the following approach $f_N = f(q_N)$, which we describe in Ch. 7.

We can now write Eq. (6.3) in the following matrix-vector form

$$M_{ij}^{(e)} \frac{dq_j^{(e)}}{dt} - \tilde{D}_{ij}^{(e)} f_j^{(e)} + F_{ij}^{(e)} f_j^{(e)} = \mathbf{0}, \quad i, j = 0, \ldots, 1 \tag{6.4}$$

where the superscript (e) denotes that the matrices are defined at the element level. Note that Eq. (6.4) is not yet well-posed. We define the well-posed formulation in Sec. 6.5[1]. Let us now describe the construction of the mass $(M^{(e)})$, differentiation $(\tilde{D}^{(e)})$, and flux $(F^{(e)})$ matrices.

6.2 Mass Matrix

As in the CG method, the mass matrix for linear polynomials $(N = 1)$ is

$$M_{ij}^{(e)} = \frac{\Delta x^{(e)}}{6} \begin{pmatrix} 2 & 1 \\ 1 & 2 \end{pmatrix}. \tag{6.5}$$

Details of this derivation can be found in Sec. 5.4.

6.3 Differentiation Matrix

The differentiation matrix in computational space becomes

$$\tilde{D}_{ij}^{(e)} = \int_{x_0}^{x_1} \frac{d}{dx}\psi_i(x)\,\psi_j(x)dx = \int_{-1}^{+1} \left(\frac{d}{d\xi}\psi_i(\xi)\frac{d\xi}{dx} \right) \psi_j(\xi)\frac{dx}{d\xi}d\xi$$

where $\tilde{D}$ is the weak form of the strong form differentiation matrix that we used in CG (see D in Sec. 5.5) - the difference between D and $\tilde{D}$ results from integration by parts where we move the differential operator from the flux function f (strong form D) to the basis function ψ (weak form $\tilde{D}$). Substituting ψ with linear basis functions, yields the following expression

$$\tilde{D}_{ij}^{(e)} = \frac{2}{\Delta x^{(e)}} \frac{\Delta x^{(e)}}{2} \int_{-1}^{+1} \begin{pmatrix} \frac{1}{2}(-1)\frac{1}{2}(1-\xi) & \frac{1}{2}(-1)\frac{1}{2}(1+\xi) \\ \frac{1}{2}(+1)\frac{1}{2}(1-\xi) & \frac{1}{2}(+1)\frac{1}{2}(1+\xi) \end{pmatrix} d\xi.$$

---

[1] The reason why Eq. (6.4) is not well-posed is because the flux term (third term) is different at the interface of two neighboring elements, depending on which element we are referring to. To make the discrete problem well-posed we must ensure that the flux leaving one element be equal to the flux entering its neighbor, i.e., that the flux is continuous on the trace of the element.

Expanding and then integrating yields

$$\tilde{D}_{ij}^{(e)} = \frac{1}{4} \left. \begin{pmatrix} -\xi + \frac{1}{2}\xi^2 & -\xi - \frac{1}{2}\xi^2 \\ \xi - \frac{1}{2}\xi^2 & \xi + \frac{1}{2}\xi^2 \end{pmatrix} \right|_{-1}^{+1}$$

and evaluating at the integration limits results in

$$\tilde{D}_{ij}^{(e)} = \frac{1}{2} \begin{pmatrix} -1 & -1 \\ 1 & 1 \end{pmatrix}. \tag{6.6}$$

6.3.1 General Form of the Weak Differentiation Matrix

The general case for the weak form differentiation matrix is written as follows

$$\tilde{D}_{ij}^{(e)} = \frac{\Delta x^{(e)}}{2} \sum_{k=0}^{Q} w_k \frac{d\psi_i}{dx}(\xi_k)\psi_j(\xi_k) \tag{6.7}$$

where $i, j = 0, \ldots, N$ and $\frac{d\psi_i}{dx}(\xi_k) \equiv \frac{d\psi_{ik}}{d\xi} \frac{2}{\Delta x^{(e)}}$. Algorithm 6.1 describes the recipe for constructing a general differentiation matrix.

Algorithm 6.1 Construction of the 1D element weak form differentiation matrix.

function ELEMENT_WEAK_DIFFERENTIATION_MATRIX
$\quad \tilde{D}^{(e)} = 0$
$\quad$**for** $e = 1 : N_e$ **do**   ▷ loop over elements
$\quad\quad$**for** $k = 0 : Q$ **do**   ▷ loop over integration points
$\quad\quad\quad$**for** $j = 0 : N$ **do**   ▷ loop over columns
$\quad\quad\quad\quad$**for** $i = 0 : N$ **do**   ▷ loop over rows
$\quad\quad\quad\quad\quad \tilde{D}_{ij}^{(e)} \mathrel{+}= w_k \frac{d\psi_{ik}}{d\xi} \psi_{jk}$
$\quad\quad\quad\quad$**end for**
$\quad\quad\quad$**end for**
$\quad\quad$**end for**
$\quad$**end for**
end function

6.4 Flux Matrix

The flux matrix (which comes from the boundary integral term) is given as

$$F_{ij}^{(e)} = \left[\psi_i(x)\psi_j(x) \right]_{x_0}^{x_1} = \left[\psi_i(\xi)\psi_j(\xi) \right]_{-1}^{+1} = \left. \begin{pmatrix} \psi_0\psi_0 & \psi_0\psi_1 \\ \psi_1\psi_0 & \psi_1\psi_1 \end{pmatrix} \right|_{-1}^{+1}.$$

Replacing ψ with the linear basis functions gives

$$F_{ij}^{(e)} = \left. \begin{pmatrix} \frac{1}{2}(1-\xi)\frac{1}{2}(1-\xi) & \frac{1}{2}(1-\xi)\frac{1}{2}(1+\xi) \\ \frac{1}{2}(1+\xi)\frac{1}{2}(1-\xi) & \frac{1}{2}(1+\xi)\frac{1}{2}(1+\xi) \end{pmatrix} \right|_{-1}^{+1}.$$

Evaluating at the endpoints gives

$$F_{ij}^{(e)} = \frac{1}{4}\begin{pmatrix} -4 & 0 \\ 0 & 4 \end{pmatrix}$$

and simplifying yields the final form

$$F_{ij}^{(e)} = \begin{pmatrix} -1 & 0 \\ 0 & 1 \end{pmatrix}. \tag{6.8}$$

6.4.1 General Form of the Flux Matrix

The general case for the flux matrix is written as follows

$$F_{ij}^{(e)} = \left[\psi_i(\xi)\psi_j(\xi) \right]_{-1}^{+1} \tag{6.9}$$

where $i, j = 0, \ldots, N$. Algorithm 6.2 describes the recipe for constructing a general flux matrix.

Algorithm 6.2 Construction of the 1D element flux matrix.

---

function ELEMENT_FLUX_MATRIX
   for $e = 1 : N_e$ **do**   ▷ loop over elements
     for $j = 0 : N$ **do**   ▷ loop over columns
       for $i = 0 : N$ **do**   ▷ loop over rows
         $F_{ij}^{(e)} = \left[\psi_i(+1)\psi_j(+1) - \psi_i(-1)\psi_j(-1) \right]$ ▷ evaluate at endpoints
       end for
     end for
   end for
end function

6.5 Resulting Element Equations

Substituting the mass, differentiation, and flux matrices from Eqs. (6.5), (6.6), and (6.8) into the DG element equation in Eq. (6.4) yields

$$\frac{\Delta x^{(e)}}{6}\begin{pmatrix} 2 & 1 \\ 1 & 2 \end{pmatrix}\frac{d}{dt}\begin{pmatrix} q_0^{(e)} \\ q_1^{(e)} \end{pmatrix} - \frac{1}{2}\begin{pmatrix} -1 & -1 \\ 1 & 1 \end{pmatrix}\begin{pmatrix} f_0^{(e)} \\ f_1^{(e)} \end{pmatrix} + \begin{pmatrix} -1 & 0 \\ 0 & 1 \end{pmatrix}\begin{pmatrix} f_0^{(*)} \\ f_1^{(*)} \end{pmatrix} = 0$$

where the subscripts 0 and 1 denote the element's local gridpoints. This, in fact, is nothing more than the matrix problem

$$M_{ij}^{(e)} \frac{dq_j^{(e)}}{dt} - \tilde{D}_{ij}^{(e)} f_j^{(e)} + F_{ij}^{(e)} f_j^{(*)} = 0, \quad i, j = 0, \dots, 1 \tag{6.10}$$

where $f^{(*)}$ denotes the numerical flux function. First, we assume the numerical flux function to be $f^{(e)}$, that is, the flux within the element; we address different numerical flux functions in Sec. 6.5.5.

Figure 6.1 illustrates the contribution of the left and right elements to the center element. The interface gridpoints between the left and center at $I - 1$ and those for the right and center at I are shown to be uniquely defined but they do not necessarily have to be as illustrated. Figure 6.1 shows one possible DG configuration where linear polynomials based on Lobatto points are used; however, we could just as well have used linear polynomials based on Legendre points which then would mean that the interface points between elements are unique but not touching. The numbers in red font are used to indicate that at each gridpoint, each element has a unique degree of freedom. Therefore, for three elements and linear polynomials we get 6 total degrees of freedom where in general we get $N_p = N_e(N + 1)$.

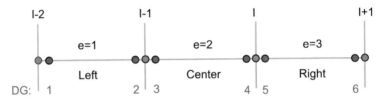

Fig. 6.1: The one-dimensional stencil for the DG method with three elements (e=1,2,3) labeled as Left, Center, and Right with $N = 1$ polynomials.

6.5.1 Left Element

Using the grid configuration illustrated in Fig. 6.1 we note that the element equations for the left element are written as follows

$$\frac{\Delta x^{(L)}}{6} \begin{pmatrix} 2 & 1 \\ 1 & 2 \end{pmatrix} \frac{d}{dt} \begin{pmatrix} q_{I-2}^{(L)} \\ q_{I-1}^{(L)} \end{pmatrix} - \frac{1}{2} \begin{pmatrix} -1 & -1 \\ 1 & 1 \end{pmatrix} \begin{pmatrix} f_{I-2}^{(L)} \\ f_{I-1}^{(L)} \end{pmatrix} + \begin{pmatrix} -1 & 0 \\ 0 & 1 \end{pmatrix} \begin{pmatrix} f_{I-2}^{(L)} \\ f_{I-1}^{(L)} \end{pmatrix} = 0 \tag{6.11}$$

where the contribution to $I-2$ is

$$\frac{1}{3}\left(2\frac{dq_{I-2}^{(L)}}{dt}+\frac{dq_{I-1}^{(L)}}{dt}\right)+\frac{1}{\Delta x^{(L)}}\left(f_{I-1}^{(L)}-f_{I-2}^{(L)}\right)=0$$

and

$$\frac{1}{3}\left(\frac{dq_{I-2}^{(L)}}{dt}+2\frac{dq_{I-1}^{(L)}}{dt}\right)+\frac{1}{\Delta x^{(L)}}\left(f_{I-1}^{(L)}-f_{I-2}^{(L)}\right)=0$$

is the contribution to $I-1$, where $\Delta x^{(L)}$ denotes the length of the left element.

Exercise Derive the difference equations for the contribution of the left element to the gridpoints $I-2$ and $I-1$. □

6.5.2 Center Element

We can write the center element equations as follows

$$\frac{\Delta x^{(C)}}{6}\begin{pmatrix}2&1\\1&2\end{pmatrix}\frac{d}{dt}\begin{pmatrix}q_{I-1}^{(C)}\\q_I^{(C)}\end{pmatrix}-\frac{1}{2}\begin{pmatrix}-1&-1\\1&1\end{pmatrix}\begin{pmatrix}f_{I-1}^{(C)}\\f_I^{(C)}\end{pmatrix}+\begin{pmatrix}-1&0\\0&1\end{pmatrix}\begin{pmatrix}f_{I-1}^{(C)}\\f_I^{(C)}\end{pmatrix}=0 \quad (6.12)$$

where the contribution to $I-1$ is

$$\frac{1}{3}\left(2\frac{dq_{I-1}^{(C)}}{dt}+\frac{dq_I^{(C)}}{dt}\right)+\frac{1}{\Delta x^{(C)}}\left(f_I^{(C)}-f_{I-1}^{(C)}\right)=0$$

and the contribution to I is

$$\frac{1}{3}\left(\frac{dq_{I-1}^{(C)}}{dt}+2\frac{dq_I^{(C)}}{dt}\right)+\frac{1}{\Delta x^{(C)}}\left(f_I^{(C)}-f_{I-1}^{(C)}\right)=0$$

where $\Delta x^{(C)}$ denotes the length of the center element.

6.5.3 Right Element

For the right element, we can write

$$\frac{\Delta x^{(R)}}{6}\begin{pmatrix}2&1\\1&2\end{pmatrix}\frac{d}{dt}\begin{pmatrix}q_I^{(R)}\\q_{I+1}^{(R)}\end{pmatrix}-\frac{1}{2}\begin{pmatrix}-1&-1\\1&1\end{pmatrix}\begin{pmatrix}f_I^{(R)}\\f_{I+1}^{(R)}\end{pmatrix}+\begin{pmatrix}-1&0\\0&1\end{pmatrix}\begin{pmatrix}f_I^{(R)}\\f_{I+1}^{(R)}\end{pmatrix}=0 \quad (6.13)$$

where the contribution to I is

$$\frac{1}{3}\left(2\frac{dq_I^{(R)}}{dt} + \frac{dq_{I+1}^{(R)}}{dt}\right) + \frac{1}{\Delta x^{(R)}}\left(f_{I+1}^{(R)} - f_I^{(R)}\right) = 0$$

and the contribution to $I + 1$ is

$$\frac{1}{3}\left(\frac{dq_I^{(R)}}{dt} + 2\frac{dq_{I+1}^{(R)}}{dt}\right) + \frac{1}{\Delta x^{(R)}}\left(f_{I+1}^{(R)} - f_I^{(R)}\right) = 0$$

where $\Delta x^{(R)}$ denotes the length of the right element.

6.5.4 Total Contribution

By looking at the left, center, and right contributions we see immediately that all three elements are completely decoupled from each other. We can see this by looking at the contribution of the left element to the gridpoint $I - 1$ given as follows:

$$\frac{1}{3}\left(\frac{dq_{I-2}^{(L)}}{dt} + 2\frac{dq_{I-1}^{(L)}}{dt}\right) + \frac{1}{\Delta x^{(L)}}\left(f_{I-1}^{(L)} - f_{I-2}^{(L)}\right) = 0 \qquad (6.14)$$

and the contribution of the center element to the gridpoint $I - 1$ given as

$$\frac{1}{3}\left(2\frac{dq_{I-1}^{(C)}}{dt} + \frac{dq_I^{(C)}}{dt}\right) + \frac{1}{\Delta x^{(C)}}\left(f_I^{(C)} - f_{I-1}^{(C)}\right) = 0. \qquad (6.15)$$

To better visualize the very local nature of DG let us collect the contributions of all three elements into a single global-vector problem using Eqs. (6.11), (6.12), and (6.13). However, before we do this, let us renumber the degrees of freedom as follows: let $q_{I-2}^{(L)} = q_1$, $q_{I-1}^{(L)} = q_2$, ..., $q_{I+1}^{(R)} = q_6$; in other words, using the red numbering in Fig. 6.1. Furthermore, let us assume that $\Delta x = \Delta x^{(L)} = \Delta x^{(C)} = \Delta x^{(R)}$ and summing the contributions from the left, center, and right elements results in the global matrix-vector problem:

$$\frac{\Delta x}{6}\begin{pmatrix} 2 & 1 & 0 & 0 & 0 & 0 \\ 1 & 2 & 0 & 0 & 0 & 0 \\ 0 & 0 & 2 & 1 & 0 & 0 \\ 0 & 0 & 1 & 2 & 0 & 0 \\ 0 & 0 & 0 & 0 & 2 & 1 \\ 0 & 0 & 0 & 0 & 1 & 2 \end{pmatrix}\frac{d}{dt}\begin{pmatrix} q_1 \\ q_2 \\ q_3 \\ q_4 \\ q_5 \\ q_6 \end{pmatrix} - \frac{1}{2}\begin{pmatrix} -1 & -1 & 0 & 0 & 0 & 0 \\ 1 & 1 & 0 & 0 & 0 & 0 \\ 0 & 0 & -1 & -1 & 0 & 0 \\ 0 & 0 & 1 & 1 & 0 & 0 \\ 0 & 0 & 0 & 0 & -1 & -1 \\ 0 & 0 & 0 & 0 & 1 & 1 \end{pmatrix}\begin{pmatrix} f_1 \\ f_2 \\ f_3 \\ f_4 \\ f_5 \\ f_6 \end{pmatrix}$$

$$+\begin{pmatrix} -1 & 0 & 0 & 0 & 0 & 0 \\ 0 & 1 & 0 & 0 & 0 & 0 \\ 0 & 0 & -1 & 0 & 0 & 0 \\ 0 & 0 & 0 & 1 & 0 & 0 \\ 0 & 0 & 0 & 0 & -1 & 0 \\ 0 & 0 & 0 & 0 & 0 & 1 \end{pmatrix}\begin{pmatrix} f_1 \\ f_2 \\ f_3 \\ f_4 \\ f_5 \\ f_6 \end{pmatrix} = \mathbf{0}. \tag{6.16}$$

Equation (6.16) is nothing more than the DSS operation of the DG method and reveals the highly localized nature of DG since all of the matrices are comprised of 2×2 sub-matrices that form the global block-diagonal matrices.

To construct the global matrix-vector problem we simply apply the DSS operation as outlined in Alg. 6.3 where M represents any of the DG matrices such as those we have discussed so far including the mass, differentiation, and flux matrices. Algorithm 6.3 is, in fact, identical to the DSS operation described for the mass

---

Algorithm 6.3 DSS operation for DG element matrices.

---

function GLOBAL_MATRIX($M^{(e)}$)
    $M = 0$
    for $e = 1 : N_e$ **do**   ▷ loop over elements
        for $j = 0 : N$ **do**   ▷ loop over columns of $M^{(e)}$
            $J = intma(j, e)$ ▷ point to global gridpoint number
            for $i = 0 : N$ **do**   ▷ loop over rows of $M^{(e)}$
                $I = intma(i, e)$ ▷ point to global gridpoint number
                $M_{IJ} \mathrel{+}= M_{ij}^{(e)}$ ▷ Gather data to global storage
            end for
        end for
    end for
end function

---

matrix in CG given in Alg. 5.4. The difference is that the *intma* pointer does not impose C^0 on the solution (see Table 6.1). Once we invoke Alg. 6.3 for all of the element matrices, the local matrix problem defined in Eq. (6.10) becomes the global matrix-vector problem

$$M_{IJ}\frac{dq_J}{dt} - \tilde{D}_{IJ}f_J + F_{IJ}f_J^{(*)} = 0, \quad I, J = 1, \ldots, N_p \tag{6.17}$$

where we have not yet defined the explicit construction of the global numerical flux $f_J^{(*)}$.

Table 6.1: intma(i,e) array for $N = 1$ associated with Fig. 6.1.

i	e=1 (Left)	e=2 (Center)	e=3 (Right)
0	1	3	5
1	2	4	6

Going back to the original element-wise numbering that we began with, we note that in general

$$q_{I-1}^{(C)} \neq q_{I-1}^{(L)} \text{ and } f_{I-1}^{(C)} \neq f_{I-1}^{(L)}$$

which tells us that Eqs. (6.14) and (6.15) are completely independent from each other and so they will evolve in time with no knowledge about what is happening at neighboring elements. However, while it is physically possible to have discontinuities it is not possible for, e.g., parcels of air (i.e., elements) to be completely decoupled from the rest of the domain [2]; fortunately, we have only arrived at this decoupling due to our treatment of the flux term. In the above equations we used the interface values specific to the element we were evaluating without regard for its neighbors - we did this to show how decoupled the elements can be from each other in the DG method. However, in reality we want information from contiguous elements to propagate across neighbors. Since at the element interfaces we have discontinuities (from the contribution of the left, center, and right elements having different solutions) then we need to use "averaged" [3] values at the interface. We call these "averaged" values a *numerical flux*.

The mathematical argument for using averaged values stems from well-posedness conditions. For example, in the extreme case where we only have one DG element, ignoring the boundary conditions would violate well-posedness. Analogously, for a multi-element problem we must impose the neighbor values as boundary conditions via the flux terms to ensure well-posedness and, consequently, avoid the inconsistency described previously in which elements have no knowledge of the solution of their neighbors.

Remark 6.2 In the DG method, elements communicate with contiguous elements only through the numerical flux terms (via the flux integrals). The remainder of the operations occur completely locally to the element because, as we have seen, the element integrals only require information specific to that element. This gives DG an advantage regarding how to lay out the communication across compute-devices[4].

---

[2] Completely decoupled parcels of air would violate the domain of influence of classical partial differential equations, such as hyperbolic systems which we describe here. For further reading on this topic, the reader is referred to the following textbooks [213, 418, 419].

[3] We use quotes here to emphasize that we have not exactly defined what we mean by *averaged*.

[4] We use the term *compute-device* to denote any type of computing processor; it can mean a classical CPU-based chip or an accelerator chip such as those found in graphics processing units (GPUs).

6.5.5 Centered Numerical Flux

Let us now consider numerical fluxes that make the problem well-posed. The simplest one possible to consider is to define the numerical flux $f^{(*)}$ as the mean value of the left and right elements. The mean value flux, or centered flux, is written as follows

$$f^{(*)} = \frac{1}{2}\left(f^{(e)} + f^{(k)}\right)$$

where the superscripts e and k denote the element and its neighbor, respectively. Using this centered flux in Eq. (6.12) yields the following equations for the center element

$$\frac{\Delta x^{(C)}}{6}\begin{pmatrix}2 & 1\\1 & 2\end{pmatrix}\frac{d}{dt}\begin{pmatrix}q_{I-1}^{(C)}\\q_I^{(C)}\end{pmatrix} - \frac{1}{2}\begin{pmatrix}-1 & -1\\+1 & +1\end{pmatrix}\begin{pmatrix}f_{I-1}^{(C)}\\f_I^{(C)}\end{pmatrix} + \begin{pmatrix}-1 & 0\\0 & 1\end{pmatrix}\begin{pmatrix}\frac{1}{2}\left(f_{I-1}^{(L)} + f_{I-1}^{(C)}\right)\\\frac{1}{2}\left(f_I^{(C)} + f_I^{(R)}\right)\end{pmatrix} = \mathbf{0}$$

$$(6.18)$$

whereas for the right element, Eq. (6.13) becomes

$$\frac{\Delta x^{(R)}}{6}\begin{pmatrix}2 & 1\\1 & 2\end{pmatrix}\frac{d}{dt}\begin{pmatrix}q_I^{(R)}\\q_{I+1}^{(R)}\end{pmatrix} - \frac{1}{2}\begin{pmatrix}-1 & -1\\+1 & +1\end{pmatrix}\begin{pmatrix}f_I^{(R)}\\f_{I+1}^{(R)}\end{pmatrix} + \begin{pmatrix}-1 & 0\\0 & 1\end{pmatrix}\begin{pmatrix}\frac{1}{2}\left(f_I^{(C)} + f_I^{(R)}\right)\\\frac{1}{2}\left(f_{I+1}^{(R)} + f_{I-2}^{(L)}\right)\end{pmatrix} = \mathbf{0}$$

$$(6.19)$$

where, for convenience, we have used periodic boundary conditions in the last term of the second row. Left-multiplying Eqs. (6.18) and (6.19) by the inverse mass matrix

$$\left(M^{(e)}\right)^{-1} = \frac{2}{\Delta x^{(e)}}\begin{pmatrix}2 & -1\\-1 & 2\end{pmatrix}$$

yields, for the center element

$$\frac{d}{dt}\begin{pmatrix}q_{I-1}^{(C)}\\q_I^{(C)}\end{pmatrix} - \frac{3}{\Delta x^{(C)}}\begin{pmatrix}-1 & -1\\+1 & +1\end{pmatrix}\begin{pmatrix}f_{I-1}^{(C)}\\f_I^{(C)}\end{pmatrix} + \frac{1}{\Delta x^{(C)}}\begin{pmatrix}-2 & -1\\1 & 2\end{pmatrix}\begin{pmatrix}f_{I-1}^{(L)} + f_{I-1}^{(C)}\\f_I^{(C)} + f_I^{(R)}\end{pmatrix} = \mathbf{0} \quad (6.20)$$

and for the right element

$$\frac{d}{dt}\begin{pmatrix}q_I^{(R)}\\q_{I+1}^{(R)}\end{pmatrix} - \frac{3}{\Delta x^{(R)}}\begin{pmatrix}-1 & -1\\+1 & +1\end{pmatrix}\begin{pmatrix}f_I^{(R)}\\f_{I+1}^{(R)}\end{pmatrix} + \frac{1}{\Delta x^{(R)}}\begin{pmatrix}-2 & -1\\1 & 2\end{pmatrix}\begin{pmatrix}f_I^{(C)} + f_I^{(R)}\\f_{I+1}^{(R)} + f_{I-2}^{(L)}\end{pmatrix} = \mathbf{0}. \quad (6.21)$$

The equation for the gridpoint I of the center element is given by the second row of Eq. (6.20) as follows

$$\frac{dq_I^{(C)}}{dt} - \frac{3}{\Delta x^{(C)}}\left(f_{I-1}^{(C)} + f_I^{(C)}\right) + \frac{1}{\Delta x^{(C)}}\left(f_{I-1}^{(L)} + f_{I-1}^{(C)} + 2f_I^{(C)} + 2f_I^{(R)}\right) = 0.$$

Rearranging this equation allows us to write

$$\frac{dq_I^{(C)}}{dt} + \frac{1}{\Delta x^{(C)}} \left[\left(2f_I^{(R)} - f_I^{(C)} \right) - \left(2f_{I-1}^{(C)} - f_{I-1}^{(L)} \right) \right] = 0$$

where the term in square brackets shows the jumps at the I and $I - 1$ gridpoint interfaces, respectively. In a similar fashion, we can obtain the solution for the right element at the gridpoint I to be

$$\frac{dq_I^{(R)}}{dt} + \frac{1}{\Delta x^{(R)}} \left[\left(2f_{I+1}^{(R)} - f_{I-2}^{(L)} \right) - \left(2f_I^{(C)} - f_I^{(R)} \right) \right] = 0.$$

Assuming $f \in C^0$, we obtain the following solution at the gridpoint I, for the center element

$$\frac{dq_I^{(C)}}{dt} + \frac{f_I - f_{I-1}}{\Delta x^{(C)}} = 0$$

and for the right element

$$\frac{dq_I^{(R)}}{dt} + \frac{f_{I+1} - f_I}{\Delta x^{(R)}} = 0$$

which result in upwinding (center element) and downwinding (right element) methods, respectively. Unfortunately, this simple DG method is not guaranteed to be stable. We discuss this point in detail in Sec. 6.7. To be clear, the problem is not with the DG method itself but rather the specific numerical flux function $f^{(*)}$ that we chose. In Ch. 18, upwinding methods are discussed in depth but for now we end the discussion on flux functions by illustrating the resulting differencing stencil when an upwind-biased numerical flux is chosen. In this case, we are guaranteed stability[5].

Before discussing the upwinding flux, let us write down the resulting matrix-vector problem as given by Eq. (6.16) but now with the centered flux included. We show this below where we only use three elements (as shown in Fig. 6.1) and use periodicity at the end points which results in the global matrix-vector problem given by

---

[5] We define stability here to mean the stability solely due to the spatial discretization method. When time-integrators are used we must be mindful of satisfying the stability condition for that specific time-integrator.

$$
\frac{\Delta x}{6}
\begin{pmatrix}
2 & 1 & 0 & 0 & 0 & 0 \\
1 & 2 & 0 & 0 & 0 & 0 \\
0 & 0 & 2 & 1 & 0 & 0 \\
0 & 0 & 1 & 2 & 0 & 0 \\
0 & 0 & 0 & 0 & 2 & 1 \\
0 & 0 & 0 & 0 & 1 & 2
\end{pmatrix}
\frac{d}{dt}
\begin{pmatrix}
q_1 \\ q_2 \\ q_3 \\ q_4 \\ q_5 \\ q_6
\end{pmatrix}
- \frac{1}{2}
\begin{pmatrix}
-1 & -1 & 0 & 0 & 0 & 0 \\
1 & 1 & 0 & 0 & 0 & 0 \\
0 & 0 & -1 & -1 & 0 & 0 \\
0 & 0 & 1 & 1 & 0 & 0 \\
0 & 0 & 0 & 0 & -1 & -1 \\
0 & 0 & 0 & 0 & 1 & 1
\end{pmatrix}
\begin{pmatrix}
f_1 \\ f_2 \\ f_3 \\ f_4 \\ f_5 \\ f_6
\end{pmatrix}
$$

$$
+ \frac{1}{2}
\begin{pmatrix}
-1 & 0 & 0 & 0 & 0 & -1 \\
0 & 1 & 1 & 0 & 0 & 0 \\
0 & -1 & -1 & 0 & 0 & 0 \\
0 & 0 & 0 & 1 & 1 & 0 \\
0 & 0 & 0 & -1 & -1 & 0 \\
1 & 0 & 0 & 0 & 0 & 1
\end{pmatrix}
\begin{pmatrix}
f_1 \\ f_2 \\ f_3 \\ f_4 \\ f_5 \\ f_6
\end{pmatrix}
= \mathbf{0}
\tag{6.22}
$$

where the changes to the flux matrix due to the centered flux are marked in red. In this form, it becomes evident that the only coupling of neighboring elements in the DG method occurs because of the numerical flux. Equation (6.22) shows that the flux term (third term) is the only term in the equations without a block-diagonal matrix.

The algorithm for modifying the flux matrix with a centered flux is described in Alg. 6.4. After constructing the flux matrix from this algorithm, we can write the global matrix-vector problem as follows

$$
M_{IJ}\frac{dq_J}{dt} - \tilde{D}_{IJ} f_J + F_{IJ}^{(*)} f_J = 0, \quad I, J = 1, \ldots, N_p.
\tag{6.23}
$$

Note that in Alg. 6.4, $\hat{\mathbf{n}}^{(e,L)} = -1$ and $\hat{\mathbf{n}}^{(e,R)} = +1$ for one-dimensional problems.

6.5.6 Rusanov Numerical Flux

The single most common numerical flux function for hyperbolic equations is the Rusanov (or local Lax-Friedrichs) flux which is a generalized upwinding method. The Rusanov flux is defined as

$$
f^{(*,e,k)} = \frac{1}{2}\left[f^{(e)} + f^{(k)} - \hat{\mathbf{n}}^{(e,k)}|\lambda_{max}|\left(q^{(k)} - q^{(e)} \right) \right]
$$

where $\hat{\mathbf{n}}^{(e,k)}$ denotes the outward pointing normal to the interface of the element e and its neighbor k, and λ_{max} is the maximum wave speed of the governing partial differential equations (PDEs). In standard DG notation, we write the Rusanov flux as follows

$$
f^{(*,e,k)} = \left\{ f^{(e,k)} \right\} - \hat{\mathbf{n}}^{(e,k)}\frac{|\lambda_{max}|}{2}[\![q^{(e,k)}]\!]
$$

where $\left\{ f^{(e,k)} \right\} = \frac{1}{2}\left(f^{(e)} + f^{(k)} \right)$ denotes the averaging operator and $[\![q^{(e,k)}]\!] = \left(q^{(k)} - q^{(e)} \right)$ denotes the jump operator.

Algorithm 6.4 Construction of the global flux matrix with centered flux.

function GLOBAL_FLUX_MATRIX
    $F^{(*)} = 0$
    for $e = 1 : N_e$ **do**   ▷ loop over elements
      $L = e - 1$ ▷ identify the left element of this face
      if $e = 1$ **then**
        $L = N_e$ ▷ apply periodicity
      end if
      $I = intma(0, e)$ ▷ global gridpoint on the right
      $J = intma(N, L)$ ▷ global gridpoint on the left
      $F_{II}^{(*)} = \frac{1}{2}\hat{\mathbf{n}}^{(e,L)}$ ▷ update flux matrix
      $F_{IJ}^{(*)} = \frac{1}{2}\hat{\mathbf{n}}^{(e,L)}$ ▷ update flux matrix
      $R = e + 1$ ▷ identify the right element of this face
      if $e = N_e$ **then**
        $R = 1$ ▷ apply periodicity
      end if
      $I = intma(N, e)$ ▷ global gridpoint on the left
      $J = intma(0, R)$ ▷ global gridpoint on the right
      $F_{II}^{(*)} = \frac{1}{2}\hat{\mathbf{n}}^{(e,R)}$ ▷ update flux matrix
      $F_{IJ}^{(*)} = \frac{1}{2}\hat{\mathbf{n}}^{(e,R)}$ ▷ update flux matrix
    end for
end function

For the linear wave equation, $\lambda_{max} = u$ but in general it represents the maximum eigenvalue of the Jacobian matrix of the governing equations of motion (e.g., see [159], [157], [162], and [163] for examples for various systems of equations including compressible Navier-Stokes, shallow water on the plane, and shallow water on the sphere). The Rusanov flux is just the average value between the two elements sharing a face with the addition of a dissipation term. The dissipation term allows the flux function to modify itself based on the flow conditions in order to construct an upwind-biased method (we assume flow is from left to right and so $u > 0$ in this particular example).

Looking at Fig. 6.1 let us now see what the Rusanov flux looks like for the center element. Since $f = qu$ we can rewrite the Rusanov flux as

$$f^{(*,e,k)} = \frac{1}{2}\left[f^{(e)} + f^{(k)} - \hat{\mathbf{n}}^{(e,k)}\left(f^{(k)} - f^{(e)} \right) \right]$$

where we have used $|\lambda_{max}| = u$ and simplified the dissipation term as follows

$$|\lambda_{max}|\left(q^{(k)} - q^{(e)} \right) = \left(f^{(k)} - f^{(e)} \right).$$

At the interface $(e, k) = (C, L)$, i.e., between the center and left elements, the outward pointing normal vector from C to L is $\hat{\mathbf{n}}^{(e,k)} = -1$. This gives

$$f^{(*,C,L)} = \frac{1}{2}\left[f^{(C)} + f^{(L)} - (-1)\left(f^{(L)} - f^{(C)} \right) \right]$$

that can be simplified to

$$f^{(*,C,L)} = f^{(L)}.$$

At the interface $(e, k) = (C, R)$, i.e., between the center and right elements, the outward pointing normal vector from C to R is $\hat{n}^{(e,k)} = +1$. This gives

$$f^{(*,C,R)} = \frac{1}{2}\left[f^{(C)} + f^{(R)} - (+1)\left(f^{(R)} - f^{(C)}\right)\right]$$

that can be simplified to

$$f^{(*,C,R)} = f^{(C)}.$$

Using these flux values gives for the center element equations

$$\frac{\Delta x^{(C)}}{6}\begin{pmatrix} 2 & 1 \\ 1 & 2 \end{pmatrix}\frac{d}{dt}\begin{pmatrix} q_{I-1}^{(C)} \\ q_I^{(C)} \end{pmatrix} - \frac{1}{2}\begin{pmatrix} -1 & -1 \\ +1 & +1 \end{pmatrix}\begin{pmatrix} f_{I-1}^{(C)} \\ f_I^{(C)} \end{pmatrix} + \begin{pmatrix} -1 & 0 \\ 0 & 1 \end{pmatrix}\begin{pmatrix} f_{I-1}^{(L)} \\ f_I^{(C)} \end{pmatrix} = 0 \quad (6.24)$$

and for the right element

$$\frac{\Delta x^{(R)}}{6}\begin{pmatrix} 2 & 1 \\ 1 & 2 \end{pmatrix}\frac{d}{dt}\begin{pmatrix} q_I^{(R)} \\ q_{I+1}^{(R)} \end{pmatrix} - \frac{1}{2}\begin{pmatrix} -1 & -1 \\ +1 & +1 \end{pmatrix}\begin{pmatrix} f_I^{(R)} \\ f_{I+1}^{(R)} \end{pmatrix} + \begin{pmatrix} -1 & 0 \\ 0 & 1 \end{pmatrix}\begin{pmatrix} f_I^{(C)} \\ f_{I+1}^{(R)} \end{pmatrix} = 0. \quad (6.25)$$

Left-multiplying these equations by the inverse mass matrix

$$\left(M^{(e)}\right)^{-1} = \frac{2}{\Delta x^{(e)}}\begin{pmatrix} 2 & -1 \\ -1 & 2 \end{pmatrix}$$

yields, for the center element

$$\frac{d}{dt}\begin{pmatrix} q_{I-1}^{(C)} \\ q_I^{(C)} \end{pmatrix} - \frac{3}{\Delta x^{(C)}}\begin{pmatrix} -1 & -1 \\ +1 & +1 \end{pmatrix}\begin{pmatrix} f_{I-1}^{(C)} \\ f_I^{(C)} \end{pmatrix} + \frac{2}{\Delta x^{(C)}}\begin{pmatrix} -2 & -1 \\ 1 & 2 \end{pmatrix}\begin{pmatrix} f_{I-1}^{(L)} \\ f_I^{(C)} \end{pmatrix} = 0 \quad (6.26)$$

and for the right element

$$\frac{d}{dt}\begin{pmatrix} q_I^{(R)} \\ q_{I+1}^{(R)} \end{pmatrix} - \frac{3}{\Delta x^{(R)}}\begin{pmatrix} -1 & -1 \\ +1 & +1 \end{pmatrix}\begin{pmatrix} f_I^{(R)} \\ f_{I+1}^{(R)} \end{pmatrix} + \frac{2}{\Delta x^{(R)}}\begin{pmatrix} -2 & -1 \\ 1 & 2 \end{pmatrix}\begin{pmatrix} f_I^{(C)} \\ f_{I+1}^{(R)} \end{pmatrix} = 0. \quad (6.27)$$

The equation for the gridpoint I of the center element is given by the second row of Eq. (6.26) as follows

$$\frac{dq_I^{(C)}}{dt} - \frac{3}{\Delta x^{(C)}}\left(f_{I-1}^{(C)} + f_I^{(C)}\right) + \frac{2}{\Delta x^{(C)}}\left(f_{I-1}^{(L)} + 2f_I^{(C)}\right) = 0.$$

Rearranging this equation allows us to write

$$\frac{dq_I^{(C)}}{dt} + \frac{1}{\Delta x^{(C)}} \left[f_I^{(C)} - \left(3f_{I-1}^{(C)} - 2f_{I-1}^{(L)} \right) \right] = 0$$

where the second and third terms in the square brackets give the jump at the $I-1$ gridpoint interface. In a similar fashion, we can obtain the solution for the right element at the gridpoint I to be

$$\frac{dq_I^{(R)}}{dt} + \frac{1}{\Delta x^{(R)}} \left[f_{I+1}^{(R)} - \left(4f_I^{(C)} - 3f_I^{(R)} \right) \right] = 0.$$

Remark 6.3 A clear difference between the CG and DG methods is that in the former the surrounding elements to a gridpoint contribute a portion of their solution to that gridpoint. In the DG method, surrounding elements contribute fluxes through the interfaces to a specific element. Therefore, DG is more local because the solution only exists in an element sense while in CG the solution exists at global gridpoints that are shared by neighboring elements.

Exercise Prove that the weak form differentiation matrix is the transpose of the strong form differentiation matrix. That is:

$$\tilde{D} = D^T.$$

Exercise Construct the weak form DG discretization for the center element when using $N = 2$ interpolation with exact integration. To do this problem, simply build the local element mass, weak-form differentiation, and flux matrices. Use the Rusanov numerical flux. You can use the results from previous homework problems from Ch. 5 if it helps. Use the fact that $\tilde{D} = D^T$ to use the results for $N = 2$ we found for CG in Ch. 5.   □

Before ending this section, let us write the global matrix-vector problem with the Rusanov numerical flux which, in one-dimension, yields a purely upwinding approach that we write as follows:

$$\frac{\Delta x}{6} \begin{pmatrix} 2 & 1 & 0 & 0 & 0 & 0 \\ 1 & 2 & 0 & 0 & 0 & 0 \\ 0 & 0 & 2 & 1 & 0 & 0 \\ 0 & 0 & 1 & 2 & 0 & 0 \\ 0 & 0 & 0 & 0 & 2 & 1 \\ 0 & 0 & 0 & 0 & 1 & 2 \end{pmatrix} \frac{d}{dt} \begin{pmatrix} q_1 \\ q_2 \\ q_3 \\ q_4 \\ q_5 \\ q_6 \end{pmatrix} - \frac{1}{2} \begin{pmatrix} -1 & -1 & 0 & 0 & 0 & 0 \\ 1 & 1 & 0 & 0 & 0 & 0 \\ 0 & 0 & -1 & -1 & 0 & 0 \\ 0 & 0 & 1 & 1 & 0 & 0 \\ 0 & 0 & 0 & 0 & -1 & -1 \\ 0 & 0 & 0 & 0 & 1 & 1 \end{pmatrix} \begin{pmatrix} f_1 \\ f_2 \\ f_3 \\ f_4 \\ f_5 \\ f_6 \end{pmatrix}$$

$$+ \begin{pmatrix} 0 & 0 & 0 & 0 & 0 & -1 \\ 0 & 1 & 0 & 0 & 0 & 0 \\ 0 & -1 & 0 & 0 & 0 & 0 \\ 0 & 0 & 0 & 1 & 0 & 0 \\ 0 & 0 & 0 & -1 & 0 & 0 \\ 0 & 0 & 0 & 0 & 0 & 1 \end{pmatrix} \begin{pmatrix} f_1 \\ f_2 \\ f_3 \\ f_4 \\ f_5 \\ f_6 \end{pmatrix} = \mathbf{0}. \qquad (6.28)$$

where the numbers marked in red show the differences between the Rusanov flux from the centered flux. Equation (6.28) shows that, as in the case with the centered flux, the Rusanov flux also does not yield a block-diagonal flux matrix. Once again, it is the numerical flux that couples adjacent elements together.

The algorithm for modifying the flux matrix to include either the centered or Rusanov flux is described in Alg. 6.5. In Alg. 6.5, $\hat{n}^{(e,L)} = -1$ and $\hat{n}^{(e,R)} = +1$, where we have now included an additional term δ_{diss} which is a switch (0 or 1). If $\delta_{diss} = 0$ then we recover the centered flux and if it is $\delta_{diss} = 1$ then we impose the Rusanov flux (which is fully upwinding in the one-dimensional case).

Algorithm 6.5 Construction of the global flux matrix with Rusanov flux.

function GLOBAL_FLUX_MATRIX
  $F^{(*)} = 0$
  for $e = 1 : N_e$ **do**   ▷ loop over elements
    $L = e - 1$ ▷ identify the left element of this face
    if $e = 1$ **then**
      $L = N_e$ ▷ apply periodicity
    end if
    $I = intma(0, e)$ ▷ global gridpoint on the right
    $J = intma(N, L)$ ▷ global gridpoint on the left
    $F^{(*)}_{II} = \frac{1}{2}\hat{n}^{(e,L)}\left(1 + \hat{n}^{(e,L)}\delta_{diss}\right)$ ▷ update flux matrix
    $F^{(*)}_{IJ} = \frac{1}{2}\hat{n}^{(e,L)}\left(1 - \hat{n}^{(e,L)}\delta_{diss}\right)$ ▷ update flux matrix
    $R = e + 1$ ▷ identify the right element of this face
    if $e = N_e$ **then**
      $R = 1$ ▷ apply periodicity
    end if
    $I = intma(N, e)$ ▷ global gridpoint on the left
    $J = intma(0, R)$ ▷ global gridpoint on the right
    $F^{(*)}_{II} = \frac{1}{2}\hat{n}^{(e,R)}\left(1 + \hat{n}^{(e,R)}\delta_{diss}\right)$ ▷ update flux matrix
    $F^{(*)}_{IJ} = \frac{1}{2}\hat{n}^{(e,R)}\left(1 - \hat{n}^{(e,R)}\delta_{diss}\right)$ ▷ update flux matrix
  end for
end function

The algorithm for just modifying the flux vector with either the centered or Rusanov flux is described in Alg. 6.6 where the term $\delta_{diss} = 0$ recovers the centered flux and $\delta_{diss} = 1$ the Rusanov flux. We use this algorithm in Sec. 6.9.5 to derive a simple global solution for the DG method that looks rather like the CG global solution.

6.6 High-Order Approximation

Now that we know how to build the element-wise equations for DG for the specific case of $N = 1$ let us extend this to the case $N = 2$. As we saw in Ch. 5 the basis functions are $\psi_0(\xi) = -\frac{1}{2}\xi(1-\xi), \psi_1(\xi) = (1-\xi)(1+\xi), \psi_2(\xi) = \frac{1}{2}\xi(1+\xi)$, where the

Algorithm 6.6 Construction of the global numerical flux vector $f^{(*)}$ with the centered or Rusanov flux.

```
function GLOBAL_FLUX_VECTOR
    f^(*) = 0
    for e = 1 : N_e do                          ▷ loop over elements
        L = e                                   ▷ identify the left element of this face
        R = e + 1                               ▷ identify the right element of this face
        if e = N_e then
            R = 1                               ▷ apply periodicity
        end if
        I = intma(N, L)                         ▷ global gridpoint on the left
        J = intma(0, R)                         ▷ global gridpoint on the right
        f^(*) = ½ [f_I + f_J − uδ_diss (q_J − q_I)]
        f_I^(*) = f^(*)
        f_J^(*) = f^(*)
    end for
end function
```

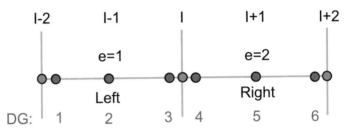

Fig. 6.2: The one-dimensional stencil for the DG method with two elements labeled as Left (e=1) and Right (e=2) with $N = 2$ polynomials.

Table 6.2: intma(i,e) array for $N = 2$ associated with Fig. 6.2.

i	e=1 (Left)	e=2 (Right)
0	1	4
1	2	5
2	3	6

subscripts 0, 1, 2 refer to the left, middle, and right points for each element. Referring to Fig. 6.2 we find that the left element equations are

$$\frac{\Delta x^{(L)}}{30} \begin{pmatrix} 4 & 2 & -1 \\ 2 & 16 & 2 \\ -1 & 2 & 4 \end{pmatrix} \frac{d}{dt} \begin{pmatrix} q_{I-2}^{(L)} \\ q_{I-1}^{(L)} \\ q_I^{(L)} \end{pmatrix} - \frac{1}{6} \begin{pmatrix} -3 & -4 & 1 \\ 4 & 0 & -4 \\ -1 & 4 & 3 \end{pmatrix} \begin{pmatrix} f_{I-2}^{(L)} \\ f_{I-1}^{(L)} \\ f_I^{(L)} \end{pmatrix} + \begin{pmatrix} -1 & 0 & 0 \\ 0 & 0 & 0 \\ 0 & 0 & 1 \end{pmatrix} \begin{pmatrix} f_{I+2}^{(R)} \\ f_{I-1}^{(L)} \\ f_I^{(L)} \end{pmatrix} = \mathbf{0} \quad (6.29)$$

where we have used the upwind Rusanov flux (with $u > 0$) and have assumed periodic boundary conditions.

Note that the corresponding equations for the right element are

$$
\frac{\Delta x^{(R)}}{30} \begin{pmatrix} 4 & 2 & -1 \\ 2 & 16 & 2 \\ -1 & 2 & 4 \end{pmatrix} \frac{d}{dt} \begin{pmatrix} q_I^{(R)} \\ q_{I+1}^{(R)} \\ q_{I+2}^{(R)} \end{pmatrix} - \frac{1}{6} \begin{pmatrix} -3 & -4 & 1 \\ 4 & 0 & -4 \\ -1 & 4 & 3 \end{pmatrix} \begin{pmatrix} f_I^{(R)} \\ f_{I+1}^{(R)} \\ f_{I+2}^{(R)} \end{pmatrix} + \begin{pmatrix} -1 & 0 & 0 \\ 0 & 0 & 0 \\ 0 & 0 & 1 \end{pmatrix} \begin{pmatrix} f_I^{(L)} \\ f_{I+1}^{(R)} \\ f_{I+2}^{(R)} \end{pmatrix} = \mathbf{0}. \quad (6.30)
$$

The global matrix-vector problem for $N = 2$ with two elements with constant element spacing Δx is given as follows

$$
\frac{\Delta x}{30} \begin{pmatrix} 4 & 2 & -1 & 0 & 0 & 0 \\ 2 & 16 & 2 & 0 & 0 & 0 \\ -1 & 2 & 4 & 0 & 0 & 0 \\ 0 & 0 & 0 & 4 & 2 & -1 \\ 0 & 0 & 0 & 2 & 16 & 2 \\ 0 & 0 & 0 & -1 & 2 & 4 \end{pmatrix} \frac{d}{dt} \begin{pmatrix} q_1 \\ q_2 \\ q_3 \\ q_4 \\ q_5 \\ q_6 \end{pmatrix} - \frac{1}{6} \begin{pmatrix} -3 & -4 & 1 & 0 & 0 & 0 \\ 4 & 0 & -4 & 0 & 0 & 0 \\ -1 & 4 & 3 & 0 & 0 & 0 \\ 0 & 0 & 0 & -3 & -4 & 1 \\ 0 & 0 & 0 & 4 & 0 & -4 \\ 0 & 0 & 0 & -1 & 4 & 3 \end{pmatrix} \begin{pmatrix} f_1 \\ f_2 \\ f_3 \\ f_4 \\ f_5 \\ f_6 \end{pmatrix}
$$

$$
+ \begin{pmatrix} 0 & 0 & 0 & 0 & 0 & -1 \\ 0 & 0 & 0 & 0 & 0 & 0 \\ 0 & 0 & 1 & 0 & 0 & 0 \\ 0 & 0 & -1 & 0 & 0 & 0 \\ 0 & 0 & 0 & 0 & 0 & 0 \\ 0 & 0 & 0 & 0 & 0 & 1 \end{pmatrix} \begin{pmatrix} f_1 \\ f_2 \\ f_3 \\ f_4 \\ f_5 \\ f_6 \end{pmatrix} = \mathbf{0} \qquad (6.31)
$$

where the gridpoint numbering $1, \ldots, 6$ is illustrated by the red gridpoints in Fig. 6.2. The *intma* pointer for this configuration is given in Table 6.2.

Exercise Derive Eq. (6.29) which is the element approximation for DG using $N = 2$ with exact integration, Rusanov flux, and periodic boundary conditions.   □

Exercise Check if the flux derivative for the I gridpoint is consistent and show the order of accuracy of the derivative.   □

Exercise Note that in Eqs. (6.28) and (6.31), the columns of the numerical flux matrix sum to zero. Explain why this is so.   □

6.7 Analysis of the Matrix Properties of the Spatial Operators

Recall that the DG global matrix-vector problem for the 1D wave equation is given by Eq. (6.23). To get the gridpoint representation, we left-multiply Eq. (6.23) by M^{-1} to get

$$
\frac{dq_I}{dt} - M_{IK}^{-1} \tilde{D}_{KJ} f_J + M_{IK}^{-1} F_{KJ}^{(*)} f_J = \mathbf{0}, \quad I, J, K = 1, \ldots, N_p \qquad (6.32)
$$

where we assume that the flux matrix $F^{(*)}$ is defined, e.g., as in Alg. 6.5. Defining

$$
\hat{F}^{(*)} = M^{-1} F^{(*)} \quad \text{and} \quad \hat{D} = M^{-1} \tilde{D}
$$

allows us to write

$$\frac{dq_I}{dt} = \left(\hat{\tilde{D}}_{IJ} - \hat{F}_{IJ}^{(*)} \right) f_J, \quad I, J = 1, \ldots, N_p \tag{6.33}$$

that can be further simplified to

$$\frac{dq_I}{dt} = \hat{D}_{IJ}^{DG} f_J, \quad I, J, K = 1, \ldots, N_p \tag{6.34}$$

where $\hat{D}^{DG} = M^{-1} D^{DG}$ represents the right-hand side matrix for DG with $D^{DG} = \tilde{D} - F^{(*)}$.

Let us now analyze the properties of these matrices. For solving the 1D wave equation, we assume periodic boundary conditions and use a total of $N_p = N_e(N + 1) = 20$ gridpoints (for $N_e = 4$ and $N = 4$) to completely cover the domain; we define the initial value problem explicitly in the following section, although it does not matter for the purposes of analyzing the matrices.

6.7.1 Sparsity Pattern of the Mass Matrix

Figure 6.3 shows the sparsity pattern for the mass matrix M for both exact ($Q = N+1$) and inexact ($Q = N$) integration using Lobatto points. Figure 6.3 shows that the mass

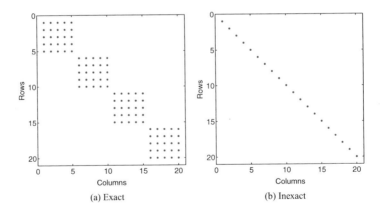

(a) Exact   (b) Inexact

Fig. 6.3: Sparsity patterns for the mass matrix M with $N_e = 4$, $N = 4$ for (a) exact and (b) inexact integration.

matrix is fully decoupled for each element, i.e., it is block-diagonal. For inexact integration, the matrix becomes diagonal.

6.7.2 Sparsity Pattern of the Differentiation Matrix

Figure 6.4 shows the sparsity pattern for the differentiation matrix D^{DG} with Rusanov flux (i.e., upwind-biased numerical flux). In Fig. 6.4 we see the outline of the elements

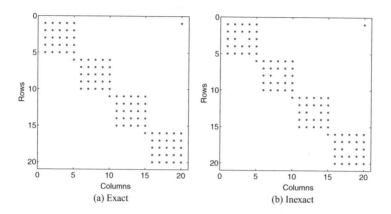

Fig. 6.4: Sparsity patterns for the differentiation matrix D^{DG} using Rusanov flux with $N_e = 4$, $N = 4$ for (a) exact and (b) inexact integration.

in the differentiation matrix. There is no change in the structure of this matrix going from exact to inexact integration; any differences visible are due to round-off error whereby small quantities that are non-zero are visible as non-zero matrix entries. The non-zero entries for D^{DG} at the top right of Fig. 6.4 (the row 1 column 20 entry) are due to the periodic boundary condition. This simply connects the first point of the first element with the last point of the last element. The reason why there is no corresponding entry at the bottom left of D^{DG} is due to the upwinding flux used (since it will only require the last point of the last element which is the bottom right entry in D^{DG}). It is important to understand that we defined the matrix D^{DG} to be $D^{DG} = \tilde{D} - F^{(*)}$ where $\tilde{D}$ is the element differentiation matrix (which is block-diagonal) and $F^{(*)}$ is the flux matrix. It is only through the flux matrix $F^{(*)}$ that we experience a coupling of neighboring elements and is the reason why D^{DG} is not block-diagonal.

6.7.3 Eigenvalue Analysis of Spatial Operator

Let us rewrite Eq. (6.34) as follows

$$\frac{dq_I}{dt} = R_{IJ}q_J \tag{6.35}$$

where $R = \hat{D}^{DG}$, with u absorbed into R, and I denotes all the gridpoints in the domain such that $I = 1, \ldots, N_e(N + 1)$ and where we assume that the gridpoint $I^{(C)}$ is distinct from $I^{(R)}$, i.e., at the interface of two elements (C and R) we use different gridpoint identifiers for the solution on the center and right elements. This is in fact the numbering described in Figs. 6.1 and 6.2.

We can now replace the matrix R by its eigenvalues as follows

$$Rx = \lambda x$$

to arrive at

$$\frac{dq_I}{dt} = \lambda q_I. \tag{6.36}$$

This equation has the analytic solution

$$q_I = q_0 \exp(\lambda t)$$

that will be bounded (less than infinity) for $Re(\lambda) \leq 0$.

Let us look at the eigenvalues of R for the example 1D wave equation problem that we defined earlier ($N_p = 20$ gridpoints). Figure 6.5 shows the eigenvalues of the right-hand side matrix R. This matrix is a representation of the complete

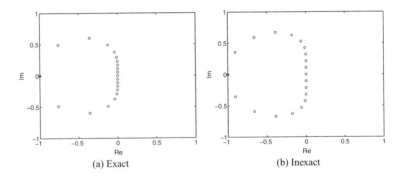

(a) Exact   (b) Inexact

Fig. 6.5: Eigenvalues of the right-hand side matrix R using Rusanov flux with $N_e = 4$, $N = 4$ for (a) exact and (b) inexact integration.

discretization of the spatial operators. Note that the real parts of the eigenvalues are less than zero with the maximum at -1.00×10^{-15}; the rest of the eigenvalues are located away from the imaginary axis and in the negative real axis (left-hand plane). Therefore, this method is perfectly stable since only a few eigenvalues are near $Re = 0$ but less than zero.

For comparison, let us now plot the eigenvalues of R using a centered numerical flux. Figure 6.6 shows that the eigenvalues in this case look very similar to those for CG (shown in Ch. 5) with the maximum real eigenvalue slightly greater than zero

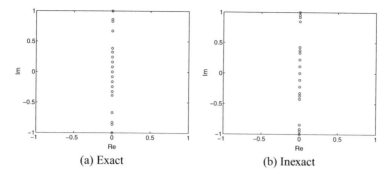

Fig. 6.6: Eigenvalues of the right-hand side matrix R using centered flux with $N_e = 4$, $N = 4$ for (a) exact and (b) inexact integration.

(5.32×10^{-15}). This does mean, however, that this method will eventually become unstable.

Before closing this section, let us look at the structure of the matrix R. Note that

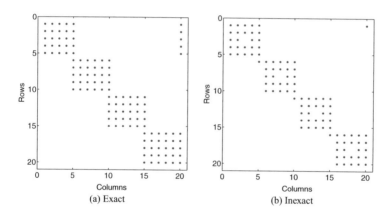

Fig. 6.7: Sparsity patterns for the right-hand side matrix R with $N_e = 4$, $N = 4$ for (a) exact and (b) inexact integration.

unlike CG, the DG right-hand side matrix is not full (even for exact integration). This is because the DG method is much more local in nature; that is, the governing equations are satisfied element-wise and the resulting matrix problem is indeed a local one as is evident from Fig. 6.7, at least for explicit time-integration. The only change in R going from exact to inexact integration is in the boundary conditions (top right corner of the matrix). This is due to the differences in the structure of the mass matrices. In other words, for exact integration, the element mass matrix is full and so the inverse of this matrix provides a *lifting* operation whereby the solution at the interfaces is *lifted* and scattered throughout the element. In this particular example, the periodic boundary conditions at the far right boundary affects all the points of that element by virtue of the lifting operation of the inverse mass matrix.

6.8 Conservation Property of DG

6.8.1 Exact Integration

Let us discuss the *conservation property* of the DG method. Local conservation of the variable q implies that the first integral in Eq. (6.1) vanishes, that is

$$\int_{\Omega_e} \psi_i \frac{\partial q_N^{(e)}}{\partial t}^{\,0} d\Omega_e + \int_{\Omega_e} \psi_i \frac{\partial f_N^{(e)}}{\partial x} d\Omega_e = 0.$$

From Eq. (6.2) we saw that using integration by parts on the second integral requires satisfying the following equality

$$\int_{\Omega_e} \psi_i \frac{\partial f_N^{(e)}}{\partial x} d\Omega_e = -\int_{\Omega_e} \frac{d\psi_i}{dx} f_N^{(e)} d\Omega_e + \left[\psi_i f_N^{(e)} \right]_{\Gamma_e}. \tag{6.37}$$

We can rewrite Eq. (6.37) in the following matrix-vector form

$$D_{i,j}^{(e)} f_j^{(e)} = -\tilde{D}_{i,j}^{(e)} f_j^{(e)} + F_{i,j}^{(e)} f_j^{(e)}$$

which we rearrange as follows

$$\left(D_{i,j}^{(e)} + \tilde{D}_{i,j}^{(e)} \right) f_j^{(e)} = F_{i,j}^{(e)} f_j^{(e)} \tag{6.38}$$

where we have treated $f_N^{(e)}$ as an $O(N)$ polynomial.

For the linear polynomial case of Eq. (6.38), replacing D, $\tilde{D}$, and F using Eqs. (5.11), (6.6), and (6.8), respectively, and expanding $f_N^{(e)}$ as follows

$$f_N^{(e)}(x,t) = \sum_{j=0}^{1} f_j^{(e)}(t)\psi_j(x) \tag{6.39}$$

yields

$$\begin{pmatrix} -1 & 0 \\ 0 & 1 \end{pmatrix} \begin{pmatrix} f_0^{(e)} \\ f_1^{(e)} \end{pmatrix} = \begin{pmatrix} -1 & 0 \\ 0 & 1 \end{pmatrix} \begin{pmatrix} f_0^{(e)} \\ f_1^{(e)} \end{pmatrix} \tag{6.40}$$

which shows that integration by parts is satisfied discretely.

Let us now consider the case when $f^{(e)} = q^{(e)}u^{(e)}$ where we now represent

$$f_N^{(e)} = f\left(q_N^{(e)}, u_N^{(e)}\right),$$
(6.41)

with

$$q_N^{(e)}(x, t) = \sum_{j=0}^{1} q_j^{(e)}(t)\psi_j(x) \quad \text{and} \quad u_N^{(e)}(x, t) = \sum_{j=0}^{1} u_j^{(e)}(t)\psi_j(x).$$
(6.42)

In order to incorporate this representation of $f_N^{(e)}$ into Eq. (6.37) requires using the product rule as follows

$$\frac{\partial f_N^{(e)}}{\partial x} = u_N^{(e)}\frac{\partial q_N^{(e)}}{\partial x} + q_N^{(e)}\frac{\partial u_N^{(e)}}{\partial x}.$$
(6.43)

Starting with the equality

$$\int_{\Omega_e} \psi_i \frac{\partial f_N^{(e)}}{\partial x} d\Omega_e = \int_{\Omega_e} \psi_i \frac{\partial f_N^{(e)}}{\partial x} d\Omega_e$$
(6.44)

and using Eq. (6.43) on the right-hand side of Eq. (6.44) yields

$$\int_{\Omega_e} \psi_i \frac{\partial f_N^{(e)}}{\partial x} d\Omega_e = \int_{\Omega_e} \psi_i \left(u_N^{(e)}\frac{\partial q_N^{(e)}}{\partial x} + q_N^{(e)}\frac{\partial u_N^{(e)}}{\partial x}\right) d\Omega_e.$$
(6.45)

We can now write Eq. (6.37) as follows

$$\int_{\Omega_e} \psi_i \left(u_N^{(e)}\frac{\partial q_N^{(e)}}{\partial x} + q_N^{(e)}\frac{\partial u_N^{(e)}}{\partial x}\right) d\Omega_e = -\int_{\Omega_e} \frac{d\psi_i}{dx} q_N^{(e)} u_N^{(e)} d\Omega_e + \left[\psi_i q_N^{(e)} u_N^{(e)}\right]_{\Gamma_e}.$$
(6.46)

Substituting the linear basis function expansion given by Eq. (6.42) and integrating exactly yields the following matrix-vector problem

$$\frac{1}{6}\begin{pmatrix} -2u_0 - u_1 & 2u_0 + u_1 \\ -u_0 - 2u_1 & u_0 + 2u_1 \end{pmatrix}\begin{pmatrix} q_0 \\ q_1 \end{pmatrix} + \frac{1}{6}\begin{pmatrix} 2(u_1 - u_0) & u_1 - u_0 \\ u_1 - u_0 & 2(u_1 - u_0) \end{pmatrix}\begin{pmatrix} q_0 \\ q_1 \end{pmatrix}$$
$$= \frac{1}{6}\begin{pmatrix} 2u_0 + u_1 & u_0 + 2u_1 \\ -2u_0 - u_1 & -u_0 - 2u_1 \end{pmatrix}\begin{pmatrix} q_0 \\ q_1 \end{pmatrix} + \begin{pmatrix} -u_0 & 0 \\ 0 & u_1 \end{pmatrix}\begin{pmatrix} q_0 \\ q_1 \end{pmatrix}$$
(6.47)

which simplifies to

$$\frac{1}{6}\begin{pmatrix} -4u_0 + u_1 & u_0 + 2u_1 \\ -2u_0 - u_1 & -u_0 + 4u_1 \end{pmatrix}\begin{pmatrix} q_0 \\ q_1 \end{pmatrix} = \frac{1}{6}\begin{pmatrix} -4u_0 + u_1 & u_0 + 2u_1 \\ -2u_0 - u_1 & -u_0 + 4u_1 \end{pmatrix}\begin{pmatrix} q_0 \\ q_1 \end{pmatrix}.$$
(6.48)

which shows that, once again, integration by parts is satisfied discretely.

Exercise Derive the matrix-vector problem given by Eq. (6.48). ☐

6.8.2 Inexact Integration

It should not be surprising that integration by parts is satisfied discretely when we use exact integration, regardless of how we treat the flux term $f_N^{(e)}$, whether it is treated as an $O(N)$ polynomial (yields the equality given by Eq. (6.40)) or as the product of two $O(N)$ polynomials (yields the equality given by Eq. (6.48)). However, as the flux term becomes more complicated, it may be impractical to use exact integration so let us see what happens when we use inexact integration.

Using inexact integration and the *group finite element* representation given by Eq. (6.39) satisfies integration by parts discretely as given in Eq. (6.40). This is the case because N Lobatto points integrate exactly $O(2N - 1)$ polynomials which is what we have in Eq. (6.37) when we treat $f_N^{(e)}$ as an $O(N)$ polynomial as defined by Eq. (6.39). Now, let us see what happens when we use N Lobatto points to integrate the representation of $f_N^{(e)}$ by the product of two functions given by Eq. (6.41).

In this case, we get the follow matrix-vector form for Eq. (6.46)

$$
\frac{1}{2}\begin{pmatrix} -u_0 & u_0 \\ -u_1 & u_1 \end{pmatrix}\begin{pmatrix} q_0 \\ q_1 \end{pmatrix} + \frac{1}{2}\begin{pmatrix} u_1 - u_0 & 0 \\ 0 & u_1 - u_0 \end{pmatrix}\begin{pmatrix} q_0 \\ q_1 \end{pmatrix}
$$
$$
= \frac{1}{2}\begin{pmatrix} u_0 & u_1 \\ -u_0 & -u_1 \end{pmatrix}\begin{pmatrix} q_0 \\ q_1 \end{pmatrix} + \begin{pmatrix} -u_0 & 0 \\ 0 & u_1 \end{pmatrix}\begin{pmatrix} q_0 \\ q_1 \end{pmatrix} \tag{6.49}
$$

that simplifies to

$$
\frac{1}{2}\begin{pmatrix} -2u_0 + u_1 & u_0 \\ -u_1 & -u_0 + 2u_1 \end{pmatrix}\begin{pmatrix} q_0 \\ q_1 \end{pmatrix} \neq \frac{1}{2}\begin{pmatrix} -u_0 & u_1 \\ -u_0 & u_1 \end{pmatrix}\begin{pmatrix} q_0 \\ q_1 \end{pmatrix}. \tag{6.50}
$$

The reason why integration by parts is not satisfied discretely is because we used N Lobatto points to integrate the $O(3N - 1)$ polynomials in Eq. (6.46).

Exercise Derive the matrix-vector problem given by Eq. (6.50). □

To conclude this section, we remind the reader that to satisfy integration by parts (IBPs) discretely requires grouping the flux terms and representing them as an $O(N)$ polynomial, and that satisfying a discrete IBPs (called summation-by-parts in the finite difference community, see [355, 125, 97]) is necessary to achieve conservation.

6.8.3 Conservation Property of CG

The proof of the *conservation property* of the CG method follows exactly the analysis shown here for the DG method. The only difference is that the element equations (Eqs. (6.37) and (6.46)) must be summed following the DSS operator outlined in Sec. 5.8. Doing so will make the flux terms vanish everywhere except at the physical boundary which simplifies the analysis. However, the fact that we have to first invoke

DSS means that the CG method only satisfies global conservation whereas DG satisfies local (element-wise) conservation and, by extension, global conservation.

6.9 Example of 1D Wave Equation Problem

Suppose we wish to solve the continuous partial differential equation

$$\frac{\partial q}{\partial t} + \frac{\partial f}{\partial x} = 0 \qquad \forall x \in [-1, +1]$$

where $f = qu$ and $u = 2$ is a constant. Thus, an initial wave $q(x, 0)$ will take exactly $t = 1$ units of time in order to complete one full revolution across the domain.

6.9.1 Initial Condition

Since the governing PDE is a hyperbolic system, it represents an initial value problem. We, therefore, need an initial condition. Let it be the following Gaussian

$$q(x, 0) = \exp(-64x^2)$$

which is the same initial condition that we used for CG in Ch. 5.

6.9.2 Boundary Condition

This problem also requires a boundary condition: let us impose periodic boundary conditions, meaning that the domain at $x = +1$ wraps around and back to $x = -1$.

6.9.3 Error Norm

To compute the error, we use the L^2 norm presented in Eqs. (5.33) or (5.34).

6.9.4 Time-Integrator

To solve the time-dependent portion of the problem we can use the 3rd order SSP Runge-Kutta (RK) method that we introduced in Ch. 5, where the Courant number must satisfy

$$C = u \frac{\Delta t}{\Delta x} \leq \frac{1}{3}.$$

6.9.5 Construction of the DG Solution

Algorithms 6.7, 6.8, and 6.9 highlight the main steps in constructing the DG solution. Algorithm 6.7 is the easiest one to explain since it is laid out in a similar fashion to the CG recipe. This approach requires that we build global matrices with the caveat that we need to build both the global gridpoint flux vector (f) as well as the numerical flux vector ($f^{(*)}$) using Alg. 6.6. Another approach is shown in Alg. 6.8 which uses the global matrix DG differentiation matrix $\hat{D}^{DG}$ which is more complicated to build although the resulting approach is more concise. However, these are not the most efficient ways of constructing the DG solution and so we also show Alg. 6.9 which is an element-wise construction of the DG solution and is the preferred approach. Let us explain the element-wise approach (Alg. 6.9).

First we build the element matrices. Then we loop through the time-integration where, at each time-step we loop through all the elements and construct the right-hand side vector R in an element-wise fashion; this step represents the volume integrals that for each element e is completely decoupled from any of its neighbors. Next we loop through all the interfaces between elements (which we refer to as *face* and denote the number of such faces by N_f) and include the neighbor information of each element via the flux integrals. In the loop $s = 1 : N_f$ the face s knows which element e and neighbor k it is related to and for this reason instead of writing the numerical flux as $f^{(*,e,k)}$ we write $f^{(*,s)}$. We enforce all boundary conditions via the numerical flux. Once we have included this contribution, we then multiply by the inverse of the local mass matrix to project the solution to a local gridpoint-wise solution. At this point, we can evolve the equations forward to the next time-step.

Algorithm 6.7 Recipe 1: DG solution algorithm for the 1D wave equation - global view 1.

---
function DG(q)
  Construct element matrices $M^{(e)}$, $\tilde{D}^{(e)}$, and $F^{(e)}$ ▷ use Algs. 5.1, 6.1, 6.2
  Construct global matrices M, $\tilde{D}$ and F via DSS   ▷ use Alg. 6.3
  Compute $\hat{D} = M^{-1}\tilde{D}$ and $\hat{F} = M^{-1}F$
  for $n = 1 : N_t$ **do**   ▷ perform time-integration loop
    for $I = 1 : N_p$ **do**   ▷ loop over points
      $R_I = \hat{D}_{IJ}f_J(q^n) - \hat{F}_{IJ}f_J^{(*)}(q^n)$
    end for
    $\frac{d}{dt}q_I = R_I$ ▷ evolve equations in time to get q^{n+1}
  end for
end function

Algorithm 6.8 Recipe 2: DG solution algorithm for the 1D wave equation - global view 2.

---

function DG(q)

    Construct $M_{ij}^{(e)}$ and $\tilde{D}_{ij}^{(e)}$ ▷ use Algs. Algs. 5.1, 6.1

    Construct M_{IJ} and $F_{IJ}^{(*)}$ via DSS   ▷ use Algs. 6.3 and 6.5

    Construct global matrix $\hat{D}^{DG}$

    for $n = 1 : N_t$ **do**   ▷ perform time-integration loop

        for $I = 1 : N_p$ **do**   ▷ loop over points

            $R_I = \hat{D}_{IJ}^{DG} f_J(q^n)$

        end for

        $\frac{d}{dt} q_I = R_I$ ▷ evolve equations in time to get q^{n+1}

    end for

end function

Algorithm 6.9 Recipe 3: DG solution algorithm for the 1D wave equation - local view.

---

function DG(q)

    Construct $M_{ij}^{(e)}$, $F_{ij}^{(e)}$ and $\tilde{D}_{ij}^{(e)}$ ▷ Algs. 5.1, 6.1, 6.2

    for $n = 1 : N_t$ **do**   ▷ perform time-integration loop

        for $e = 1 : N_e$ **do**   ▷ loop over elements

            for $i = 0 : N$ **do**   ▷ loop over interpolation points

                $R_i^{(e)} = \tilde{D}_{ij}^{(e)} f_j^{(e)}(q^n)$ ▷ volume integrals

            end for

        end for

        for $s = 1 : N_f$ **do**   ▷ loop over faces between elements e and k

            $R_i^{(e)} \mathrel{-}= F_{ij}^{(e)} f_j^{(*,s)}(q^n)$ ▷ use Alg. 6.6 to build $f_j^{(*,s)}$

        end for

        $R_i^{(e)} = \left(M_{ij}^{(e)} \right)^{-1} R_j^{(e)}$

        $\frac{d}{dt} q_i^{(e)} = R_i^{(e)}$ ▷ evolve equations in time to get q^{n+1}

    end for

end function

6.9.6 Solution Accuracy for a Smooth Problem

Unlike CG, DG can use either Lobatto or Legendre points to do both the interpolation and integration because DG does not require C^0 continuity at the gridpoints on the element faces. Let us now analyze the results obtained using both point sets.

In Fig. 6.8 we show the snapshot of the exact and DG numerical solutions (with Rusanov flux) after one revolution ($t = 1$) using $N = 8$ order polynomials and $N_e = 8$ elements for a total of $N_p = 72$ gridpoints. Figure 6.8 shows that there is no noticeable difference between the exact and numerical solutions. This is corroborated by the convergence rates shown below where we use a Courant number of $C = 0.25$.

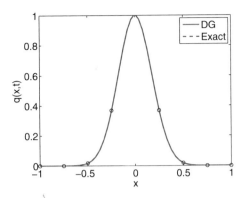

Fig. 6.8: Snapshot of the exact and DG numerical solution for the 1D wave equation using $N = 8$ order polynomials with $N_e = 8$ elements with exact integration after one revolution ($t = 1$ time unit).

Lobatto Points

Let us now see how well our model behaves by varying the number of elements and the polynomial order of the basis functions based on Lobatto points. Figure 6.9 shows the convergence rates using Lagrange polynomials with Lobatto points for a total number of gridpoints $N_p = N_e(N + 1)$. Figure 6.9 shows that there is little

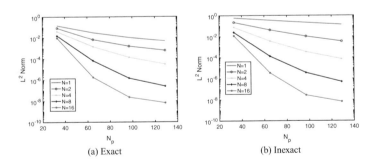

(a) Exact   (b) Inexact

Fig. 6.9: Convergence rates of DG with Rusanov flux with nodal functions using Lobatto points for the 1D wave equation for polynomial orders $N = 1, N = 2, N = 4$, $N = 8$, and $N = 16$ using a total number of gridpoints N_p for (a) exact integration ($Q = N + 1$) and (b) inexact integration ($Q = N$).

difference between exact and inexact integration. The main differences occur for low values of N (≤ 4), however, as N increases the difference between exact and inexact integration all but disappears; this behavior is similar to that shown for CG in Ch. 5

and seems to adhere to the analysis discussed in [151]. Note that the aliasing error that CG exhibited for $N = 2$ does not appear here for DG. The reason why this does not happen is due to the numerical flux function used in DG; the numerical flux is a dissipation mechanism and it is sufficient to avoid the aliasing error in this case. In fact, if we take $N = 0$ (which reduces to a cell-centered finite volume method) through a Taylor series expansion of the resulting differencing stencil it can be shown that the flux term is a Laplacian-like operator (see Ch. 18 for this analysis).

Once again, the question we need to answer is whether using high-order is more efficient than low-order. Figure 6.10 shows the L^2 error norm as a function of wallclock time in seconds (work-precision diagram). Figure 6.10 shows that the high-order methods are, in fact, more efficient to reach a certain level of accuracy than the low-order methods. For example, to achieve an accuracy of 10^{-4} is most efficiently reached with $N = 8$; $N = 1$ and $N = 2$ would require prohibitively large computational times to achieve these levels of accuracy. This result holds for both

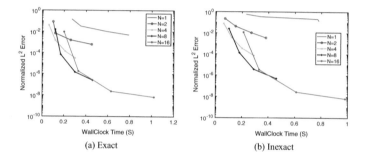

Fig. 6.10: Work-Precision diagram for DG with Rusanov flux with nodal functions using Lobatto points for the 1D wave equation for nodal polynomial orders $N = 1$, $N = 2$, $N = 4$, $N = 8$, and $N = 16$ as a function of wallclock time (in seconds) for (a) exact integration ($Q = N + 1$) and (b) inexact integration ($Q = N$).

exact and inexact integration.

In Fig. 6.11 the convergence rates for DG with Lobatto points are shown, except that now we use the centered numerical flux instead of Rusanov. The convergence rates for this approach look acceptable, and in fact, as will be shown in Ch. 7 the DG method with centered fluxes behaves rather similarly to the CG method. This is the case because both methods are non-dissipative; the DG method allows for a simple dissipation mechanism through upwind fluxes (such as Rusanov).

Legendre Points

Figure 6.12 shows the convergence rates for Lagrange polynomials based on Legendre points. Figure 6.12 shows that there is absolutely no difference between exact

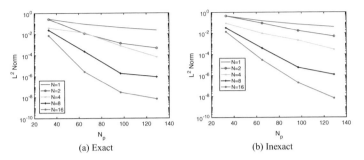

(a) Exact   (b) Inexact

Fig. 6.11: Convergence rates of DG with centered flux with nodal functions using Lobatto points for the 1D wave equation for polynomial orders $N = 1$, $N = 2$, $N = 4$, $N = 8$, and $N = 16$ using a total number of gridpoints N_p for (a) exact integration ($Q = N + 1$) and (b) inexact integration ($Q = N$).

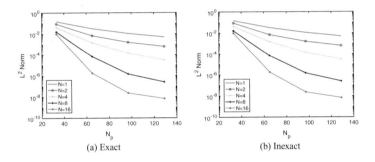

(a) Exact   (b) Inexact

Fig. 6.12: Convergence rates of DG with Rusanov flux with nodal functions using Legendre points for the 1D wave equation for polynomial orders $N = 1$, $N = 2$, $N = 4$, $N = 8$, and $N = 16$ using a total number of gridpoints N_p for (a) exact integration ($Q = N + 1$) and (b) inexact integration ($Q = N$).

($Q = N + 1$) and inexact ($Q = N$) integration; this is because, unlike the Lobatto points that only integrate $O(2N - 1)$, the Legendre points integrate $O(2N + 1)$. This means that even for $Q = N$ the integration is still exact.

Figure 6.13 shows the L^2 error norm as a function of wallclock time in seconds. Figure 6.13 shows that the high-order methods are, in fact, more efficient to reach a certain level of accuracy than the low-order methods, just like we saw for the Lobatto points.

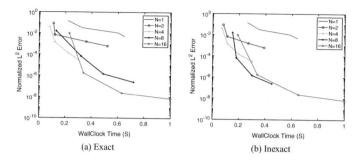

(a) Exact   (b) Inexact

Fig. 6.13: Work-Precision diagram for DG with Rusanov flux with nodal functions using Legendre points for the 1D wave equation for polynomial orders $N = 1$, $N = 2$, $N = 4$, $N = 8$, and $N = 16$ as a function of wallclock time (in seconds) for (a) exact integration ($Q = N + 1$) and (b) inexact integration ($Q = N$).

Modal Basis with Legendre Points

As a final example, we now run the same problem using Legendre points but instead of using the Lagrange polynomials, we use the Legendre polynomials (modal functions) as our basis functions. Recall that in this approach, no interpolation points are required, we only need integration points which we use Legendre points of order $Q = N$. Figure 6.14 shows that the modal function approach yields qualitatively

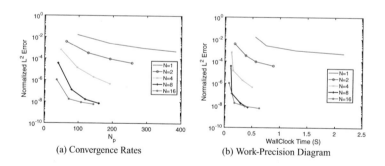

(a) Convergence Rates   (b) Work-Precision Diagram

Fig. 6.14: Results for DG with Rusanov flux with modal functions using Legendre points for the 1D wave equation for polynomial orders $N = 1$, $N = 2$, $N = 4$, $N = 8$, and $N = 16$ using a total number of gridpoints N_p showing (a) convergence rates and (b) work-precision diagram (cost).

similar results to the nodal function approach. We should expect to get the exact same answer when we use the same polynomial order and integration accuracy, regardless of type of basis function used. The differences exhibited between the exact integration nodal and modal approaches is purely due to differences in implementation.

6.9.7 Solution Accuracy for a Non-Smooth Problem

In Fig. 6.8 we showed the solution after one revolution for a smooth problem (i.e. infinitely differentiable solution). For this class of problems, any high-order method will do very well. Now let us see how DG performs for a non-smooth initial condition. On the domain $x \in [-1, 1]$, let the initial condition be defined as

$$q(x, 0) = \begin{cases} 1 \text{ for } |x| \leq R_c \\ 0 \text{ for } |x| > R_c \end{cases}$$

where $R_c = \frac{1}{8}$.

Figure 6.15 shows the exact and DG solutions after one revolution $(t = 1)$ using $N = 8$ order polynomials and $N_e = 16$ elements for a total of $N_p = 144$ gridpoints for a non-smooth initial condition (a square wave). Figure 6.15 shows

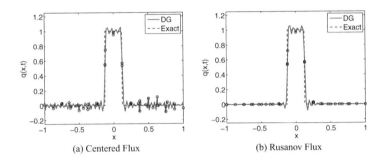

(a) Centered Flux   (b) Rusanov Flux

Fig. 6.15: Snapshot of the exact and DG numerical solution for the 1D wave equation with a non-smooth initial condition using $N = 8$ order polynomials with $N_e = 16$ elements with exact integration after one revolution ($t = 1$ time unit) for (a) centered flux (no dissipation) and (b) Rusanov flux (upwind-biased with dissipation).

that for discontinuous solutions the DG method with a centered flux is able to track the solution but it is just as noisy as the CG method presented in Ch. 5. On the other hand, by simply replacing the centered flux with the Rusanov flux, which is an upwind-biased flux with a dissipation mechanism, the solution becomes much less noisier. In fact, away from the discontinuity the solution is very smooth; the noise near the discontinuity is telling us that we have still not reached the well-resolved scale. Through the use of filters and flux limiters, we can clean up the DG solution further. This topic we reserve for Ch. 18.

Chapter 7
1D Unified Continuous and Discontinuous Galerkin Methods for Systems of Hyperbolic Equations

7.1 Introduction

In Chapters 5 and 6 we described the discretization of the 1D wave equation using both continuous and discontinuous Galerkin methods, respectively. However, in those chapters, the methods were described in the traditional way which made them appear to be quite different although similarities could be observed. In this chapter, we construct these methods in a unified way so that they can co-exist within the same piece of software. In order to achieve this unification of CG and DG requires analyzing the storage of the methods; this is the objective of Sec. 7.2. Using this information, we then build a unified CG/DG approximation for the one-dimensional wave equation in Sec. 7.3. We then extend the unified approach for discretizing systems of equations first for the shallow water equations in Sec. 7.4 and then for the Euler equations in Sec. 7.5. For any numerical method being used to approximate differential equations, it is important to understand the dissipation and dispersion characteristics. In Sec. 7.6 we analyze the dissipation and dispersion properties of both CG and DG methods for the one-dimensional wave equation. We then test some of the conjectures extracted from that analysis on multiscale problems in Sec. 7.7 to show the effects of the interplay between dissipation and dispersion on such problems. For that study we use a linearized system of equations that can be characterized as either an acoustic problem or a linearized shallow water system.

7.2 CG and DG Storage of Data

One key difference that can be observed between the CG and DG methods is in the different ways in which the two methods store the solution vector. In the case

F. X. Giraldo, *An Introduction to Element-Based Galerkin Methods on Tensor-Product Bases*, Texts in Computational Science and Engineering 24, https://doi.org/10.1007/978-3-030-55069-1_7

of CG, we use a global gridpoint numbering approach whereas in DG we use a local element-wise approach; these different storage schemes are illustrated in Fig. 7.1. The way information is stored in both methods is an effect of the conditions that both methods require. For example, since CG requires continuity at the element interfaces then it makes sense to store the solution gridpoint-wise since this ensures that the solution remains continuous at element interfaces (since the solution can only have one value at the interface if it is stored in the same location). The global gridpoint (GGP) storage approach is shown in Fig. 7.1: for example, the gridpoint 2 of element 1 (e=1) is the same as gridpoint 1 of element 2 (e=2) and so it has the global identification number 2 (blue). The *intma* matrix for this setup can be deduced from the figure to be

$$intma\_CG = \begin{pmatrix} 1 & 2 & 3 \\ 2 & 3 & 4 \end{pmatrix}$$

where the two rows denote the number of points per element and the three columns the number of elements; let us add the suffix *CG* to emphasize that this is the connectivity matrix for the CG method. In classical CG, we store the solution vector q as a vector with dimension N_p, where $N_p = N_e N + 1$ denotes the total number of unique global gridpoints.

Looking at Fig. 7.1 again, we see that we can store the DG solution vector q in one of two ways. One way is derived by noting that the *intma* matrix for DG can be written as

$$intma\_DG = \begin{pmatrix} 1 & 3 & 5 \\ 2 & 4 & 6 \end{pmatrix}$$

where we now have $N_p = N_e(N + 1)$ gridpoints. The second way to store the DG solution vector is discussed below.

7.2.1 From DG to CG Storage

In classical DG, we do not use this *intma_DG* matrix but rather embed it directly into the solution vector q as follows: let $q(N + 1, N_e)$ be the solution matrix that stores the data for all gridpoints $(N + 1)$ within all the elements (N_e).

However, if we wish to write one piece of software that can handle both types of methods requires storing the data similarly for both methods. We can choose to do this one of two ways: the first way requires storing the CG solution as $q(N+1, N_e)$ but then use *intma_CG* to perform the DSS operation. This CG method now looks very much like DG and we shall refer to this method as CGd where the "d" denotes that we are using a *discontinuous* storage scheme (the data is stored as if the *intma_DG* matrix is embedded within the solution vector). This strategy permits the unification of both CG and DG within the same code and is the preferred method to use if you already have a working DG code.

7.2.2 From CG to DG Storage

What if you already have a working CG code that is in classical CG storage whereby the solution is stored as the vector $q(N_p)$. Let us now show that this is, in fact, the preferred way to construct general storage schemes. Starting from this approach, we can now include the DG method simply by using *intma_DG* to map from local element-wise (LEW) storage to global gridpoint (GGP) storage as follows: $(i, e) \rightarrow I$ just as we did for the CG method in Ch. 5 [1]. We can also use *intma_DG* for CG with the addition of *intma_CG* to handle the DSS operation. This is the CGd method described in the previous section. Finally, we can use *intma_CG* exclusively and recover the classical CG method; we call this method CGc where the "c" represents that we are assuming a *continuous* storage scheme. All three methods (CGc, CGd, and DG) can now be used within the same piece of software if we define the solution vector to be $q(N_p)$, where we keep track of the map $(i, e) \rightarrow I$ by virtue of the connectivity matrix *intma* where it can be used to store either *intma_CG* or *intma_DG*, with the proper dimension of N_p. Let us refer to this storage scheme as the augmented global gridpoint (AGGP) storage .

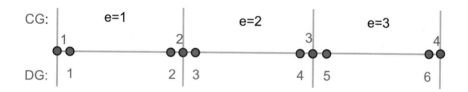

Fig. 7.1: CG and DG storage for 3 linear elements. The CG storage is in blue and the DG storage in red.

   The advantage of using either the local element-wise storage or the augmented global gridpoint storage is that we can construct unified CG/DG methods within the same computer code; this approach is described in detail in [2]. In addition, using the CGd method allows us to use many of the attractive features of the DG method with CG. Examples of such features originally designed for DG that can be used with CGd include non-conforming adaptive mesh refinement (see Ch. 19) and limiters (see Ch. 18). Let us begin by describing the unified CG/DG method for the 1D wave equation.

---

[1] LEW storage will sometimes be referred to as DG storage in this text, while GGP is sometimes called CG storage.

7.3 1D Wave Equation

Let us consider the 1D wave equation in conservation form

$$\frac{\partial q}{\partial t} + \frac{\partial f}{\partial x} = 0 \quad \text{where} \quad f = qu$$

and u is the velocity. Representing the element-wise discrete solution as

$$q_N^{(e)}(x,t) = \sum_{j=0}^{N} q_j^{(e)}(t)\psi_j(x) \tag{7.1a}$$

$$f_N^{(e)}(x,t) = f\left(q_N^{(e)}\right) \tag{7.1b}$$

and taking the integral form within a single element yields

$$\int_{\Omega_e} \psi_i \left(\frac{\partial q_N^{(e)}}{\partial t} + \frac{\partial f_N^{(e)}}{\partial x}\right) d\Omega_e = 0.$$

Using the product rule to replace the second term in the above equation yields

$$\int_{\Omega_e} \psi_i \frac{\partial q_N^{(e)}}{\partial t} d\Omega_e + \int_{\Omega_e} \frac{\partial}{\partial x}\left(\psi_i f_N^{(e)}\right) d\Omega_e - \int_{\Omega_e} \frac{d\psi_i}{dx} f_N^{(e)} d\Omega_e = 0.$$

Applying the Fundamental Theorem of Calculus to the second integral yields

$$\int_{\Omega_e} \psi_i \frac{\partial q_N^{(e)}}{\partial t} d\Omega_e + \sum_{k=1}^{N_{FN}} \left[\hat{n}^{(e,k)}\psi_i f_N^{(*,e,k)}\right]_{\Gamma_e} - \int_{\Omega_e} \frac{d\psi_i}{dx} f_N^{(e)} d\Omega_e = 0 \tag{7.2}$$

where $f^{(*,e,k)}$ denotes the numerical flux (Riemann solver) evaluated between elements e and k, $\hat{n}^{(e,k)}$ is the normal vector pointing from the element e to k, and N_{FN} denotes the number of face neighbors of the element e; in 1D the number is 2 (left and right faces). For the purpose of illustration let us use the Rusanov flux defined as

$$f^{(*,e,k)} = \frac{1}{2}\left[f^{(e)} + f^{(k)} - \hat{n}^{(e,k)}|u|\left(q^{(k)} - q^{(e)}\right)\right].$$

Equation (7.2) is nothing more than the 1D wave equation discretized using the weak form DG that we described in Ch. 6 that we wrote in matrix-vector form as

$$M_{ij}^{(e)}\frac{dq_j^{(e)}}{dt} + \sum_{k=1}^{N_{FN}} F_{ij}^{(e,k)} f_j^{(*,e,k)} - \tilde{D}_{ij}^{(e)} f_j^{(e)} = 0 \tag{7.3}$$

with $i,j = 0,\dots,N$, $e = 1,\dots,N_e$, where M is the mass matrix, F is the flux matrix, and $\tilde{D}$ is the weak form differentiation matrix. The key to a unified CG/DG representation resides in Eq. (7.3). Note that the flux term disappears at element interfaces if the solution is continuous. This means that the only difference between CG and DG in Eq. (7.3) is in the flux term; however, for CG we still need to apply DSS.

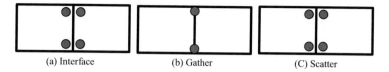

| (a) Interface | (b) Gather | (C) Scatter |

Fig. 7.2: Graphical representation of the Direct Stiffness Summation operator. The process begins with the (a) local element-wise (LEW) storage of the *interface* that is then (b) *gathered* to form a continuous solution (GGP), which is then (c) *scattered* to define a new LEW-based solution vector that satisfies the continuity condition of CG.

7.3.1 Communicator

Let us now introduce the concept of a *communicator*. The communicator in *element-based Galerkin* methods is the operator that allows the solution to be communicated from one element to its neighbors. Referring to Figs. 7.1 and 7.2a we start with the LEW storage (in red) and then gather the solution as in Fig. 7.2b yielding the GGP storage (in blue) of Fig. 7.1. The gathered solution is then scattered to its original LEW storage (in red) of Fig. 7.1 and 7.2c.

Rewriting Eq. (7.3) as

$$M_{ij}^{(e)} \frac{dq_j^{(e)}}{dt} = -\sum_{k=1}^{N_{FN}} F_{ij}^{(e,k)} f_j^{(*,e,k)} + \tilde{D}_{ij}^{(e)} f_j^{(e)} \equiv R_i^{(e,k)} \tag{7.4}$$

we can refer to the communicator for DG as

$$C_{DG} = \left(M^{(e)}\right)^{-1} R^{(e,k)} \tag{7.5}$$

that now yields a solution for DG that has taken the solution of its neighbors into account.

Similarly, we can define the communicator for CG as the following operations

$$C_{CG} = M^{-1} S\left(\mathcal{G}\left(R^{(e,k)}\right)\right) \tag{7.6}$$

where $\mathcal{G}$ is the gather operation (Fig. 7.2b), S is the scatter operation (Fig. 7.2c), and M is the global mass matrix defined as

$$M = \bigwedge_{e=1}^{N_e} M^{(e)}$$

with N_e denoting the number of elements in the grid. The DSS operator is described in Sec. 5.8. Algorithms 7.1 and 7.2 define the gather and scatter operations. Note that the gather and scatter operations presented in Algs. 7.1 and 7.2 look very similar

---

Algorithm 7.1 Gather operation for CG.

---

 function $\mathcal{G}(R_i^{(e)})$
  $R = 0$
  for $e = 1 : N_e$ **do**   ▷ loop over elements
   for $i = 0 : N$ **do**
    $I = intma\_CG(i, e)$ ▷ get unique global gridpoint ID
    $R_I \mathrel{+}= R_i^{(e)}$
   end for
  end for
 end function

---

to the DSS operator defined in Alg. 5.7. In Alg. 5.7 there is only a gather operation because the data is stored in a global gridpoint fashion and therefore it need not be scattered to the local element-wise storage (i.e., it is using CGc storage).

Algorithm 7.2 Scatter operation for CG.

---

 function $\mathcal{S}(R_I)$
  for $e = 1 : N_e$ **do**   ▷ loop over elements
   for $i = 0 : N$ **do**
    $I = intma\_DG(i, e)$ ▷ get redundant global gridpoint ID
    $R_i^{(e)} = R_I$
   end for
  end for
 end function

Finally, we can represent both the CG and DG solutions in the following way

$$\frac{dq_i^{(e)}}{dt} = C_i \tag{7.7}$$

where C is either the CG or DG communicator that can be determined at runtime.

7.3.2 Construction of the Unified CG/DG Solution

7.3.2.1 Recipe 1

Algorithm 7.3 highlights the main steps in constructing the unified CG/DG solution. First we construct all the element matrices including the global mass matrix - the mass matrix is the only global matrix that needs to be constructed. Then we begin the time-integration loop where at each time step we perform the following steps. First we loop through all the elements and construct the right-hand side vector for the volume integral terms (loop $e = 1 : N_e$ and $i = 0 : N$) - this step is required by both CG and DG and is completely local in nature since no neighboring element information is

Algorithm 7.3 Recipe 1: unified CG/DG solution algorithm for the 1D wave equation.

---

function CG/DG(q)

  Construct $M_{ij}^{(e)}$, $\tilde{D}_{ij}^{(e)}$, $F_{ij}^{(e)}$ ▷ use Algs. 5.1, 6.1, 6.2

  Construct M_{IJ} via DSS   ▷ use Alg. 5.4

  for $n = 1 : N_t$ **do**   ▷ time-integration loop

    for $e = 1 : N_e$ **do**   ▷ volume integral loop

      for $i = 0 : N$ **do**

        $R_i^{(e)} = \tilde{D}_{ij}^{(e)} f_j^{(e)}(q^n)$

      end for

    end for

    if CG **then**   ▷ CG communicator

      (CGC1) Gather $R_I = \mathcal{G}(R_i^{(e)})$ ▷ use Alg. 7.1

      (CGC2) $R_I = (M_{IJ})^{-1} R_J$ ▷ apply boundary conditions

      (CGC3) Scatter $R_i^{(e)} = \mathcal{S}(R_I)$ ▷ use Alg. 7.2

    else if DG **then**   ▷ DG communicator

      for $s = 1 : N_f$ **do**   ▷ flux integral loop

        (DGC1) $R_i^{(e)} \mathrel{-}= F_{ij}^{(e)} f_j^{(*,e,k)}(q^n)$ ▷ i, e, k are functions of s

      end for

      (DGC2) $R_i^{(e)} = \left(M_{ij}^{(e)}\right)^{-1} R_j^{(e)}$

    end if

    $\frac{d}{dt} q_i^{(e)} = R_i^{(e)}$ ▷ evolve equations in time to get q^{n+1}

  end for

end function

---

necessary to complete this step. Next we invoke the *communicator* for both CG and DG which links the decoupled solutions together; the labels (CGC1,CGC2,CGC3) and (DGC1,DGC2) in Alg. 7.3 denote the different steps of the communicator for both CG and DG.

For CG, we need to gather the solution to make it C^0 (CGC1) and then multiply by the inverse of the global mass matrix (CGC2); this step yields a global gridpoint solution. Here, we enforce boundary conditions. Now we can scatter (CGC3) the solution back to local element-wise storage . This step completes the communicator for CG.

For DG, we loop over all the faces (N_f) which are defined as the interfaces between two elements (s represents the face that links the interfaces of the elements e and k; we define this part of the algorithm precisely in Sec. 7.3.3) and include the flux terms (DGC1) to the RHS vector R; this step includes the contribution of the flux integral terms. Next, we multiply by the inverse local mass matrix (DGC2) which now yields a local gridpoint solution at the interpolation points of each element. This step now completes the communicator for DG.

Once all of these steps are complete, we can now march the equations forward one time-step.

Algorithm 7.4 Recipe 2: unified CG/DG solution algorithm for the 1D wave equation.

---

function CG/DG(q)

    Construct $M_{ij}^{(e)}, \tilde{D}_{ij}^{(e)}, F_{ij}^{(e)}$ ▷ use Algs. 5.1, 6.1, 6.2

    Construct M_{IJ} via DSS   ▷ use Alg. 5.4

    for $n = 1 : N_t$ **do**   ▷ time-integration loop

        for $e = 1 : N_e$ **do**   ▷ volume integral loop

            for $i = 0 : N$ **do**

$$R_i^{(e)} = \tilde{D}_{ij}^{(e)} f_j^{(e)}(q^n)$$

            end for

        end for

        for $s = 1 : N_f$ **do**   ▷ *flux integral loop*

$$R_i^{(e)} \mathrel{-}= F_{ij}^{(e)} f_j^{(*,e,k)}(q^n)$$ ▷ *apply boundary conditions*

        end for

        if CG **then**   ▷ CG communicator

            (CGC1) Gather $R_I = \mathcal{G}(R_i^{(e)})$

            (CGC2) $R_I = (M_{IJ})^{-1} R_J$

            (CGC3) Scatter $R_i^{(e)} = \mathcal{S}(R_I)$

        else if DG **then**   ▷ DG communicator

            (DGC1) $R_i^{(e)} = \left(M_{ij}^{(e)}\right)^{-1} R_j^{(e)}$

        end if

$$\frac{d}{dt} q_i^{(e)} = R_i^{(e)}$$ ▷ evolve equations in time to get q^{n+1}

    end for

end function

7.3.2.2 Recipe 2

We can modify Algorithm 7.3 further to make the CG and DG constructions more similar. Algorithm 7.4 shows the modification in italics (see the comments in the algorithm); all that has been done here is to move the flux integral loop outside the *communicator* step which means that this step will be executed by both CG and DG. This way, we can impose boundary conditions for both CG and DG in exactly the same way and only at the element level. Note that because we now apply boundary conditions via the flux integrals, we no longer require imposing them at the global gridpoint level (step CGC2 in the CG communicator). The only concern in adding the flux terms to CG is whether we indeed will recover the same solution as in Algorithm 7.3. To show that the solutions are identical, all we need to show is what happens to the element flux matrix after we invoke DSS.

In Sec. 6.4 we found that the element flux matrix for linear polynomials ($N = 1$) is

$$F_{ij}^{(e)} = \begin{pmatrix} -1 & 0 \\ 0 & 1 \end{pmatrix}$$

for Lobatto points. In fact, we can follow that derivation to generalize the flux matrix for any order N to be

$$F_{ij}^{(e)} = \begin{pmatrix} -1 & 0 & \cdots & 0 \\ 0 & 0 & \cdots & 0 \\ \vdots & \vdots & \vdots & \vdots \\ 0 & \cdots & 0 & 1 \end{pmatrix},$$

in other words, only the first and last elements of the matrix are populated (for one-dimensional geometries). However, let us consider applying DSS to the $N = 1$ element flux matrix because the same arguments follow for any order flux matrix. Let us now define

$$F_{IJ} = \bigwedge_{e=1}^{N_e} F_{ij}^{(e)}$$

which, as an example, let us assume $N_e = 4$ elements. For this particular element configuration, we would arrive at the following global flux matrix

$$F = \begin{pmatrix} -1 & 0 & 0 & 0 & 0 \\ 0 & 0 & 0 & 0 & 0 \\ 0 & 0 & 0 & 0 & 0 \\ 0 & 0 & 0 & 0 & 0 \\ 0 & 0 & 0 & 0 & 1 \end{pmatrix}$$

which demonstrates that the contribution of the flux matrix to all internal degrees of freedom is exactly zero and non-zero only at the physical boundary of the domain. However, in Alg. 7.4 it is a waste of floating point operations to have the CG method compute internal flux values that, in the end, sum to zero. Instead, we can make the algorithm faster by simply skipping all internal flux computations for CG and only perform boundary terms.

7.3.2.3 Recipe 3

There is at least one more simplification that we can make to the unified CG/DG algorithm which will bring these two methods even closer. Note that in order to do this requires performing a gather-scatter on the mass matrix for CG[2]. Once this operation is completed, the CG and DG algorithms are almost identical. The only difference now is that for CG, we still require DSS (gather-scatter) on the RHS vector (where we need to perform the multiplication of the inverse mass matrix within the DSS operation). The new algorithm is shown in Alg. 7.5. This algorithm implicitly assumes that we are using local element-wise (i.e., DG) storage where the solutions exist element-wise. This is evident by the fact that we are using the indices (i, e) to define the solution; this is also the case for Recipes 1 and 2.

---

[2] This operation only needs to be done once for a static grid. However, if using adaptive mesh refinement, then the mass matrix has to be recomputed.

Algorithm 7.5 Recipe 3: unified CG/DG solution algorithm for the 1D wave equation.

---

function CG/DG(q)

    Construct $M_{ij}^{(e)}, \tilde{D}_{ij}^{(e)}, F_{ij}^{(e)}$ ▷ use Algs. 5.1, 6.1, 6.2

    if CG **then**   ▷ CG communicator

        Gather $M_{IJ} = \mathcal{G}(M_{ij}^{(e)})$

        Scatter $M_{ij}^{(e)} = \mathcal{S}(M_{IJ})$

    end if

    for $n = 1 : N_t$ **do**   ▷ time-integration loop

        for $e = 1 : N_e$ **do**   ▷ volume integral Loop

             for $i = 0 : N$ **do**

                $R_i^{(e)} = \tilde{D}_{ij}^{(e)} f_j^{(e)}(q^n)$

             end for

        end for

        for $s = 1 : N_f$ **do**   ▷ flux integral loop

             $R_i^{(e)} \mathrel{-}= F_{ij}^{(e)} f_j^{(*,e,k)}(q^n)$ ▷ apply boundary conditions

        end for

        if CG **then**   ▷ CG communicator

             (CGC1) Gather $R_I = \mathcal{G}(R_i^{(e)})$

             (CGC2) Scatter $R_i^{(e)} = \mathcal{S}(R_I)$

        end if

        $R_i^{(e)} = \left(M_{ij}^{(e)}\right)^{-1} R_j^{(e)}$ ▷ construct gridpoint solution

        $\frac{d}{dt} q_i^{(e)} = R_i^{(e)}$ ▷ evolve equations in time to get q^{n+1}

    end for

    end function

7.3.2.4 Recipe 4

As the final recipe, let us start with Recipe 3 and generalize it using the augmented global gridpoint storage, which we show in Alg. 7.6. The difference between Alg. 7.6 and all the other recipes described in this chapter is that the augmented global gridpoint (AGGP) storage used here can accommodate CGd, DG, and CGc, making AGGP the storage method that allows the largest class of methods. The main difference between Recipe 4 and the previous three recipes is that Recipe 4 uses a gridpoint storage. The advantage of using this storage scheme is that CGc can also be included into the algorithm. Note that in the gridpoint storage scheme used here, e.g., the right-hand side vector R is not referenced by the element indices i, e but rather by the gridpoint index I, where I is pointed to by the element connectivity matrix $intma(i, e)$. This simple approach allows us to use all the methods described so far within the same piece of software.

Algorithm 7.6 Recipe 4: unified CG/DG solution algorithm for the 1D wave equation.

---

function CG/DG(q)
    Construct $M_{ij}^{(e)}, \tilde{D}_{ij}^{(e)}, F_{ij}^{(e)}$ ▷ use Algs. 5.1, 6.1, 6.2
    Construct global M_{IJ} ▷ for CG use DSS; for DG store block-diagonal matrix
    for $n = 1 : N_t$ **do**   ▷ time-integration loop
        for $e = 1 : N_e$ **do**   ▷ volume integral loop
            for $i = 0 : N$ **do**
                $I = intma(i, e)$ ▷ for CGd, DG use $intma\_DG$; for CGc $intma\_CG$
                $R_I \mathrel{+}= \tilde{D}_{ij}^{(e)} f_j^{(e)}(q^n)$
            end for
        end for
        for $s = 1 : N_f$ **do**   ▷ flux integral loop
            $R_I \mathrel{-}= F_{ij}^{(e)} f_j^{(*,e,k)}(q^n)$ ▷ apply boundary conditions
        end for
        Construct global R_I ▷ for CGc and DG do nothing; for CGd do DSS
        $R_I = M_{IJ}^{-1} R_J$ ▷ construct gridpoint solution
        $\frac{d}{dt} q_I = R_I$ ▷ evolve equations in time to get q^{n+1}
    end for
end function

7.3.3 Face Data Structure

The flux integral loops ($s = 1 : N_f$) in Algs. 7.3 - 7.6 need to be explained in detail before these algorithms can be used. We omitted the details to allow for a simpler description of the algorithms. Let us now describe what this specific operation is doing. Let us refer to Fig. 7.1 and note that there are 4 distinct element interfaces (light blue vertical lines). Therefore, the total number of faces in the grid is $N_f = 4$. Next, let us define the following face data structure

$$face(1:4, N_f) = \begin{pmatrix} 0 & 1 & 1 & 1 \\ -1 & 0 & 0 & -1 \\ 1 & 1 & 2 & 3 \\ -1 & 2 & 3 & -1 \end{pmatrix}$$

with the corresponding normal vector $\hat{\mathbf{n}}^{(L)} = (-1, 1, 1, 1)^T$, with $\hat{\mathbf{n}}^{(R)} = -\hat{\mathbf{n}}^{(L)}$, where the superscripts L and R denote the left and right elements of the face s and the columns of the array $face$ denote the specific faces of the grid; the first face has normal vector $\hat{\mathbf{n}}^{(L)} = -1$ because, by convention, we assume that the left element is always defined as interior to a boundary. The columns of the face data structure contain the information for each of the four faces (in this figure, we can think of them as edges) shown in Fig. 7.1. For each column, the first two rows contain the left and right local gridpoint numbering (position 0 corresponds to the left and position 1 to the right). Rows 3 and 4 contain the left and right element identifiers. The negative number -1 indicates that no value is associated with this position. This face data structure configuration could represent, e.g., *impermeable wall* boundary conditions[3]. However, if we imposed periodic boundary conditions then we would require the following face data structure

$$face(1:4, N_f) = \begin{pmatrix} 0 & 1 & 1 & 1 \\ 1 & 0 & 0 & 0 \\ 1 & 1 & 2 & 3 \\ 3 & 2 & 3 & 1 \end{pmatrix}$$

where columns 1 and 4 are connected. For further discussions on boundary conditions, the reader is referred to Appendix C.

7.4 1D Shallow Water Equations

Let us consider the 1D shallow water equations written as the following balance law (conservation form plus a source)

$$\frac{\partial h_S}{\partial t} + \frac{\partial U}{\partial x} = 0 \tag{7.8a}$$

$$\frac{\partial U}{\partial t} + \frac{\partial}{\partial x}\left(\frac{U^2}{h} + \frac{1}{2}gh_S^2 \right) = -gh_B \frac{\partial h_S}{\partial x} \tag{7.8b}$$

where $h = h_S + h_B$ is the total height of the fluid, h_S is the surface height measured from the mean level, and h_B is the distance from the mean level to the bottom, g is the gravitational constant, and $U = hu$ is the momentum where u is the velocity. The reason why we discuss the shallow water equations is because these equations are typically used as the prototype equations before atmosphere and ocean models are developed. These equations are derived from the 3D Navier-Stokes equations and are depth-integrated in order to reduce the spatial dependence. In addition to being used for testing numerical methods, the shallow water equations are also used to simulate storm-surges and tsunami wave propagation (e.g., see [6, 121, 40, 266, 258]).

The shallow water equations can be rewritten in the following compact vector form

$$\frac{\partial \mathbf{q}}{\partial t} + \frac{\partial \mathbf{f(q)}}{\partial x} = \mathbf{S(q)} \tag{7.9}$$

where

$$\mathbf{q} = \begin{pmatrix} h_S \\ U \end{pmatrix} \tag{7.10a}$$

$$\mathbf{f(q)} = \begin{pmatrix} U \\ \frac{U^2}{h} + \frac{1}{2}gh_S^2 \end{pmatrix} \tag{7.10b}$$

$$\mathbf{S(q)} = \begin{pmatrix} 0 \\ -gh_B \frac{\partial h_S}{\partial x} \end{pmatrix}. \tag{7.10c}$$

---

[3] Impermeable wall boundary conditions are also known as *no flux* or *reflecting*.

Inserting the basis function expansion in Eq. (7.1) into the compact vector form of the equations given in Eq. (7.9), multiplying by a test function ψ and integrating within each element Ω_e yields

$$\int_{\Omega_e} \psi_i \left(\frac{\partial \mathbf{q}_N^{(e)}}{\partial t} + \frac{\partial \mathbf{f}\left(\mathbf{q}_N^{(e)}\right)}{\partial x} \right) d\Omega_e = \int_{\Omega_e} \psi_i \mathbf{S}\left(\mathbf{q}_N^{(e)}\right) d\Omega_e.$$

Using the product rule to modify the second term on the left-hand side gives

$$\int_{\Omega_e} \psi_i \frac{\partial \mathbf{q}_N^{(e)}}{\partial t} d\Omega_e + \int_{\Omega_e} \frac{d}{dx}\left[\psi_i \mathbf{f}\left(\mathbf{q}_N^{(e)}\right)\right] d\Omega_e - \int_{\Omega_e} \frac{d\psi_i}{dx}\mathbf{f}\left(\mathbf{q}_N^{(e)}\right) d\Omega_e$$
$$= \int_{\Omega_e} \psi_i \mathbf{S}\left(\mathbf{q}_N^{(e)}\right) d\Omega_e.$$

Using the Fundamental Theorem of Calculus to simplify the second integral yields

$$\int_{\Omega_e} \psi_i \frac{\partial \mathbf{q}_N^{(e)}}{\partial t} d\Omega_e + \sum_{k=1}^{N_{FN}} \left[\hat{n}^{(e,k)}\psi_i \mathbf{f}\left(\mathbf{q}_N^{(e,k)}\right)\right]\Big|_{\Gamma_e} - \int_{\Omega_e} \frac{d\psi_i}{dx}\mathbf{f}\left(\mathbf{q}_N^{(e)}\right) d\Omega_e$$
$$= \int_{\Omega_e} \psi_i \mathbf{S}\left(\mathbf{q}_N^{(e)}\right) d\Omega_e \quad (7.11)$$

where $\mathbf{f}\left(\mathbf{q}_N^{(e,k)}\right)$ denotes the numerical flux; for simplicity we can assume that it is the Rusanov flux defined as in the 1D wave equation.

Writing Eq. (7.11) in matrix-vector form yields

$$M_{ij}^{(e)} \frac{d\mathbf{q}_j^{(e)}}{dt} + \sum_{k=1}^{N_{FN}} F_{ij}^{(e,k)}\mathbf{f}\left(\mathbf{q}_j^{(e,k)}\right) - \tilde{D}_{ij}^{(e)}\mathbf{f}\left(\mathbf{q}_j^{(e)}\right) = \mathbf{S}_i^{(e)} \quad (7.12)$$

with $i, j = 0, \ldots, N$, $e = 1, \ldots, N_e$, where $\mathbf{S}_i^{(e)} = \int_{\Omega_e} \psi_i \mathbf{S}\left(\mathbf{q}_N^{(e)}\right) d\Omega_e$ is the source function vector and where we define $\mathbf{q}_N^{(e)} = \sum_{j=0}^{N} \psi_j \mathbf{q}_j^{(e)}$. At this point we have already seen every term in Eq. (7.11) except the terms $\mathbf{q}_j^{(e)}$, $\mathbf{f}\left(\mathbf{q}_j^{(e)}\right)$, $\mathbf{f}\left(\mathbf{q}_j^{(e,k)}\right)$, and $\mathbf{S}\left(\mathbf{q}_j^{(e)}\right)$; let us now explicitly write these terms. Beginning with the vector $\mathbf{q}_j^{(e)}$ we note that it is nothing more than the expansion coefficients from Eq. (7.1a) but now defined for the 1D shallow water equations which are defined as

$$\mathbf{q}_j^{(e)} = \begin{pmatrix} h_{S,j}^{(e)} \\ U_j^{(e)} \end{pmatrix}. \quad (7.13)$$

The term $\mathbf{S}\left(\mathbf{q}_i^{(e)}\right)$ is expressed as follows

$$\mathbf{S}\left(\mathbf{q}_i^{(e)}\right) = -\begin{pmatrix} 0 \\ gh_{B,i}^{(e)}\left(\sum_{k=0}^{N}\left(\frac{d\psi_k}{d\xi}\frac{d\xi}{dx}\right)h_{S,k}^{(e)}\right) \end{pmatrix}. \tag{7.14}$$

The term $\mathbf{f}\left(\mathbf{q}_j^{(e)}\right)$ is a bit more complicated and is expressed as follows

$$\mathbf{f}\left(\mathbf{q}_j^{(e)}\right) = \begin{pmatrix} U_j^{(e)} \\ U_j^{(e)}\left(\sum_{k=0}^{N}\psi_k U_k^{(e)}\right)/\left(\sum_{k=0}^{N}\psi_k h_k^{(e)}\right) + \frac{1}{2}gh_{S,j}^{(e)}\left(\sum_{k=0}^{N}\psi_k h_{S,k}^{(e)}\right) \end{pmatrix}. \tag{7.15}$$

Comparing Eq. (7.15) with Eq. (7.10b) we can now see where these terms are coming from. To define the numerical flux term let us first write the flux using the following notation

$$\mathbf{f}\left(\mathbf{q}_j^{(e)}\right) = \begin{pmatrix} f_{h_S} \\ f_U \end{pmatrix} \tag{7.16}$$

which now allows us to define the numerical flux (Rusanov) as follows

$$\mathbf{f}\left(\mathbf{q}_j^{(e,k)}\right) = \begin{pmatrix} \{f_{h_S}\}^{(e,k)} - \hat{n}^{(e,k)}\frac{|\lambda_{max}|}{2}[\![h_{S,j}]\!]^{(e,k)} \\ \{f_U\}^{(e,k)} - \hat{n}^{(e,k)}\frac{|\lambda_{max}|}{2}[\![U_j]\!]^{(e,k)} \end{pmatrix} \tag{7.17}$$

where we have used classical DG notation with the above delimiters defined as follows:

$$\{f\}^{(e,k)} = \frac{1}{2}\left(f^{(e)} + f^{(k)}\right), \tag{7.18}$$

$$[\![\mathbf{q}]\!]^{(e,k)} = \mathbf{q}^{(k)} - \mathbf{q}^{(e)} \tag{7.19}$$

and λ_{max} is the maximum eigenvalue of the 1D shallow water equations which is $|u| + \sqrt{gh}$; this term represents the maximum propagation speed of all possible waves in the system.

7.4.1 Example of Linearized 1D Shallow Water Equations

Suppose we wish to solve the one-dimensional linearized shallow water equations

$$\frac{\partial}{\partial t}\begin{pmatrix} h_S \\ U \end{pmatrix} + \frac{\partial}{\partial x}\begin{pmatrix} U \\ gh_B h_S \end{pmatrix} = \begin{pmatrix} 0 \\ gh_S\frac{\partial h_B}{\partial x} \end{pmatrix} \tag{7.20}$$

where $h = h_S + h_B$ is the total height of the water column, and h_S is the height of the fluid from mean (sea) level to the surface of the wave, h_B is the depth of the bathymetry, g is the gravitational constant, and $U = hu$ is the momentum.

7.4.2 Analytic Solution and Initial Condition

Since the governing PDE is a hyperbolic system, this problem represents an initial value problem. We, therefore, need an initial condition. Setting $g = h_B = 1$, the following relations

$$h_S(x, t) = \frac{1}{2} \cos c\pi x \cos c\pi t \quad U(x, t) = \frac{1}{2} \sin c\pi x \sin c\pi t \qquad (7.21)$$

satisfy an analytic solution to the system described in Eq. (7.20) for any constant c where the domain is defined to be $(x, t) \in [0, 1] \times [0, 1]$. From the analytic solution we can produce the following initial condition

$$h_S(x, 0) = \frac{1}{2} \cos c\pi x \quad \text{and} \quad U(x, t) = 0 \qquad (7.22)$$

that we can use to begin the simulation.

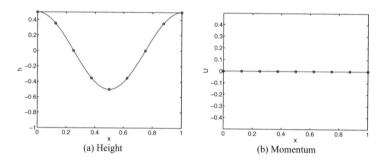

(a) Height   (b) Momentum

Fig. 7.3: Analytic solution at time $t = 1$ using $c = 2$ for (a) height of the water column (h) and (b) momentum (U).

7.4.3 Boundary Condition

This problem also requires a boundary condition: let us impose impermeable wall boundary conditions which are satisfied by the analytic solution given in Eq. (7.21) for all integer values of c. In other words, at $x = 0$ and $x = 1$, the momentum is zero.

7.4.4 Error Norm

We use the integral form of the L^2 norm given by Eq. (5.34). In addition, let us define the mass conservation measure as follows

$$\Delta M = |\text{Mass(t)} - \text{Mass}(0)|$$

where Mass(t) is the total mass at time t and M(0) the mass at the initial time. The mass is defined as follows:

$$\text{Mass(t)} = \int_{\Omega} \left(h_{S,i}^{(e)}(t) + h_{B,i}^{(e)} \right) d\Omega$$

where the integral is computed by the same quadrature used for the L^2 error norm as given in Ch. 5.

Exercise Write an algorithm for solving the 1D shallow water equations. Use the algorithm for the 1D wave equation as a template.   □

7.4.5 Time-Integrator

Any of the time-integrators presented in Ch. 20 could be used. In the results presented below, we use the 4th order Runge Kutta method which can be written for the ODE $\frac{dq}{dt} = R(q)$ as follows

$$q^{(0)} = q^n,$$

$$q^{(1)} = q^n + \frac{\Delta t}{2} R\left(q^{(0)}\right),$$

$$q^{(2)} = q^n + \frac{\Delta t}{2} R\left(q^{(1)}\right),$$

$$q^{(3)} = q^n + \Delta t R\left(q^{(2)}\right),$$

$$q^{n+1} = q^n + \frac{\Delta t}{6} \left[R\left(q^{(0)}\right) + 2R\left(q^{(1)}\right) + 2R\left(q^{(2)}\right) + R\left(q^{(3)}\right) \right].$$

7.4.6 CG Solution Accuracy

Figure 7.4 shows the convergence rates for various polynomial orders, N, for a total number of gridpoints $N_p = N_e N + 1$. In this example, an RK4 time step of $\Delta t = 1 \times 10^{-3}$ is used for all the simulations and the norms are computed at a final time of $t = 1$ using exact integration with Lobatto points. The CG method achieves

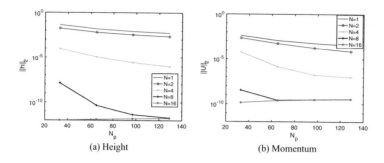

(a) Height   (b) Momentum

Fig. 7.4: Convergence rates of CG for the 1D linearized shallow water equations for polynomial orders $N = 1$, $N = 2$, $N = 4$, $N = 8$, and $N = 16$ using a total number of gridpoints N_p for (a) height variable h_S and (b) momentum $U = hu$.

accuracy of 10^{-12} for water height and 10^{-10} for momentum very quickly, even with fewer than 100 gridpoints. Figure 7.5 shows the mass conservation for the various simulations. Note that all simulations achieve excellent mass conservation (10^{-15}) which demonstrates an important property of the CG method.

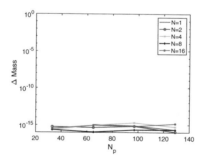

Fig. 7.5: Mass loss of CG for the 1D linearized shallow water equations for polynomial orders $N = 1$, $N = 2$, $N = 4$, $N = 8$, and $N = 16$ using a total number of gridpoints N_p.

Let us now look at one specific simulation comprised of $N_e = 8$, $N = 8$, $Q = 9$, and $\Delta t = 1 \times 10^{-4}$. Figure 7.6a shows the time history of the h_S and U L^2-error norms. The error oscillates somewhat but this is preferrable to a steady rise in the error which is not desirable. Figure 7.6b shows the change in mass conservation as a function of time. Once again, the oscillatory behavior is preferred to a steady rise in the error because this would mean that the error would be increasing. This oscillatory behavior may be due to truncation error that has very little significance at the machine precision level[4].

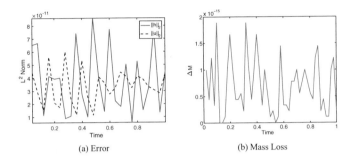

(a) Error   (b) Mass Loss

Fig. 7.6: Time history characteristics of an exact integration CG simulation comprised of $N_e = 8$ and $N = 8$. The left panel (a) shows the L^2 errors for the height and momentum variables and the right panel (b) shows the change in mass.

7.4.7 DG Solution Accuracy

Figure 7.7 shows the convergence rates for various polynomial orders, N, for a total number of gridpoints $N_p = N_e(N+1)$. Once again, an RK4 time step of $\Delta t = 1 \times 10^{-3}$ is used for all the simulations and the norms are computed at a final time of $t = 1$ using exact integration with Lobatto points.

Let us look at the specific simulation comprised of $N_e = 8$, $N = 8$, $Q = 9$, and $\Delta t = 1 \times 10^{-4}$, just as we did for CG. Figure 7.9a shows the time history of the h_S and U L^2-error norms while Figure 7.9b shows the change in mass conservation. The results are qualitatively similar with those for CG.

---

[4] Although machine double precision is $O(10^{-16})$ an error of $O(10^{-15})$ is in that range because it clearly oscillates near $O(10^{-16})$.

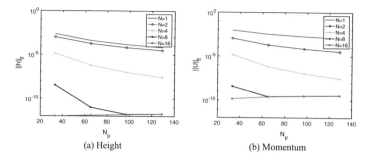

Fig. 7.7: Convergence rates of DG for the 1D linearized shallow water equations for polynomial orders $N = 1$, $N = 2$, $N = 4$, $N = 8$, and $N = 16$ using a total number of gridpoints N_p for (a) height variable h_S and (b) momentum $U = hu$.

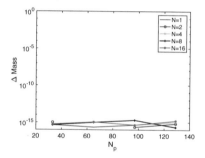

Fig. 7.8: Mass loss of DG for the 1D linearized shallow water equations for polynomial orders $N = 1$, $N = 2$, $N = 4$, $N = 8$, and $N = 16$ using a total number of gridpoints N_p.

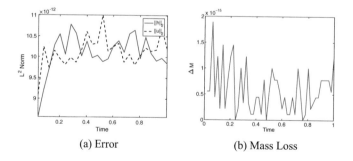

Fig. 7.9: Time history characteristics of an exact integration DG simulation comprised of $N_e = 8$ and $N = 8$. The left panel (a) shows the L^2 errors for the height and momentum variables and the right panel (b) shows the change in mass.

7.5 1D Euler Equations

Let us next consider the 1D Euler equations written in conservation form

$$\frac{\partial \rho}{\partial t} + \frac{\partial U}{\partial x} = 0 \tag{7.23a}$$

$$\frac{\partial U}{\partial t} + \frac{\partial}{\partial x}\left(\frac{U^2}{\rho} + P\right) = 0 \tag{7.23b}$$

$$\frac{\partial E}{\partial t} + \frac{\partial}{\partial x}\left(\frac{(E+P)U}{\rho}\right) = 0 \tag{7.23c}$$

$$P = \rho R T \tag{7.23d}$$

where ρ is the density, $U = \rho u$ is the momentum where u is the velocity, $E = \rho\left(c_v T + \frac{1}{2}u^2\right)$ is the total (internal plus kinetic) energy, T is the temperature, c_v is the specific heat at constant volume, P is the pressure, and R is the gas constant. We discuss the Euler equations because they are one of the most ubiquitous equations solved in various fields of science and engineering. For example, these equations are used in *computational fluid dynamics* for simulating flow over airfoils (e.g., see [11, 232, 233, 149, 234, 238, 23, 22, 131, 29, 126, 239, 189, 124, 1, 109, 285, 142]) and also for constructing *nonhydrostatic weather prediction* models[5] (e.g., see [316, 116, 367, 112, 319, 345, 303, 203, 38, 374, 410, 413, 212, 375, 344, 378, 328, 5, 139, 225, 162, 315, 165, 137, 221, 160, 278, 78, 381, 417, 220, 416]).

The Euler equations can be rewritten in the following compact vector form

$$\frac{\partial \mathbf{q}}{\partial t} + \frac{\partial \mathbf{f}}{\partial x} = \mathbf{0} \tag{7.24}$$

where

$$\mathbf{q} = \begin{pmatrix} \rho \\ U \\ E \end{pmatrix} \tag{7.25a}$$

$$\mathbf{f} = \begin{pmatrix} U \\ \frac{U^2}{\rho} + P \\ \frac{(E+P)U}{\rho} \end{pmatrix} \tag{7.25b}$$

Inserting the basis function expansion in Eq. (7.1) into the compact vector form of the equations given in Eq. (7.24), multiplying by a test function ψ and integrating within each element Ω_e yields

$$\int_{\Omega_e} \psi_i \left(\frac{\partial \mathbf{q}_N^{(e)}}{\partial t} + \frac{\partial \mathbf{f}\left(\mathbf{q}_N^{(e)}\right)}{\partial x} \right) d\Omega_e = 0.$$

---

[5] Nonhydrostatic models do not make the hydrostatic assumption, which implies that vertical accelerations are diagnostic. See, e.g., [204].

Integrating by parts gives

$$\int_{\Omega_e} \psi_i \frac{\partial \mathbf{q}_N^{(e)}}{\partial t} d\Omega_e + \int_{\Omega_e} \frac{d}{dx}\left[\psi_i \mathbf{f}\left(\mathbf{q}_N^{(e)}\right)\right] d\Omega_e - \int_{\Omega_e} \frac{d\psi_i}{dx}\mathbf{f}\left(\mathbf{q}_N^{(e)}\right) d\Omega_e = 0.$$

Evaluating the second term yields

$$\int_{\Omega_e} \psi_i \frac{\partial \mathbf{q}_N^{(e)}}{\partial t} d\Omega_e + \sum_{k=1}^{N_{FN}} \left[\hat{\mathbf{n}}^{(e,k)}\psi_i \mathbf{f}\left(\mathbf{q}_N^{(e,k)}\right)\right]_{\Gamma_e} - \int_{\Omega_e} \frac{d\psi_i}{dx}\mathbf{f}\left(\mathbf{q}_N^{(e)}\right) d\Omega_e = 0 \quad (7.26)$$

where $\mathbf{f}\left(\mathbf{q}_N^{(e,k)}\right)$ denotes the numerical flux (Riemann solver); as for the 1D shallow water equations, we can assume that it is the Rusanov flux.

Writing Eq. (7.26) in matrix-vector form yields

$$M_{ij}^{(e)}\frac{d\mathbf{q}_j^{(e)}}{dt} + \sum_{k=1}^{N_{FN}} F_{ij}^{(e,k)}\mathbf{f}\left(\mathbf{q}_j^{(e,k)}\right) - \tilde{D}_{ij}^{(e)}\mathbf{f}\left(\mathbf{q}_j^{(e)}\right) = 0. \quad (7.27)$$

The vector $\mathbf{q}_j^{(e)}$ is defined as

$$\mathbf{q}_j^{(e)} = \begin{pmatrix} \rho_j^{(e)} \\ U_j^{(e)} \\ E_j^{(e)} \end{pmatrix} \quad (7.28)$$

and the flux term $\mathbf{f}\left(\mathbf{q}_j^{(e)}\right)$ is expressed as follows

$$\mathbf{f}\left(\mathbf{q}_j^{(e)}\right) = \begin{pmatrix} U_j^{(e)} \\ U_j^{(e)}\left(\sum_{k=0}^{N}\psi_k U_k^{(e)}\right)/\left(\sum_{k=0}^{N}\psi_k \rho_k^{(e)}\right) + \left(\sum_{k=0}^{N}\psi_k P_k^{(e)}\right) \\ \left(E_j^{(e)} + P_j^{(e)}\right)\left(\sum_{k=0}^{N}\psi_k U_k^{(e)}\right)/\left(\sum_{k=0}^{N}\psi_k \rho_k^{(e)}\right) \end{pmatrix}. \quad (7.29)$$

Comparing Eq. (7.29) with Eq. (7.25b) we can now see where these terms are coming from. Defining the flux term as follows

$$\mathbf{f}\left(\mathbf{q}_j^{(e)}\right) = \begin{pmatrix} f_\rho \\ f_U \\ f_E \end{pmatrix} \quad (7.30)$$

allows us to define the numerical flux (Rusanov) as follows

$$\mathbf{f}\left(\mathbf{q}_j^{(e,k)}\right) = \begin{pmatrix} \{f_\rho\}^{(e,k)} - \hat{\mathbf{n}}^{(e,k)}\frac{|\lambda_{max}|}{2}[\![\rho_j]\!]^{(e,k)} \\ \{f_U\}^{(e,k)} - \hat{\mathbf{n}}^{(e,k)}\frac{|\lambda_{max}|}{2}[\![U_j]\!]^{(e,k)} \\ \{f_E\}^{(e,k)} - \hat{\mathbf{n}}^{(e,k)}\frac{|\lambda_{max}|}{2}[\![E_j]\!]^{(e,k)} \end{pmatrix} \quad (7.31)$$

where $\lambda_{max} = |u| + \sqrt{\gamma RT}$, with the second term being the speed of sound (speed of the acoustic waves).

Exercise Write an algorithm for solving the 1D Euler equations. Use the algorithm for the 1D shallow water equations as a template.   □

7.6 Dissipation and Dispersion Analysis

Using the eigenvalue analysis presented in Sec. 5.9.3 we can see that the semi-discrete (in space) method for either CG or DG can be written as

$$\frac{dq}{dt} = Rq.$$

Therefore, to analyze the stability of the method requires computing the eigenvalues of R because we can write the term on the right as $Rq = \lambda q$ where the eigenvalues λ of R have both a real part $(Re(\lambda))$ and an imaginary part $(Im(\lambda))$. To maintain stability, we require $Re(\lambda) < 0$, otherwise the solution increases without bound for increasing time (since the analytic solution is $q(x,t) = q_0 \exp(\lambda t)$). Note that $Im(\lambda)$ is not problematic because (via Euler's formula) the complex values represent a bounded oscillatory behavior (due to the sinusoidals). Therefore, we can represent the analytic solution as

$$q(x,t) = q_0 \exp(Re(\lambda)t) \left[\cos\left(Im(\lambda)\right)t + i \sin\left(Im(\lambda)\right)t \right]$$

which reveals that the exponential term expresses the amplitudes of the solution while the trigonometric functions dictate the phase angle of the solution. The increase or decrease in the amplitudes is predicted by the dissipation mechanisms of the numerical method (too little dissipation and the amplitudes may grow without bound) while the phase angle of the solution is nothing more than the dispersion of the method[6]. Therefore, to discern the dissipation and dispersion of the method, we only need to look at the real and imaginary parts of the eigenvalues of the matrix R. By constructing the matrix R for various polynomial orders, we can study the influence of high-order on both dissipation and dispersion.

To perform this analysis we first construct the eigenvalue decomposition

$$R = V \Lambda V^{-1}$$

where V, Λ, $V^{-1} \in C^{N \times N}$, N are the number of gridpoints in the domain, and the ith column of V contains the eigenvector associated with the ith eigenvalue of the diagonal matrix Λ, and C denotes complex space. At this point, we cannot be certain which eigenvalue is associated with which wavenumber. To determine the associated wavenumbers, we need to transform the eigenvectors from the time-domain to the

---

[6] Dispersion essentially measures how out of phase the numerical solution is from the exact solution.

frequency-domain via the discrete Fourier transform

$$\tilde{V}_k = \sum_{n=1}^{N} V_n \exp\left(-2\pi i \frac{(k-1)(n-1)}{N}\right), \quad \forall k = 1, \ldots, N.$$

However, we note that the resulting Fourier vector $\tilde{V}_k$ needs to be shifted in the proper form such that the zero frequency is in the middle of the vector[7]. At this point, we can extract the wavenumber from the Fourier vector by taking the magnitude of the vector and finding its position in the vector. Let us call this the k_{max} position. Once all k_{max} values are found for each frequency-domain vector we then map k_{max} from $n = 1, \ldots, N$ to $n = -\frac{N}{2}, \ldots, +\frac{N}{2}$, being careful to do it properly for both odd and even N. Once we know which wavenumber is associated with which eigenvalue, we can plot the dissipation and dispersion relations as a function of wavenumber.

7.6.1 Continuous Galerkin Method

In Fig. 7.10 we show the results for various polynomial orders for the CG method for the 1D wave equation (see Sec. 7.3) using Lobatto points with exact integration.

The results in Fig. 7.10 show that the CG method is completely non-dissipative (blue crosses). On the other hand, depending on the polynomial order, the CG method can be dispersive at the highest wave numbers[8]. For $N = 1$ and $N = 2$, the CG method exhibits perfect non-dispersive behavior for normalized wave numbers less than 50% of the total wave numbers. At 75% of the waves, CG begins to exhibit lagging waves. Increasing the polynomial order delays the onset of the lagging phase error; however, one alarming feature is that the dispersion plot exhibits a gap. These gaps in the dispersion relation worsen for increasing N.

In Fig. 7.11 we show the same dissipation-dispersion plots as shown in Fig. 7.10 except that now we include a diffusion operator with dissipation coefficient $\mu = 0.0051$[9]. Figure 7.11 shows that by adding just a small amount of dissipation makes the spectral gaps vanish. Clearly, the addition of the dissipation mechanism will alter the non-dissipative nature of the CG method. We note that the dissipation mainly affects the high-frequency waves (large wavenumbers) which are not well-resolved anyway. This results in a better numerical solution since we damp the waves that cannot be resolved. This can be seen quite clearly, e.g., for $N = 8$ where we see that the lines comprised of the blue-crosses (dissipation curve) and red-circles (dispersion) intersect near 70% of the highest waves where the dispersion relation moves above the ideal line. From 70% to 100%, the region where the dispersion is worst, the dissipation is the strongest. This is a highly desirable trait to have in a numerical method. We have not addressed, though, how one needs to select

---

[7] Many FFT packages have an *fftshift* command to perform this operation.

[8] The black solid line with a slope of one shows what to expect for a non-dispersive solution.

[9] We reserve the complete discussion of this approach to Ch. 18.

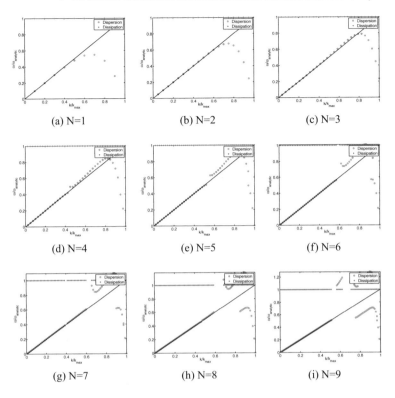

Fig. 7.10: The dissipation and dispersion analysis for CG with no dissipation for exact integration for various orders ($N = 1, \ldots, 9$) with Lobatto points.

the dissipation mechanism. We reserve this topic for Ch. 18. From this dispersion analysis, we can understand why one should not use the CG method without any form of dissipation. However, the dissipation-dispersion analysis shown in Fig. 7.11 is not representative of the method actually used in practice. The reason for this is because the method presented in those figures is for exact integration. Recall that for exact integration, one must contend with inverting a global sparse mass matrix which would be prohibitive for very large systems. Instead, we rely on high-order polynomials with inexact integration; this method is often referred to as the *spectral element method* [296]. The dissipation-dispersion curves for inexact integration and with dissipation are shown in Fig. 7.12. Besides gaining efficiency, the inexact integration approach also offers better dispersion relations as seen by comparing the results for $N \leq 8$ between Figs. 7.11 and 7.12. Note how the dispersion curves for inexact integration in Fig. 7.12 are well behaved and non-dispersive for many more wavenumbers than the exact integration results shown in Fig. 7.11. However, this difference in the dispersion relations, in fact, does not make any difference because

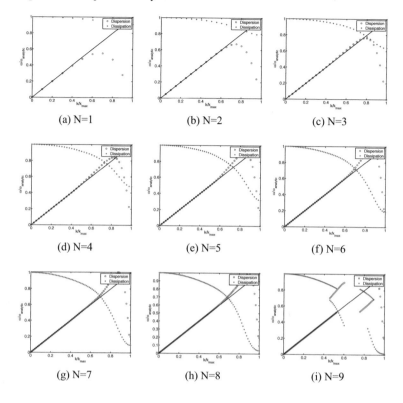

Fig. 7.11: The dissipation and dispersion analysis for CG with dissipation for exact integration for various orders ($N = 1, \ldots, 9$) with Lobatto points.

for these wavenumbers, the method dissipates the dispersion resulting in those waves being damped.

7.6.2 Discontinuous Galerkin Method

Let us now present the dissipation-dispersion analysis for the discontinuous Galerkin (DG) method. Figure 7.13 shows the dissipation and dispersion results for the DG method with no dissipation and using Lobatto points with exact integration. No dissipation in this context means that we have used a centered numerical flux function with no upwinding. These results show that the DG method without dissipation should never be used because the dispersion for the method is not desirable even at low-order (e.g., $N = 2$). As we increase the polynomial order N the dispersion gets worse. Note that the vertical scale for each of the subfigures increases due to

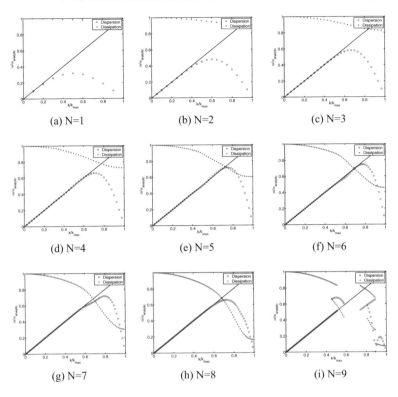

Fig. 7.12: The dissipation and dispersion analysis for CG with dissipation for inexact integration for various orders ($N = 1, \ldots, 9$) with Lobatto points.

the increasing phase lead errors of the method. Note, however, that this method is completely non-dissipative as is evident from the values of the blue-crosses in the figures that are at a value of unity (no damping).

Let us now see how the dissipation and dispersion characteristics of the DG method change when we add some dissipation via the upwind numerical flux (this is achieved by the Rusanov flux, in this case). Figure 7.14 shows the dissipation and dispersion characteristics of the DG method with dissipation (Rusanov flux) using Lobatto points with exact integration. Notice that the dispersion plots do not change very much, however, the dissipation plots do. In fact, one can see that the dissipation is strongest precisely where the dispersion is worst. Therefore, the DG method with dissipation actually controls the dispersion error by effectively dampening waves that are too dispersive. Although these plots look far messier than those for CG, DG is still quite an effective method for discretizing wave-like behavior. Although the method presented in Fig. 7.14 is in fact used for DG (exact integration) one may also be inclined to consider a DG method with inexact integration. The difference in

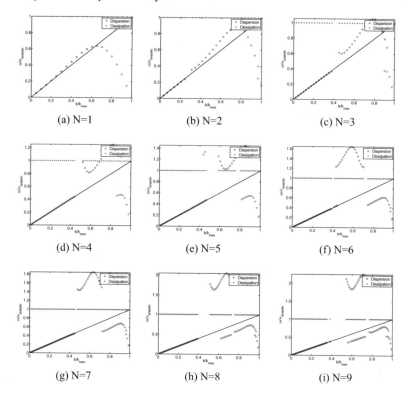

Fig. 7.13: The dissipation and dispersion analysis for DG with no dissipation (centered flux) for exact integration for various orders ($N = 1, \ldots, 9$) with Lobatto points.

cost between exact and inexact integration is not as great for DG as it is for CG. The reason for this is due to the fact that for DG the mass matrix is block-diagonal, i.e., the mass matrix corresponding to each element does not extend beyond the element. This makes it quite easy and inexpensive to invert. However, for high polynomial order it will be necessary to either invert and left-multiply or solve a system of dimensions $(N + 1)^{2d}$ for each element, where N is the order of the polynomial and d is the spatial dimension (assuming tensor-product bases). For high polynomial orders, this cost can be significant although certainly not prohibitive. For the sake of increased efficiency, it is desirable to use inexact integration that results in a diagonal mass matrix which makes the cost of inverting a mass matrix all but vanish; this method is very popular and is called the *discontinuous spectral element method* (see, e.g., [146], [35]). In Fig. 7.15, we present the dissipation-dispersion plots for DG with dissipation and inexact integration for polynomial orders $N = 1, \ldots, 9$. As in the CG case, we see that the dispersion curves for DG with inexact integration

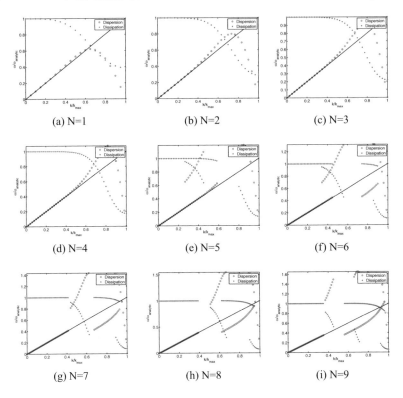

Fig. 7.14: The dissipation and dispersion analysis for DG with dissipation (Rusanov flux) for exact integration for various orders ($N = 1, \ldots, 9$) with Lobatto points.

are better behaved than those for exact integration. However, it is important to understand that although the dispersion curves for inexact integration appear better, the shape of the curves are inconsequential because the highly dispersive regions of the wave spectrum are in fact highly damped so (in principle) the odd appearance of the dispersion curves should not matter. It is prudent, however, to perform basic dispersion tests on simplified problems before using any method on large complex systems.

7.7 Dispersion and High-Frequency Waves

In Chs. 5 and 6 we saw that both CG and DG achieve impressive levels of accuracy for simple scalar wave equations. In this chapter we also saw that both CG and DG achieve equally impressive levels of accuracy for systems of hyperbolic equations

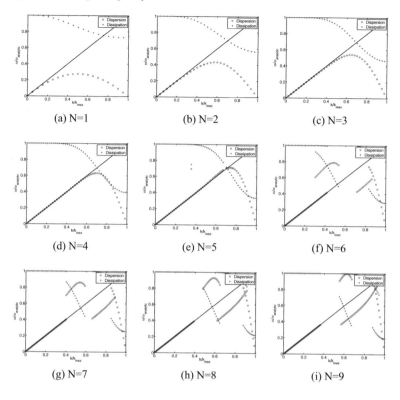

Fig. 7.15: The dissipation and dispersion analysis for DG with dissipation (Rusanov flux) for inexact integration for various orders ($N = 1, \ldots, 9$) with Lobatto points.

such as the shallow water equations. However, the dispersion analysis conducted in Sec. 7.6 might make the reader wonder whether it is a good idea at all to use high-order in light of the seemingly disturbing dispersion relations obtained for orders $N = 8$ and $N = 9$. To test whether these dispersion results are of concern, we apply both CG and DG to a test problem that has both long (low-frequency) and short (high-frequency) waves. The reason for using this test problem is to see what happens when the numerical method cannot resolve the high-frequency waves. We observed in Sec. 7.6 that the highly dispersive part of the spectrum should in fact be dissipated in some sense. Let us see if dissipation really cures the ills of dispersion errors.

7.7.1 Multi-scale Test Problem

The test problem that we use consists of the linearized shallow water equations which we write in the following form:

$$\frac{\partial h}{\partial t} + \frac{\partial U}{\partial x} = 0 \qquad (7.32a)$$

$$\frac{\partial U}{\partial t} + \frac{\partial h}{\partial x} = 0 \qquad (7.32b)$$

where (h, U) denote the fluid height (h) and momentum (U). Let us use the initial conditions defined in [389] as follows

$$h(x, 0) = h_{long}(x) + h_{short}(x)$$

$$h_{long}(x) = \exp\left[-\left(\frac{x - x_L}{\sigma}\right)^2\right]$$

$$h_{short}(x) = \exp\left[-\left(\frac{x - x_S}{\sigma}\right)^2\right] \cos\left(2\pi k \frac{x - x_S}{\sigma}\right)$$

with $U(x, 0) = h(x, 0)$. We use the same values as in [389]: $x \in [-1, +1]$, $x_L = 0.75$, $x_S = 0.25$, $\sigma = 0.1$, $k = 7$, with periodic boundary conditions. Integrating the equations to $t = 1$ time units is equivalent to having the wave move one full revolution through the domain.

7.7.2 CG and DG Solutions

If we run this test case with sufficient resolution, the two signals (low-frequency and high-frequency) propagate perfectly without one affecting the other. Better said, at sufficiently high resolution, the high-frequency signal does not pollute the low-frequency signal.

Figure 7.16 shows the solution after one full revolution ($t = 1$ time unit) when the grid is sufficiently fine to resolve both the low and high-frequency waves. The solution after one full revolution appears very similar to the initial condition. In Fig. 7.16, one can see the multiscale nature of the initial condition which has a long (low-frequency) wave on the right and short (high-frequency) waves on the left. We use both the CG and DG methods without any dissipation. Let us now see what happens when the grid is under-resolved.

Figure 7.17 shows the solution for a coarser resolution which under-resolves the high-frequency waves. Because the grid resolution is not sufficiently fine to capture the high-frequency waves, the numerical methods will incorrectly interpret them and thereby manifest themselves as Gibbs phenomena that pollutes the entire solution, including the well-resolved long (low-frequency) waves. Looking back at the results

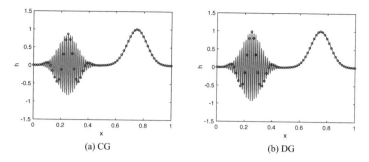

Fig. 7.16: Well-resolved low- and high-frequency waves for the *multiscale* problem for (a) CG and (b) DG after one revolution with $N = 8$ and $N_e = 64$ elements with no dissipation.

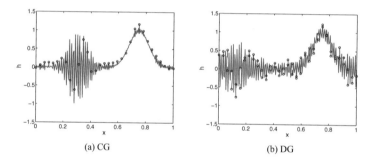

Fig. 7.17: Under-resolved high-frequency waves for the *multiscale* problem for (a) CG and (b) DG after one revolution with $N = 8$ and $N_e = 32$ elements with no dissipation.

of Sec. 7.6 we can now begin to understand what those plots were saying and see the effects of not resolving well part of the spectrum. The short waves that are not resolved (high wavenumbers) will be improperly interpreted by the numerical method. To avoid letting these unresolved waves pollute the solution (including the long waves that should be well-resolved) we must dampen the unresolved high-frequency waves.

Figure 7.18 shows the effect of dampening the under-resolved high-frequency waves. Precisely because we acknowledge that we do not have sufficient resolution to resolve all high-frequency waves, we introduce some dissipation that will eliminate the unresolved high-frequency waves and thereby protect the well-resolved portion of the solution (the long waves) so that we can represent it properly. We use two different forms of dissipation mechanisms for CG and DG. For DG, we use the Rusanov flux

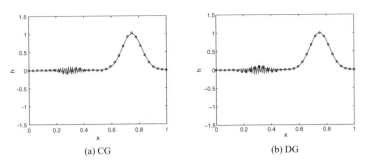

(a) CG   (b) DG

Fig. 7.18: The effect of dampening the under-resolved high-frequency waves for the *multiscale* problem for (a) CG and (b) DG after one revolution with $N = 8$ and $N_e = 32$ elements.

while for CG we use a simple low-pass filter (see Sec. 18.3) [10]. Regardless of the type of dissipation method used, the point is that by introducing some dissipation we are able to eliminate undesirable waves. We should point out that without these additional dissipative mechanisms both CG and DG are completely non-dissipative. The results in Fig. 7.18 show that the added dissipation ameliorates the pollution of the resolved regime caused by the unresolved waves (e.g., high-frequency). Figure 7.18 shows that the numerical methods are not resolving the high-frequency waves but this is less important than having them pollute the resolvable scales. Due to the dissipation at these scales, these unresolved waves will eventually disappear (for longer time simulations).

---

[10] The filter used for CG is the erfc-log filter with strength $\mu = 0.002$. Similar results could also be obtained with CG and Laplacian artificial viscosity of strength $\mu = 1 \times 10^{-5}$; however, we would need to add the operator $\mu \frac{\partial^2 h}{\partial x^2}$ to Eq. (7.32a) and $\mu \frac{\partial^2 U}{\partial x^2}$ to Eq. (7.32b).

Chapter 8
1D Continuous Galerkin Methods for Elliptic Equations

8.1 Introduction

In Chs. 5 and 6 we discussed how to use CG and DG to discretize a scalar conservation law. In Ch. 7 we extended these ideas to discretize systems of conservation laws for both CG and DG. However, these types of equations are entirely hyperbolic (first order equations in these cases). In this chapter we learn how to use the CG method to discretize second order equations that are elliptic.

This chapter begins with the definition of the elliptic operators in Sec. 8.2 that we use to describe the CG method for second order operators. Section 8.3 describes the finite difference approach for constructing second order operators which we use to compare and contrast against the CG approach which we introduce in Sec. 8.4. To better understand the construction of the CG elliptic operators, we first compare strong and weak forms of first derivatives in Sec. 8.5. This leads us to Sec. 8.6 where we describe the construction of the weak form of second order operators for use in CG discretizations. In Sec. 8.7 we perform an analysis of the matrix properties of these spatial operators. We end the chapter with Sec. 8.8 where we show results for a CG discretization of the 1D Poisson equation.

8.2 Elliptic Equations

If one is interested in solving not only first order systems, such as classical systems of hyperbolic equations, then we need to have a means to discretize first and second derivatives. It is also possible to go to arbitrarily high derivatives but we reserve

© The Editor(s) (if applicable) and The Author(s), under exclusive license
to Springer Nature Switzerland AG 2020
F. X. Giraldo, *An Introduction to Element-Based Galerkin Methods on Tensor-Product Bases*, Texts in Computational Science and Engineering 24,
https://doi.org/10.1007/978-3-030-55069-1_8

this for Ch. 18. For the moment let us assume that we are interested in solving the following elliptic equation

$$\nabla^2 q(\mathbf{x}) = f(\mathbf{x}) \tag{8.1}$$

in some domain $\mathbf{x} \in \Omega$ with the Dirichlet boundary condition

$$q|_{\Gamma_D} = g(x)$$

and Neumann boundary condition

$$(\mathbf{n} \cdot \nabla q)_{\Gamma_N} = h(x)$$

where $\Gamma = \Gamma_D \bigcup \Gamma_N$ is the boundary of Ω. The wave equations that we have already seen are examples of Cauchy or initial value problems whereas Eq. (8.1) defines a boundary value problem and is in fact a Poisson equation. The reason why the physical meaning of the operator is important (i.e., whether it is an initial versus a boundary value problem) is that this will dictate how we approach the solution of the governing equation. For example, recall that for the wave equation, we had an initial value that we then used to start the simulation and the Lagrangian derivative $\frac{d}{dt} = \frac{\partial}{\partial t} + u\frac{\partial}{\partial x}$ moves the fluid parcel along a certain path along the characteristics of this hyperbolic operator.

In contrast, the Poisson equation is a boundary value problem and has no such initial condition - furthermore, it has no time-dependence. What this means is that we will have to solve this equation in one step, i.e., as a global system. Let us now describe the solution of this operator using the finite difference method in order to facilitate the comparison with the continuous Galerkin method in this chapter and with the discontinuous Galerkin method in Ch. 9.

8.3 Finite Difference Method

In order to compare and contrast the Galerkin formulations of the Laplacian operator, let us first derive the finite difference approximation to this operator. Looking at Fig. 8.1 we see that if we wish to construct the second partial derivative, q_{xx}, at the

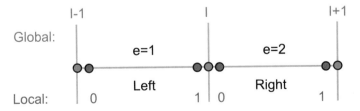

Fig. 8.1: Contribution to the gridpoint I from the left $(I - 1, I)$ and right $(I, I + 1)$ elements.

gridpoint I we must take Taylor series expansions about I for the points $I + 1$ and $I - 1$ as follows

$$q_{I+1} = q_I + \Delta x \frac{dq_I}{dx} + \frac{\Delta x^2}{2} \frac{d^2 q_I}{dx^2} + \frac{\Delta x^3}{6} \frac{d^3 q_I}{dx^3} + O\left(\Delta x^4\right)$$

and

$$q_{I-1} = q_I - \Delta x \frac{dq_I}{dx} + \frac{\Delta x^2}{2} \frac{d^2 q_I}{dx^2} - \frac{\Delta x^3}{6} \frac{d^3 q_I}{dx^3} + O\left(\Delta x^4\right).$$

Adding the above two relations yields

$$q_{I+1} + q_{I-1} = 2q_I + \Delta x^2 \frac{d^2 q_I}{dx^2} + O\left(\Delta x^4\right).$$

Next, isolating the second derivative, yields

$$\frac{d^2 q_I}{dx^2} = \frac{q_{I+1} - 2q_I + q_{I-1}}{\Delta x^2} + O\left(\Delta x^2\right) \tag{8.2}$$

which is the second order, centered finite difference approximation of the second derivative. We will use this expression to compare and contrast the derivatives that we obtain with the Galerkin methods with linear polynomial approximations.

8.4 Continuous Galerkin Method

Looking at Eq. (8.1), it is easy to believe that the solution q must lie in the space $C^2(\Omega)$; that is, that the solution q, and its first and second derivatives must be continuous everywhere. If this were in fact true then this would be a very restrictive space; e.g., it would not be possible to approximate this PDE with either linear or quadratic basis functions. The fact of the matter is that the solution q does not even need to live in $C^1(\Omega)$. Fortunately, the solution must be in $H^1(\Omega)$ which is the Sobolev space with semi-norm 1 that we defined in Ch. 5. The reason why this is very good news is because this space is much less restrictive (and hence much larger) than the spaces C^2 and C^1. This is the reason why Galerkin methods have been so successful for solving self-adjoint operators (such as elliptic PDEs); in other words, since the basis functions used to discretize the governing equations satisfy the analytic solutions to these equations, the method is optimal. Recall from Ch. 3 that the generating polynomials of the basis functions are the eigenfunctions of the singular Sturm-Liouville operator which do indeed satisfy the Laplacian operator.

The fact that the CG solutions live in $H^1(\Omega)$ means that the natural way to solve Eq. (8.1) is by multiplying it by a test function ψ that lives in $H^1(\Omega)$ that, for short, we will write simply as H^1. This yields the first step in the discretization

$$\int_{\Omega_e} \psi_i \nabla^2 q_N \, d\Omega_e = \int_{\Omega_e} \psi_i f_N \, d\Omega_e \tag{8.3}$$

where we always will take a local (element-wise) view of the problem, assuming that we will apply DSS later in order to form the global problem. Let us now use the product rule

$$\nabla \cdot (\psi_i \nabla q_N) = \nabla \psi_i \cdot \nabla q_N + \psi_i \nabla^2 q_N$$

that, upon integrating, using the divergence theorem, and rearranging, allows us to rewrite Eq. (8.3) as

$$\int_{\Gamma_e} \psi_i \mathbf{n} \cdot \nabla q_N \, d\Gamma_e - \int_{\Omega_e} \nabla \psi_i \cdot \nabla q_N \, d\Omega_e = \int_{\Omega_e} \psi_i f_N \, d\Omega_e, \qquad (8.4)$$

which is *Green's first identity*.

If we define the problem statement of Eq. (8.1) with homogeneous Dirichlet boundary conditions then we can write the function space of our solution as $q \in H_0^1(\Omega)$. Furthermore, if we impose the additional boundary condition $(\mathbf{n} \cdot \nabla q_N)_{\Gamma_N} = 0$, known as the natural boundary condition, then we can write the weak formulation of the original PDE as

$$-\int_{\Omega_e} \nabla \psi_i \cdot \nabla q_N \, d\Omega_e = \int_{\Omega_e} \psi_i f_N \, d\Omega_e \qquad (8.5)$$

which is the integral form we will use with the continuous Galerkin method. To ensure well-posedness we need to impose at least one Dirichlet boundary condition.

For general boundary conditions we require satisfying the integral form

$$\int_{\Gamma_N} \psi_i h(x) \, d\Gamma_N - \int_{\Omega_e} \nabla \psi_i \cdot \nabla q_N \, d\Omega_e = \int_{\Omega_e} \psi_i f_N \, d\Omega_e, \qquad (8.6)$$

where $h(x)$ denotes the Neumann boundary condition and the Dirichlet boundary condition

$$q|_{\Gamma_D} = g(x)$$

will need to be enforced in a strong sense. That is, we construct the global matrix problem of Eq. (8.5) and then modify the matrix problem such that it then satisfies the Dirichlet boundary conditions exactly.

Note that another option for enforcing Dirichlet boundary conditions is to impose *lifting* conditions by constructing a linear decomposition of the solution as follows

$$q(x) = q^I(x) + q^D(x)$$

where q^I and q^D are the solutions at the interior and at the Dirichlet boundaries. This equation is then substituted into Eq. (8.6) to yield

$$-\int_{\Omega_e} \nabla \psi_i \cdot \nabla q_N^I \, d\Omega_e = \int_{\Omega_e} \psi_i f_N \, d\Omega_e - \int_{\Gamma_N} \psi_i h(x) \, d\Gamma_N \qquad (8.7)$$

$$-\int_{\Gamma_D} \psi_i \mathbf{n} \cdot \nabla q_N^D \, d\Gamma_D + \int_{\Omega_e} \nabla \psi_i \cdot \nabla q_N^D \, d\Omega_e.$$

This approach is known as lifting because the known Dirichlet boundary condition $q^{\mathcal{D}}$ is lifted across the interior solution due to the basis function expansion approximation of $q^{\mathcal{D}}$ away from the boundary. The advantage of this approach is that the degrees of freedom associated to this known Dirichlet condition are removed from the system on the left-hand side. In addition, this approach will ensure that the resulting matrix problem remains symmetric positive-definite, which is critical if, e.g., the conjugate gradient method is used for solving the resulting system of equations.

Remark 8.1 Note that the reason we choose to use Green's first identity is to ensure that our solution space is in the Sobolev space H^1 which we saw in Ch. 2 and state once again:

$$(\psi, q)_{H^1} = \int_{\Omega} (\nabla \psi \cdot \nabla q + \psi q) \, d\Omega.$$

If we did not use Green's first identity, then we would not be guaranteed to be in H^1. This would not be disastrous, it would just restrict the space of possible solutions. The reason why H^1 is a good choice is because we are able to use linear functions (ψ) to represent second order derivatives. Otherwise, we would be restricted to use higher than linear polynomials. Building second order derivatives explicitly (instead of using Green's first identity) increases the complexity and memory of a computer model.

8.5 First Derivatives in their Strong and Weak Forms

Before moving forward to the discrete approximation of the Laplacian operator, let us discuss further the role of Green's identity that will be pivotal in the construction of the Laplacian matrix. Let us discuss the role of Green's identities using the first derivative operator because, for this operator, the approximation using the strong and weak forms both live in H^1.

8.5.1 Strong Form

To approximate the derivative $\frac{dq}{dx}$ we can expand the function q in the following way

$$q_N^{(e)}(x) = \sum_{j=0}^{N} \psi_j(x) q_j^{(e)}$$

that, after differentiation yields

$$\frac{dq_N^{(e)}(x)}{dx} = \sum_{j=0}^{N} \frac{d\psi_j(x)}{dx} q_j^{(e)}. \tag{8.8}$$

In what follows we will make use of the following observation: if we knew the derivative at all gridpoints (q_x) then we could represent the derivative everywhere by the basis function expansion

$$\frac{dq_N^{(e)}(x)}{dx} = \sum_{j=0}^{N} \psi_j(x) q_{x,j}^{(e)}, \tag{8.9}$$

where $q_{x,j}^{(e)} = \frac{dq^{(e)}(x_j)}{dx}$. We first map from physical (x) to computational space (ξ), i.e., we map each element defined in physical space to the reference element $\xi \in [-1, +1]$. The mapping that we use is the usual 1D map

$$\xi = \frac{2(x - x_0)}{x_N - x_0} - 1 \tag{8.10}$$

that gives $\xi = -1$ when $x = x_0$ and $\xi = +1$ when $x = x_N$. Since we wish to define $\psi = \psi(\xi)$ then, by the chain rule, $\frac{d\psi(x)}{dx} = \frac{d\psi(\xi)}{d\xi} \frac{d\xi}{dx}$ where from Eq. (8.10) we get $\frac{d\xi}{dx} = \frac{2}{\Delta x}$ with $\Delta x = x_N - x_0$ being the size of the element in physical space. Using all of this information allows us to rewrite Eq. (8.8) as follows

$$\frac{dq_N^{(e)}(x)}{dx} = \sum_{j=0}^{N} \frac{d\psi_j(\xi)}{d\xi} \frac{d\xi(x)}{dx} q_j^{(e)}. \tag{8.11}$$

Next, if we multiply both sides of Eq. (8.11) with ψ_i and integrate from x_0 to x_N, yields

$$\int_{x_0}^{x_N} \psi_i \frac{dq_N^{(e)}(x)}{dx} \, dx = \int_{x_0}^{x_N} \psi_i \sum_{j=0}^{N} \frac{d\psi_j(\xi)}{d\xi} \frac{d\xi(x)}{dx} q_j^{(e)} \, dx \tag{8.12}$$

where $i = 0, \ldots, N$. Note that to transform this integral from physical space to the reference element requires us to change the bounds of integration and change the variable such that $dx = \frac{dx}{d\xi} d\xi$, where from Eq. (8.10) we can write $\frac{dx}{d\xi} = \frac{\Delta x}{2}$. Using this information as well as Eq. (8.9) allows us to write Eq. (8.12) as follows

$$\int_{-1}^{+1} \psi_i \psi_j \frac{dx}{d\xi} \, d\xi \, q_{x,j}^{(e)} = \int_{-1}^{+1} \psi_i \frac{d\psi_j(\xi)}{d\xi} \frac{d\xi(x)}{dx} \frac{dx}{d\xi} \, d\xi \, q_j^{(e)} \tag{8.13}$$

with $i, j = 0, \ldots, N$, where the summation symbol has disappeared because we have written this equation implying a matrix-vector form. In addition, the term $q_x^{(e)}$ represents the expansion coefficients of the first derivative of $q^{(e)}$ and is in fact what we seek. Simplifying yields,

$$\frac{\Delta x^{(e)}}{2} \int_{-1}^{+1} \psi_i \psi_j \, d\xi \, q_{x,j}^{(e)} = \int_{-1}^{+1} \psi_i \frac{d\psi_j(\xi)}{d\xi} \, d\xi \, q_j^{(e)} \tag{8.14}$$

where we can further simplify by writing

$$M_{ij}^{(e)} q_{x,j}^{(e)} = D_{ij}^{(e)} q_j^{(e)} \tag{8.15}$$

where

$$M_{ij}^{(e)} = \frac{\Delta x^{(e)}}{2} \int_{-1}^{+1} \psi_i \psi_j \, d\xi$$

is the mass matrix and

$$D_{ij}^{(e)} = \int_{-1}^{+1} \psi_i \frac{d\psi_j(\xi)}{d\xi} \, d\xi$$

is the *strong form* differentiation matrix. In Ch. 5 we saw that for linear elements (N=1), the mass and differentiation matrices are given as follows

$$M_{ij}^{(e)} = \frac{\Delta x^{(e)}}{6} \begin{pmatrix} 2 & 1 \\ 1 & 2 \end{pmatrix},$$

$$D_{ij}^{(e)} = \frac{1}{2} \begin{pmatrix} -1 & 1 \\ -1 & 1 \end{pmatrix}.$$

Let us now see how these element matrices allow us to construct the derivative at the gridpoint I shown in Fig. 8.1.

Resulting Element Equations

Substituting linear basis functions into Eq. (8.15) results in the element equations

$$\frac{\Delta x^{(e)}}{6} \begin{pmatrix} 2 & 1 \\ 1 & 2 \end{pmatrix} \begin{pmatrix} q_{x,0}^{(e)} \\ q_{x,1}^{(e)} \end{pmatrix} = \frac{1}{2} \begin{pmatrix} -1 & 1 \\ -1 & 1 \end{pmatrix} \begin{pmatrix} q_0^{(e)} \\ q_1^{(e)} \end{pmatrix}.$$

Element Contribution to Gridpoint I

From Fig. 8.1 we can see that both the left and right elements contribute a portion of their solutions to the global gridpoint I.

Left Element

Since the following is the left element equation

$$\frac{\Delta x^{(L)}}{6} \begin{pmatrix} 2 & 1 \\ 1 & 2 \end{pmatrix} \begin{pmatrix} q_{x,I-1} \\ q_{x,I} \end{pmatrix} = \frac{1}{2} \begin{pmatrix} -1 & 1 \\ -1 & 1 \end{pmatrix} \begin{pmatrix} q_{I-1} \\ q_I \end{pmatrix}$$

then the contribution to the global gridpoint I is given by the second row as follows

$$\frac{\Delta x^{(L)}}{6} (q_{x,I-1} + 2q_{x,I}) = \frac{1}{2} (-q_{I-1} + q_I)$$

where

$$\frac{\Delta x^{(L)}}{6} (2q_{x,I-1} + q_{x,I}) = \frac{1}{2} (-q_{I-1} + q_I)$$

is the contribution to the global gridpoint $I - 1$.

Right Element

For the right element we have the equation

$$\frac{\Delta x^{(R)}}{6} \begin{pmatrix} 2 & 1 \\ 1 & 2 \end{pmatrix} \begin{pmatrix} q_{x,I} \\ q_{x,I+1} \end{pmatrix} = \frac{1}{2} \begin{pmatrix} -1 & 1 \\ -1 & 1 \end{pmatrix} \begin{pmatrix} q_I \\ q_{I+1} \end{pmatrix}$$

then the contribution to the global gridpoint I is given by the first row as follows

$$\frac{\Delta x^{(R)}}{6} (2q_{x,I} + q_{x,I+1}) = \frac{1}{2}(-q_I + q_{I+1})$$

where

$$\frac{\Delta x^{(R)}}{6} (q_{x,I} + 2q_{x,I+1}) = \frac{1}{2}(-q_I + q_{I+1})$$

is the contribution to the global gridpoint $I + 1$.

Total Contribution

Assuming $\Delta x^{(L)} = \Delta x^{(R)} = \Delta x$, we can write the total contribution for both elements given in Fig. 8.1 by the following matrix-vector problem:

$$\frac{\Delta x}{6} \begin{pmatrix} 2 & 1 & 0 \\ 1 & 4 & 1 \\ 0 & 1 & 2 \end{pmatrix} \begin{pmatrix} q_{x,I-1} \\ q_{x,I} \\ q_{x,I+1} \end{pmatrix} = \frac{1}{2} \begin{pmatrix} -1 & 1 & 0 \\ -1 & 0 & 1 \\ 0 & -1 & 1 \end{pmatrix} \begin{pmatrix} q_{I-1} \\ q_I \\ q_{I+1} \end{pmatrix}. \tag{8.16}$$

Multiplying by the inverse mass matrix given by Eq. (5.17) yields

$$\begin{pmatrix} q_{x,I-1} \\ q_{x,I} \\ q_{x,I+1} \end{pmatrix} = \frac{1}{4\Delta x} \begin{pmatrix} -5 & 6 & -1 \\ -2 & 0 & 2 \\ 1 & -6 & 5 \end{pmatrix} \begin{pmatrix} q_{I-1} \\ q_I \\ q_{I+1} \end{pmatrix}. \tag{8.17}$$

The equation for the global gridpoint I is given by the second row:

$$q_{x,I} = \frac{q_{I+1} - q_{I-1}}{2\Delta x}$$

which is the same approximation we got for the 2nd order finite difference approximation.

8.5.2 Weak Form

To approximate the derivative $\frac{dq}{dx}$ using the weak form, we begin with the equality

$$\frac{dq_N^{(e)}(x)}{dx} = \frac{dq_N^{(e)}(x)}{dx}.$$

Multiplying both sides by a test function and integrating yields

$$\int_{x_0}^{x_N} \psi_i \frac{dq_N^{(e)}(x)}{dx}\, dx = \int_{x_0}^{x_N} \psi_i \frac{dq_N^{(e)}(x)}{dx}\, dx. \tag{8.18}$$

Using the product rule

$$\frac{d}{dx}\left(\psi_i(x)q_N^{(e)}(x)\right) = \frac{d\psi_i}{dx}q_N^{(e)}(x) + \psi_i(x)\frac{dq_N^{(e)}(x)}{dx}$$

we can write Eq. (8.18) as follows

$$\int_{x_0}^{x_N} \psi_i \frac{dq_N^{(e)}(x)}{dx}\, dx = \int_{x_0}^{x_N} \frac{d}{dx}\left(\psi_i(x)q_N^{(e)}(x)\right)\, dx - \int_{x_0}^{x_N} \frac{d\psi_i}{dx}q_N^{(e)}(x)\, dx. \tag{8.19}$$

Mapping this equation to the reference element yields

$$\int_{-1}^{+1} \psi_i\psi_j \frac{dx}{d\xi}\, d\xi\, q_{x,j}^{(e)} = [\psi_i(\xi)q_N^{(e)}(\xi)]_{\hat{\Gamma}} - \int_{-1}^{+1} \frac{d\psi_i(\xi)}{d\xi}\frac{d\xi(x)}{dx}\psi_j(\xi) \frac{dx}{d\xi}\, d\xi\, q_j^{(e)} \tag{8.20}$$

with $i, j = 0, \ldots, N$ and $e = 1, \ldots, N_e$, where we have used the Fundamental Theorem of Calculus to simplify the first term on the right-hand side and where $\hat{\Gamma}$ denotes the boundary in terms of the computational space of the reference element. Simplifying yields,

$$M_{ij}^{(e)} q_{x,j}^{(e)} = -\tilde{D}_{ij}^{(e)} q_j^{(e)} \tag{8.21}$$

where

$$M_{ij}^{(e)} = \frac{\Delta x^{(e)}}{2} \int_{-1}^{+1} \psi_i\psi_j\, d\xi$$

is the mass matrix,

$$\tilde{D}_{ij}^{(e)} = \int_{-1}^{+1} \frac{d\psi_i(\xi)}{d\xi}\psi_j(\xi)\, d\xi$$

is the *weak form* differentiation matrix and (for simplicity) we have assumed that the boundary term $[\psi_i(x)q_N(x)]|_{\Gamma_\mathcal{D}} = g(x)$ vanishes.

We already know from the previous section that for $N = 1$ the mass matrix is

$$M_{ij}^{(e)} = \frac{\Delta x^{(e)}}{6} \begin{pmatrix} 2 & 1 \\ 1 & 2 \end{pmatrix}.$$

We also know the weak form differentiation matrix, $\tilde{D}$, from Ch. 6 to be

$$\tilde{D}_{ij}^{(e)} = \frac{1}{2}\begin{pmatrix} -1 & -1 \\ 1 & 1 \end{pmatrix}.$$

Let us now see how these element matrices allow us to construct the derivative at the gridpoint I.

Resulting Element Equations

Substituting linear basis functions into Eq. (8.21) results in the element equations

$$\frac{\Delta x^{(e)}}{6} \begin{pmatrix} 2 & 1 \\ 1 & 2 \end{pmatrix} \begin{pmatrix} q_{x,0} \\ q_{x,1} \end{pmatrix} = -\frac{1}{2} \begin{pmatrix} -1 & -1 \\ 1 & 1 \end{pmatrix} \begin{pmatrix} q_0 \\ q_1 \end{pmatrix}.$$

Element Contribution to Gridpoint I

Just as we did for the strong form, let us see how the left and right elements contribute to the solution of the global gridpoint I.

Left Element

Since the following is the left element equation

$$\frac{\Delta x^{(L)}}{6} \begin{pmatrix} 2 & 1 \\ 1 & 2 \end{pmatrix} \begin{pmatrix} q_{x,I-1} \\ q_{x,I} \end{pmatrix} = -\frac{1}{2} \begin{pmatrix} -1 & -1 \\ 1 & 1 \end{pmatrix} \begin{pmatrix} q_{I-1} \\ q_I \end{pmatrix}$$

then the contribution to the gridpoint I is given by the second row as follows

$$\frac{\Delta x^{(L)}}{6} \left(q_{x,I-1} + 2q_{x,I} \right) = -\frac{1}{2} \left(q_{I-1} + q_I \right)$$

where

$$\frac{\Delta x^{(L)}}{6} \left(2q_{x,I-1} + q_{x,I} \right) = -\frac{1}{2} \left(-q_{I-1} - q_I \right)$$

is the contribution to $I - 1$.

Right Element

For the right element we have the equation

$$\frac{\Delta x^{(R)}}{6} \begin{pmatrix} 2 & 1 \\ 1 & 2 \end{pmatrix} \begin{pmatrix} q_{x,I} \\ q_{x,I+1} \end{pmatrix} = -\frac{1}{2} \begin{pmatrix} -1 & -1 \\ 1 & 1 \end{pmatrix} \begin{pmatrix} q_I \\ q_{I+1} \end{pmatrix}$$

then the contribution to the gridpoint I is given by the first row as follows

$$\frac{\Delta x^{(R)}}{6} \left(2q_{x,I} + q_{x,I+1} \right) = -\frac{1}{2} \left(-q_I - q_{I+1} \right)$$

where

$$\frac{\Delta x^{(R)}}{6}\left(q_{x,I} + 2q_{x,I+1}\right) = -\frac{1}{2}\left(q_I + q_{I+1}\right)$$

is the contribution to $I + 1$.

Total Contribution

Assuming $\Delta x^{(L)} = \Delta x^{(R)} = \Delta x$, we can write the total contribution for both elements given in Fig. 8.1 by the following matrix-vector problem:

$$\frac{\Delta x}{6}\begin{pmatrix} 2 & 1 & 0 \\ 1 & 4 & 1 \\ 0 & 1 & 2 \end{pmatrix}\begin{pmatrix} q_{x,I-1} \\ q_{x,I} \\ q_{x,I+1} \end{pmatrix} = \frac{1}{2}\begin{pmatrix} -1 & -1 & 0 \\ 1 & 0 & -1 \\ 0 & 1 & 1 \end{pmatrix}\begin{pmatrix} q_{I-1} \\ q_I \\ q_{I+1} \end{pmatrix}. \qquad (8.22)$$

The second row of this matrix-vector problem denotes the solution at the global gridpoint I which is the following

$$\frac{\Delta x}{6}\left(q_{x,I-1} + 4q_{x,I} + q_{x,I+1}\right) = \frac{1}{2}\left(-q_{I-1} + q_{I+1}\right).$$

Diagonalizing the mass matrix yields

$$q_{x,I} = \frac{q_{I+1} - q_{I-1}}{2\Delta x}$$

which is, once again, the same expression we obtained for the 2nd order finite difference approximation and is identical to the *strong* form representation.

8.6 Second Derivatives in their Weak Form

Let us now see how we can use the continuous Galerkin machinery to approximate a second derivative.

8.6.1 Laplacian Matrix

Let us define the weak form Laplacian operator defined on the left-hand side of Eq. (8.5) as follows

$$L_{ij}^{(e)} = \int_{x_0}^{x_1} \frac{d\psi_i(x)}{dx}\frac{d\psi_j(x)}{dx}\,dx = \int_{-1}^{+1} \frac{d\psi_i(\xi)}{d\xi}\frac{d\xi}{dx}\frac{d\psi_j(\xi)}{d\xi}\frac{d\xi}{dx}\frac{\Delta x^{(e)}}{2}\,d\xi$$

since

$$\frac{d\psi_j}{dx}(x) = \frac{d\psi_j}{d\xi}(\xi)\frac{d\xi}{dx} = \frac{1}{2}\xi_j\frac{2}{\Delta x^{(e)}}$$

because the basis functions for linear elements are

$$\psi_j = \frac{1}{2}(1 + \xi_j \xi)$$

and their derivatives are

$$\frac{d\psi_j}{d\xi} = \frac{1}{2}\xi_j.$$

Substituting the differentiated terms for ψ we get in matrix form

$$L_{ij}^{(e)} = \frac{2}{\Delta x^{(e)}} \int_{-1}^{+1} \begin{pmatrix} \frac{1}{4} & -\frac{1}{4} \\ -\frac{1}{4} & \frac{1}{4} \end{pmatrix} d\xi.$$

Integrating yields

$$L_{ij}^{(e)} = \frac{1}{\Delta x^{(e)}} \begin{pmatrix} 1 & -1 \\ -1 & 1 \end{pmatrix}.$$

Exercise Derive the element Laplacian matrix for quadratic (N=2) polynomials with exact integration. ☐

Exercise Does using exact or inexact integration matter? Explain your answer. ☐

Next, assume that if we indeed knew the second derivative, $\frac{d^2 q}{dx^2}$ then we could expand it as follows

$$\frac{d^2 q_N^{(e)}}{dx^2}(x) = \sum_{i=0}^{N} \psi_i(x) q_{xx,i}^{(e)}$$

where $q_{xx}^{(e)}$ are the expansion coefficients of the second derivative and is in fact what we seek. Using the product rule

$$\psi \frac{d^2 q_N^{(e)}}{dx^2} = \frac{d}{dx}\left(\psi \frac{dq_N^{(e)}}{dx}\right) - \frac{d\psi}{dx}\frac{dq_N^{(e)}}{dx}$$

where we now ignore the boundary term (first term on the right) [1], we get

$$\int_{x_0}^{x_1} \psi_i(x)\psi_j(x)\, dx\, q_{xx,j}^{(e)} = -\int_{x_0}^{x_1} \frac{d\psi_i(x)}{dx}\frac{d\psi_j(x)}{dx}\, dx\, q_j^{(e)}. \qquad (8.23)$$

Equation (8.23) can then be written in matrix-vector form as follows

$$M_{ij}^{(e)} q_{xx,j}^{(e)} = -L_{ij}^{(e)} q_j^{(e)}. \qquad (8.24)$$

---

[1] The physical justification for ignoring the boundary term is that we are assuming that the element in question is far away from a boundary. Of course, if we were imposing boundary conditions in a strong sense, then this boundary integral would also vanish or if the natural boundary condition (that the flux is zero) were imposed.

8.6.2 Resulting Element Equations

Substituting linear basis functions into Eq. (8.24) yields the following element equations

$$\frac{\Delta x^{(e)}}{6} \begin{pmatrix} 2 & 1 \\ 1 & 2 \end{pmatrix} \begin{pmatrix} q_{xx,0}^{(e)} \\ q_{xx,1}^{(e)} \end{pmatrix} = -\frac{1}{\Delta x^{(e)}} \begin{pmatrix} 1 & -1 \\ -1 & 1 \end{pmatrix} \begin{pmatrix} q_0^{(e)} \\ q_1^{(e)} \end{pmatrix}.$$

8.6.3 Element Contribution to Gridpoint I

Let us now see what the discrete linear Laplacian operator for the gridpoint I in Fig. 8.1 will be.

Left Element

Since the following is the left element equation

$$\frac{\Delta x^{(L)}}{6} \begin{pmatrix} 2 & 1 \\ 1 & 2 \end{pmatrix} \begin{pmatrix} q_{xx,I-1} \\ q_{xx,I} \end{pmatrix} = -\frac{1}{\Delta x^{(L)}} \begin{pmatrix} 1 & -1 \\ -1 & 1 \end{pmatrix} \begin{pmatrix} q_{I-1} \\ q_I \end{pmatrix}$$

then the contribution to the gridpoint I is given by the second row as follows

$$\frac{\Delta x^{(L)}}{6} \left(q_{xx,I-1} + 2 q_{xx,I} \right) = -\frac{1}{\Delta x^{(L)}} \left(-q_{I-1} + q_I \right)$$

where

$$\frac{\Delta x^{(L)}}{6} \left(2 q_{xx,I-1} + q_{xx,I} \right) = -\frac{1}{\Delta x^{(L)}} \left(q_{I-1} - q_I \right)$$

is the contribution to $I - 1$.

Right Element

For the right element we have the equation

$$\frac{\Delta x^{(R)}}{6} \begin{pmatrix} 2 & 1 \\ 1 & 2 \end{pmatrix} \begin{pmatrix} q_{xx,I} \\ q_{xx,I+1} \end{pmatrix} = -\frac{1}{\Delta x^{(R)}} \begin{pmatrix} 1 & -1 \\ -1 & 1 \end{pmatrix} \begin{pmatrix} q_I \\ q_{I+1} \end{pmatrix}$$

then the contribution to the gridpoint I is given by the first row as follows

$$\frac{\Delta x^{(R)}}{6} \left(2 q_{xx,I} + q_{xx,I+1} \right) = -\frac{1}{\Delta x^{(R)}} \left(q_I - q_{I+1} \right)$$

where

$$\frac{\Delta x^{(R)}}{6}\left(q_{xx,I} + 2q_{xx,I+1}\right) = -\frac{1}{\Delta x^{(R)}}\left(-q_I + q_{I+1}\right)$$

is the contribution of the right element to $I + 1$.

Total Contribution

Summing up the contributions from the left and right elements and assuming that $\Delta x^{(L)} = \Delta x^{(R)} = \Delta x$ gives the global matrix-vector problem:

$$\frac{\Delta x}{6}\begin{pmatrix} 2 & 1 & 0 \\ 1 & 4 & 1 \\ 0 & 1 & 2 \end{pmatrix}\begin{pmatrix} q_{xx,I-1} \\ q_{xx,I} \\ q_{xx,I+1} \end{pmatrix} = -\frac{1}{\Delta x}\begin{pmatrix} 1 & -1 & 0 \\ -1 & 2 & -1 \\ 0 & -1 & 1 \end{pmatrix}\begin{pmatrix} f_{I-1} \\ f_I \\ f_{I+1} \end{pmatrix}.$$

Diagonalizing the mass matrix yields

$$\frac{\Delta x}{6}\begin{pmatrix} 3 & 0 & 0 \\ 0 & 6 & 0 \\ 0 & 0 & 3 \end{pmatrix}\begin{pmatrix} q_{xx,I-1} \\ q_{xx,I} \\ q_{xx,I+1} \end{pmatrix} = -\frac{1}{\Delta x}\begin{pmatrix} 1 & -1 & 0 \\ -1 & 2 & -1 \\ 0 & -1 & 1 \end{pmatrix}\begin{pmatrix} f_{I-1} \\ f_I \\ f_{I+1} \end{pmatrix}.$$

The total contribution to the gridpoint I is given by the second row which, after simplifying, yields

$$\Delta x q_{xx,I} = \frac{1}{\Delta x}\left(q_{I-1} - 2q_I + q_{I+1}\right)$$

which is identical to a centered 2nd order finite difference method

$$q_{xx,I} = \frac{q_{I+1} - 2q_I + q_{I-1}}{\Delta x^2}.$$

Exercise Recompute the second order derivative but this time do not lump the mass matrix. ☐

Exercise Prove that this derivative is consistent and show the order of accuracy. ☐

8.6.4 1D Elliptic Equation

Now that we know how to construct a second derivative with the CG method we can use this information to solve elliptic equations. For the Poisson problem in one dimension we write: find $q_N^{(e)} \in H^1(\Omega)$ such that

$$\int_{\Omega_e} \psi_i \frac{d^2 q_N^{(e)}}{dx^2}\, d\Omega_e = \int_{\Omega_e} \psi_i f_N^{(e)}\, d\Omega_e \quad \forall\, \psi \in H^1(\Omega_e). \tag{8.25}$$

Using the product rule and the Fundamental Theorem of Calculus (i.e., Green's first identity), we get

$$\left[\psi_i \frac{dq_N^{(e)}}{dx} \right]_{\Gamma_e} - \int_{\Omega_e} \frac{d\psi_i}{dx} \frac{dq_N^{(e)}}{dx} \, d\Omega_e = \int_{\Omega_e} \psi_i f_N^{(e)} \, d\Omega_e. \tag{8.26}$$

Assuming that we defined the problem statement of Eq. (8.25) as having homogeneous Dirichlet boundary conditions then we can write the function space of our solution as $q_N^{(e)} \in H_0^1(\Omega_e)$. Equation (8.26) then becomes

$$-\int_{\Omega_e} \frac{d\psi_i}{dx} \frac{dq_N^{(e)}}{dx} \, d\Omega_e = \int_{\Omega_e} \psi_i f_N^{(e)} \, d\Omega_e \tag{8.27}$$

with the Dirichlet boundary condition

$$q|_{\Gamma_D} = g(x)$$

where $g(x = x_L) = a$ and $g(x = x_R) = b$ and x_L and x_R are the left and right boundaries (Γ) of the domain Ω; only if $a = b = 0$ will the Dirichlet boundary conditions be homogeneous.

Introducing the basis function expansion

$$q_N^{(e)}(x) = \sum_{j=0}^{N} \psi_j(x) q_j^{(e)}$$

and

$$f_N^{(e)}(x) = \sum_{j=0}^{N} \psi_j(x) f_j^{(e)}$$

into Eq. (8.26) yields

$$\left[\psi_i \frac{dq_N^{(e)}}{dx} \right]_{\Gamma_e} - \int_{\Omega_e} \frac{d\psi_i(x)}{dx} \frac{d\psi_j(x)}{dx} \, d\Omega_e \, q_j^{(e)} = \int_{\Omega_e} \psi_i(x) \psi_j(x) \, d\Omega_e \, f_j^{(e)} \tag{8.28}$$

where we have left the Neumann boundary condition terms without expanding. We discuss this term later.

Introducing the mapping from the physical element Ω_e to the computational element $\widehat{\Omega}$ and invoking numerical integration gives

$$\left[\psi_i \frac{dq_N^{(e)}}{dx} \right]_{\Gamma_e} - \sum_{k=0}^{Q} w_k \left(\frac{d\psi_i(\xi_k)}{d\xi} \frac{d\xi_k}{dx} \right) \left(\frac{d\psi_j(\xi_k)}{d\xi} \frac{d\xi_k}{dx} \right) \frac{dx}{d\xi} q_j^{(e)}$$

$$= \sum_{k=0}^{Q} w_k \psi_i(\xi_k) \psi_j(\xi_k) \frac{dx}{d\xi} f_j^{(e)}. \tag{8.29}$$

Introducing the metric and Jacobian terms

$$\frac{d\xi}{dx} = \frac{2}{\Delta x} \quad \text{and} \quad \frac{dx}{d\xi} = \frac{\Delta x}{2}$$

allows us to simplify Eq. (8.29) as follows

$$\left[\psi_i \frac{dq_N^{(e)}}{dx} \right]_{\Gamma_e} - \frac{2}{\Delta x^{(e)}} \sum_{k=0}^{Q} w_k \frac{d\psi_i(\xi_k)}{d\xi} \frac{d\psi_j(\xi_k)}{d\xi} q_j^{(e)} = \frac{\Delta x^{(e)}}{2} \sum_{k=0}^{Q} w_k \psi_i(\xi_k) \psi_j(\xi_k) f_j^{(e)}.$$

$$(8.30)$$

Equation (8.30) has the following corresponding element matrix form

$$B_i^{(e)} - L_{ij}^{(e)} q_j^{(e)} = M_{ij}^{(e)} f_j^{(e)} \tag{8.31}$$

where

$$B_i^{(e)} = \left[\psi_i \frac{dq_N^{(e)}}{dx} \right]_{\Gamma_e},$$

$$L_{ij}^{(e)} = \frac{2}{\Delta x^{(e)}} \sum_{k=0}^{Q} w_k \frac{d\psi_i(\xi_k)}{d\xi} \frac{d\psi_j(\xi_k)}{d\xi}, \tag{8.32}$$

and

$$M_{ij}^{(e)} = \frac{\Delta x^{(e)}}{2} \sum_{k=0}^{Q} w_k \psi_i(\xi_k) \psi_j(\xi_k).$$

Applying the *direct stiffness summation* (DSS) operator yields the following global matrix problem

$$B_I - L_{IJ} \, q_J = M_{IJ} \, f_J \tag{8.33}$$

where q_J, f_J are the values of q and f at the global gridpoints $J = 1, ..., N_p$ and

$$M_{IJ} = \bigwedge_{e=1}^{N_e} M_{ij}^{(e)}, \qquad L_{IJ} = \bigwedge_{e=1}^{N_e} L_{ij}^{(e)}, \qquad B_I = \bigwedge_{e=1}^{N_e} B_i^{(e)},$$

where the DSS operator performs the summation via the mapping $(i, e) \rightarrow (I)$ where $(i, j) = 0, ..., N$, $e = 1, ..., N_e$, $(I, J) = 1, ..., N_p$, and N_e, N_p are the total number of elements and gridpoints.

Algorithm 8.1 highlights the steps required to build a solution to an elliptic equation. As in Algorithm 5.5 for the 1D wave equation, we first build the element matrices. In this case they are the mass and Laplacian matrices as well as the boundary vector. These element matrices are then used to construct the global matrices via the DSS operator given by Algorithm 5.4. At this point, we build the right-hand side vector R_I. Since we require Dirichlet boundary conditions to make this problem well-posed[2] we now modify the global matrix L and the global RHS vector $R - B$ such that the solution satisfies the boundary condition. For example, assume that we wish to impose homogeneous Dirichlet boundary conditions on the right boundary.

---

[2] Without Dirichlet boundary conditions, the solution is non-unique.

In this case, we modify $L(N_p, 1 : N_p - 1) = R(N_p) - B(N_p) = 0$ and $L(N_p, N_p) = 1$ which then enforces $q(N_p) = 0$. Once the boundary conditions have been imposed on the system, we can then solve the linear algebra problem $Ax = b$ where in this case we have $A = -L$, $x = q$ and $b = R - B$. Note that for general Dirichlet boundary conditions we set $R(N_p) - B(N_p) = g(x_R) = b$. In Ch. 14 we cover a different approach for imposing Dirichlet boundary conditions for elliptic problems using a unified CG/DG method.

Algorithm 8.1 CG solution algorithm for the 1D elliptic equation.

---

function CG(q)

    Construct $M_{ij}^{(e)}$, $L_{ij}^{(e)}$, and $B_i^{(e)}$ ▷ use Algs. 5.2 and 8.2

    Construct M_{IJ}, L_{IJ}, and B_I via DSS   ▷ use Algs. 5.4 and 5.7

    Construct $R_I = M_{IJ} f_J$

    Modify L_{IJ} and $R_I - B_I$ to satisfy Dirichlet boundary conditions

    Solve $-L_{IJ} q_J = R_I - B_I$ ▷ $Ax = b$ linear system

end function

Algorithm 8.2 Construction of the 1D element Laplacian matrix.

---

function ELEMENT_LAPLACIAN_MATRIX

    $L^{(e)} = \mathbf{0}$

    for $e = 1 : N_e$ **do**   ▷ loop over elements

        for $k = 0 : Q$ **do**   ▷ loop over integration points

            for $j = 0 : N$ **do**   ▷ loop over columns

                for $i = 0 : N$ **do**   ▷ loop over rows

$$L_{ij}^{(e)} \mathrel{+}= \frac{2}{\Delta x^{(e)}} w_k \frac{d\psi_{ik}}{d\xi} \frac{d\psi_{jk}}{d\xi}$$

                end for

            end for

        end for

    end for

end function

A Comment on the Boundary Vector B

Although we have constructed the boundary vector B_I by first constructing the element vector and then applying DSS, it is better to think about constructing B_I directly since we know it only affects the end points of the domain. Since there are no Neumann boundary conditions imposed between the interfaces of the elements (interior faces) the contribution at interior faces must vanish. The boundary term in element form is written as

$$B_i^{(e)} = \left[\psi_i \frac{dq_N^{(e)}}{dx} \right]_{\Gamma_e}$$

which, in fact, means that we need to evaluate the functions at the end points as
follows

$$
B_i^{(e)} = \begin{pmatrix} \psi_0(\xi)\frac{dq_N^{(e)}}{d\xi} \\ \psi_1(\xi)\frac{dq_N^{(e)}}{d\xi} \\ \vdots \\ \psi_N(\xi)\frac{dq_N^{(e)}}{d\xi} \end{pmatrix}_{-1}^{+1}
$$

and since ψ is cardinal then all but the end point terms disappear to yield

$$
B_i^{(e)} = \begin{pmatrix} -\frac{dq_N^{(e)}}{d\xi}(\xi = -1) \\ 0 \\ \vdots \\ 0 \\ +\frac{dq_N^{(e)}}{d\xi}(\xi = +1) \end{pmatrix}.
$$

Exercise Show why the terms in the vector $B_i^{(e)}$ vanish everywhere except at the end
points.   □

However, if we apply DSS to this element vector, we will not get the interior
face contributions to vanish because $q \in C^0$ only and so the derivatives are not
continuous. Therefore, if we wish to include the boundary vector in the element
matrix form then we need to follow one of two choices.

The first choice is to recognize that the boundary term $\frac{dq}{dx} = h(x)$ is something
known that we are imposing on the system. Therefore we can replace $\frac{dq_N}{dx}$ above by
the function $h(x)$ which may or may not be non-zero at the end points but it *will* be
zero in the interior. This leads to

$$
B_i^{(1)} = \begin{pmatrix} -h(x_L) \\ 0 \\ \vdots \\ 0 \end{pmatrix},
$$

$$
B_i^{(e)} = \begin{pmatrix} 0 \\ \vdots \\ 0 \end{pmatrix} \quad \forall e = 2, ..., N_e - 1,
$$

and

$$
B_i^{(N_e)} = \begin{pmatrix} 0 \\ \vdots \\ 0 \\ h(x_R) \end{pmatrix},
$$

where we assume that the elements are numbered from left to right and so $e = 1$ is the far left element (with boundary x_L) and $e = N_e$ is the far right element (with boundary x_R). With this notation, we can now apply DSS correctly for this vector.

The second choice is to recognize that since $\frac{dq}{dx} = h(x)$ at Neumann boundaries then we replace $\frac{dq_N}{dx} = \sum_{j=0}^{N} \psi_j h_j$ and rewrite the boundary term as follows

$$\left[\psi_i \frac{dq_N}{dx} \right] = [\psi_i \psi_j] h_j.$$

We may now rewrite the boundary element vector as follows

$$B_i^{(e)} = F_{ij}^{(e)} h_j(x)$$

where F is the exact same flux matrix given in Chs. 6 and 7 which we rewrite for convenience here

$$F_{ij}^{(e)} = \begin{pmatrix} -1 & 0 & \cdots & 0 & 0 \\ 0 & 0 & \cdots & 0 & 0 \\ \vdots & \vdots & \cdots & \vdots & \vdots \\ 0 & 0 & \cdots & 0 & 0 \\ 0 & 0 & \cdots & 0 & 1 \end{pmatrix}.$$

Now we can rewrite the element problem as follows

$$F_{ij}^{(e)} h_j(x) - L_{ij}^{(e)} q_j^{(e)} = M_{ij}^{(e)} f_j^{(e)} \tag{8.34}$$

with the corresponding global statement

$$F_{IJ} h_J - L_{IJ} q_J = M_{IJ} f_J \tag{8.35}$$

where

$$F_{IJ} = \bigwedge_{e=1}^{N_e} F_{ij}^{(e)}$$

and h_J only needs to be defined at h_1 (left boundary) and h_{N_p} (right boundary) for $J = 1, ..., N_p$.

8.7 Analysis of the Matrix Properties of the Spatial Operators

Let us now analyze the sparsity pattern of both the mass M and Laplacian matrix L. In Sec. 5.9.1 we showed the sparsity pattern of the mass matrix but for periodic boundary conditions; which we do not use below. In addition, we also look at the eigenvalues of the Laplacian matrix because this will tell us if we should expect to be able to obtain a stable numerical solution; for example, is the matrix L well-conditioned or singular. For the analysis given below, we assume homogeneous Dirichlet boundary conditions.

8.7.1 Sparsity Pattern of the Mass Matrix

Figure 8.2 shows the sparsity pattern for the mass matrix M using both exact and inexact integration for four elements ($N_e = 4$) each using 4th order polynomials ($N = 4$) based on Lobatto points. Figure 8.2 shows that the mass matrix is not full

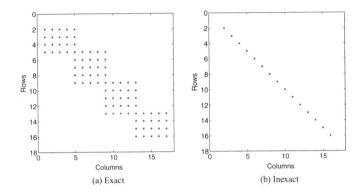

(a) Exact   (b) Inexact

Fig. 8.2: Sparsity patterns for the mass matrix M with $N_e = 4$, $N = 4$ for (a) exact and (b) inexact integration for CG.

but rather sparse; one can see the overlap highlighting the interfaces of the four elements. This figure shows that the mass matrix is tightly banded, i.e., it has a small bandwidth[3] and that the matrix becomes diagonal when inexact integration is used.

8.7.2 Sparsity Pattern of the Laplacian Matrix

Figure 8.3 shows the sparsity pattern for the Laplacian matrix L for exact and inexact integration. Similarly to the mass matrix, Fig. 8.3 shows that the Laplacian matrix is not completely full, i.e., it has a small bandwidth (also of dimension $2N + 1$). For the Laplacian matrix, there is no difference between exact and inexact integration since for $2N - 2$ polynomials, N Lobatto points evaluate the integrals exactly.

---

[3] Recall that in general the bandwidth of the mass matrix is $2N + 1$ which, for $N = 4$ yields a bandwidth of 9. This can be clearly seen, e.g., at row 5 which represents an interface between elements 1 and 2.

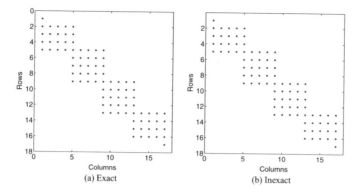

Fig. 8.3: Sparsity patterns for the Laplacian matrix L with $N_e = 4$, $N = 4$ for (a) exact and (b) inexact integration for CG.

8.7.3 Eigenvalue Analysis of the Laplacian Operator

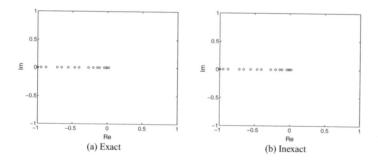

Fig. 8.4: Eigenvalues of the Laplacian matrix L with $N_e = 4$, $N = 4$ for (a) exact and (b) inexact integration for CG.

Figure 8.4 shows that the eigenvalues of the Laplacian are all on the real axis and, although we have normalized the values, we can state that the condition number of this matrix is rather small (i.e., it is well-conditioned)[4]. This is very good news because in order to construct the numerical solution requires the inversion of L;

---

[4] Matrices representing elliptic operators are symmetric positive-definite, meaning that the eigenvalues are all real and greater than zero. In Fig. 8.4 we are actually showing the eigenvalues of the operator ∇^2 and not $-\nabla^2$.

in practice we do not construct the inverse but it is important to show that the inverse does indeed exist and is finite (i.e., the determinant of L must be non-zero). One could solve the resulting linear algebra matrix problem using either direct or iterative methods but the success of both of these methods depends on the matrix L being well-conditioned. Thankfully, we can see that this matrix indeed is well-conditioned. In fact, a good method for solving this matrix problem is the family of Krylov methods; examples of this family of methods include the conjugate gradient method and the Generalized Minimal Residual (GMRES) (see [326] and [380] for details).

8.8 Example of 1D Poisson Equation Problem

Suppose we wish to solve the continuous second order differential equation

$$\frac{d^2 q}{dx^2} = f(x) \qquad \forall x \in [-1, 1]$$

where the exact solution is given by

$$q(x) = \sin(c\pi x)$$

and, therefore, the right-hand side forcing function is

$$f(x) = -(c\pi)^2 \sin(c\pi x)$$

where c is any constant. For integer values of c, the Dirichlet boundary conditions are homogeneous. For all values of c the Dirichlet boundary conditions are defined as

$$g(-1) = \sin(-c\pi) \qquad \text{and} \qquad g(+1) = \sin(c\pi)$$

where $g(x) = q|_{\Gamma_\mathcal{D}}$. In what follows we use $c = 2$.

8.8.1 Error Norm

We use the integral form of the L^2 norm given by Eq. (5.34).

8.8.2 Solution Accuracy

Figure 8.5 shows the exact and CG solutions using $N = 16$ order polynomials and $N_e = 4$ elements for a total of $N_p = 65$ gridpoints. Figure 8.5 shows that there is no difference between the exact and numerical solutions. Let us now see how well our

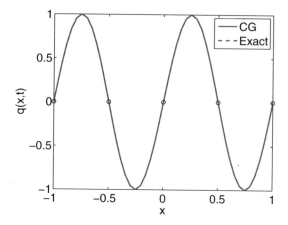

Fig. 8.5: Exact and CG numerical solution for the 1D Poisson equation using $N = 16$ order polynomials with $N_e = 4$ elements with exact integration.

model behaves by varying the number of elements and the polynomial order of the basis functions.

Figure 8.6 shows the convergence rates for various polynomial orders N (based on Lobatto points), for a total number of gridpoints N_p where, for 1D, $N_p = N_e N + 1$. Figure 8.6 shows that there is little difference between exact ($Q = N + 1$) and inexact

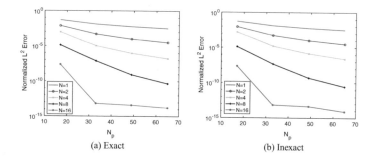

(a) Exact   (b) Inexact

Fig. 8.6: Convergence rates of CG for the 1D Poisson equation for polynomial orders $N = 1, 2, 4, 8, 16$ using a total number of gridpoints N_p for (a) exact integration and (b) inexact integration.

($Q = N$) integration at least in the rates of convergence. The main differences occur for low values of N (≤ 4), however, as N increases the differences between exact and inexact integration all but disappear. Furthermore, the accuracy of the CG method, especially for the high-order polynomials, approaches machine double precision (1×10^{-16}) rather quickly.

Chapter 9
1D Discontinuous Galerkin Methods for Elliptic Equations

9.1 Introduction

In Ch. 8 we saw that the CG method can be applied in a relatively straightforward manner to solve elliptic equations. All that was required was to know how to compute first derivatives. A judicious use of Green's first identity then permits a simple discretization of the Laplacian operator. We learn in this chapter that DG cannot use the same representation of the Laplacian operator. Rather, we need to revisit first order derivatives and construct an auxiliary variable and extend this approach to approximate the Laplacian operator (using the so-called flux formulation) or we can use the original Laplacian operator (using the so-called primal formulation) but with the addition of penalty terms to stabilize the numerical solution. In Sec. 9.2 we introduce the elliptic operator that we solve in this chapter and discuss it in terms of the CG formulation that was described in Ch. 8. The literature on DG methods for elliptic equations is quite extensive. In Sec. 9.3 we introduce some of the key works from the literature on tackling elliptic operators with DG and choose some powerful, yet simple, methods that we have found to represent a good balance between accuracy and simplicity. Because we focus on the flux formulation of the DG method, we discuss the construction of first derivatives with DG in Sec. 9.4. In Sec. 9.5 we describe one possibility for constructing a consistent DG discretization for a second order operator using the flux formulation. We then analyze the properties of the resulting DG spatial operators for approximating second order spatial operators in Sec. 9.6. We end the chapter with an example of the 1D Poisson equation using the DG method in Sec. 9.7.

© The Editor(s) (if applicable) and The Author(s), under exclusive license to Springer Nature Switzerland AG 2020
F. X. Giraldo, *An Introduction to Element-Based Galerkin Methods on Tensor-Product Bases*, Texts in Computational Science and Engineering 24, https://doi.org/10.1007/978-3-030-55069-1_9

9.2 Elliptic Equations

In Ch. 8 we saw that a multi-dimensional elliptic equation is defined as follows

$$\mathbf{\nabla}^2 q(\mathbf{x}) = f(\mathbf{x}) \tag{9.1}$$

where $\mathbf{x} \in \Omega$ with the Dirichlet boundary condition

$$q|_{\Gamma_{\mathcal{D}}} = g(\mathbf{x})$$

and Neumann boundary condition

$$(\mathbf{n} \cdot \mathbf{\nabla} q)_{\Gamma_N} = h(\mathbf{x})$$

where $\Gamma = \Gamma_{\mathcal{D}} \bigcup \Gamma_N$ is the boundary of Ω. This operator represents a boundary value problem and, in fact, can be solved using the DG method by virtue of solving a linear matrix problem.

We saw in Ch. 8 that for CG, the linear matrix problem is defined as follows

$$B_I - L_{IJ}\, q_J = M_{IJ} f_J$$

where q_J, f_J are the values of q and f at the global gridpoints $J = 1, ..., N_P$. With the appropriate Dirichlet boundary conditions, the numerical solution of the Poisson problem is well-posed and hence solvable. In the next few sections, we construct the DG solution to this elliptic equation.

9.3 Discontinuous Galerkin Method

The numerical solution of elliptic partial differential equations (PDEs) has a long history and, to this day, remains a rich field of study. In Ch. 8 we discussed the classical approach for solving the PDE

$$\mathbf{\nabla}^2 q(x) = f(x)$$

using continuous methods, with appropriate boundary conditions. To conform to the literature on elliptic operators, let us rewrite it as follows

$$-\mathbf{\nabla}^2 q(x) = f(x).$$

In a classical continuous Galerkin approach, we choose $\psi \in S$, where S is a predefined vector space, and multiply by a test function and integrate to arrive at the form

$$\int_{\Omega} \mathbf{\nabla}\psi_i \cdot \mathbf{\nabla} q_N \, d\Omega - \int_{\Gamma} \psi_i \left(\hat{\mathbf{n}} \cdot \mathbf{\nabla} q_N\right) d\Gamma = \int_{\Omega} \psi_i f_N \, d\Omega$$

where $f = f(x)$ and $q = q(x)$. In 1968 Lions [12] sought to satisfy the solution to this equation for the case when Dirichlet boundary conditions need to be satisfied weakly. Lions wrote the discretized equations as follows

$$\int_{\Omega} \nabla \psi_i \cdot \nabla q_N \, d\Omega - \int_{\Gamma} \psi_i \, (\hat{\mathbf{n}} \cdot \nabla q_N) \, d\Gamma + \int_{\Gamma} \mu \psi_i \, (q_N - g) \, d\Gamma = \int_{\Omega} \psi_i f_N \, d\Omega$$

where the Dirichlet boundary condition imposed is $q(x)|_{\Gamma_D} = g(x)$, and μ is a penalty term that enforces the solution to satisfy the boundary condition. However, this approach suffers from an inconsistency since the exact solution does not satisfy the above PDE. Nitsche [12] later proposed the following modification (in red font)

$$\int_{\Omega} \nabla \psi_i \cdot \nabla q_N \, d\Omega - \int_{\Gamma} \psi_i \, (\hat{\mathbf{n}} \cdot \nabla q_N) \, d\Gamma - \int_{\Gamma} q_N \, (\hat{\mathbf{n}} \cdot \nabla \psi_i) \, d\Gamma$$

$$+ \int_{\Gamma} \mu \psi_i \, (q_N - g) \, d\Gamma = \int_{\Omega} \psi_i f_N \, d\Omega$$

that now makes the numerical discretization consistent. We now use this form to write general element-based interior penalty methods [110, 404] for solving elliptic PDEs

$$\int_{\Omega_e} \nabla \psi_i \cdot \nabla q_N \, d\Omega_e - \int_{\Gamma_e} \psi_i \, (\hat{\mathbf{n}} \cdot \nabla \hat{q}_N) \, d\Gamma_e - \tau \int_{\Gamma_e} q_N \, (\hat{\mathbf{n}} \cdot \nabla \hat{\psi}_i) \, d\Gamma_e$$

$$+ \int_{\Gamma_e} \mu \psi_i q_N \, d\Gamma_e = \int_{\Omega_e} \psi_i f_N \, d\Omega_e \tag{9.2}$$

where, for simplicity, we assume homogeneous Dirichlet boundary conditions (i.e., $g = 0$), introduce a parameter $\tau = \{-1, 0, 1\}$, and use $\hat{q}$ and $\hat{\psi}$ to represent the numerical flux functions used for both the solution variable q and the basis functions ψ. We can now derive various DG methods for elliptic equations based on specific values of τ. For example, $\tau = 1$ yields the *Symmetric Interior Penalty Galerkin* (SIPG) method [357], whereas $\tau = -1$ yields the *Non-Symmetric Interior Penalty Galerkin* (NIPG) method, and $\tau = 0$ yields the *Incomplete Interior Penalty Galerkin* (IIPG) method. We describe the SIPG method in detail in Ch. 14.

Although the *primal formulation* used in Eq. (9.2) is useful for discussing various DG methods, let us begin with the *flux formulation*

$$\mathbf{Q} = \nabla q \text{ and } -\nabla \cdot \mathbf{Q} = f$$

where the strategy is to recognize that the Laplacian operator is defined as $\Delta q \equiv \nabla^2 q = \nabla \cdot \nabla q$. Therefore, we recognize that a second order operator is decomposed into two first order operators and then we are able to apply standard DG methods such as those discussed in Ch. 6 but we must choose the numerical flux function differently.

Since most DG methods for elliptic operators rely on the decomposition of the second order operator to a system of first order operators, let us first look at the approximation of first derivatives with DG.

9.4 First Derivatives in Weak Form

When we discuss the discrete approximation of continuous operators with DG we need to discuss the weak form since we need integration by parts in order to introduce the flux terms. We could apply integration by parts twice to derive a strong form derivative with flux terms but we reserve this approach to Ch. 13. In what follows, we only use weak form derivatives.

Beginning with the equality

$$\frac{dq^{(e)}(x)}{dx} = \frac{dq^{(e)}(x)}{dx}$$

we then write the integral problem statement

$$\int_{x_0}^{x_N} \psi_i(x) \frac{dq_N^{(e)}(x)}{dx} \, dx = \int_{x_0}^{x_N} \psi_i(x) \frac{dq_N^{(e)}(x)}{dx} \, dx \tag{9.3}$$

where $i = 1, \dots, N$ and $e = 1, \dots, N_e$. Using the product rule

$$\frac{d}{dx}\left(\psi_i(x) q_N^{(e)}(x)\right) = \frac{d\psi_i(x)}{dx} q_N^{(e)}(x) + \psi_i(x) \frac{dq_N^{(e)}(x)}{dx}$$

we can write Eq. (9.3) as follows

$$\int_{x_0}^{x_N} \psi_i(x) \frac{dq_N^{(e)}(x)}{dx} \, dx = \int_{x_0}^{x_N} \frac{d}{dx}\left(\psi_i(x) q_N^{(e)}(x)\right) \, dx - \int_{x_0}^{x_N} \frac{d\psi_i(x)}{dx} q_N^{(e)}(x) \, dx. \tag{9.4}$$

Let us now introduce the basis function expansions

$$q_N^{(e)}(x) = \sum_{j=0}^{N} \psi_j(x) q_j^{(e)}$$

and

$$\frac{dq_N^{(e)}}{dx}(x) = \sum_{j=0}^{N} \psi_j(x) q_{x,j}^{(e)}$$

where $q_{x,j}^{(e)} = \frac{dq^{(e)}}{dx}(x_j)$ is the derivative that we seek. Substituting these expansions into Eq. (9.4) and mapping to the reference element yields

$$\int_{-1}^{+1} \psi_i(\xi) \psi_j(\xi) \frac{dx}{d\xi} \, d\xi \, q_{x,j}^{(e)} = \int_{-1}^{+1} \frac{d}{d\xi}\left(\psi_i(\xi)\psi_j(\xi)\right) \frac{d\xi(x)}{dx} \frac{dx}{d\xi} \, d\xi \, q_j^{(e)}$$

$$- \int_{-1}^{+1} \frac{d\psi_i(\xi)}{d\xi} \frac{d\xi(x)}{dx} \psi_j(\xi) \frac{dx}{d\xi} \, d\xi \, q_j^{(e)} \tag{9.5}$$

with $i, j = 0, \ldots, N$ and $e = 1, \ldots, N_e$. Simplifying yields,

$$M_{ij}^{(e)} q_{x,j}^{(e)} = F_{ij}^{(e)} q_j^{(*,e)} - \tilde{D}_{ij}^{(e)} q_j^{(e)} \tag{9.6}$$

where

$$M_{ij}^{(e)} = \frac{\Delta x^{(e)}}{2} \int_{-1}^{+1} \psi_i(\xi)\psi_j(\xi) \, d\xi$$

is the mass matrix,

$$\tilde{D}_{ij}^{(e)} = \int_{-1}^{+1} \frac{d\psi_i(\xi)}{d\xi} \psi_j(\xi) \, d\xi$$

is the *weak form* differentiation matrix, and

$$F_{ij}^{(e)} = \int_{-1}^{+1} \frac{d}{d\xi} \left(\psi_i(\xi)\psi_j(\xi) \right) \, d\xi \equiv \left[\psi_i(\xi)\psi_j(\xi) \right]_{-1}^{+1}$$

is the flux matrix, where we have replaced $q^{(e)}$ in the flux term by $q^{(*,e)}$ which we define shortly.

We already know from Ch. 6 that for $N = 1$ the mass matrix is

$$M_{ij}^{(e)} = \frac{\Delta x^{(e)}}{6} \begin{pmatrix} 2 & 1 \\ 1 & 2 \end{pmatrix}$$

the weak form differentiation matrix is

$$\tilde{D}_{ij}^{(e)} = \frac{1}{2} \begin{pmatrix} -1 & -1 \\ 1 & 1 \end{pmatrix},$$

and the flux matrix is

$$F_{ij}^{(e)} = \begin{pmatrix} -1 & 0 \\ 0 & 1 \end{pmatrix},$$

where we have not yet defined $q^{(*,e)}$.

9.4.1 Resulting Element Equations

For linear elements ($N = 1$), substituting the mass, differentiation, and flux matrices into Eq. (9.6) yields the following element equations defining the weak form first derivative in the DG method

$$\frac{\Delta x^{(e)}}{6} \begin{pmatrix} 2 & 1 \\ 1 & 2 \end{pmatrix} \begin{pmatrix} q_{x,0}^{(e)} \\ q_{x,1}^{(e)} \end{pmatrix} = \begin{pmatrix} -1 & 0 \\ 0 & 1 \end{pmatrix} \begin{pmatrix} q_0^{(*,e)} \\ q_1^{(*,e)} \end{pmatrix} - \frac{1}{2} \begin{pmatrix} -1 & -1 \\ 1 & 1 \end{pmatrix} \begin{pmatrix} q_0^{(e)} \\ q_1^{(e)} \end{pmatrix}. \tag{9.7}$$

Exercise Derive the resulting element equations similar to Eq. (9.7) but for quadratic (N=2) polynomials.   □

Diagonalizing the mass matrix in Eq. (9.7) results in the following element equations

$$\frac{\Delta x^{(e)}}{2} \begin{pmatrix} 1 & 0 \\ 0 & 1 \end{pmatrix} \begin{pmatrix} q_{x,0}^{(e)} \\ q_{x,1}^{(e)} \end{pmatrix} = \begin{pmatrix} -1 & 0 \\ 0 & 1 \end{pmatrix} \begin{pmatrix} q_{0}^{(*,e)} \\ q_{1}^{(*,e)} \end{pmatrix} - \frac{1}{2} \begin{pmatrix} -1 & -1 \\ 1 & 1 \end{pmatrix} \begin{pmatrix} q_{0}^{(e)} \\ q_{1}^{(e)} \end{pmatrix}. \tag{9.8}$$

Recall that diagonalizing the mass matrix as such means that we are using $Q = N$ integration formulas. Although this is a perfectly reasonable quadrature choice, we make this choice here to simplify the discussion.

9.4.2 Element Derivative at the Gridpoint I

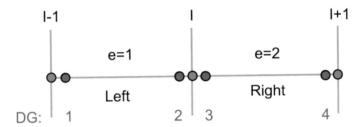

Fig. 9.1: Contribution to the gridpoint I from the left $(I - 1, I)$ and right $(I, I + 1)$ elements.

Figure 9.1 shows that the left and right elements contribute a portion of their solution to the gridpoint I; however, since the solution is allowed to be discontinuous, there exist two values at the gridpoint I which we denote as $(q_I^{(L)}, q_I^{(R)}) = (q_2, q_3)$.

Left Element

From Eq. (9.7) we write the left element equation as follows

$$\frac{\Delta x^{(L)}}{6} \begin{pmatrix} 2 & 1 \\ 1 & 2 \end{pmatrix} \begin{pmatrix} q_{x,I-1}^{(L)} \\ q_{x,I}^{(L)} \end{pmatrix} = \begin{pmatrix} -1 & 0 \\ 0 & 1 \end{pmatrix} \begin{pmatrix} q_{I-1}^{(*,L)} \\ q_{I}^{(*,L)} \end{pmatrix} - \frac{1}{2} \begin{pmatrix} -1 & -1 \\ 1 & 1 \end{pmatrix} \begin{pmatrix} q_{I-1}^{(L)} \\ q_{I}^{(L)} \end{pmatrix}$$

where the first row denotes the derivative at the gridpoint $I - 1$ and the second row the derivative at the gridpoint I.

Using the diagonal form of the mass matrix results in

$$\frac{\Delta x^{(L)}}{2} q_{x,I}^{(L)} = q_I^{(*,L)} - \frac{1}{2} \left(q_{I-1}^{(L)} + q_I^{(L)} \right) \tag{9.9}$$

for the derivative at the gridpoint I for the left element.

Right Element

Similarly, from Eq. (9.7) we can write the right element equation as follows

$$\frac{\Delta x^{(R)}}{6}\begin{pmatrix}2 & 1\\1 & 2\end{pmatrix}\begin{pmatrix}q_{x,I}^{(R)}\\q_{x,I+1}^{(R)}\end{pmatrix}=\begin{pmatrix}-1 & 0\\0 & 1\end{pmatrix}\begin{pmatrix}q_I^{(*,R)}\\q_{I+1}^{(*,R)}\end{pmatrix}-\frac{1}{2}\begin{pmatrix}-1 & -1\\1 & 1\end{pmatrix}\begin{pmatrix}q_I^{(R)}\\q_{I+1}^{(R)}\end{pmatrix}$$

where the first row gives the derivative at the gridpoint I and the second row the derivative at the gridpoint $I+1$.

Using the diagonal form of the mass matrix yields the derivative of the right element at the gridpoint I as

$$\frac{\Delta x^{(R)}}{2}q_{x,I}^{(R)}=-q_I^{(*,R)}-\frac{1}{2}\left(-q_I^{(R)}-q_{I+1}^{(R)}\right).\tag{9.10}$$

Numerical Flux Function

Let us now define the numerical flux function $q^{(*,e)}$. Recall that this numerical flux function is constructed as a measure of the preferred propagation direction of the continuous partial differential equation. In the case of a hyperbolic system, one obtains the eigenvalues of the Jacobian matrix as we saw in Ch. 6. However, since in this section we are interested in elliptic equations then it makes sense to use the average value at the interface between two elements[1]. In other words, for the function at the gridpoint I we define the following numerical flux

$$q_I^{(*,e)}=\frac{1}{2}\left(q_I^{(e)}+q_I^{(k)}\right)\tag{9.11}$$

where the superscript (k) denotes the contiguous neighbor of (e) that shares the gridpoint I. It should be understood, however, that this is not the only possible numerical flux. There is a plethora of numerical fluxes available, that when used in the flux formulation of second order operators, results in a large family of DG methods for elliptic equations. The interested reader is referred to, e.g., the work of Arnold et al. [12] and the books by Riviere [318] and DiPietro and Ern [108] for other possibilities.

Resulting First Derivative at the Gridpoint I

Using Eq. (9.11) for the numerical flux for the interface between the left and right elements at the gridpoint I yields

$$q_I^{(*,L)}=\frac{1}{2}\left(q_I^{(L)}+q_I^{(R)}\right).\tag{9.12}$$

---

[1] The average value makes sense because, here, there is no preferred direction as in a hyperbolic system.

Substituting Eq. (9.12) for the numerical flux into the left element equation given in Eq. (9.9) yields

$$\frac{\Delta x^{(L)}}{2} q_{x,I}^{(L)} = \frac{1}{2}\left(q_I^{(L)} + q_I^{(R)}\right) - \frac{1}{2}\left(q_{I-1}^{(L)} + q_I^{(L)}\right)$$

that, after simplifying results in

$$q_{x,I}^{(L)} = \frac{q_I^{(R)} - q_{I-1}^{(L)}}{\Delta x^{(L)}}.$$

To see what this statement means, let us assume that $q \in C^0$ which results in $q_I^{(R)} = q_I^{(L)}$ and yields

$$q_{x,I}^{(L)} = \frac{q_I^{(L)} - q_{I-1}^{(L)}}{\Delta x^{(L)}}. \qquad (9.13)$$

For the right element we get

$$q_{x,I}^{(R)} = \frac{q_{I+1}^{(R)} - q_I^{(R)}}{\Delta x^{(R)}}. \qquad (9.14)$$

Equations (9.13) and (9.14) are first order approximations to the first derivative at two different elements surrounding the interface gridpoint I. This is true only when $q_I^{(R)} = q_I^{(L)}$ which, in general, will not be the case. Generally speaking, $q_I^{(R)}$ and $q_I^{(L)}$ are two distinct solutions at different gridpoints.

Exercise Show that the derivatives in Eqs. (9.13) and (9.14) represent first order approximations. ☐

9.5 Second Derivatives

Recall that the DG method was developed with first order equations in mind (see [310]). On that note, it makes sense to write the second order operator as a coupled system of two first order operators, the so-called flux formulation [22]. That is, let us replace the second order operator

$$\frac{d^2 q}{dx^2}(x)$$

with the first order system

$$\frac{dq}{dx}(x) = Q(x)$$

$$\frac{d^2q}{dx^2}(x) = \frac{dQ}{dx}(x).$$

With this system in mind, we now apply the DG machinery to the first order operator in order to determine the auxiliary variable Q that is then used to construct the second order operator $\frac{d^2q}{dx^2}$.

9.5.1 First Derivative: Auxiliary Variable

From the differential equation

$$Q(x) = \frac{dq}{dx}(x)$$

we can construct an approximation to the auxiliary variable Q via the integral form

$$\int_{x_0}^{x_1} \psi_i(x) Q_N^{(e)}(x)\, dx = \int_{x_0}^{x_1} \psi_i(x) \frac{dq_N^{(e)}(x)}{dx}\, dx$$

and applying integration by parts to the right-hand side gives

$$\int_{x_0}^{x_1} \psi_i(x) Q_N^{(e)}(x)\, dx = \int_{x_0}^{x_1} \frac{d}{dx}\left(\psi_i(x) q_N^{(e)}(x)\right) dx - \int_{x_0}^{x_1} \frac{d\psi_i(x)}{dx} q_N^{(e)}(x)\, dx$$

where $i = 0, \ldots, N$ and $e = 1, \ldots, N_e$. For linear elements we get

$$\frac{d\psi_j}{dx}(x) = \frac{d\psi_j}{d\xi}(\xi)\frac{d\xi}{dx} = \frac{1}{2}\xi_j \frac{2}{\Delta x}.$$

Expanding the solution variables as usual with

$$Q_N^{(e)}(x) = \sum_{i=0}^{N} \psi_i(x) Q_i^{(e)}$$

$$q_N^{(e)}(x) = \sum_{i=0}^{N} \psi_i(x) q_i^{(e)}$$

allows us to write the integrals as follows

$$\int_{x_0}^{x_1} \psi_i(x)\psi_j(x)\, dx\, Q_j^{(e)} = \left[\psi_i(x)\psi_j(x)\right]_{x_0}^{x_1} q_j^{(*,e)} - \int_{x_0}^{x_1} \frac{d\psi_i(x)}{dx}\psi_j(x)\, dx\, q_j^{(e)}$$

where $i, j = 0, \ldots, N$ and $e = 1, \ldots, N_e$. This equation takes the following matrix-vector form

$$M_{ij}^{(e)} Q_j^{(e)} = F_{ij}^{(e)} q_j^{(*,e)} - \tilde{D}_{ij}^{(e)} q_j^{(e)} \tag{9.15}$$

where the mass $M^{(e)}$, weak form differentiation matrix $\tilde{D}^{(e)}$, and flux $F^{(e)}$ matrices have been previously defined.

9.5.2 Resulting Element Equations

Assuming linear elements and substituting the element matrices into Eq. (9.15) results in the following equations for the auxiliary variable Q

$$\frac{\Delta x^{(e)}}{6} \begin{pmatrix} 2 & 1 \\ 1 & 2 \end{pmatrix} \begin{pmatrix} Q_0^{(e)} \\ Q_1^{(e)} \end{pmatrix} = \begin{pmatrix} -1 & 0 \\ 0 & 1 \end{pmatrix} \begin{pmatrix} q_0^{(*,e)} \\ q_1^{(*,e)} \end{pmatrix} - \frac{1}{2} \begin{pmatrix} -1 & -1 \\ 1 & 1 \end{pmatrix} \begin{pmatrix} q_0^{(e)} \\ q_1^{(e)} \end{pmatrix} \tag{9.16}$$

where diagonalizing the mass matrix results in

$$\frac{\Delta x^{(e)}}{2} \begin{pmatrix} 1 & 0 \\ 0 & 1 \end{pmatrix} \begin{pmatrix} Q_0^{(e)} \\ Q_1^{(e)} \end{pmatrix} = \begin{pmatrix} -1 & 0 \\ 0 & 1 \end{pmatrix} \begin{pmatrix} q_0^{(*,e)} \\ q_1^{(*,e)} \end{pmatrix} - \frac{1}{2} \begin{pmatrix} -1 & -1 \\ 1 & 1 \end{pmatrix} \begin{pmatrix} q_0^{(e)} \\ q_1^{(e)} \end{pmatrix} . \tag{9.17}$$

9.5.3 Element Derivative at the Gridpoint I

Let us now write the element equations for both the left and right elements (see Fig. 9.1).

Left Element

From Eq. (9.16), the left element equation is given by

$$\frac{\Delta x^{(L)}}{6} \begin{pmatrix} 2 & 1 \\ 1 & 2 \end{pmatrix} \begin{pmatrix} Q_{I-1}^{(L)} \\ Q_I^{(L)} \end{pmatrix} = \begin{pmatrix} -1 & 0 \\ 0 & 1 \end{pmatrix} \begin{pmatrix} q_{I-1}^{(*,L)} \\ q_I^{(*,L)} \end{pmatrix} - \frac{1}{2} \begin{pmatrix} -1 & -1 \\ 1 & 1 \end{pmatrix} \begin{pmatrix} q_{I-1}^{(L)} \\ q_I^{(L)} \end{pmatrix}$$

where from Eq. (9.17) the derivative at the gridpoint I is given by the second row as follows

$$\frac{\Delta x^{(L)}}{2} \left(Q_I^{(L)} \right) = q_I^{(*,L)} - \frac{1}{2} \left(q_{I-1}^{(L)} + q_I^{(L)} \right)$$

where we have diagonalized the mass matrix for simplicity.

Right Element

From Eq. (9.16), for the right element we have the equation

$$\frac{\Delta x^{(R)}}{6} \begin{pmatrix} 2 & 1 \\ 1 & 2 \end{pmatrix} \begin{pmatrix} Q_I^{(R)} \\ Q_{I+1}^{(R)} \end{pmatrix} = \begin{pmatrix} -1 & 0 \\ 0 & 1 \end{pmatrix} \begin{pmatrix} q_I^{(*,R)} \\ q_{I+1}^{(*,R)} \end{pmatrix} - \frac{1}{2} \begin{pmatrix} -1 & -1 \\ 1 & 1 \end{pmatrix} \begin{pmatrix} q_I^{(R)} \\ q_{I+1}^{(R)} \end{pmatrix}$$

where from Eq. (9.17) the derivative at the gridpoint I is given by the first row as follows

$$\frac{\Delta x^{(R)}}{2} \left(Q_I^{(R)} \right) = -q_I^{(*,R)} - \frac{1}{2} \left(-q_I^{(R)} - q_{I+1}^{(R)} \right).$$

Total Contribution

Let us now write the entire matrix-vector problem for the two-element configuration given in Fig. 9.1. Assuming that $\Delta x = \Delta x^{(L)} = \Delta x^{(R)}$ and using the global gridpoint numbering for DG (red numbers in Fig. 9.1) we write

$$\frac{\Delta x}{6} \begin{pmatrix} 2 & 1 & 0 & 0 \\ 1 & 2 & 0 & 0 \\ 0 & 0 & 2 & 1 \\ 0 & 0 & 1 & 2 \end{pmatrix} \begin{pmatrix} Q_1 \\ Q_2 \\ Q_3 \\ Q_4 \end{pmatrix} = \begin{pmatrix} -1 & 0 & 0 & 0 \\ 0 & 1 & 0 & 0 \\ 0 & 0 & -1 & 0 \\ 0 & 0 & 0 & 1 \end{pmatrix} \begin{pmatrix} q_1^{(*)} \\ q_2^{(*)} \\ q_3^{(*)} \\ q_4^{(*)} \end{pmatrix} - \frac{1}{2} \begin{pmatrix} -1 & -1 & 0 & 0 \\ 1 & 1 & 0 & 0 \\ 0 & 0 & -1 & -1 \\ 0 & 0 & 1 & 1 \end{pmatrix} \begin{pmatrix} q_1 \\ q_2 \\ q_3 \\ q_4 \end{pmatrix} \qquad (9.18)$$

where we now need to introduce the numerical flux values.

9.5.4 Possible Choices for the Numerical Flux

To make this analysis simpler to follow, let us make a specific choice for the numerical flux. Following the *local discontinuous Galerkin* method [86, 72] let us define the numerical flux function to be

$$q^{(*,e,k)} = \alpha q^{(k)} + \beta q^{(e)} \qquad (9.19a)$$

$$Q^{(*,e,k)} = \beta Q^{(k)} + \alpha Q^{(e)} \qquad (9.19b)$$

where the superscript $(*, e, k)$ represents the numerical flux $(*)$ at the interface of the two elements (e, k). However, in what follows we only use either the superscript $(*, e)$ or $(*)$ because we are referring to the interface at a specific gridpoint (I). Regarding the scalars (α, β) the choice $(\alpha, \beta) = (\frac{1}{2}, \frac{1}{2})$ leads to the centered flux of Bassi and Rebay [22], while the choice $\alpha = 1 - \beta$ leads to the flip-flop numerical flux of Cockburn and Shu [86]. To satisfy Dirichlet boundary conditions we require choosing $\beta = 0$ and for satisfying Neumann boundary conditions we require $\beta = 1$. For convenience, below we only consider the centered flux although it should be understood that the flux given in Eq. (9.19) can be used or any of the fluxes given in Arnold et al. [12].

9.5.5 Resulting Auxiliary Variable at the Gridpoint I

Using the centered flux, Eq. (9.18) becomes

$$
\frac{\Delta x}{6}
\begin{pmatrix} 2 & 1 & 0 & 0 \\ 1 & 2 & 0 & 0 \\ 0 & 0 & 2 & 1 \\ 0 & 0 & 1 & 2 \end{pmatrix}
\begin{pmatrix} Q_1 \\ Q_2 \\ Q_3 \\ Q_4 \end{pmatrix}
= \frac{1}{2}
\begin{pmatrix} -1 & 0 & 0 & 0 \\ 0 & 1 & 1 & 0 \\ 0 & -1 & -1 & 0 \\ 0 & 0 & 0 & 1 \end{pmatrix}
\begin{pmatrix} q_1 \\ q_2 \\ q_3 \\ q_4 \end{pmatrix}
+ \frac{1}{2}
\begin{pmatrix} -g_1 \\ 0 \\ 0 \\ g_4 \end{pmatrix}
$$

$$
- \frac{1}{2}
\begin{pmatrix} -1 & -1 & 0 & 0 \\ 1 & 1 & 0 & 0 \\ 0 & 0 & -1 & -1 \\ 0 & 0 & 1 & 1 \end{pmatrix}
\begin{pmatrix} q_1 \\ q_2 \\ q_3 \\ q_4 \end{pmatrix}
\tag{9.20}
$$

where g_i denotes the Dirichlet boundary condition at the gridpoint i. Adding the first and third terms on the right-hand side and diagonalizing the mass matrix yields the following simplified form

$$
\frac{\Delta x}{2}
\begin{pmatrix} Q_1 \\ Q_2 \\ Q_3 \\ Q_4 \end{pmatrix}
= \frac{1}{2}
\begin{pmatrix} 0 & 1 & 0 & 0 \\ -1 & 0 & 1 & 0 \\ 0 & -1 & 0 & 1 \\ 0 & 0 & -1 & 0 \end{pmatrix}
\begin{pmatrix} q_1 \\ q_2 \\ q_3 \\ q_4 \end{pmatrix}
+ \frac{1}{2}
\begin{pmatrix} -g_1 \\ 0 \\ 0 \\ g_4 \end{pmatrix}.
\tag{9.21}
$$

From Eq. (9.21) we see that the Q value at the gridpoint I has two distinct values depending on whether we are considering the left or right elements. The value of $Q_I^{(L)} = Q_2$ is obtained from the second row which yields

$$
Q_2 = \frac{q_3 - q_1}{\Delta x}
$$

and the value of $Q_I^{(R)} = Q_3$ is given by the third row which yields

$$
Q_3 = \frac{q_4 - q_2}{\Delta x}.
$$

To see what these statements mean, let us assume that $q \in C^0$ which results in $q_I^{(R)} = q_I^{(L)}$ and yields

$$
Q_I = \frac{q_I - q_{I-1}}{\Delta x}
$$

from the viewpoint of the left element (Q_2). From the viewpoint of the right element (Q_3) we get

$$
Q_I = \frac{q_{I+1} - q_I}{\Delta x}.
$$

Both of these equations are consistent first order difference approximations to the first order derivative.

Exercise Construct the derivative for the auxiliary variable using the flip-flop numerical flux with $(\alpha, \beta) = (1, 0)$ with linear polynomials (N=1). □

9.5.6 Second Derivative

Now that we have computed the auxiliary variable, $Q_N^{(e)}(x)$, we can then compute the second derivative from the equality

$$\frac{d^2q}{dx^2}(x) = \frac{dQ}{dx}(x)$$

that, in integral form, is written as follows

$$\int_{x_0}^{x_1} \psi_i(x)\frac{d^2 q_N^{(e)}}{dx^2}\, dx = \int_{x_0}^{x_1} \psi_i(x)\frac{dQ_N^{(e)}(x)}{dx}\, dx.$$

Applying integration by parts to the right-hand side gives

$$\int_{x_0}^{x_1} \psi_i(x)\frac{d^2 q_N^{(e)}}{dx^2}\, dx = \int_{x_0}^{x_1} \frac{d}{dx}\left(\psi_i(x)Q_N^{(e)}(x)\right)\, dx - \int_{x_0}^{x_1} \frac{d\psi_i(x)}{dx}Q_N^{(e)}(x)\, dx.$$

Using linear elements results in the following basis function derivative

$$\frac{d\psi_j}{dx}(x) = \frac{d\psi_j}{d\xi}(\xi)\frac{d\xi}{dx} = \frac{1}{2}\xi_j\frac{2}{\Delta x}.$$

Expanding the auxiliary variable as

$$Q_N^{(e)}(x) = \sum_{i=0}^{N} \psi_i(x)Q_i^{(e)}$$

and the second derivative of q as

$$\frac{d^2 q_N^{(e)}}{dx^2}(x) = \sum_{i=0}^{N} \psi_i(x)q_{xx,i}^{(e)}$$

where $q_{xx,i}^{(e)} = \frac{d^2 q^{(e)}}{dx^2}(x_j)$ is the derivative that we seek, allows us to write the integral in the following way

$$\int_{x_0}^{x_1} \psi_i(x)\psi_j(x)\, dx\, q_{xx,j}^{(e)} = \left[\psi_i(x)\psi_j(x)\right]_{x_0}^{x_1} Q_j^{(*,e)} - \int_{x_0}^{x_1} \frac{d\psi_i(x)}{dx}\psi_j(x)\, dx\, Q_j^{(e)}$$

where $i, j = 0, \ldots, N$ and $e = 1, \ldots, N_e$, This expression then simplifies to the matrix-vector form

$$M^{(e)}_{ij}\, q^{(e)}_{xx,j} = F^{(e)}_{ij}\, Q^{(*,e)}_{j} - \tilde{D}^{(e)}_{ij}\, Q^{(e)}_{j} \tag{9.22}$$

which is identical to the expression we obtained for the auxiliary variable Q given by Eq. (9.15).

9.5.7 Resulting Element Equations

The resulting element equations defining the second derivative in the DG method is, in fact, identical to Eq. (9.16) which we now write as follows

$$\frac{\Delta x^{(e)}}{6}\begin{pmatrix} 2 & 1 \\ 1 & 2 \end{pmatrix}\begin{pmatrix} q^{(e)}_{xx,0} \\ q^{(e)}_{xx,1} \end{pmatrix} = \begin{pmatrix} -1 & 0 \\ 0 & 1 \end{pmatrix}\begin{pmatrix} Q^{(*,e)}_0 \\ Q^{(*,e)}_1 \end{pmatrix} - \frac{1}{2}\begin{pmatrix} -1 & -1 \\ 1 & 1 \end{pmatrix}\begin{pmatrix} Q^{(e)}_0 \\ Q^{(e)}_1 \end{pmatrix}. \tag{9.23}$$

9.5.8 Element Second Derivative at the Gridpoint I

We now consider the contributions of the left and right elements to the gridpoint I.

Left Element

From Eq. (9.23), the left element equation is

$$\frac{\Delta x^{(L)}}{6}\begin{pmatrix} 2 & 1 \\ 1 & 2 \end{pmatrix}\begin{pmatrix} q^{(L)}_{xx,I-1} \\ q^{(L)}_{xx,I} \end{pmatrix} = \begin{pmatrix} -1 & 0 \\ 0 & 1 \end{pmatrix}\begin{pmatrix} Q^{(*,L)}_{I-1} \\ Q^{(*,L)}_{I} \end{pmatrix} - \frac{1}{2}\begin{pmatrix} -1 & -1 \\ 1 & 1 \end{pmatrix}\begin{pmatrix} Q^{(L)}_{I-1} \\ Q^{(L)}_{I} \end{pmatrix}.$$

Right Element

From Eq. (9.23), for the right element we have the equation

$$\frac{\Delta x^{(R)}}{6}\begin{pmatrix} 2 & 1 \\ 1 & 2 \end{pmatrix}\begin{pmatrix} q^{(R)}_{xx,I} \\ q^{(R)}_{xx,I+1} \end{pmatrix} = \begin{pmatrix} -1 & 0 \\ 0 & 1 \end{pmatrix}\begin{pmatrix} Q^{(*,R)}_{I} \\ Q^{(*,R)}_{I+1} \end{pmatrix} - \frac{1}{2}\begin{pmatrix} -1 & -1 \\ 1 & 1 \end{pmatrix}\begin{pmatrix} Q^{(R)}_{I} \\ Q^{(R)}_{I+1} \end{pmatrix}.$$

Total Contribution

Let us now write the entire matrix-vector problem for the two-element configuration given in Fig. 9.1. Assuming that $\Delta x = \Delta x^{(L)} = \Delta x^{(R)}$ and using the global gridpoint numbering for DG we write

$$\frac{\Delta x}{6}\begin{pmatrix}2&1&0&0\\1&2&0&0\\0&0&2&1\\0&0&1&2\end{pmatrix}\begin{pmatrix}q_{xx,1}\\q_{xx,2}\\q_{xx,3}\\q_{xx,4}\end{pmatrix}=\begin{pmatrix}-1&0&0&0\\0&1&0&0\\0&0&-1&0\\0&0&0&1\end{pmatrix}\begin{pmatrix}Q_1^{(*)}\\Q_2^{(*)}\\Q_3^{(*)}\\Q_4^{(*)}\end{pmatrix}-\frac{1}{2}\begin{pmatrix}-1&-1&0&0\\1&1&0&0\\0&0&-1&-1\\0&0&1&1\end{pmatrix}\begin{pmatrix}Q_1\\Q_2\\Q_3\\Q_4\end{pmatrix}\quad(9.24)$$

which is, in fact, identical to Eq. (9.18) for the auxiliary variable Q.

9.5.9 Resulting Second Derivative at the Gridpoint I

Using the centered flux for $Q^{(*)}$, Eq. (9.24) becomes

$$\frac{\Delta x}{6}\begin{pmatrix}2&1&0&0\\1&2&0&0\\0&0&2&1\\0&0&1&2\end{pmatrix}\begin{pmatrix}q_{xx,1}\\q_{xx,2}\\q_{xx,3}\\q_{xx,4}\end{pmatrix}=\frac{1}{2}\begin{pmatrix}-1&0&0&0\\0&1&1&0\\0&-1&-1&0\\0&0&0&1\end{pmatrix}\begin{pmatrix}Q_1\\Q_2\\Q_3\\Q_4\end{pmatrix}+\frac{1}{2}\begin{pmatrix}-h_1\\0\\0\\h_4\end{pmatrix}$$

$$-\frac{1}{2}\begin{pmatrix}-1&-1&0&0\\1&1&0&0\\0&0&-1&-1\\0&0&1&1\end{pmatrix}\begin{pmatrix}Q_1\\Q_2\\Q_3\\Q_4\end{pmatrix}\quad(9.25)$$

where h_i denotes the Neumann boundary condition at the gridpoint i. Adding the first and third terms on the right-hand side and diagonalizing the mass matrix yields the simplified form

$$\frac{\Delta x}{2}\begin{pmatrix}q_{xx,1}\\q_{xx,2}\\q_{xx,3}\\q_{xx,4}\end{pmatrix}=\frac{1}{2}\begin{pmatrix}0&1&0&0\\-1&0&1&0\\0&-1&0&1\\0&0&-1&0\end{pmatrix}\begin{pmatrix}Q_1\\Q_2\\Q_3\\Q_4\end{pmatrix}+\frac{1}{2}\begin{pmatrix}-h_1\\0\\0\\h_4\end{pmatrix}.\quad(9.26)$$

From Eq. (9.26) we can see that the second derivative of q at the gridpoint I has two distinct values depending on whether we are considering the left or right elements. The value of $q_{xx,I}^{(L)} = q_{xx,2}$ is obtained from the second row which yields

$$q_{xx,I}^{(L)} = \frac{Q_3 - Q_1}{\Delta x} \equiv \frac{Q_I^{(R)} - Q_{I-1}^{(L)}}{\Delta x}$$

and the value of $q_{xx,I}^{(R)} = q_{xx,3}$ is given by the third row which yields

$$q_{xx,I}^{(R)} = \frac{Q_4 - Q_2}{\Delta x} \equiv \frac{Q_{I+1}^{(R)} - Q_I^{(L)}}{\Delta x}.$$

Note that the auxiliary variables are given as follows

$$Q^{(R)}_{I+1} \approx \frac{q_{I+1} - q_I}{\Delta x}$$

and

$$Q^{(L)}_I \approx \frac{q_I - q_{I-1}}{\Delta x};$$

this allows us to write

$$q^{(R)}_{xx,I} = \frac{q_{I+1} - 2q_I + q_{I-1}}{\Delta x^2} \qquad (9.27)$$

which is a valid second order derivative and, in fact, identical to the centered finite difference approximation.

Exercise Show that the second derivative from the viewpoint of $q^{(L)}_{xx,I}$ is identical to that given by Eq. (9.27). □

Exercise Construct the second derivative using the flip-flop numerical flux with $(\alpha, \beta) = (1, 0)$ for linear polynomials (N=1). □

Exercise Approximate the second order operator at the gridpoint I

$$\frac{d^2 q}{dx^2}(x)$$

using the two-step approach (in strong form)

$$Q(x) = \frac{dq}{dx}(x)$$
$$\frac{d^2 q}{dx^2}(x) = \frac{dQ(x)}{dx}$$

for $N = 1$. *Hint*: It will have a stencil size spanning the gridpoints $(I - 2, I - 1, I, I + 1, I + 2)$. □

Exercise Show that this second derivative is indeed second order accurate. □

9.5.10 1D Elliptic Equation

Now that we know how to construct first and second derivatives with the DG method we can use this information to solve elliptic equations. The continuous differential form of the Poisson problem is

$$\frac{d^2 q(x)}{dx^2} = f(x)$$

with the Dirichlet boundary condition

$$q(x)_\Gamma = g(x).$$

Recall that we will not build second order derivatives directly (i.e., the primal formulation) but rather will do it via the two-step process (i.e., the flux formulation)

$$\frac{dq(x)}{dx} = Q(x) \tag{9.28a}$$

$$\frac{dQ(x)}{dx} = f(x). \tag{9.28b}$$

Multiplying Eqs. (9.28) by a test function and integrating yields the DG problem statement: find $\left(q_N^{(e)}, Q_N^{(e)}\right) \in L^2(\Omega_e)$ such that

$$\int_{\Omega_e} \psi_i \frac{dQ_N^{(e)}}{dx} \, d\Omega_e = \int_{\Omega_e} \psi_i f_N^{(e)} \, d\Omega_e \tag{9.29a}$$

$$\int_{\Omega_e} \psi_i \frac{dq_N^{(e)}}{dx} \, d\Omega_e = \int_{\Omega_e} \psi_i Q_N^{(e)} \, d\Omega_e \tag{9.29b}$$

$\forall \, \psi \in L^2(\Omega_e)$ where we satisfy the integrals at the element level only. Using the product rule and the Fundamental Theorem of Calculus, results in

$$\left[\psi_i Q_N^{(*,e)}\right]_{\Gamma_e} - \int_{\Omega_e} \frac{d\psi_i}{dx} Q_N^{(e)} \, d\Omega_e = \int_{\Omega_e} \psi_i f_N^{(e)} \, d\Omega_e \tag{9.30a}$$

$$\left[\psi_i q_N^{(*,e)}\right]_{\Gamma_e} - \int_{\Omega_e} \frac{d\psi_i}{dx} q_N^{(e)} \, d\Omega_e = \int_{\Omega_e} \psi_i Q_N^{(e)} \, d\Omega_e. \tag{9.30b}$$

Introducing the basis function expansions

$$q_N^{(e)}(x) = \sum_{j=0}^{N} \psi_j(x) q_j^{(e)}$$

$$Q_N^{(e)}(x) = \sum_{j=0}^{N} \psi_j(x) Q_j^{(e)}$$

$$f_N^{(e)}(x) = \sum_{j=0}^{N} \psi_j(x) f_j^{(e)}$$

into Eqs. (9.30) yields

$$F_{ij}^{(e)} Q_j^{(*,e)} - \tilde{D}_{ij}^{(e)} Q_j^{(e)} = M_{ij}^{(e)} f_j^{(e)} \tag{9.31a}$$

$$F_{ij}^{(e)} q_j^{(*,e)} - \tilde{D}_{ij}^{(e)} q_j^{(e)} = M_{ij}^{(e)} Q_j^{(e)} \tag{9.31b}$$

where the element mass, flux, and weak differentiation matrices are defined as follows:

$$M_{ij}^{(e)} = \int_{\Omega_e} \psi_i \psi_j \, d\Omega_e,$$

$$F_{ij}^{(e)} = \left[\psi_i \psi_j\right]_{\Gamma_e},$$

and

$$\tilde{D}_{ij}^{(e)} = \int_{\Omega_e} \frac{d\psi_i}{dx} \psi_j \, d\Omega_e.$$

Introducing the mapping from the physical element Ω_e to the reference element $\widehat{\Omega}$ with the following metric and Jacobian terms

$$\frac{d\xi}{dx} = \frac{2}{\Delta x} \quad \text{and} \quad \frac{dx}{d\xi} = \frac{\Delta x}{2}$$

allows us to write the element matrices as follows

$$M_{ij}^{(e)} = \frac{\Delta x^{(e)}}{2} \sum_{k=0}^{Q} w_k \psi_i(\xi_k) \psi_j(\xi_k),$$

$$F_{ij}^{(e)} = \left[\psi_i(\xi) \psi_j(\xi)\right]_{\widehat{\Gamma}},$$

and

$$\tilde{D}_{ij}^{(e)} = \sum_{k=0}^{Q} w_k \frac{d\psi_i(\xi_k)}{d\xi} \psi_j(\xi_k).$$

Next, we define the numerical fluxes ($Q^{(*,e)}$ and $q^{(*,e)}$) using the centered fluxes

$$Q^{(*,e)} = \frac{1}{2}\left(Q^{(e)} + Q^{(k)}\right)$$

and

$$q^{(*,e)} = \frac{1}{2}\left(q^{(e)} + q^{(k)}\right)$$

where the superscript (k) denotes the neighbor of the element (e).

Since Eqs. (9.31) can be viewed as a global matrix problem (since the numerical fluxes connect adjacent elements), then we can use *direct stiffness summation* to construct the global matrix problem

$$B_I^{(Q)} + F_{IJ}Q_J - \tilde{D}_{IJ}Q_J = M_{IJ}f_J \qquad (9.32a)$$

$$B_I^{(q)} + F_{IJ}q_J - \tilde{D}_{IJ}q_J = M_{IJ}Q_J \qquad (9.32b)$$

where the vectors B_I are due to the Dirichlet boundary conditions (with respect to Q and q) that are imposed for this problem and $I, J = 1, ..., N_p$ where $N_p = N_e(N+1)$. For example, B only has two contributions each (one at each endpoint of the physical domain) with $B^{(Q)} = \frac{1}{2}n_\Gamma \frac{dq}{dx}(x)|_\Gamma$ and $B^{(q)} = \frac{1}{2}n_\Gamma q(x)|_\Gamma$ where n_Γ is the outward pointing unit normal vector of the physical boundary Γ, and the $\frac{1}{2}$ comes from the fact that the numerical flux function that we use is the average value and so this boundary condition is derived from the ghost cell (element k) for the element e.

Rearranging Eqs. (9.32) allows us to write

$$\hat{D}_{IJ}Q_J = M_{IJ}f_J - B_I^{(Q)} \tag{9.33a}$$

$$\hat{D}_{IJ}q_J = M_{IJ}Q_J - B_I^{(q)} \tag{9.33b}$$

where

$$\hat{D} = F - \tilde{D}.$$

Note that the solution of the elliptic equation requires us to find q, therefore, we now seek to isolate Q in Eq. (9.33a) as follows: let

$$Q_J = \hat{D}_{JK}^{-1}\left(M_{KL}f_L - B_K^{(Q)}\right)$$

which we now substitute into Eq. (9.33b) as such

$$\hat{D}_{IJ}q_J = M_{IJ}\hat{D}_{JK}^{-1}\left(M_{KL}f_L - B_K^{(Q)}\right) - B_I^{(q)}$$

where $I, J, K, L = 1, ..., N_p$. Rearranging this equation allows us to write the *local discontinuous Galerkin* (LDG) representation of the Poisson problem as

$$\hat{D}_{IK}M_{KL}^{-1}\hat{D}_{LJ}q_J = \left(M_{IJ}f_J - B_I^{(Q)}\right) - \hat{D}_{IK}M_{KJ}^{-1}B_J^{(q)} \tag{9.34}$$

which we can write as follows

$$L_{IJ}q_J = \left(M_{IJ}f_J - B_I^{(Q)}\right) - \hat{D}_{IK}M_{KJ}^{-1}B_J^{(q)} \tag{9.35}$$

where

$$L_{IJ} = \hat{D}_{IK}M_{KL}^{-1}\hat{D}_{LJ}$$

is the DG representation of the Laplacian operator. Algorithm 9.1 describes the steps required to solve an elliptic equation with LDG. In Alg. 9.1 we include the effect of the boundary condition directly into the global flux matrices in order to avoid constructing the boundary vector B.

Algorithm 9.1 LDG solution algorithm for the 1D elliptic equation.

---

function DG(q)
    Construct $M_{ij}^{(e)}, \tilde{D}_{ij}^{(e)}$ ▷ use Algs. 5.1, 6.1
    Construct global M and $\tilde{D}$ via DSS   ▷ use Alg. 6.3
    Construct global $F^{(Q)}$ and $F^{(q)}$ ▷ use Alg. 6.4
    Include the action of $B^{(Q)}$ and $B^{(q)}$ into $F^{(Q)}$ and $F^{(q)}$
    Construct global $\hat{D}^{(Q)} = \tilde{D} - F^{(Q)}$ and $\hat{D}^{(q)} = \tilde{D} - F^{(q)}$
    Construct global Laplacian $L = \hat{D}^{(Q)}M^{-1}\hat{D}^{(q)}$
    Invert $L q = M f$ ▷ use LU factorization or conjugate gradient
end function

9.6 Analysis of the Matrix Properties of the Spatial Operators

Let us now analyze the sparsity patterns of both M and L, that is, the mass and Laplacian matrices. In addition, we also look at the eigenvalues of the Laplacian matrix. For the analysis given below, we assume homogeneous Dirichlet boundary conditions.

9.6.1 Sparsity Pattern of the Mass Matrix

Figure 9.2 shows the sparsity pattern for the mass matrix M using both exact and inexact integration using Lobatto points. Figure 9.2 shows that the mass matrix is

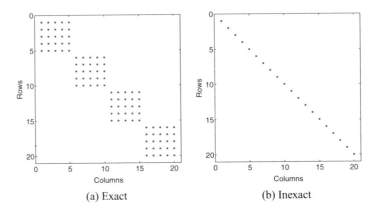

(a) Exact   (b) Inexact

Fig. 9.2: Sparsity patterns for the mass matrix M with $N_e = 4$, $N = 4$ for (a) exact and (b) inexact integration for DG.

block diagonal and that it becomes completely diagonal when inexact integration is used.

9.6.2 Sparsity Pattern of the Laplacian Matrix

Figure 9.3 shows the sparsity pattern for the Laplacian matrix L for exact and inexact integration using Lobatto points. Figure 9.3 shows that the Laplacian matrix is not completely full, however, unlike the mass matrix, it is not block diagonal but is actually a global matrix (with bandwidth $2N + 2$). For the Laplacian matrix, there is no difference between exact and inexact integration since for $O(2N-2)$ polynomials, N Lobatto integration points evaluate the integrals exactly.

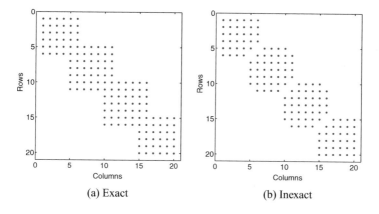

(a) Exact   (b) Inexact

Fig. 9.3: Sparsity patterns for the Laplacian matrix L with $N_e = 4$, $N = 4$ for (a) exact and (b) inexact integration for DG.

9.6.3 Eigenvalue Analysis of the Laplacian Operator

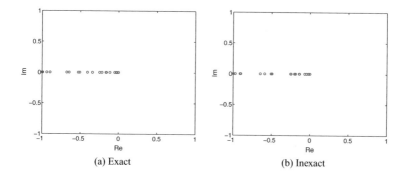

(a) Exact   (b) Inexact

Fig. 9.4: Eigenvalues of the Laplacian matrix L with $N_e = 4$, $N = 4$ for (a) exact and (b) inexact integration for DG.

Figure 9.4 shows that the eigenvalues of the Laplacian are all on the real axis and, although we have normalized the values, we can state that the condition number of this matrix is rather small (i.e., it is well-conditioned)[2]. This is very good news because in order to construct the numerical solution requires the inversion of L; in

---

[2] Matrices representing elliptic operators are symmetric positive-definite, meaning that the eigenvalues are all real and greater than zero. In Fig. 9.4 we show the eigenvalues of the operator ∇^2 and not $-\nabla^2$.

practice we do not construct the inverse but it is important to show that the inverse does indeed exist (i.e., the determinant of L must not be zero). The resulting linear algebra matrix problem can be solved with either direct or iterative methods but the success of both of these methods depends on the matrix L being well-conditioned. Thankfully, this matrix indeed is well-conditioned and, similarly to the CG version, can be solved by methods such as conjugate gradient (see [326] and [380] for details).

9.7 Example of 1D Poisson Equation Problem

Suppose we wish to solve the continuous second order differential equation

$$\frac{d^2 q}{dx^2} = f(x) \qquad \forall x \in [-1, 1]$$

where the exact solution is given as

$$q(x) = \sin(c\pi x)$$

and

$$f(x) = -(c\pi)^2 \sin(c\pi x)$$

where c is any constant. For integer values of c, the Dirichlet boundary conditions are homogeneous. For all values of c the Dirichlet boundary conditions are defined as

$$g(-1) = \sin(-c\pi) \qquad \text{and} \qquad g(+1) = \sin(c\pi).$$

In what follows we use $c = 2$.

9.7.1 Error Norm

We use the integral form of the L^2 norm given by Eq. (5.34).

9.7.2 Solution Accuracy

Figure 9.5 shows the exact and DG solutions using $N = 16$ order Lobatto polynomials and $N_e = 4$ elements for a total of $N_p = 68$ gridpoints. Figure 9.5 shows that there is no difference between the exact and numerical solutions. Let us now see how well our model behaves by varying the number of elements and the polynomial order of the basis functions.

Figure 9.6 shows the convergence rates for various polynomial orders, N, for a total number of gridpoints $N_p = N_e(N + 1)$. Figure 9.6 shows that there is little

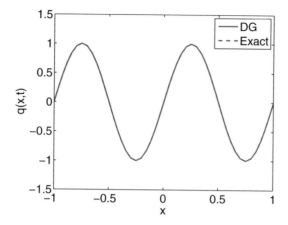

Fig. 9.5: Exact and DG numerical solution for the 1D Poisson equation using $N = 16$ order Lobatto polynomials with $N_e = 4$ elements with exact integration.

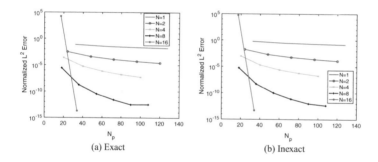

(a) Exact   (b) Inexact

Fig. 9.6: Convergence rates of DG for the 1D Poisson equation for Lobatto polynomial orders $N = 1, 2, 4, 8, 16$ using a total number of gridpoints N_p for (a) exact and (b) inexact integration.

difference between exact and inexact integration at least in the rates of convergence. The main differences occur for low values of N (≤ 4), however, as N increases the differences between exact and inexact integration all but disappear. Furthermore, the accuracy of the DG method, especially for the high-order polynomials, approaches machine double precision (1×10^{-16}).

Part III
Multi-Dimensional Problems

Part III focuses on solving multi-dimensional problems. We begin with Chs. 10 and 11 that discuss interpolation and integration in two and three dimensions. These are then followed by Ch. 12 which describes the CG discretization of elliptic equations in two dimensions. This chapter also describes the construction of metric terms for general unstructured grids and extends these ideas to three dimensions. Chapter 13 describes the construction of the LDG method for elliptic equations. The treatment of elliptic equations ends with Ch. 14 which introduces the symmetric interior penalty method used to unify the CG and DG methods. The next three chapters (15, 16, and 17) treat the construction of hyperbolic equations with the CG, DG, and unified CG/DG methods, respectively. It is in these chapters that efficiency of the methods are described in detail, which includes the discussion of the sum factorization approach that allows efficient implementation of tensor-product based methods.

Chapter 10
Interpolation in Multiple Dimensions

10.1 Introduction

In Chapter 3 we discussed interpolation in one dimension (1D). In this chapter we use some of the ideas introduced there to extend interpolation to two and three dimensions. In one dimension, there is no room to choose the shape of the domain. That is, in the domain $x \in [-1, +1]$ we are constrained to line elements. However, in two dimensions this door is flung wide open and we are now free to choose all sorts of polygons as the basic building blocks of our interpolation. For example, let us assume that we wish to interpolate a function on the unit square $(x, y) \in [-1, +1]^2$. Clearly, we could tile the unit square using quadrilaterals, triangles, and even pentagons, hexagons, octagons, etc. (albeit with the need to chop up some of these more elaborate polygons in order to fit them exactly in the unit square). However, let us restrict our attention to quadrilateral and hexahedral (i.e., cubes) decompositions of our two-dimensional (2D) and three-dimensional (3D) domains since tensor-product bases are the focus of this text. For interpolation on triangular domains the reader is referred to the following works [201, 400, 398, 369, 368, 399, 396, 87, 229, 276, 199, 114, 168, 157, 167, 371, 397, 169, 34, 191, 141, 247, 164], which only comprise a small subset of this rather large literature.

This chapter is organized as follows. Section 10.2 describes interpolation on the quadrilateral. In Secs. 10.2.1 and 10.2.2 we describe interpolation on the quadrilateral using both modal and nodal functions, respectively. In Sec. 10.2.4 we show examples of quadrilateral elements for various polynomial orders. We end this chapter with a description of interpolation on hexahedra in Secs. 10.3.1 and Secs. 10.3.2 for both modal and nodal representations, respectively.

© The Editor(s) (if applicable) and The Author(s), under exclusive license
to Springer Nature Switzerland AG 2020
F. X. Giraldo, *An Introduction to Element-Based Galerkin Methods on
Tensor-Product Bases*, Texts in Computational Science and Engineering 24,
https://doi.org/10.1007/978-3-030-55069-1_10

10.2 Interpolation on the Quadrilateral

10.2.1 Modal Interpolation

In 1D, we already saw that interpolation can be defined in either modal or nodal space. When modal functions are used this means that we are representing any function as the sum of multiple kinds of waves having $N + 1$ distinct frequencies and amplitudes. For example, the modal interpolation of a function $f(\xi)$ is given as

$$f(\xi) = \sum_{i=0}^{N} \phi_i(\xi) \tilde{f}_i.$$

The modal functions can be defined to be either the simple monomials $\phi_i^{1D}(\xi) = \xi^i \ \forall \ i = 0, ..., N$ or the Jacobi polynomials, which we introduced in Ch. 3 and are described in Appendix B, that are orthogonal in the interval $\xi \in [-1, +1]$, $\phi_i^{1D}(\xi) = P_i^{(\alpha,\beta)}(\xi) \ \forall \ i = 0, ..., N$. In 2D, we simply apply a tensor-product of the 1D modal functions as follows

$$\phi_{ij}^{2D}(\xi, \eta) = \phi_i^{1D}(\xi) \otimes \phi_j^{1D}(\eta), \ \forall \ (i, j) = 0, ..., N$$

where the symbol $\otimes$ is known as the *Kronecker*, tensor, or outer product.

10.2.2 Nodal Interpolation

Recall that when nodal interpolation is used, we use the value of the function at specific interpolation points as our expansion coefficients. For example, the nodal interpolation of a function $f(\xi)$ is defined as follows

$$f(\xi) = \sum_{i=0}^{N} L_i(\xi) f_i$$

where L are Lagrange polynomials and $f_i = f(\xi_i)$ is the value of the function $f(\xi)$ evaluated at $\xi = \xi_i$. Using the same idea that we just introduced for the construction of 2D modal interpolation functions on the quadrilateral, we can write the tensor-product of the 1D Lagrange polynomials as follows

$$L_{ij}^{2D}(\xi, \eta) = L_i^{1D}(\xi) \otimes L_j^{1D}(\eta), \ \forall \ (i, j) = 0, ..., N.$$

It becomes immediately obvious why the quadrilateral is the element of choice in most high-order element-based Galerkin models.

10.2.3 Popularity of Quadrilateral Elements

The popularity of quadrilateral elements is very much a direct consequence of the ease with which interpolation functions can be constructed. However, another reason for their ubiquity is due to their computational efficiency. Methods for multi-dimensions that are based on tensor-products tend to be very fast because a very nice trick (called sum factorization) reduces the cost of the method to $O(N^{d+1})$ where N is the order of the interpolation functions and d is the dimension of the space. A naive approach costs $O(N^{2d})$ that arises from performing both interpolation $O(N^d)$ and integration $O(N^d)$ directly, which is unnecessary but can be found in the literature.

10.2.4 Example of Quadrilateral Basis Functions

Let us take a close look at nodal basis functions on the quadrilateral based on Lobatto and Legendre points.

10.2.4.1 Lobatto Points

To simplify the exposition, let us renumber the 2D basis functions as follows: let

$$\psi_i(\xi, \eta) = L_j^{1D}(\xi) \otimes L_k^{1D}(\eta), \ \forall \ (j, k) = 0, ..., N, \ i = 1, ..., (N + 1)^2$$

where i is associated with j, k as follows:

$$i = j + 1 + k(N + 1). \tag{10.1}$$

Note that the order in each direction can be different which requires modifying the map as follows

$$i = j + 1 + k(N_\xi + 1) \tag{10.2}$$

where $j = 0, ..., N_\xi$, $k = 0, ..., N_\eta$, and $i = 1, ..., (N_\xi + 1)(N_\eta + 1)$. Figure 10.1 shows the Lobatto interpolation points for elements of orders $N = 1, 2$, and 3. Let us now show the basis functions for each of these three elements. In Figs. 10.2, 10.3, 10.4 the basis functions associated with the $4, 9, 16$ interpolation points of polynomial orders $N = 1, 2, 3$, respectively, are displayed.

10.2.4.2 Legendre Points

Let us now look at the Lagrange polynomial (nodal) basis functions when we use Legendre points to construct the Lagrange polynomials. Once again, the 2D basis functions are defined as follows: let

$$\psi_i(\xi, \eta) = L_j^{1D}(\xi) \otimes L_k^{1D}(\eta), \ \forall \ (j, k) = 0, ..., N, \ i = 1, ..., (N + 1)^2$$

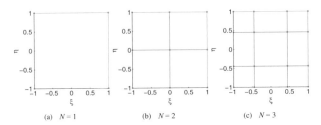

Fig. 10.1: Quadrilateral elements of orders (a) $N = 1$, (b) $N = 2$, and (c) $N = 3$ based on Lobatto points.

where i is associated with j, k as given by Eqs. (10.1) or (10.2). Figure 10.5 shows the Legendre interpolation points for elements of orders $N = 1, 2$, and 3. Let us now show the basis functions for each of these three elements.

In Figs. 10.6, 10.7, 10.8 the basis functions associated with the 4, 9, 16 interpolation points of polynomial orders $N = 1, 2, 3$, respectively, are displayed. The one clear difference between the Lobatto and Legendre points is that the Lagrange polynomials based on the Lobatto points have extremal values far smaller than those for the Legendre points. For the Legendre-based polynomials, the large extremal values occur at the element edges (specifically, at the corners). The Lagrange polynomials are well-behaved for the points internal to the space within the minimum and maximum Legendre points; the boundaries are not defined in this space. Fortunately, these large values at the boundaries do not have an adverse effect on the quality of the numerical solution. These points cannot be used in a CG formulation because the boundaries are not included and so one cannot satisfy C^0 continuity. In a DG formulation, these points can be used because, for this method, we do not require that the solution be C^0 continuous. For DG formulations, the boundaries are not so critical since we only need them to evaluate flux integrals. Furthermore, the corner points are never used since one typically uses Legendre-Gauss quadrature (Legendre points) that do not include the endpoints [-1,+1].

10.3 Interpolation on the Hexahedron

In Sec. 10.2 we saw that when using tensor-product basis functions it is rather trivial to construct two-dimensional basis functions from the one-dimensional functions. This is the case for either modal or nodal bases. In this section, we show that extending these ideas to three dimensions is just as straightforward.

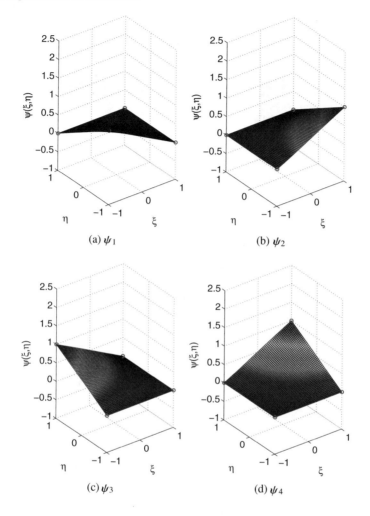

Fig. 10.2: Quadrilateral Lagrange basis functions of order $N = 1$ at the four interpolation points at which the functions satisfy cardinality for Lobatto points.

10.3.1 Modal Interpolation

To construct a general d-dimensional approximation to the function $f(\xi)$ via modal interpolation, we can write

$$f(\boldsymbol{\xi}) = \sum_{l=1}^{M_N} \phi_l(\boldsymbol{\xi}) \tilde{f}_l \tag{10.3}$$

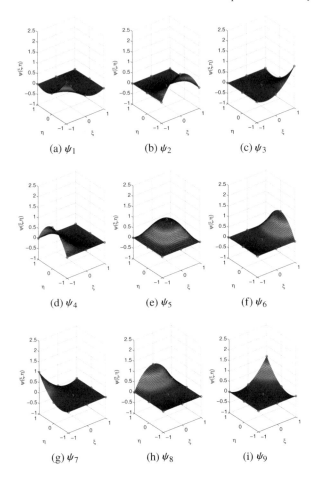

Fig. 10.3: Quadrilateral Lagrange basis functions of order $N = 2$ at the nine inter-polation points at which the functions satisfy cardinality for Lobatto points.

where $l = 1, \ldots, M_N$ and $M_N = (N + 1)^d$ denotes the number of terms in the expansion (for uniform polynomial order in all directions) and $\boldsymbol{\xi} = (\xi, \eta, \zeta)$.

If we let $\phi^{1D}(\xi)$ represent Jacobi polynomials of order N defined for a variable $\xi \in [-1, +1]$ then we can use tensor-products to write the three-dimensional basis functions as follows

$$\phi_l(\boldsymbol{\xi}) \equiv \phi_{ijk}^{3D}(\xi, \eta, \zeta) = \phi_i^{1D}(\xi) \otimes \phi_j^{1D}(\eta) \otimes \phi_k^{1D}(\zeta), \ \forall \ (i, j, k) = 0, \ldots, N \quad (10.4)$$

where the map $(i, j, k) \rightarrow l$ is described in Sec. 10.3.3.

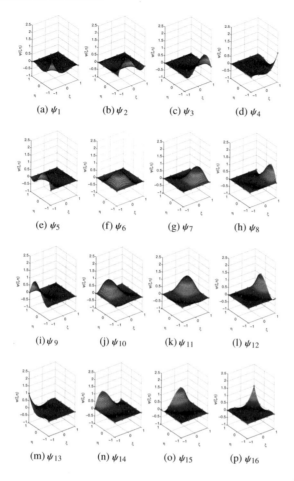

Fig. 10.4: Quadrilateral Lagrange basis functions of order $N = 3$ at the sixteen interpolation points at which the functions satisfy cardinality for Lobatto points.

10.3.2 Nodal Interpolation

On the other hand, the nodal interpolation of the function $f(\xi)$ is defined as follows

$$f(\xi) = \sum_{l=1}^{M_N} L_l(\xi) f_l \tag{10.5}$$

where L are multi-dimensional Lagrange polynomials and $f_l = f(\xi_l)$ is the value of the function $f(\xi)$ evaluated at ξ. Using the same idea that we just introduced for the

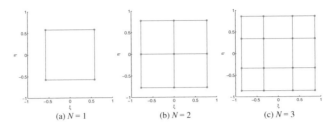

Fig. 10.5: Quadrilateral elements of orders (a) $N = 1$, (b) $N = 2$, and (c) $N = 3$ based on Legendre points.

construction of 3D modal interpolation functions on the hexahedron, we can write the tensor-product of the 1D Lagrange polynomials as follows

$$L_I(\xi) \equiv L_{ijk}^{3D}(\xi, \eta, \zeta) = L_i^{1D}(\xi) \otimes L_j^{1D}(\eta) \otimes L_k^{1D}(\zeta), \ \forall \ (i, j, k) = 0, ..., N. \quad (10.6)$$

10.3.3 Indexing of the Basis Functions

In Eqs. (10.3) and (10.5), we wrote the basis functions in monolithic form whereby they are represented as a vector of length M_N for a specific position ξ as such $\phi_I(\xi)$. While this allows us to define the interpolation problem compactly (only using one index) it does not represent the most efficient way of performing operations such as interpolation and differentiation. In Chs. 15 and 16 we discuss *sum factorization* in detail which allows us to construct efficient differentiation and integration when we write the basis functions as tensor-products. This means that in most CG and DG software, we define the basis functions in the following form $\phi_{ijk}(\xi)$. However, in the special case when long vectors are required then one should revert to the monolithic approach.

Before we close this chapter, it should be emphasized that the approach for constructing basis functions that we presented is sufficiently flexible to allow for basis functions of different polynomial orders across all d-dimensions. More specifically, in Eq. (10.3) we stated that $M_N = (N + 1)^d$ which implies that the polynomial order is uniform across all d-dimensions. However, we could also construct polynomials of different orders across all d-dimensions as follows:

$$M_N = (N_\xi + 1)(N_\eta + 1)(N_\zeta + 1)$$

where (N_ξ, N_η, N_ζ) can be different. This then changes our tensor-product basis to the following:

$$L_I(\xi) \equiv L_{ijk}^{3D}(\xi, \eta, \zeta) = L_i^{1D}(\xi) \otimes L_j^{1D}(\eta) \otimes L_k^{1D}(\zeta) \quad (10.7)$$

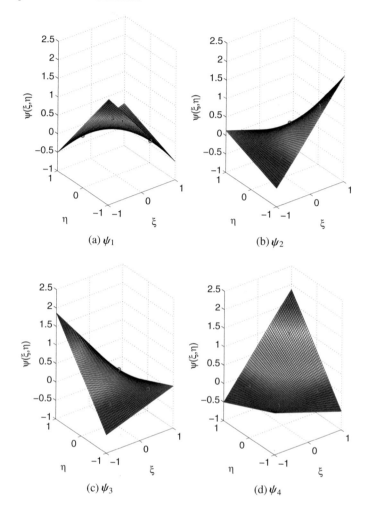

Fig. 10.6: Quadrilateral Lagrange basis functions of order $N = 1$ at the four interpolation points at which the functions satisfy cardinality for Legendre points.

$\forall i = 0, \ldots, N_\xi;\ j = 0, \ldots, N_\eta;\ k = 0, \ldots, N_\zeta$ where the map from the tensor-product space to the monolithic space is given by

$$l = i + 1 + j(N_\xi + 1) + k(N_\xi + 1)(N_\eta + 1).$$

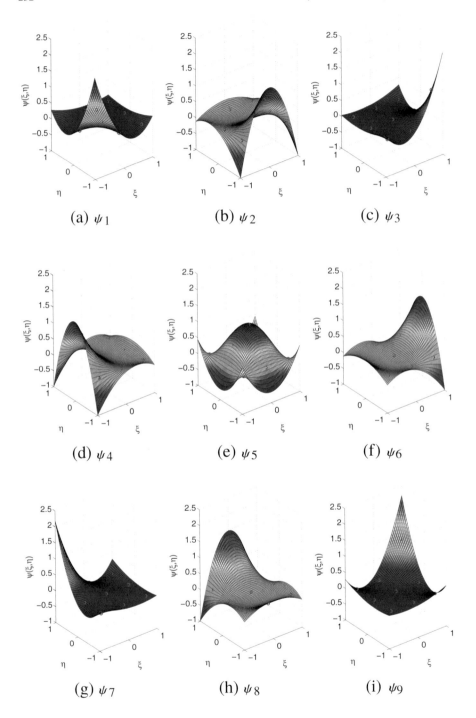

Fig. 10.7: Quadrilateral basis functions of order $N = 2$ at the nine interpolation points at which the functions satisfy cardinality for Legendre points.

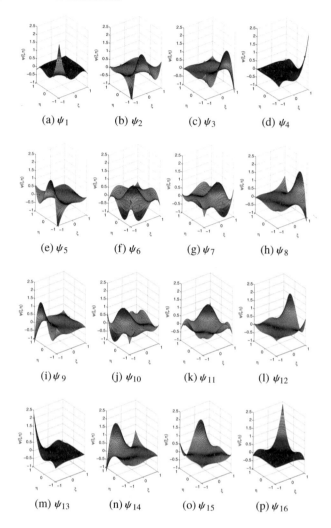

Fig. 10.8: Quadrilateral basis functions of order $N = 3$ at the sixteen interpolation points at which the functions satisfy cardinality for Legendre points.

Chapter 11
Numerical Integration in Multiple Dimensions

11.1 Introduction

Galerkin methods require the evaluation of integrals of the type

$$A = \int_{\Omega} f(\mathbf{x})d\Omega \text{ and } B = \int_{\Gamma} f(\mathbf{x})d\Gamma \qquad (11.1)$$

where Ω is the domain that we wish to integrate within and Γ are its boundaries. We saw in Chs. 5 and 6 that, in one dimension, A is a line integral while B denotes a jump value. In two dimensions, A is an area integral while B is a line integral. Finally, in three dimensions, A is a volume integral and B is a surface area integral. For brevity we always refer to A as a volume integral and B as either a boundary or flux integral.

In Eq. (11.1) we cannot integrate $f(\mathbf{x})$ analytically because f is an unknown function that we, unfortunately, do not know in closed form. Instead, we only have f sampled at certain selected spatial positions. This then motivates the need for numerical integration which in one dimension is referred to as *quadrature* or in more dimensions as *cubature*. We saw in Ch. 4 that the idea behind numerical integration is very much tied to the construction of interpolating functions by using polynomial representations. Thus if we assume that a function takes the form of a well-known polynomial then we can determine the best way to integrate that polynomial within a given *a priori* error bound. One of the interesting issues in numerical integration is that it can be viewed as an operation that is independent of the type of basis function (e.g., modal or nodal) used to represent functions within an element. This is certainly true for tensor-product elements (such as quadrilaterals in 2D and hexahedra in 3D). However, on the triangle, the choice of basis function will affect the way that a function is represented within an element. For a discussion on triangular elements,

F. X. Giraldo, *An Introduction to Element-Based Galerkin Methods on Tensor-Product Bases*, Texts in Computational Science and Engineering 24, https://doi.org/10.1007/978-3-030-55069-1_11

the reader is referred to [356, 260, 111, 94, 259, 93, 392, 200]. In this chapter, we only discuss tensor-product elements such as quadrilaterals and hexahedra.

This chapter is organized as follows. Section 11.2 discusses integration on the quadrilateral and in Sec. 11.3 we discuss integration on the hexahedron.

11.2 Numerical Integration on the Quadrilateral

Numerical integration on the quadrilateral is straightforward since we use the one-dimensional numerical integration formulas discussed in Ch. 4. In a similar manner that the two-dimensional (2D) interpolation functions are constructed using a tensor-product of the one-dimensional (1D) interpolation functions, we construct the 2D integration formulas using a tensor-product of the 1D integration formulas.

For example, assume that we want to evaluate the area integral

$$\int_{\Omega_e} f(\mathbf{x}) \, d\Omega_e = \int_{x_0}^{x_1} \int_{y_0}^{y_1} f(x, y) \, dx \, dy$$

where (x_i, y_i) for $i = 0, 1$ are the bounds of integration in physical space. Mapping the physical space (physical element Ω_e) to the computational space (reference element $\hat{\Omega}$) gives

$$\int_{\Omega_e} f(\mathbf{x}) \, d\Omega_e = \int_{\hat{\Omega}} f(\xi, \eta) J(\xi, \eta) \, d\hat{\Omega}$$

where $J(\xi, \eta)$ is the determinant of the Jacobian of the mapping from physical (x, y) to computational (ξ, η) space that is derived in Chs. 15 and 16. The term $J(\xi, \eta)$ ensures that the physical area is represented within the integration along the reference element. Rewriting this continuous (definite) integral as a Riemann sum[1] with carefully chosen weights and integration points yields

$$\int_{\hat{\Omega}} f(\xi, \eta) J(\xi, \eta) \, d\hat{\Omega} = \int_{-1}^{+1} \int_{-1}^{+1} f(\xi, \eta) \, J(\xi, \eta) \, d\xi \, d\eta \qquad (11.2)$$

$$\approx \sum_{i=0}^{Q_\xi} \sum_{j=0}^{Q_\eta} w_i w_j f(\xi_i, \eta_j) J(\xi_i, \eta_j)$$

where w_i and ξ_i are the 1D quadrature weights and points of degree Q_ξ described in Ch. 4. For example, we can use the roots of the Legendre or Lobatto polynomials for ξ_i and their corresponding weights w_i. Similarly, the weights and points w_j and η_j are points of degree Q_η that represent the second spatial dimension of the quadrilateral element; note that Q_ξ and Q_η can be equal but they need not be. For the quadrilateral, the theory of interpolation and integration relies entirely on the tensor-product in 1D on a line segment; this means that everything we learned about integration in 1D can be applied directly to the quadrilateral including inexact integration. In other

---

[1] We call it a Riemann sum here to make the connection between integration and a sum. However, in element-based Galerkin methods we are interested in viewing all integrals as Lebesgue integrals. We discuss this point in Ch. 2.

words, the interpolation points can also be used for integration thereby resulting in a co-located approach. For CG, we need Lobatto points because they include the endpoints, but for DG we can choose Lobatto or Legendre points.

Note that we did not discuss the type of basis function used on the quadrilateral. In other words, we could have chosen Lagrange polynomials or Legendre polynomials, for example. Both of these basis functions are constructed as a tensor-product of 1D basis functions and so there is no advantage in writing the quadrature formulas in another (monolithic) form; that is, the natural way to represent an integral on the quadrilateral is as a tensor-product of 1D integrals. For the triangle, this is not necessarily the case[2]; however, triangular domains are beyond the scope of this text and will not be discussed further.

11.3 Numerical Integration on the Hexahedron

In Sec. 11.2 we saw that we can apply the quadrature formulas we learned in 1D to quadrilaterals in 2D by using tensor-products. In a similar fashion, we can extend this approach to three dimensions for constructing numerical integration formulas on the hexahedron (i.e., numerical cubature).

Let us consider evaluating the volume integral

$$\int_{\Omega_e} f(\mathbf{x}) \, d\Omega_e = \int_{x_0}^{x_1} \int_{y_0}^{y_1} \int_{z_0}^{z_1} f(x, y, z) \, dx \, dy \, dz$$

where (x_i, y_i, z_i) for $i = 0, 1$ are the bounds of integration in physical space. Mapping from physical (Ω_e) to computational $(\hat{\Omega})$ space gives

$$\int_{\Omega_e} f(\mathbf{x}) \, d\Omega_e = \int_{\hat{\Omega}} f(\xi, \eta, \zeta) J(\xi, \eta, \zeta) \, d\hat{\Omega}$$

where $J(\xi, \eta, \zeta)$ is the determinant of the Jacobian of the mapping from physical (x, y, z) to computational (ξ, η, ζ) space and ensures that the physical volume is represented within the reference element. Rewriting this continuous integral as a Riemann sum with carefully chosen weights and integration points yields

$$\int_{\hat{\Omega}} f(\xi, \eta, \zeta) J(\xi, \eta, \zeta) \, d\hat{\Omega} = \int_{-1}^{+1} \int_{-1}^{+1} \int_{-1}^{+1} f(\xi, \eta, \zeta) J(\xi, \eta, \zeta) \, d\xi \, d\eta \, d\zeta$$

and finally

$$\int_{-1}^{+1} \int_{-1}^{+1} \int_{-1}^{+1} f(\xi, \eta, \zeta) J(\xi, \eta, \zeta) \, d\xi \, d\eta \, d\zeta \qquad (11.3)$$

$$\approx \sum_{i=0}^{Q_\xi} \sum_{j=0}^{Q_\eta} \sum_{k=0}^{Q_\zeta} w_i w_j w_k f(\xi_i, \eta_j, \zeta_k) J(\xi_i, \eta_j, \zeta_k)$$

where (w_i, w_j, w_k) are the quadrature weights associated with the reference coordinates (ξ_i, η_j, ζ_k) with integration formulas of degree (Q_ξ, Q_η, Q_ζ), respectively.

---

[2] Regarding triangles, using a modal representation permits a tensor-product approach whereas using a nodal representation only allows the monolithic approach.

Therefore, like for the quadrilateral, the hexahedron is constructed directly on the theory of interpolation and integration of tensor-products for 1D on the line segment. Similar to 2D, in 3D we can use Lobatto points for CG and either Lobatto or Legendre points for DG.

11.4 Types of Integrals Required

Now that we have defined the integration formulas for elements defined on the line, quadrilateral, and hexahedron, we can now discuss the types of integrals required by both the CG and DG method in all three dimensions. In 1D, we only require line integrals of the type described in Ch. 4. In 2D, we require area integrals of the type describe in Eq. (11.3) in addition to line integrals described in Ch. 4 for the flux integrals. Finally, in 3D we require volume integrals of the type describe in Eq. (11.4) in addition to area integrals described in Eq. (11.3) for the flux integrals.

Chapter 12
2D Continuous Galerkin Methods for Elliptic Equations

12.1 Introduction

So far we have only discussed the implementation of continuous Galerkin (CG) and discontinuous Galerkin (DG) methods in one dimension. We have done so for the advection and diffusion equations, and by extension the advection-diffusion equation. In addition, in Ch. 7 we discussed the application of both methods for systems of nonlinear equations in one dimension. In this chapter we now extend the ideas that we have learned so far in one dimension and extend it to two dimensions where we focus on the discretization of elliptic equations using the CG method. In what follows, we shall use the global gridpoint (GPP) storage. In Ch. 14 we describe the local element-wise (LEW) storage.

The remainder of this chapter is organized as follows. In Sec. 12.2 we describe the continuous form of the elliptic equation that we solve with the CG method and in Sec. 12.3 we describe its integral form. In Sec. 12.4 we describe the construction of the basis functions which leads to the element equations defined in Sec. 12.8. Understanding the procedure for constructing the element equations will allow us to better understand the construction of the global matrix-vector problem which we discuss in Sec. 12.9. To fully appreciate the role of integration by parts, we outline in Sec. 12.10 the direct discretization of second order operators. In Sec. 12.11 we give specific examples for the relevant matrices for solving elliptic equations for the specific case of linear elements. This then leads us to the discretization of elliptic equations which we discuss in Sec. 12.12. In Sec. 12.13 we analyze the sparsity and eigenvalues of the global matrices. This leads us to Sec. 12.14 where we show an example for solving a Poisson problem. We end this chapter with a discussion on the computational cost of using high-order in Sec. 12.15.

F. X. Giraldo, *An Introduction to Element-Based Galerkin Methods on Tensor-Product Bases*, Texts in Computational Science and Engineering 24, https://doi.org/10.1007/978-3-030-55069-1_12

12.2 Problem Statement for the Elliptic Equation

Let us begin our discussion with the Poisson equation

$$\nabla^2 q = f(\mathbf{x}) \tag{12.1}$$

where $q = q(\mathbf{x})$ is our scalar solution variable, $\nabla = \left(\frac{\partial}{\partial x}\mathbf{i} + \frac{\partial}{\partial y}\mathbf{j}\right)$ is the two-dimensional (2D) gradient operator and $\nabla^2 = \nabla \cdot \nabla$ is the Laplacian operator, and f is a forcing function. To complete the mathematical description of this partial differential equation requires introducing boundary conditions. The domain Ω in which we solve Eq. (12.1) has a corresponding boundary $\Gamma = \{\Gamma_{\mathcal{D}} \bigcup \Gamma_N\}$. On the boundary we require one or both of the following types of boundary conditions:

$$q \mid_{\Gamma_{\mathcal{D}}} = g(\mathbf{x}) \qquad \text{and/or} \qquad \hat{\mathbf{n}} \cdot \nabla q \mid_{\Gamma_N} = h(\mathbf{x})$$

where $g(\mathbf{x})$ and $h(\mathbf{x})$ denote the Dirichlet and Neumann boundary conditions, respectively, with $\hat{\mathbf{n}}$ denoting the outward pointing unit normal vector of the boundary. The interested reader is referred to Appendix C for further reading on boundary conditions.

12.3 Integral Form

To construct the integral form of Eq. (12.1), we begin by expanding the solution variable as follows

$$q_N^{(e)}(\mathbf{x}) = \sum_{i=1}^{M_N} \psi_i(\mathbf{x}) q_i^{(e)}$$

with

$$f_N^{(e)}(\mathbf{x}) = \sum_{i=1}^{M_N} \psi_i(\mathbf{x}) f_i^{(e)}.$$

The integer M_N is a function of N which determines the number of points inside each element Ω_e which for quadrilaterals is $M_N = (N+1)^2$. However, it is important to recall that each direction can use a different polynomial order as we described in Ch. 10. In the more general case, we would define $M_N = (N_\xi + 1)(N_\eta + 1)$ where N_ξ and N_η are the polynomial orders along the two directions of the reference element.

Next, we substitute $q_N^{(e)}$ and $f_N^{(e)}$ into the partial differential equation, multiply by a test function and integrate within each element yielding the problem statement: find $q_N^{(e)} \in H^1$ such that

$$\int_{\Omega_e} \psi_i \nabla^2 q_N^{(e)} \, d\Omega_e = \int_{\Omega_e} \psi_i f_N^{(e)} \, d\Omega_e \ \forall \ \psi \in H^1 \tag{12.2}$$

with $i = 1, \ldots, M_N$, where ψ are the locally defined basis functions in each element Ω_e such that

$$\Omega = \bigcup_{e=1}^{N_e} \Omega_e$$

with N_e denoting the number of elements. To be clear, we do not solve Eq. (12.2) as written. Instead, we use integration by parts in order to include boundary data. We discuss this in Sec. 12.8.

12.4 Basis Functions and the Reference Element

12.5 Basis Functions in Two Dimensions

Crucial to the construction of CG approximations is the basis functions used to approximate all the variables in the model. For tensor-product grids (quadrilaterals and hexahedra) we saw in Ch. 10 that we can use the same machinery used in 1D to extend the methods to multiple dimensions. For example, to construct 2D basis functions on the reference element $(\xi, \eta) \in [-1, +1]^2$ we use the tensor-product notation

$$\psi_i(\xi, \eta) = h_j(\xi) \otimes h_k(\eta) \tag{12.3}$$

where h are the 1D basis functions described in Ch. 3, $\otimes$ is the tensor-product, and the 1D indices vary as follows: $j = 0, \ldots, N_\xi$, $k = 0, \ldots, N_\eta$ with the 2D index varying as $i = 1, \ldots, M_N$ where $M_N = (N_\xi + 1)(N_\eta + 1)$. To get from the 1D local indices (j, k) to the 2D local index i requires the mapping $i = j + 1 + k(N_\xi + 1)$. In what follows, we use the indices (i, j) to refer to either the tensor-product or monolithic numbering - when it is not immediately obvious, we will specify which index we are referring to.

We can now expand the solution variable q as follows

$$q_N^{(e)}(\mathbf{x}) = \sum_{i=1}^{M_N} \psi_i(\mathbf{x}) q_i^{(e)} \tag{12.4}$$

which implies the approximation of the gradient operator to be

$$\nabla q_N^{(e)}(\mathbf{x}) = \sum_{i=1}^{M_N} \nabla \psi_i(\mathbf{x}) q_i^{(e)} \tag{12.5}$$

where the partial derivatives are defined as follows

$$\frac{\partial q_N^{(e)}(\mathbf{x})}{\partial x} = \sum_{i=1}^{M_N} \frac{\partial \psi_i(\mathbf{x})}{\partial x} q_i^{(e)} \tag{12.6a}$$

$$\frac{\partial q_N^{(e)}(\mathbf{x})}{\partial y} = \sum_{i=1}^{M_N} \frac{\partial \psi_i(\mathbf{x})}{\partial y} q_i^{(e)}. \tag{12.6b}$$

Since we perform all of our computations in the reference element with coordinates (ξ, η) then we must transform all derivatives from (x, y) to (ξ, η). Using the chain rule, we write the derivatives of the basis functions as

$$\frac{\partial \psi(x(\xi, \eta), y(\xi, \eta))}{\partial x} = \frac{\partial \psi(\xi, \eta)}{\partial \xi} \frac{\partial \xi(x, y)}{\partial x} + \frac{\partial \psi(\xi, \eta)}{\partial \eta} \frac{\partial \eta(x, y)}{\partial x} \tag{12.7a}$$

$$\frac{\partial \psi(x(\xi, \eta), y(\xi, \eta))}{\partial y} = \frac{\partial \psi(\xi, \eta)}{\partial \xi} \frac{\partial \xi(x, y)}{\partial y} + \frac{\partial \psi(\xi, \eta)}{\partial \eta} \frac{\partial \eta(x, y)}{\partial y}. \tag{12.7b}$$

The derivatives $\frac{\partial \psi}{\partial \xi}$ are computed directly from Eq. (12.3) as follows

$$\frac{\partial \psi_i(\xi, \eta)}{\partial \xi} = \frac{\partial h_j(\xi)}{\partial \xi} \otimes h_k(\eta) \tag{12.8a}$$

$$\frac{\partial \psi_i(\xi, \eta)}{\partial \eta} = h_j(\xi) \otimes \frac{\partial h_k(\eta)}{\partial \eta} \tag{12.8b}$$

where the 1D derivatives $\frac{\partial h_j(\xi)}{\partial \xi}$ and $\frac{\partial h_k(\eta)}{\partial \eta}$ were defined in Ch. 3. The metric terms $\frac{\partial \xi}{\partial \mathbf{x}}$ are defined in Sec. 12.7.

12.6 Basis Functions in Three Dimensions

The 3D basis functions on the reference element $(\xi, \eta, \zeta) \in [-1, +1]^3$ are written as

$$\psi_i(\xi, \eta, \zeta) = h_j(\xi) \otimes h_k(\eta) \otimes h_l(\zeta) \tag{12.9}$$

where h are the 1D basis functions described in Ch. 3 and the 1D indices vary as follows: $j = 0, \ldots, N_\xi$, $k = 0, \ldots, N_\eta$, $l = 0, \ldots, N_\zeta$ with the 3D index varying as $i = 1, \ldots, M_N$ where $M_N = (N_\xi + 1)(N_\eta + 1)(N_\zeta + 1)$. To get from the 1D local indices (j, k, l) to the 3D local index i requires the mapping $i = j + 1 + k(N_\xi + 1) + l(N_\xi + 1)(N_\eta + 1)$.

We can now expand the solution variable q and the gradient operator using Eqs. (12.4) and (12.5), respectively, where the partial derivatives are defined as follows

$$\frac{\partial q_N^{(e)}(\mathbf{x})}{\partial x} = \sum_{i=1}^{M_N} \frac{\partial \psi_i(\mathbf{x})}{\partial x} q_i^{(e)} \tag{12.10a}$$

$$\frac{\partial q_N^{(e)}(\mathbf{x})}{\partial y} = \sum_{i=1}^{M_N} \frac{\partial \psi_i(\mathbf{x})}{\partial y} q_i^{(e)} \tag{12.10b}$$

$$\frac{\partial q_N^{(e)}(\mathbf{x})}{\partial z} = \sum_{i=1}^{M_N} \frac{\partial \psi_i(\mathbf{x})}{\partial z} q_i^{(e)}. \tag{12.10c}$$

To transform all derivatives from (x, y, z) to (ξ, η, ζ) we write the derivatives of the basis functions as

$$\frac{\partial \psi(x(\boldsymbol{\xi}), y(\boldsymbol{\xi}), z(\boldsymbol{\xi}))}{\partial x} = \frac{\partial \psi(\boldsymbol{\xi})}{\partial \xi} \frac{\partial \xi(\mathbf{x})}{\partial x} + \frac{\partial \psi(\boldsymbol{\xi})}{\partial \eta} \frac{\partial \eta(\mathbf{x})}{\partial x} + \frac{\partial \psi(\boldsymbol{\xi})}{\partial \zeta} \frac{\partial \zeta(\mathbf{x})}{\partial x} \tag{12.11a}$$

$$\frac{\partial \psi(x(\boldsymbol{\xi}), y(\boldsymbol{\xi}), z(\boldsymbol{\xi}))}{\partial y} = \frac{\partial \psi(\boldsymbol{\xi})}{\partial \xi} \frac{\partial \xi(\mathbf{x})}{\partial y} + \frac{\partial \psi(\boldsymbol{\xi})}{\partial \eta} \frac{\partial \eta(\mathbf{x})}{\partial y} + \frac{\partial \psi(\boldsymbol{\xi})}{\partial \zeta} \frac{\partial \zeta(\mathbf{x})}{\partial y} \tag{12.11b}$$

$$\frac{\partial \psi(x(\boldsymbol{\xi}), y(\boldsymbol{\xi}), z(\boldsymbol{\xi}))}{\partial z} = \frac{\partial \psi(\boldsymbol{\xi})}{\partial \xi} \frac{\partial \xi(\mathbf{x})}{\partial z} + \frac{\partial \psi(\boldsymbol{\xi})}{\partial \eta} \frac{\partial \eta(\mathbf{x})}{\partial z} + \frac{\partial \psi(\boldsymbol{\xi})}{\partial \zeta} \frac{\partial \zeta(\mathbf{x})}{\partial z}. \tag{12.11c}$$

The derivatives $\frac{\partial \psi}{\partial \xi}$ are computed directly from Eq. (12.9) as follows

$$\frac{\partial \psi_i(\xi, \eta)}{\partial \xi} = \frac{\partial h_j(\xi)}{\partial \xi} \otimes h_k(\eta) \otimes h_l(\zeta) \tag{12.12a}$$

$$\frac{\partial \psi_i(\xi, \eta)}{\partial \eta} = h_j(\xi) \otimes \frac{\partial h_k(\eta)}{\partial \eta} \otimes h_l(\zeta) \tag{12.12b}$$

$$\frac{\partial \psi_i(\xi, \eta)}{\partial \zeta} = h_j(\xi) \otimes h_k(\eta) \otimes \frac{\partial h_l(\zeta)}{\partial \zeta} \tag{12.12c}$$

where the 1D derivatives $\frac{\partial h_j(\xi)}{\partial \xi}$, $\frac{\partial h_k(\eta)}{\partial \eta}$, and $\frac{\partial h_k(\zeta)}{\partial \zeta}$ were defined in Ch. 3. Let us now define the metric terms $\frac{\partial \xi}{\partial x}$.

12.7 Metric Terms of the Mapping

12.7.1 Metric Terms in Two Dimensions

Since we have defined $\mathbf{x} = \mathbf{x}(\xi, \eta)$ the chain rule yields

$$d\mathbf{x} = \frac{\partial \mathbf{x}}{\partial \xi} d\xi + \frac{\partial \mathbf{x}}{\partial \eta} d\eta$$

which can be written in the following matrix form

$$\begin{pmatrix} dx \\ dy \end{pmatrix} = \begin{pmatrix} \frac{\partial x}{\partial \xi} & \frac{\partial x}{\partial \eta} \\ \frac{\partial y}{\partial \xi} & \frac{\partial y}{\partial \eta} \end{pmatrix} \begin{pmatrix} d\xi \\ d\eta \end{pmatrix}$$

where

$$
\mathbf{J} = \begin{pmatrix} \dfrac{\partial x}{\partial \xi} & \dfrac{\partial x}{\partial \eta} \\[2mm] \dfrac{\partial y}{\partial \xi} & \dfrac{\partial y}{\partial \eta} \end{pmatrix}
\tag{12.13}
$$

is the Jacobian of the transformation, with determinant

$$
\det(\mathbf{J}) \equiv J^{(e)} = \frac{\partial x}{\partial \xi}\frac{\partial y}{\partial \eta} - \frac{\partial y}{\partial \xi}\frac{\partial x}{\partial \eta}.
$$

Alternatively, we can write the derivatives of $\xi(x, y)$ to get

$$
d\xi = \frac{\partial \xi}{\partial x} dx + \frac{\partial \xi}{\partial y} dy
$$

which, in matrix form, yields

$$
\begin{pmatrix} d\xi \\[1mm] d\eta \end{pmatrix} = \begin{pmatrix} \dfrac{\partial \xi}{\partial x} & \dfrac{\partial \xi}{\partial y} \\[2mm] \dfrac{\partial \eta}{\partial x} & \dfrac{\partial \eta}{\partial y} \end{pmatrix} \begin{pmatrix} dx \\[1mm] dy \end{pmatrix}
$$

where

$$
\mathbf{J}^{-1} = \begin{pmatrix} \dfrac{\partial \xi}{\partial x} & \dfrac{\partial \xi}{\partial y} \\[2mm] \dfrac{\partial \eta}{\partial x} & \dfrac{\partial \eta}{\partial y} \end{pmatrix}
\tag{12.14}
$$

is the inverse Jacobian of the transformation. Again, using the chain rule, we can write the derivatives in the physical space in terms of the reference element space as follows

$$
\frac{\partial}{\partial \mathbf{x}} = \frac{\partial \xi}{\partial \mathbf{x}}\frac{\partial}{\partial \xi} + \frac{\partial \eta}{\partial \mathbf{x}}\frac{\partial}{\partial \eta}.
$$

Expanding into its components, we see that the derivatives are in fact given by

$$
\begin{pmatrix} \dfrac{\partial}{\partial x} \\[3mm] \dfrac{\partial}{\partial y} \end{pmatrix} = \mathbf{J}^{-1} \begin{pmatrix} \dfrac{\partial}{\partial \xi} \\[3mm] \dfrac{\partial}{\partial \eta} \end{pmatrix}
$$

where the inverse Jacobian matrix is

$$
\mathbf{J}^{-1} = \frac{1}{J^{(e)}} \begin{pmatrix} \dfrac{\partial y}{\partial \eta} & -\dfrac{\partial x}{\partial \eta} \\[2mm] -\dfrac{\partial y}{\partial \xi} & \dfrac{\partial x}{\partial \xi} \end{pmatrix}
\tag{12.15}
$$

and has been obtained by inverting Eq. (12.13).

Exercise Show that Eq. (12.15) is the inverse of Eq. (12.13). □

Let the mapping $\mathbf{x} = \Psi(\xi)$ define the transformation from the local reference coordinate system, ξ, in the reference element $\hat{\Omega}$ to the physical Cartesian coordinate system, $\mathbf{x}$, within each physical element Ω_e. This mapping $\Psi(\boldsymbol{\xi})$ (and its inverse $\boldsymbol{\xi} = \Psi^{-1}(\mathbf{x})$) now allows us to construct basis function derivatives in terms of the reference element (ξ, η) as follows

$$\frac{\partial \psi}{\partial x} = \frac{\partial \psi}{\partial \xi}\frac{\partial \xi}{\partial x} + \frac{\partial \psi}{\partial \eta}\frac{\partial \eta}{\partial x}$$

where from equating Eqs. (12.14) and (12.15) we see that the metric terms are in fact defined explicitly as

$$\frac{\partial \xi}{\partial x} = \frac{1}{J^{(e)}}\frac{\partial y}{\partial \eta}, \quad \frac{\partial \xi}{\partial y} = -\frac{1}{J^{(e)}}\frac{\partial x}{\partial \eta}, \quad \frac{\partial \eta}{\partial x} = -\frac{1}{J^{(e)}}\frac{\partial y}{\partial \xi}, \quad \frac{\partial \eta}{\partial y} = \frac{1}{J^{(e)}}\frac{\partial x}{\partial \xi}. \tag{12.16}$$

The final step required to compute these metric terms is to approximate the derivatives $\frac{\partial x}{\partial \xi}$ using the basis functions. Let us approximate the physical coordinates by the basis function expansion

$$\mathbf{x}_N^{(e)}(\xi, \eta) = \sum_{i=1}^{M_N} \psi_i(\xi, \eta)\mathbf{x}_i^{(e)}$$

which then yields the derivatives

$$\frac{\partial \mathbf{x}_N^{(e)}}{\partial \xi}(\xi, \eta) = \sum_{i=1}^{M_N} \frac{\partial \psi_i}{\partial \xi}(\xi, \eta)\mathbf{x}_i^{(e)}. \tag{12.17}$$

For 2D tensor-products, this expression is written as follows

$$\frac{\partial \mathbf{x}_N^{(e)}}{\partial \xi}(\xi, \eta) = \sum_{i=0}^{N_\xi}\sum_{j=0}^{N_\eta} \frac{\partial h_i}{\partial \xi}(\xi)h_j(\eta)\mathbf{x}_{i,j}^{(e)} \tag{12.18a}$$

$$\frac{\partial \mathbf{x}_N^{(e)}}{\partial \eta}(\xi, \eta) = \sum_{i=0}^{N_\xi}\sum_{j=0}^{N_\eta} h_i(\xi)\frac{\partial h_j}{\partial \eta}(\eta)\mathbf{x}_{i,j}^{(e)}. \tag{12.18b}$$

12.7.1.2 Covariant and Contravariant Vectors

The metric terms given by Eq. (12.17) are called the *covariant* metric terms and define the tangent vectors on the physical mesh ($\mathbf{x}$) along the reference element space ($\boldsymbol{\xi}$). In contrast, Eq. (12.16) denote the contravariant metric terms which allow us to map from the covariant vectors to the *contravariant* vectors. To better explain this, let us assume that we have a vector defined in the two-dimensional Cartesian space

$$\mathbf{u} = u\hat{\mathbf{i}} + v\hat{\mathbf{j}}$$

that we call the covariant vector, where $\hat{\mathbf{i}}$ and $\hat{\mathbf{j}}$ are the Cartesian coordinate unit vectors. The contravariant representation of $\mathbf{u}$ is then defined as follows

$$\mathbf{u}^{\xi} = \nabla\xi \cdot \mathbf{u} \tag{12.19}$$

where

$$\nabla\xi = \frac{\partial\xi}{\partial x}\hat{\mathbf{i}} + \frac{\partial\xi}{\partial y}\hat{\mathbf{j}}.$$

Expanding Eq. (12.19) and substituting $\mathbf{u} = \frac{d\mathbf{x}}{dt}$ yields

$$\mathbf{u}^{\xi} = \frac{\partial\xi}{\partial\mathbf{x}} \cdot \frac{d\mathbf{x}}{dt} \equiv \frac{d\xi}{dt}. \tag{12.20}$$

This analysis shows that the covariant velocities are the velocities along the physical coordinate system (in this case x-y) while the contravariant velocities are the velocities along the reference coordinate system (in this case ξ-η). This mapping is valid for all vectors, not just the velocity vectors.

The connection between covariant and contravariant metric vectors and their construction in multiple dimensions is better explained if we first write the covariant metric terms as follows

$$\mathbf{v}_i = \frac{\partial\mathbf{x}}{\partial\xi_i}$$

which, for example, gives

$$\mathbf{v}_1 = \frac{\partial\mathbf{x}}{\partial\xi} \equiv \frac{\partial x}{\partial\xi}\hat{\mathbf{i}} + \frac{\partial y}{\partial\xi}\hat{\mathbf{j}} + 0\,\hat{\mathbf{k}},$$

$$\mathbf{v}_2 = \frac{\partial\mathbf{x}}{\partial\eta} \equiv \frac{\partial x}{\partial\eta}\hat{\mathbf{i}} + \frac{\partial y}{\partial\eta}\hat{\mathbf{j}} + 0\,\hat{\mathbf{k}}, \tag{12.21}$$

$$\mathbf{v}_3 = \hat{\mathbf{k}}$$

where the directional unit vectors $(\hat{\mathbf{i}}, \hat{\mathbf{j}}, \hat{\mathbf{k}})$ are along the covariant direction (in this case, the Cartesian x-y-z space). Then, following Kopriva [236][1], we define the connection between the covariant vectors $\mathbf{v}_i$ with the contravariant vectors $\mathbf{v}^i$ as follows

$$\mathbf{v}^i = \frac{1}{J^{(e)}}\left(\mathbf{v}_j \times \mathbf{v}_k\right) \tag{12.22}$$

where the cycling order is $[(i, j, k), (j, k, i), (k, i, j)]$ and $\mathbf{v}^1 = \nabla\xi$ and $\mathbf{v}^2 = \nabla\eta$.

Exercise Using the definition given in Eq. (12.22) derive the contravariant metrics in Eq. (12.16).   □

---

[1] In my opinion, the clearest discussion on the topic of covariant and contravariant vectors is found in Chapters 5 and 6 of David Kopriva's text. The interested reader should refer to this text for the derivation of the discussion that we use in this section.

Using the connection between the covariant and contravariant metric terms defined in Eqs. (12.22) we can rewrite Eq. (12.20) as follows

$$u^i = \mathbf{v}^i \cdot \mathbf{u}, \tag{12.23}$$

which, component-wise, yields

$$u^1 \equiv u^\xi = \mathbf{v}^1 \cdot \mathbf{u} = \nabla\xi \cdot \mathbf{u}$$

and

$$u^2 \equiv u^\eta = \mathbf{v}^2 \cdot \mathbf{u} = \nabla\eta \cdot \mathbf{u}.$$

Exercise Derive the contravariant velocities (u^ξ, u^η, u^ζ) for the covariant metrics defined as $\frac{\partial \mathbf{x}}{\partial \xi}, \frac{\partial \mathbf{x}}{\partial \eta}, \frac{\partial \mathbf{x}}{\partial \zeta}$ where $\mathbf{x} = x\,\hat{\mathbf{i}} + y\,\hat{\mathbf{j}} + z\,\hat{\mathbf{k}}$. Use Eq. (12.21) with $\mathbf{v}_3 = \frac{\partial \mathbf{x}}{\partial \zeta}$. □

12.7.2 Metric Terms in Three Dimensions

12.7.2.1 Cross-Product Metric Terms in Three Dimensions

To construct the metric terms in three dimensions, we start with $\mathbf{x} = \mathbf{x}(\xi, \eta, \zeta)$ and apply the chain rule as follows

$$d\mathbf{x} = \frac{\partial \mathbf{x}}{\partial \xi}d\xi + \frac{\partial \mathbf{x}}{\partial \eta}d\eta + \frac{\partial \mathbf{x}}{\partial \zeta}d\zeta$$

which we rewrite more compactly as

$$d\mathbf{x} = \mathbf{x}_\xi d\xi + \mathbf{x}_\eta d\eta + \mathbf{x}_\zeta d\zeta.$$

This now yields the following matrix form

$$\begin{pmatrix} dx \\ dy \\ dz \end{pmatrix} = \begin{pmatrix} x_\xi & x_\eta & x_\zeta \\ y_\xi & y_\eta & y_\zeta \\ z_\xi & z_\eta & z_\zeta \end{pmatrix} \begin{pmatrix} d\xi \\ d\eta \\ d\zeta \end{pmatrix}$$

where

$$\mathbf{J} = \begin{pmatrix} x_\xi & x_\eta & x_\zeta \\ y_\xi & y_\eta & y_\zeta \\ z_\xi & z_\eta & z_\zeta \end{pmatrix} \tag{12.24}$$

is the Jacobian of the transformation, with determinant

$$J^{(e)} \equiv \mathbf{x}_\xi \cdot (\mathbf{x}_\eta \times \mathbf{x}_\zeta) = x_\xi \left(y_\eta z_\zeta - y_\zeta z_\eta \right) - y_\xi \left(x_\eta z_\zeta - x_\zeta z_\eta \right) + z_\xi \left(x_\eta y_\zeta - x_\zeta y_\eta \right).$$

Alternatively, we can write the derivatives of $\boldsymbol{\xi}(x, y, z)$ to get

$$d\boldsymbol{\xi} = \boldsymbol{\xi}_x dx + \boldsymbol{\xi}_y dy + \boldsymbol{\xi}_z dz$$

which, in matrix form, yields

$$\begin{pmatrix} d\xi \\ d\eta \\ d\zeta \end{pmatrix} = \begin{pmatrix} \xi_x & \xi_y & \xi_z \\ \eta_x & \eta_y & \eta_z \\ \zeta_x & \zeta_y & \zeta_z \end{pmatrix} \begin{pmatrix} dx \\ dy \\ dz \end{pmatrix}$$

where

$$\mathbf{J}^{-1} = \begin{pmatrix} \xi_x & \xi_y & \xi_z \\ \eta_x & \eta_y & \eta_z \\ \zeta_x & \zeta_y & \zeta_z \end{pmatrix} \tag{12.25}$$

is the inverse Jacobian of the transformation. Again, using the chain rule, we can write the derivatives in the physical space in terms of the reference element space as follows

$$\frac{\partial}{\partial \mathbf{x}} = \boldsymbol{\xi}_x \frac{\partial}{\partial \xi} + \eta_x \frac{\partial}{\partial \eta} + \zeta_x \frac{\partial}{\partial \zeta}.$$

Expanding into its components, we see that the derivatives are in fact given by

$$\begin{pmatrix} \frac{\partial}{\partial x} \\ \frac{\partial}{\partial y} \\ \frac{\partial}{\partial z} \end{pmatrix} = \mathbf{J}^{-1} \begin{pmatrix} \frac{\partial}{\partial \xi} \\ \frac{\partial}{\partial \eta} \\ \frac{\partial}{\partial \zeta} \end{pmatrix}$$

where the inverse Jacobian matrix is

$$\mathbf{J}^{-1} = \frac{1}{J^{(e)}} \begin{pmatrix} y_\eta z_\zeta - y_\zeta z_\eta & x_\zeta z_\eta - x_\eta z_\zeta & x_\eta y_\zeta - x_\zeta y_\eta \\ y_\zeta z_\xi - y_\xi z_\zeta & x_\xi z_\zeta - x_\zeta z_\xi & x_\zeta y_\xi - x_\xi y_\zeta \\ y_\xi z_\eta - y_\eta z_\xi & x_\eta z_\xi - x_\xi z_\eta & x_\xi y_\eta - x_\eta y_\xi \end{pmatrix} \tag{12.26}$$

and has been obtained by inverting Eq. (12.24).

Exercise Show that Eq. (12.26) is the inverse of Eq. (12.24).   □

Let the mapping $\mathbf{x} = \boldsymbol{\Psi}(\boldsymbol{\xi})$ define the transformation from the local reference coordinate system, $\boldsymbol{\xi}$, in the reference element $\hat{\Omega}$ to the physical Cartesian coordinate system, $\mathbf{x}$, within each physical element Ω_e. This mapping $\boldsymbol{\Psi}(\boldsymbol{\xi})$ (and its inverse $\boldsymbol{\xi} = \boldsymbol{\Psi}^{-1}(\mathbf{x})$) now allows us to construct basis function derivatives in terms of the reference element (ξ, η, ζ) as follows

$$\psi_x = \psi_\xi \xi_x + \psi_\eta \eta_x + \psi_\zeta \zeta_x$$

where from equating Eqs. (12.25) and (12.26) we see that the metric terms are in fact defined explicitly as

$$\xi_x = \frac{1}{J^{(e)}} \left(y_\eta z_\zeta - y_\zeta z_\eta \right), \quad \xi_y = \frac{1}{J^{(e)}} \left(x_\zeta z_\eta - x_\eta z_\zeta \right), \quad \xi_z = \frac{1}{J^{(e)}} \left(x_\eta y_\zeta - x_\zeta y_\eta \right),$$
$$(12.27a)$$

$$\eta_x = \frac{1}{J^{(e)}} \left(y_\zeta z_\xi - y_\xi z_\zeta \right), \quad \eta_y = \frac{1}{J^{(e)}} \left(x_\xi z_\zeta - x_\zeta z_\xi \right), \quad \eta_z = \frac{1}{J^{(e)}} \left(x_\zeta y_\xi - x_\xi y_\zeta \right),$$
$$(12.27b)$$

$$\zeta_x = \frac{1}{J^{(e)}} \left(y_\xi z_\eta - y_\eta z_\xi \right), \quad \zeta_y = \frac{1}{J^{(e)}} \left(x_\eta z_\xi - x_\xi z_\eta \right), \quad \zeta_z = \frac{1}{J^{(e)}} \left(x_\xi y_\eta - x_\eta y_\xi \right).$$
$$(12.27c)$$

The final step required to compute these metric terms is to approximate the derivatives x_ξ using the basis functions. Let us approximate the physical coordinates by the basis function expansion

$$\mathbf{x}_N^{(e)}(\xi, \eta) = \sum_{i=1}^{M_N} \psi_i(\xi, \eta) \mathbf{x}_i^{(e)}$$

which then yields the derivatives

$$\frac{\partial \mathbf{x}_N^{(e)}}{\partial \xi}(\xi, \eta) = \sum_{i=1}^{M_N} \frac{\partial \psi_i}{\partial \xi}(\xi, \eta) \mathbf{x}_i^{(e)}. \tag{12.28}$$

For 3D tensor-products, this expression is written as follows

$$\frac{\partial \mathbf{x}_N^{(e)}}{\partial \xi}(\xi, \eta, \zeta) = \sum_{i=0}^{N_\xi} \sum_{j=0}^{N_\eta} \sum_{k=0}^{N_\zeta} \frac{\partial h_i}{\partial \xi}(\xi) h_j(\eta) h_k(\zeta) \mathbf{x}_{i,j,k}^{(e)} \tag{12.29a}$$

$$\frac{\partial \mathbf{x}_N^{(e)}}{\partial \eta}(\xi, \eta, \zeta) = \sum_{i=0}^{N_\xi} \sum_{j=0}^{N_\eta} \sum_{k=0}^{N_\zeta} h_i(\xi) \frac{\partial h_j}{\partial \eta}(\eta) h_k(\zeta) \mathbf{x}_{i,j,k}^{(e)} \tag{12.29b}$$

$$\frac{\partial \mathbf{x}_N^{(e)}}{\partial \eta}(\xi, \eta, \zeta) = \sum_{i=0}^{N_\xi} \sum_{j=0}^{N_\eta} \sum_{k=0}^{N_\zeta} h_i(\xi) h_j(\eta) \frac{\partial h_k}{\partial \zeta}(\zeta) \mathbf{x}_{i,j,k}^{(e)} \tag{12.29c}$$

where N_ξ, N_η, N_ζ are the polynomial orders along each of the three directions of the reference element.

The metric terms derived in Eq. (12.27) can be written in the following more compact form

$$\nabla \xi = \frac{1}{J^{(e)}} \left(\frac{\partial \mathbf{x}}{\partial \eta} \times \frac{\partial \mathbf{x}}{\partial \zeta} \right) \tag{12.30a}$$

$$\nabla \eta = \frac{1}{J^{(e)}} \left(\frac{\partial \mathbf{x}}{\partial \zeta} \times \frac{\partial \mathbf{x}}{\partial \xi} \right) \tag{12.30b}$$

$$\nabla\zeta = \frac{1}{J^{(e)}}\left(\frac{\partial \mathbf{x}}{\partial \xi} \times \frac{\partial \mathbf{x}}{\partial \eta}\right) \tag{12.30c}$$

which, we refer to as the *cross-product* form of the metric terms. Note also that the 2D metric terms defined in Eqs. (12.16) and (12.22) are also cross-product metric terms. To derive these relations, we substitute $\frac{\partial \mathbf{x}}{\partial \zeta} = \hat{\mathbf{k}}$ and $\mathbf{x} = x\hat{\mathbf{i}} + y\hat{\mathbf{j}} + 0\hat{\mathbf{k}}$.

12.7.2.3 Curl Invariant Metric Terms in Three Dimensions

In [235], Kopriva proved that this form of the metric terms in three dimensions will not satisfy *constant-state preservation*; constant-state preservation means that constant solutions do not evolve. To see what this condition implies on the metric terms, let us consider the conservation law

$$\frac{\partial q}{\partial t} + \nabla \cdot \mathbf{f} = 0,$$

which can be written in terms of the reference element coordinates[2] as follows

$$\frac{\partial q}{\partial t} + \frac{1}{J^{(e)}}\nabla_\xi \cdot \left(J^{(e)}\mathbf{f}^i\right) = 0,$$

where $\mathbf{f}^i$ denotes the contravariant vector as described in Eq. (12.23), and

$$\nabla_\xi = \frac{\partial}{\partial \xi}\mathbf{v}^1 + \frac{\partial}{\partial \eta}\mathbf{v}^2 + \frac{\partial}{\partial \zeta}\mathbf{v}^3$$

where $\mathbf{v}^i$ is given in Eq. (12.22). Note that if we let $q =$ constant and $\mathbf{f} =$ constant then the conservation law becomes

$$\nabla_\xi \cdot \left(J^{(e)}\nabla\xi^i\right) = 0, \tag{12.31}$$

which defines the three (for each value of $i = 1, 2, 3$) discrete metric identities given in [235], where

$$\nabla\xi^1 = \nabla\xi, \quad \nabla\xi^2 = \nabla\eta, \quad \nabla\xi^3 = \nabla\zeta.$$

In [235] it is shown that, in 3D, the cross-product form of the metric terms given by Eq. (12.30) do not satisfy the discrete metric identities given by Eq. (12.31). Instead, the *curl invariant form* defined as follows [373, 235]

$$\nabla\xi^i = \frac{1}{2J^{(e)}}\left[\left(\mathbf{x}_{\xi^j} \times \mathbf{x}\right)_{\xi^k} - \left(\mathbf{x}_{\xi^k} \times \mathbf{x}\right)_{\xi^j}\right] \tag{12.32}$$

does satisfy the discrete metric identities, where (i, j, k) are cyclic as follows $(1, 2, 3)$, $(2, 3, 1)$, $(3, 1, 2)$,

$$\xi^1 = \xi, \quad \xi^2 = \eta, \quad \xi^3 = \zeta.$$

---

[2] For a discussion on the advantages of using the contravariant form of the governing equations, the reader is referred to Appendix D where the contravariant form of the 3D Euler equations in conservation form and the 2D shallow water equations in non-conservation form are discussed.

and $\mathbf{x}_{\xi^j}$ denotes the partial derivative of the physical coordinate $\mathbf{x}$ with respect to the reference element coordinate ξ^j. In Eq. (12.32), the partial derivatives $\mathbf{x}_{\xi^j}$ are obtained using Eq. (12.17) for the monolithic form or Eq. (12.29) for the tensor-product form. Terms such as $\left(\mathbf{x}_{\xi^j} \times \mathbf{x} \right)_{\xi^k}$ are computed as follows: let, e.g., $\omega = \mathbf{x}_{\xi^j} \times \mathbf{x}$ and then compute ω_{ξ^k} using Eqs. (12.17) or (12.29).

Exercise Write an algorithm for computing the metric terms using the curl invariant form for the monolithic form. Feel free to use Alg. 12.2 as a template.   □

Exercise Write an algorithm for computing the metric terms using the curl invariant form for the tensor-product form. Feel free to use Alg. 12.5 as a template. You should use Alg. 12.4 to construct the derivatives with respect to the reference element coordinates.   □

12.7.3 Normal Vectors

The remaining metric terms required in element-based Galerkin methods are the normal vectors. Fortunately, we have already derived all of the necessary components in order to construct the normal vectors. By writing the contravariant metric terms as the cross-products of the covariant metric vectors as given by Eq. (12.30) we can see that the contravariant metric terms are in fact the normal vectors. However, these normal vectors are only required along the faces of the element. The normal vectors are given by the following compact relation

$$\hat{\mathbf{n}} = \xi^i \left(\frac{\nabla \xi^i}{| \nabla \xi^i |} \right)_{\xi^i = -1,+1} \tag{12.33}$$

which, can be expanded along each direction as follows

$$\hat{\mathbf{n}}_\xi = \xi \left(\frac{\nabla \xi}{| \nabla \xi |} \right)_{\xi = -1,+1} \tag{12.34a}$$

$$\hat{\mathbf{n}}_\eta = \eta \left(\frac{\nabla \eta}{| \nabla \eta |} \right)_{\eta = -1,+1} \tag{12.34b}$$

$$\hat{\mathbf{n}}_\zeta = \zeta \left(\frac{\nabla \zeta}{| \nabla \zeta |} \right)_{\zeta = -1,+1}. \tag{12.34c}$$

The face Jacobian (which we shall see in Chs. 13 and 16 defines the length of the face in 2D and the surface area of the face in 3D and is defined as follows:

$$J^{(f)} = | \nabla \xi^i |. \tag{12.35}$$

It should be emphasized that the definition of the normal vectors defined in this section are independent of the construction of the metric terms. That is, this approach works for either the cross-product or curl invariant forms presented and hold in two and three dimensions.

12.7.4 Algorithm for the Metric Terms

Algorithm 12.1 summarizes the procedure for constructing the metric terms in two dimensions for a general polynomial order N with quadrature strength Q where $J^{(e)}$ denotes the determinant of the volume Jacobian. The only additional information required by this algorithm is the grid information, i.e., it is assumed that we know the physical gridpoints $\mathbf{x}$ for each element in the grid. We always have to know the basis functions ψ along with the total number of interpolation points M_N and quadrature points M_Q per element. Here we assume that these quantities are constant across all elements N_e.

Algorithm 12.1 describes the procedure for constructing metric terms in two dimensions using the monolithic form of the basis functions; Algorithm 12.2 describes the procedure in three dimensions. However, if we are using tensor-product basis functions then we can exploit this property and use the one-dimensional basis functions directly as described in Alg. 12.3 in two dimensions which uses the representation of the basis function derivatives given by Eq. (12.18). The algorithm for three dimensions with tensor-products is outlined in Alg. 12.5 where (N_ξ, N_η, N_ζ) and (Q_ξ, Q_η, Q_ζ) denote the polynomial and quadrature orders along each of the three directions along the reference element (ξ, η, ζ).

Algorithm 12.1 Construction of the 2D metric terms using the monolithic form.

function 2D_METRIC_TERMS
$\quad \frac{\partial \mathbf{x}^{(e)}}{\partial \xi} = \mathbf{0}$ ⊳ initialize all four arrays
$\quad$ **for** $e = 1 : N_e$ **do** ⊳ loop over elements
$\quad\quad$ **for** $k = 1 : M_Q$ **do** ⊳ loop over integration points
$\quad\quad\quad$ **for** $i = 1 : M_N$ **do** ⊳ loop over interpolation points
$$\frac{\partial x_k^{(e)}}{\partial \xi} \mathrel{+}= \frac{\partial \psi_{ik}}{\partial \xi} x_i^{(e)}; \ \frac{\partial x_k^{(e)}}{\partial \eta} \mathrel{+}= \frac{\partial \psi_{ik}}{\partial \eta} x_i^{(e)}$$
$$\frac{\partial y_k^{(e)}}{\partial \xi} \mathrel{+}= \frac{\partial \psi_{ik}}{\partial \xi} y_i^{(e)}; \ \frac{\partial y_k^{(e)}}{\partial \eta} \mathrel{+}= \frac{\partial \psi_{ik}}{\partial \eta} y_i^{(e)}$$
$\quad\quad\quad$ **end for**
$\quad\quad$ **end for**

$\quad\quad$ **for** $k = 1 : M_Q$ **do** ⊳ loop over integration points
$$J_k^{(e)} = \frac{\partial x_k^{(e)}}{\partial \xi} \cdot \frac{\partial y_k^{(e)}}{\partial \eta} - \frac{\partial x_k^{(e)}}{\partial \eta} \cdot \frac{\partial y_k^{(e)}}{\partial \xi}$$
$$\frac{\partial \xi_k^{(e)}}{\partial x} = +\frac{\partial y_k^{(e)}}{\partial \eta} / J_k^{(e)}; \ \frac{\partial \xi_k^{(e)}}{\partial y} = -\frac{\partial x_k^{(e)}}{\partial \eta} / J_k^{(e)}$$
$$\frac{\partial \eta_k^{(e)}}{\partial x} = -\frac{\partial y_k^{(e)}}{\partial \xi} / J_k^{(e)}; \ \frac{\partial \eta_k^{(e)}}{\partial y} = +\frac{\partial x_k^{(e)}}{\partial \xi} / J_k^{(e)}$$
$\quad\quad$ **end for**
$\quad$ **end for**
end function

Algorithm 12.2 Construction of the 3D metric terms using the monolithic form.

---

function 3D_METRIC_TERMS

$\dfrac{\partial \mathbf{x}^{(e)}}{\partial \xi} = \mathbf{0}$ ▷ initialize all four arrays

for $e = 1 : N_e$ **do** ▷ loop over elements

 for $k = 1 : M_Q$ **do** ▷ loop over integration points

  for $i = 1 : M_N$ **do** ▷ loop over interpolation points

$$\frac{\partial x_k^{(e)}}{\partial \xi} \mathrel{+}= \frac{\partial \psi_{ik}}{\partial \xi} x_i^{(e)};\ \frac{\partial x_k^{(e)}}{\partial \eta} \mathrel{+}= \frac{\partial \psi_{ik}}{\partial \eta} x_i^{(e)};\ \frac{\partial x_k^{(e)}}{\partial \zeta} \mathrel{+}= \frac{\partial \psi_{ik}}{\partial \zeta} x_i^{(e)}$$

$$\frac{\partial y_k^{(e)}}{\partial \xi} \mathrel{+}= \frac{\partial \psi_{ik}}{\partial \xi} y_i^{(e)};\ \frac{\partial y_k^{(e)}}{\partial \eta} \mathrel{+}= \frac{\partial \psi_{ik}}{\partial \eta} y_i^{(e)};\ \frac{\partial y_k^{(e)}}{\partial \zeta} \mathrel{+}= \frac{\partial \psi_{ik}}{\partial \zeta} y_i^{(e)}$$

$$\frac{\partial z_k^{(e)}}{\partial \xi} \mathrel{+}= \frac{\partial \psi_{ik}}{\partial \xi} z_i^{(e)};\ \frac{\partial z_k^{(e)}}{\partial \eta} \mathrel{+}= \frac{\partial \psi_{ik}}{\partial \eta} z_i^{(e)};\ \frac{\partial z_k^{(e)}}{\partial \zeta} \mathrel{+}= \frac{\partial \psi_{ik}}{\partial \zeta} z_i^{(e)}$$

  end for

 end for

 for $k = 1 : M_Q$ **do** ▷ loop over integration points

$$J_k^{(e)} = \frac{\partial x_k^{(e)}}{\partial \xi}\left(\frac{\partial y_k^{(e)}}{\partial \eta}\cdot\frac{\partial z_k^{(e)}}{\partial \zeta} - \frac{\partial y_k^{(e)}}{\partial \zeta}\cdot\frac{\partial z_k^{(e)}}{\partial \eta}\right)$$

$$J_k^{(e)} \mathrel{+}= \frac{\partial y_k^{(e)}}{\partial \xi}\left(\frac{\partial x_k^{(e)}}{\partial \zeta}\cdot\frac{\partial z_k^{(e)}}{\partial \eta} - \frac{\partial x_k^{(e)}}{\partial \eta}\cdot\frac{\partial z_k^{(e)}}{\partial \zeta}\right)$$

$$J_k^{(e)} \mathrel{+}= \frac{\partial z_k^{(e)}}{\partial \xi}\left(\frac{\partial x_k^{(e)}}{\partial \eta}\cdot\frac{\partial y_k^{(e)}}{\partial \zeta} - \frac{\partial x_k^{(e)}}{\partial \zeta}\cdot\frac{\partial y_k^{(e)}}{\partial \eta}\right)$$

$$\frac{\partial \xi_k^{(e)}}{\partial x} = \left(\frac{\partial y_k^{(e)}}{\partial \eta}\cdot\frac{\partial z_k^{(e)}}{\partial \zeta} - \frac{\partial y_k^{(e)}}{\partial \zeta}\cdot\frac{\partial z_k^{(e)}}{\partial \eta}\right)\Big/J_k^{(e)}$$

$$\frac{\partial \xi_k^{(e)}}{\partial y} = \left(\frac{\partial x_k^{(e)}}{\partial \zeta}\cdot\frac{\partial z_k^{(e)}}{\partial \eta} - \frac{\partial x_k^{(e)}}{\partial \eta}\cdot\frac{\partial z_k^{(e)}}{\partial \zeta}\right)\Big/J_k^{(e)}$$

$$\frac{\partial \xi_k^{(e)}}{\partial z} = \left(\frac{\partial x_k^{(e)}}{\partial \eta}\cdot\frac{\partial y_k^{(e)}}{\partial \zeta} - \frac{\partial x_k^{(e)}}{\partial \zeta}\cdot\frac{\partial y_k^{(e)}}{\partial \eta}\right)\Big/J_k^{(e)}$$

$$\frac{\partial \eta_k^{(e)}}{\partial x} = \left(\frac{\partial y_k^{(e)}}{\partial \zeta}\cdot\frac{\partial z_k^{(e)}}{\partial \xi} - \frac{\partial y_k^{(e)}}{\partial \xi}\cdot\frac{\partial z_k^{(e)}}{\partial \zeta}\right)\Big/J_k^{(e)}$$

$$\frac{\partial \eta_k^{(e)}}{\partial y} = \left(\frac{\partial x_k^{(e)}}{\partial \xi}\cdot\frac{\partial z_k^{(e)}}{\partial \zeta} - \frac{\partial x_k^{(e)}}{\partial \zeta}\cdot\frac{\partial z_k^{(e)}}{\partial \xi}\right)\Big/J_k^{(e)}$$

$$\frac{\partial \eta_k^{(e)}}{\partial z} = \left(\frac{\partial x_k^{(e)}}{\partial \zeta}\cdot\frac{\partial y_k^{(e)}}{\partial \xi} - \frac{\partial x_k^{(e)}}{\partial \xi}\cdot\frac{\partial y_k^{(e)}}{\partial \zeta}\right)\Big/J_k^{(e)}$$

$$\frac{\partial \zeta_k^{(e)}}{\partial x} = \left(\frac{\partial y_k^{(e)}}{\partial \xi}\cdot\frac{\partial z_k^{(e)}}{\partial \eta} - \frac{\partial y_k^{(e)}}{\partial \eta}\cdot\frac{\partial z_k^{(e)}}{\partial \xi}\right)\Big/J_k^{(e)}$$

$$\frac{\partial \zeta_k^{(e)}}{\partial y} = \left(\frac{\partial x_k^{(e)}}{\partial \eta}\cdot\frac{\partial z_k^{(e)}}{\partial \xi} - \frac{\partial x_k^{(e)}}{\partial \xi}\cdot\frac{\partial z_k^{(e)}}{\partial \eta}\right)\Big/J_k^{(e)}$$

$$\frac{\partial \zeta_k^{(e)}}{\partial z} = \left(\frac{\partial x_k^{(e)}}{\partial \xi}\cdot\frac{\partial y_k^{(e)}}{\partial \eta} - \frac{\partial x_k^{(e)}}{\partial \eta}\cdot\frac{\partial y_k^{(e)}}{\partial \xi}\right)\Big/J_k^{(e)}$$

  end for

 end for

end function

Algorithm 12.3 Construction of the 2D metric terms using tensor-products.

function 2D_METRIC_TERMS

$\quad \frac{\partial \mathbf{x}^{(e)}}{\partial \xi} = \mathbf{0}$ ▷ initialize all four arrays

$\quad$ **for** $e = 1 : N_e$ **do** ▷ loop over elements

$\quad\quad$ **for** $l = 0 : Q_\eta, \; k = 0 : Q_\xi$ **do** ▷ loop over integration points

$\quad\quad\quad$ **for** $j = 0 : N_\eta, \; i = 0 : N_\xi$ **do** ▷ loop over interpolation points

$$\frac{\partial x_{k,l}^{(e)}}{\partial \xi} \mathrel{+}= \frac{\partial h_{i,k}}{\partial \xi}(\xi)h_{j,l}(\eta)x_{i,j}^{(e)}; \quad \frac{\partial x_{k,l}^{(e)}}{\partial \eta} \mathrel{+}= h_{i,k}(\xi)\frac{\partial h_{j,l}}{\partial \eta}(\eta)x_{i,j}^{(e)}$$

$$\frac{\partial y_{k,l}^{(e)}}{\partial \xi} \mathrel{+}= \frac{\partial h_{i,k}}{\partial \xi}(\xi)h_{j,l}(\eta)y_{i,j}^{(e)}; \quad \frac{\partial y_{k,l}^{(e)}}{\partial \eta} \mathrel{+}= h_{i,k}(\xi)\frac{\partial h_{j,l}}{\partial \eta}(\eta)y_{i,j}^{(e)}$$

$\quad\quad\quad$ **end for**

$\quad\quad$ **end for**

$\quad\quad$ **for** $l = 0 : Q_\eta, \; k = 0 : Q_\xi$ **do** ▷ loop over integration points

$$J_{k,l}^{(e)} = \frac{\partial x_{k,l}^{(e)}}{\partial \xi} \cdot \frac{\partial y_{k,l}^{(e)}}{\partial \eta} - \frac{\partial x_{k,l}^{(e)}}{\partial \eta} \cdot \frac{\partial y_{k,l}^{(e)}}{\partial \xi}$$

$$\frac{\partial \xi_{k,l}^{(e)}}{\partial x} = +\frac{\partial y_{k,l}^{(e)}}{\partial \eta}/J_{k,l}^{(e)}; \quad \frac{\partial \xi_{k,l}^{(e)}}{\partial y} = -\frac{\partial x_{k,l}^{(e)}}{\partial \eta}/J_{k,l}^{(e)}$$

$$\frac{\partial \eta_{k,l}^{(e)}}{\partial x} = -\frac{\partial y_{k,l}^{(e)}}{\partial \xi}/J_{k,l}^{(e)}; \quad \frac{\partial \eta_{k,l}^{(e)}}{\partial y} = +\frac{\partial x_{k,l}^{(e)}}{\partial \xi}/J_{k,l}^{(e)}$$

$\quad\quad$ **end for**

$\quad$ **end for**

end function

Algorithm 12.4 Construction of the 3D coordinate derivatives using tensor-products.

function 3D_COORDINATE_DERIVATIVES

$\quad \frac{\partial \mathbf{x}^{(e)}}{\partial \xi} = \mathbf{0}$ ▷ initialize all nine arrays

$\quad$ **for** $e = 1 : N_e$ **do** ▷ loop over elements

$\quad\quad$ **for** $n = 0 : Q_\zeta, \; m = 0 : Q_\eta, \; l = 0 : Q_\xi$ **do** ▷ loop over integration points

$\quad\quad\quad$ **for** $k = 0 : N_\zeta, \; j = 0 : N_\eta, \; i = 0 : N_\xi$ **do** ▷ loop over interpolation points

$$\frac{\partial \mathbf{x}_{l,m,n}^{(e)}}{\partial \xi} \mathrel{+}= \frac{\partial h_{i,l}}{\partial \xi}(\xi)h_{j,m}(\eta)h_{k,n}(\zeta)\mathbf{x}_{i,j,k}^{(e)}$$

$$\frac{\partial \mathbf{x}_{l,m,n}^{(e)}}{\partial \eta} \mathrel{+}= h_{i,l}(\xi)\frac{\partial h_{j,m}}{\partial \eta}(\eta)h_{k,n}(\zeta)\mathbf{x}_{i,j,k}^{(e)}$$

$$\frac{\partial \mathbf{x}_{l,m,n}^{(e)}}{\partial \zeta} \mathrel{+}= h_{i,l}(\xi)h_{j,m}(\eta)\frac{\partial h_{k,n}}{\partial \zeta}(\zeta)\mathbf{x}_{i,j,k}^{(e)}$$

$\quad\quad\quad$ **end for**

$\quad\quad$ **end for**

$\quad$ **end for**

end function

Algorithm 12.5 Construction of the 3D metric terms using tensor-products.

---

function 3D_METRIC_TERMS

$\dfrac{\partial \mathbf{x}^{(e)}}{\partial \xi}$ =3D_COORDINATE_DERIVATIVES   ▷ use Alg. 12.4

for $e = 1 : N_e$ **do**   ▷ loop over elements

     for $n = 0 : Q_\zeta, \; m = 0 : Q_\eta, \; l = 0 : Q_\xi$ **do**   ▷ loop over integration points

$$J^{(e)}_{l,m,n} = \frac{\partial x^{(e)}_{l,m,n}}{\partial \xi}\left(\frac{\partial y^{(e)}_{l,m,n}}{\partial \eta} \cdot \frac{\partial z^{(e)}_{l,m,n}}{\partial \zeta} - \frac{\partial y^{(e)}_{l,m,n}}{\partial \zeta} \cdot \frac{\partial z^{(e)}_{l,m,n}}{\partial \eta}\right)$$

$$J^{(e)}_{l,m,n} \; += \; \frac{\partial y^{(e)}_{l,m,n}}{\partial \xi}\left(\frac{\partial x^{(e)}_{l,m,n}}{\partial \zeta} \cdot \frac{\partial z^{(e)}_{l,m,n}}{\partial \eta} - \frac{\partial x^{(e)}_{l,m,n}}{\partial \eta} \cdot \frac{\partial z^{(e)}_{l,m,n}}{\partial \zeta}\right)$$

$$J^{(e)}_{l,m,n} \; += \; \frac{\partial z^{(e)}_{l,m,n}}{\partial \xi}\left(\frac{\partial x^{(e)}_{l,m,n}}{\partial \eta} \cdot \frac{\partial y^{(e)}_{l,m,n}}{\partial \zeta} - \frac{\partial x^{(e)}_{l,m,n}}{\partial \zeta} \cdot \frac{\partial y^{(e)}_{l,m,n}}{\partial \eta}\right)$$

$$\frac{\partial \xi^{(e)}_{l,m,n}}{\partial x} = \left(\frac{\partial y^{(e)}_{l,m,n}}{\partial \eta} \cdot \frac{\partial z^{(e)}_{l,m,n}}{\partial \zeta} - \frac{\partial y^{(e)}_{l,m,n}}{\partial \zeta} \cdot \frac{\partial z^{(e)}_{l,m,n}}{\partial \eta}\right) / J^{(e)}_{l,m,n}$$

$$\frac{\partial \xi^{(e)}_{l,m,n}}{\partial y} = \left(\frac{\partial x^{(e)}_{l,m,n}}{\partial \zeta} \cdot \frac{\partial z^{(e)}_{l,m,n}}{\partial \eta} - \frac{\partial x^{(e)}_{l,m,n}}{\partial \eta} \cdot \frac{\partial z^{(e)}_{l,m,n}}{\partial \zeta}\right) / J^{(e)}_{l,m,n}$$

$$\frac{\partial \xi^{(e)}_{l,m,n}}{\partial z} = \left(\frac{\partial x^{(e)}_{l,m,n}}{\partial \eta} \cdot \frac{\partial y^{(e)}_{l,m,n}}{\partial \zeta} - \frac{\partial x^{(e)}_{l,m,n}}{\partial \zeta} \cdot \frac{\partial y^{(e)}_{l,m,n}}{\partial \eta}\right) / J^{(e)}_{l,m,n}$$

$$\frac{\partial \eta^{(e)}_{l,m,n}}{\partial x} = \left(\frac{\partial y^{(e)}_{l,m,n}}{\partial \zeta} \cdot \frac{\partial z^{(e)}_{l,m,n}}{\partial \xi} - \frac{\partial y^{(e)}_{l,m,n}}{\partial \xi} \cdot \frac{\partial z^{(e)}_{l,m,n}}{\partial \zeta}\right) / J^{(e)}_{l,m,n}$$

$$\frac{\partial \eta^{(e)}_{l,m,n}}{\partial y} = \left(\frac{\partial x^{(e)}_{l,m,n}}{\partial \xi} \cdot \frac{\partial z^{(e)}_{l,m,n}}{\partial \zeta} - \frac{\partial x^{(e)}_{l,m,n}}{\partial \eta} \cdot \frac{\partial z^{(e)}_{l,m,n}}{\partial \xi}\right) / J^{(e)}_{l,m,n}$$

$$\frac{\partial \eta^{(e)}_{l,m,n}}{\partial z} = \left(\frac{\partial x^{(e)}_{l,m,n}}{\partial \zeta} \cdot \frac{\partial y^{(e)}_{l,m,n}}{\partial \xi} - \frac{\partial x^{(e)}_{l,m,n}}{\partial \xi} \cdot \frac{\partial y^{(e)}_{l,m,n}}{\partial \zeta}\right) / J^{(e)}_{l,m,n}$$

$$\frac{\partial \zeta^{(e)}_{l,m,n}}{\partial x} = \left(\frac{\partial y^{(e)}_{l,m,n}}{\partial \xi} \cdot \frac{\partial z^{(e)}_{l,m,n}}{\partial \eta} - \frac{\partial y^{(e)}_{l,m,n}}{\partial \eta} \cdot \frac{\partial z^{(e)}_{l,m,n}}{\partial \xi}\right) / J^{(e)}_{l,m,n}$$

$$\frac{\partial \zeta^{(e)}_{l,m,n}}{\partial y} = \left(\frac{\partial x^{(e)}_{l,m,n}}{\partial \eta} \cdot \frac{\partial z^{(e)}_{l,m,n}}{\partial \xi} - \frac{\partial x^{(e)}_{l,m,n}}{\partial \xi} \cdot \frac{\partial z^{(e)}_{l,m,n}}{\partial \eta}\right) / J^{(e)}_{l,m,n}$$

$$\frac{\partial \zeta^{(e)}_{l,m,n}}{\partial z} = \left(\frac{\partial x^{(e)}_{l,m,n}}{\partial \xi} \cdot \frac{\partial y^{(e)}_{l,m,n}}{\partial \eta} - \frac{\partial x^{(e)}_{l,m,n}}{\partial \eta} \cdot \frac{\partial y^{(e)}_{l,m,n}}{\partial \xi}\right) / J^{(e)}_{l,m,n}$$

     end for

   end for

end function

12.8 Element Equations on a Single Element

Let us return to the element equation that we showed in Eq. (12.2), where we now need to employ integration by parts in order to bring down the second order operator (∇^2) to a product of first order operators, since we are using the space H^1.

12.8.1 Integration by Parts for the Diffusion Operator

To maintain the solution q in the vector space H^1 requires using the product rule as follows

$$\nabla \cdot \left(\psi_i \nabla q_N^{(e)} \right) = \nabla \psi_i \cdot \nabla q_N^{(e)} + \psi_i \nabla^2 q_N^{(e)}$$

which we rewrite as

$$\psi_i \nabla^2 q_N^{(e)} = \nabla \cdot \left(\psi_i \nabla q_N^{(e)} \right) - \nabla \psi_i \cdot \nabla q_N^{(e)}$$

where $i = 1, \ldots, M_N$ and $e = 1, \ldots, N_e$. Substituting this identity into Eq. (12.2) and using the divergence theorem leads to the following element integral equation

$$\int_{\Gamma_e} \psi_i \hat{\mathbf{n}} \cdot \nabla q_N^{(e)} \, d\Gamma_e - \int_{\Omega_e} \nabla \psi_i \cdot \nabla q_N^{(e)} \, d\Omega_e = \int_{\Omega_e} \psi_i f_N^{(e)} \, d\Omega_e. \qquad (12.36)$$

Substituting the expansions for q and f of the type given in Eqs. (12.4) and (12.5) yields

$$\int_{\Gamma_e} \psi_i \hat{\mathbf{n}} \cdot \nabla q_N^{(e)} \, d\Gamma_e - \int_{\Omega_e} \nabla \psi_i \cdot \left(\sum_{j=1}^{M_N} \nabla \psi_j q_j^{(e)} \right) d\Omega_e = \int_{\Omega_e} \psi_i \left(\sum_{j=1}^{M_N} \psi_j f_j^{(e)} \right) d\Omega_e$$
$$(12.37)$$

where, upon applying the DSS operation, the boundary integral (first term on the left) will vanish everywhere except at the domain boundary where either Dirichlet or Neumann boundary conditions are be imposed.

12.8.2 Matrix-Vector Problem Resulting from Exact Integration

At this point we have to decide whether to evaluate the integrals using exact or inexact integration. Let us first see what happens when we use exact integration. This quadrature rule yields the following matrix-vector problem

$$B_i^{(e)} - L_{ij}^{(e)} q_j^{(e)} = M_{ij}^{(e)} f_j^{(e)} \qquad (12.38)$$

where the above matrices are defined as follows

$$M_{ij}^{(e)} \equiv \int_{\Omega_e} \psi_i \psi_j \, d\Omega_e \equiv \int_{\hat{\Omega}} \psi_i \psi_j J^{(e)}(\xi) \, d\hat{\Omega} = \sum_{k=1}^{M_Q} w_k J^{(e)}(\xi_k) \psi_i(\xi_k) \psi_j(\xi_k)$$
(12.39)

is the mass matrix where w_k and $J_k^{(e)}$ are the quadrature weights and determinant of the Jacobian evaluated at the quadrature point ξ_k where, for convenience, have written the sum in monolithic form. The Laplacian matrix is defined as follows

$$L_{ij}^{(e)} \equiv \int_{\Omega_e} \nabla \psi_i \cdot \nabla \psi_j \, d\Omega_e = \sum_{k=1}^{M_Q} w_k J^{(e)}(\xi_k) \nabla \psi_i(\xi_k) \cdot \nabla \psi_j(\xi_k) \qquad (12.40)$$

and the boundary vector, resulting from the integration by parts of the Laplacian operator, is

$$B_i^{(e)} \equiv \int_{\Gamma_e} \psi_i \hat{\mathbf{n}} \cdot \nabla q_N^{(e)} \, d\Gamma_e = \sum_{k=0}^{Q} w_k^{(f)} J^{(f)}(\xi_k) \psi_i(\xi_k) \hat{\mathbf{n}}_k \cdot \nabla q_k^{(e)} \qquad (12.41)$$

where $w_k^{(f)}$ and $J^{(f)}$ denote the face quadrature weights and Jacobian (as opposed to w_k and $J^{(e)}$ which are the volume quadrature weights and Jacobian). In Eq. (12.41) we have taken some liberties mixing monolithic and tensor-product notation. For now, it suffices to understand that in 2D, a boundary integral reduces to a line integral. Although there is no problem integrating the above matrices exactly, let us look at the other choice we have at our disposal, namely, inexact integration.

Exercise Equations (12.38), (12.39), and (12.40) are all valid for either 2D or 3D; however, Eq. (12.41) is only valid for 2D. Extend Eq. (12.41) to 3D. □

12.8.3 Matrix-Vector Problem Resulting from Inexact Integration

Looking at the mass, boundary, and Laplacian matrices we see that they represent integrals of polynomials of degree 2N for mass along both directions; 2N-1 for the boundary term (assuming Lobatto points), and for the Laplacian we get a minimum of 2N-2 and a maximum of 2N. Thus if we use inexact integration (2N-1 integration) we only commit a small numerical error for both the mass and Laplacian matrices while for the boundary vector we commit no such error and, in fact, obtain it exactly!

Exercise Show that the mass matrix given in Eq. (12.39) represents a 2N degree polynomial in each direction. □

Exercise Show that the Laplacian matrix given in Eq. (12.40) represents 2N-2 and 2N degree polynomials along each direction. To show this you need to expand the basis functions into their tensor-product forms. □

Exercise Show that the boundary vector given in Eq. (12.41) represents a 2N-1 degree polynomial along the boundary. To show this, assume that the coordinate axes are aligned with the reference element coordinates and use Lobatto points. ☐

Let us now use inexact integration and see what the resulting matrix-vector problem looks like. In this case, we let $M_Q = (Q + 1)^2 = M_N$ where $Q = N$ to get the following matrix-vector problem

$$B_i^{(e)} - L_{ij}^{(e)} q_j^{(e)} = M_{ij}^{(e)} f_j^{(e)}$$

where changes in our matrix-vector problem from exact to inexact integration occur in the mass and boundary matrices. The mass matrix simplifies to

$$M_{ij}^{(e)} = w_i J_i^{(e)} \delta_{ij},$$

the boundary vector is written as follows

$$B_i^{(e)} = \sum_{i=0}^{N} w_i^{(f)} J_i^{(f)} \hat{\mathbf{n}}_i \cdot \nabla q_i^{(e)}$$

and the Laplacian matrix is

$$L_{ij}^{(e)} = \sum_{k=1}^{M_N} w_k J_k^{(e)} \nabla \psi_{ik} \cdot \nabla \psi_{jk}$$

which remains unchanged.

12.8.4 Algorithms for the Element Matrices

Let us now describe the algorithms required to construct the element mass and Laplacian matrices. To construct the local element-wise mass matrix for each element, we follow the procedure summarized in Alg. 12.6 which is valid for all dimensions. Comparing it to Alg. 5.1 we note that the procedure follows the same strategy that we used to build the one-dimensional element mass matrix. This is only the case if we use the monolithic form but not so if we exploit the tensor-product form of the basis functions. The tensor-product form is given in Alg. 12.7 which, in two-dimensions, results in a four-dimensional array with $m, n, i, j = 0, \ldots, N$. The terms $\psi_{I,K}$ and $\psi_{J,K}$ denote the 2D basis functions in monolithic numbering (I and J) which are then summed for all integration points (K). To show that it is straightforward to extend these algorithms to 3D, we outline the construction of the 3D mass matrix using the tensor-product form in Alg. 12.8.

The construction of the element Laplacian matrix in monolithic form is outlined in Alg. 12.9. The gradient of the basis function $\nabla \psi$ is obtained using Eqs. (12.7) and (12.8). In order to build the gradient of the basis functions requires the metric

---

Algorithm 12.6 Construction of the 2D/3D element mass matrix in monolithic form.

function ELEMENT_MASS_MATRIX
  $M^{(e)} = \mathbf{0}$
  for $e = 1 : N_e$ **do**   ▷ loop over elements
    for $k = 1 : M_Q$ **do**   ▷ loop over integration points
      for $j = 1 : M_N$ **do**   ▷ loop over columns
        for $i = 1 : M_N$ **do**   ▷ loop over rows
          $M^{(e)}_{ij} \mathrel{+}= w_k J^{(e)}_k \psi_{ik} \psi_{jk}$
        end for
      end for
    end for
  end for
end function

Algorithm 12.7 Construction of the 2D element mass matrix using tensor-products.

function ELEMENT_MASS_MATRIX
  $M^{(e)} = \mathbf{0}$
  for $e = 1 : N_e$ **do**   ▷ loop over elements
    for $l = 0 : Q_\eta,\ k = 0 : Q_\xi$ **do**   ▷ loop over integration points
      for $j = 0 : N_\eta,\ i = 0 : N_\xi$ **do** ▷ loop over interpolation points
        $\psi_{J,K} = h_{i,k} h_{j,l}$
        for $n = 0 : N_\eta,\ m = 0 : N_\xi$ **do** ▷ loop over interpolation points
          $\psi_{I,K} = h_{m,k} h_{n,l}$
          $M^{(e)}_{m,n,i,j} \mathrel{+}= w_{k,l} J^{(e)}_{k,l} \psi_{I,K} \psi_{J,K}$
        end for
      end for
    end for
  end for
end function

Algorithm 12.8 Construction of the 3D element mass matrix using tensor-products.

function ELEMENT_MASS_MATRIX_3D
  $M^{(e)} = \mathbf{0}$
  for $e = 1 : N_e$ **do**   ▷ loop over elements
    for $q = 0 : Q_\zeta,\ p = 0 : Q_\eta,\ o = 0 : Q_\xi$ **do** ▷ loop over integration points
      for $n = 0 : N_\zeta,\ m = 0 : N_\eta,\ l = 0 : N_\xi$ **do** ▷ loop over interpolation points
        $\psi_{J,K} = h_{l,o} h_{m,p} h_{n,q}$
        for $k = 0 : N_\zeta,\ j = 0 : N_\eta,\ i = 0 : N_\xi$ **do** ▷ loop over interpolation points
          $\psi_{I,K} = h_{i,o} h_{j,p} h_{k,q}$
          $M^{(e)}_{i,j,k,l,m,n} \mathrel{+}= w_{o,p,q} J^{(e)}_{o,p,q} \psi_{I,K} \psi_{J,K}$
        end for
      end for
    end for
  end for
end function

terms. For the metric terms we use Alg. 12.1. Algorithm 12.10 describes the steps for

---

Algorithm 12.9 Construction of the 2D element Laplacian matrix.

---

function ELEMENT_LAPLACIAN_MATRIX
  $L^{(e)} = 0$
  for $e = 1 : N_e$ **do**   ▷ loop over elements
    for $k = 1 : M_Q$ **do**   ▷ loop over integration points
      for $j = 1 : M_N$ **do**   ▷ loop over columns
        for $i = 1 : M_N$ **do**   ▷ loop over rows
          $L_{ij}^{(e)} \mathrel{+}= w_k J_k^{(e)} \boldsymbol{\nabla} \psi_{ik} \cdot \boldsymbol{\nabla} \psi_{jk}$
        end for
      end for
    end for
  end for
end function

---

constructing the Laplacian matrix using the tensor-product form. In this algorithm, although we use the one-dimensional basis functions h we also define the resulting two-dimensional basis function derivatives $\frac{\partial \psi}{\partial x}$ and $\frac{\partial \psi}{\partial y}$ to allow the reader to make the connection between Algs. 12.9 and 12.10. However, these values do not need to be stored as arrays but as scalars; we only denote them with indices to allow the reader to connect these values with those in Alg. 12.9.

Algorithm 12.10 Construction of the 2D element Laplacian matrix using tensor-products.

---

function ELEMENT_LAPLACIAN_MATRIX
  $L^{(e)} = 0$
  for $e = 1 : N_e$ **do**   ▷ loop over elements
    for $l = 0 : Q, \; k = 0 : Q$ **do**   ▷ loop over integration points
      for $j = 0 : N, \; i = 0 : N$ **do**   ▷ loop over interpolation points
        $\frac{\partial \psi_{J,K}}{\partial x} = \frac{\partial h_{i,k}}{\partial \xi} h_{j,l} \frac{\partial \xi_{k,l}}{\partial x} + h_{i,k} \frac{\partial h_{j,l}}{\partial \eta} \frac{\partial \eta_{k,l}}{\partial x}$
        $\frac{\partial \psi_{J,K}}{\partial y} = \frac{\partial h_{i,k}}{\partial \xi} h_{j,l} \frac{\partial \xi_{k,l}}{\partial y} + h_{i,k} \frac{\partial h_{j,l}}{\partial \eta} \frac{\partial \eta_{k,l}}{\partial y}$
        for $n = 0 : N, \; m = 0 : N$ **do**   ▷ loop over interpolation points
          $\frac{\partial \psi_{I,K}}{\partial x} = \frac{\partial h_{m,k}}{\partial \xi} h_{n,l} \frac{\partial \xi_{k,l}}{\partial x} + h_{m,k} \frac{\partial h_{n,l}}{\partial \eta} \frac{\partial \eta_{k,l}}{\partial x}$
          $\frac{\partial \psi_{I,K}}{\partial y} = \frac{\partial h_{m,k}}{\partial \xi} h_{n,l} \frac{\partial \xi_{k,l}}{\partial y} + h_{m,k} \frac{\partial h_{n,l}}{\partial \eta} \frac{\partial \eta_{k,l}}{\partial y}$
          $L_{m,n,i,j}^{(e)} \mathrel{+}= w_{k,l} J_{k,l}^{(e)} \left(\frac{\partial \psi_{I,K}}{\partial x} \frac{\partial \psi_{J,K}}{\partial x} + \frac{\partial \psi_{I,K}}{\partial y} \frac{\partial \psi_{J,K}}{\partial y} \right)$
        end for
      end for
    end for
  end for
end function

Exercise Write down the algorithm for the 3D Laplacian matrix using tensor products. Start with Alg. 12.10 and feel free to use the outline of Alg. 12.8 to construct the 3D Laplacian matrix algorithm.   □

12.9 Global Matrix-Vector Problem

Once all of the element matrices have been defined, the global matrices are then constructed by invoking the *direct stiffness summation* (DSS) operation which we describe in Sec. 12.9.1. That is, we construct

$$M_{IJ} = \bigwedge_{e=1}^{N_e} M_{ij}^{(e)}, \qquad B_I = \bigwedge_{e=1}^{N_e} B_i^{(e)}, \qquad L_{IJ} = \bigwedge_{e=1}^{N_e} L_{ij}^{(e)}$$

where $i, j = 1, \dots, M_N$ and $I, J = 1, \dots, N_p$, which then allows us to write the global matrix problem for exact integration as

$$B_I - L_{IJ} q_J = M_{IJ} f_J \tag{12.42}$$

and for inexact integration as

$$B_I - L_{IJ} q_J = M_I f_I$$

since for inexact integration M is diagonal. Next, we can rearrange the global matrix-vector problem as

$$-L_{IJ} q_J = M_{IJ} f_J - B_I$$

and left multiplying by $-L^{-1}$ yields

$$q_I = -L_{IK}^{-1} (M_{KJ} f_J - B_K) \tag{12.43}$$

for exact integration, and

$$q_I = -L_{IJ}^{-1} (M_J f_J - B_J) \tag{12.44}$$

for inexact integration.

12.9.1 Direct Stiffness Summation

The DSS operation for any matrix (here we take the mass matrix as an example) is described in Alg. 12.11 where we loop over all the elements N_e in the grid. Then we loop over each of the M_N interpolation points in the element. The integer array *intma* is the map which takes us from the local element space (i, e) to the global gridpoint space I.

Algorithm 12.11 DSS operation for a matrix.

function GLOBAL_MASS_MATRIX
  $M = \mathbf{0}$
  for $e = 1 : N_e$ **do**   ▷ loop over elements
    for $j = 1 : M_N$ **do**   ▷ loop over columns of M^e
      $J = intma(j, e)$ ▷ point to global grid point number
      for $i = 1 : M_N$ **do**   ▷ loop over rows of M^e
        $I = intma(i, e)$ ▷ point to global grid point number
        $M_{IJ} \mathrel{+}= M_{ij}^{(e)}$ ▷ gather data to CG storage
      end for
    end for
  end for
end function

Exercise Compute the complexity for building the matrix M in Alg. 12.11. Here we define complexity as the operation count, i.e., the number of mathematical operations.   ☐

By looking carefully at Alg. 12.11 we see that it is, in fact, virtually identical to Alg. 5.4 that we saw in Ch. 5 for the DSS operation in one dimension. The only difference is that *intma* now contains the map for a 2D or 3D element (as opposed to a 1D element used in Ch. 5). The DSS operation for a vector quantity follows in a similar fashion. Let us now describe the RHS vector R.

Algorithm 12.12 DSS operation for a vector.

function GLOBAL_RHS_VECTOR
  $R = \mathbf{0}$
  for $e = 1 : N_e$ **do**   ▷ loop over elements
    for $i = 1 : M_N$ **do**   ▷ loop over rows of $R^{(e)}$
      $I = intma(i, e)$ ▷ point to global grid point number
      $R_I \mathrel{+}= R_i^{(e)}$ ▷ gather data to CG storage
    end for
  end for
end function

Exercise Compute the complexity for building the vector R in Alg. 12.12.   ☐

Algorithm 12.12 describes the steps for constructing the DSS operator for a vector. The difference between DSS for a vector and that for a matrix is that we need to go over one less loop for a vector but the idea is the same.

So far, we have not exploited the tensor-product nature of our grids and basis functions. This means that the monolithic forms that have been emphasized so far are quite general and can be applied to any type of element shape (e.g., triangles, quadrilaterals, tetrahedra, hexahedra, triangular prisms, etc.) provided that we are able to construct the basis functions on these domains. However, if we want to use the tensor-product approach using the one-dimensional basis functions then we can use the algorithms described in Algs. 12.13 and 12.14 for 2D or Algs. 12.15 and 12.16 for 3D.

Algorithm 12.13 2D DSS operation for a matrix using tensor-products.

---

function GLOBAL_MASS_MATRIX
  $M = \mathbf{0}$
  for $e = 1 : N_e$ **do**   ▷ loop over elements
    for $j = 0 : N_\eta,\ i = 0 : N_\xi$ **do**
      $J = intma(i, j, e)$ ▷ point to global grid point number
      for $n = 0 : N_\eta,\ m = 0 : N_\xi$ **do**
        $I = intma(m, n, e)$ ▷ point to global grid point number
        $M_{IJ} \mathrel{+}= M^{(e)}_{m,n,i,j}$
      end for
    end for
  end for
end function

Algorithm 12.14 2D DSS operation for a vector using tensor-products.

---

function GLOBAL_RHS_VECTOR
  $R = \mathbf{0}$
  for $e = 1 : N_e$ **do**   ▷ loop over elements
    for $j = 0 : N_\eta,\ i = 0 : N_\xi$ **do**
      $I = intma(i, j, e)$ ▷ point to global grid point number
      $R_I \mathrel{+}= R^{(e)}_{i,j}$
    end for
  end for
end function

Algorithm 12.15 3D DSS operation for a matrix using tensor-products.

---

function GLOBAL_MASS_MATRIX
  $M = \mathbf{0}$
  for $e = 1 : N_e$ **do**   ▷ loop over elements
    for $n = 0 : N_\zeta,\ m = 0 : N_\eta,\ l = 0 : N_\xi$ **do**
      $J = intma(l, m, n, e)$ ▷ point to global grid point number
      for $k = 0 : N_\zeta,\ j = 0 : N_\eta,\ i = 0 : N_\xi$ **do**
        $I = intma(i, j, k, e)$ ▷ point to global grid point number
        $M_{IJK} \mathrel{+}= M^{(e)}_{i,j,k,l,m,n}$
      end for
    end for
  end for
end function

Algorithm 12.16 3D DSS operation for a vector using tensor-products.

---
function GLOBAL_RHS_VECTOR
   $R = \mathbf{0}$
   for $e = 1 : N_e$ **do**   ▷ loop over elements
     for $k = 0 : N_\zeta,\ j = 0 : N_\eta,\ i = 0 : N_\xi$ **do**
      $I = intma(i, j, k, e)$ ▷ point to global grid point number
      $R_I \mathrel{+}= R^{(e)}_{i,j,k}$
     end for
   end for
end function

Remark 12.1 In general, we will not solve the problem in the form given in Eqs. (12.43) and (12.44) since it is impractical to construct the inverse of L. Instead, we merely write it in this form to reveal the structure of the difference equations at the grid points. In practice, we would solve this equation using Gaussian elimination or an iterative solver. In many programming environments (e.g., Julia, Matlab, or Octave) this is easy to do especially for a small system.

12.9.2 Boundary Condition

In practice, if Dirichlet boundary conditions are required then we will not need the vector B in Eq. (12.42) but instead, will modify the resulting matrix problem $Ax = b$ such that A is zeroed out along the row for the gridpoint we wish to impose the boundary except at the diagonal which we replace by 1. Then b at that row is replaced by the Dirichlet solution at the gridpoint. This represents a strong imposition of the Dirichlet boundary condition and is summarized in Alg. 12.17. In Ch. 14 we explore a different approach which we can use for both Dirichlet and Neumann boundary conditions.

Algorithm 12.17 Strong imposition of Dirichlet boundary conditions.

---
function DIRICHLET(q_D)
   $R = M \cdot f$ ▷ form the RHS vector
   for $i = 1 : N_{boun}$ **do**   ▷ loop over boundary points
     $I = boundary(i)$ ▷ find global gridpoint for each boundary point
     for $J = 1 : N_{poin}$ **do**   ▷ loop over columns
      $L_{IJ} = 0$
     end for
     $L_{II} = 1$ ▷ next two lines apply Dirichlet boundary
     $R_I = q_D(I)$
   end for
end function

Another approach for imposing Dirichlet boundary conditions is to decompose the solution as follows

$$q = q_I + q_D$$

where q_I and q_D denote the interior and Dirichlet boundary values, respectively. Since q_D are known, then we can rewrite Eq. (12.1) as follows

$$\nabla^2 q_I = f(\mathbf{x}) - \nabla^2 q_D$$

where the right-hand side is known and we only seek the degrees of freedom for q_I. An advantage of using this *lift* operator approach is that: (i) the degrees of freedom of the matrix problem are reduced and (ii) the linear system remains symmetric positive definite. This second point is important when iterative solvers are used to solve the linear system; e.g., the conjugate gradient method can be used effectively to invert this system.

12.10 Second Order Operators without Integration by Parts

If we were interested in approximating the Laplacian operator directly as follows

$$\nabla^2 q_N^{(e)}(x, y) = \sum_{j=1}^{M_N} \nabla^2 \psi_j q_j^{(e)}$$

then we would need to construct

$$\nabla^2 \psi = \frac{\partial^2 \psi}{\partial x^2} + \frac{\partial^2 \psi}{\partial y^2}$$

where $\psi = \psi(\xi(x, y), \eta(x, y))$. This means that we need to construct the first derivative by the chain rule as follows

$$\psi_x = \psi_\xi \xi_x + \psi_\eta \eta_x$$

and

$$\psi_y = \psi_\xi \xi_y + \psi_\eta \eta_y.$$

Taking the second derivative of these functions give

$$\psi_{xx} = \psi_{\xi\xi} \xi_x^2 + 2\psi_{\xi\eta} \xi_x \eta_x + \psi_{\eta\eta} \eta_x^2 + \psi_\xi \xi_{xx} + \psi_\eta \eta_{xx}$$

and

$$\psi_{yy} = \psi_{\xi\xi} \xi_y^2 + 2\psi_{\xi\eta} \xi_y \eta_y + \psi_{\eta\eta} \eta_y^2 + \psi_\xi \xi_{yy} + \psi_\eta \eta_{yy}.$$

Although we can continue in this fashion, the challenge is that we will require constructing (and storing) the derivatives of the basis functions

$$\psi_\xi, \psi_\eta, \psi_{\xi\xi}, \psi_{\xi\eta}, \psi_{\eta\eta},$$

which are only $M_N \times M_Q$ matrices but we also need to store the metric terms

$$\xi_x, \xi_y, \xi_{xx}, \xi_{yy}, \eta_x, \eta_y, \eta_{xx}, \eta_{yy}$$

which are $N_e \times M_Q$ matrices and so becomes prohibitive. This approach requires the construction and storage of matrices with dimensions $5M_N M_Q + 8N_e M_Q$ where M_N are the number of interpolation points, M_Q are the number of integration points, and N_e are the number of elements.

Exercise Compute the storage requirements for constructing a Laplacian operator without integration by parts in three dimensions. □

In 2D for the integration by parts approach, all we need are the following matrices

$$\psi_\xi, \psi_\eta$$

and

$$\xi_x, \xi_y, \eta_x, \eta_y$$

which requires the construction and storage of matrices with dimensions $2M_N M_Q + 4N_e M_N$. This represents a significant savings in computer memory especially as the number of elements, N_e, can be on the order of millions and the integration points, M_Q, can be quite large especially for high-order methods (think of $M_Q = (N + 2)^2$ points for exact integration of a $2N$ degree polynomial). For example, for eighth order polynomials, $N = 8$, we need $M_Q = 100$ integration points per element! Another disadvantage of using the direct discretization of the Laplacian operator is that it becomes not so obvious how to incorporate Neumann boundary conditions; Dirichlet boundary conditions can be imposed either strongly or with the lifting operator approach. Note that in 3D with integration by parts, we need the following matrices

$$\psi_\xi, \psi_\eta, \psi_\zeta$$

and

$$\xi_x, \xi_y, \xi_z, \eta_x, \eta_y, \eta_z, \zeta_x, \zeta_y, \zeta_z.$$

12.11 Example of 2D CG for Linear Elements

Let us construct the matrix problem for the 2D Poisson problem using linear elements $N = 1$. To simplify the discussion let us assume that we are far away from boundaries such that the global matrix problem reads

$$-L_{IJ} q_J = M_{IJ} f_J$$

where we now have to define the element matrices $L_{ij}^{(e)}$ and $M_{ij}^{(e)}$ that we then use to construct the global matrices L_{IJ} and M_{IJ} via direct stiffness summation (DSS).

12.11.1 2D Basis Functions on Quadrilaterals

Recall that for linear elements, the 1D basis function defined at the element grid point i is

$$h_i(\xi) = \frac{1}{2}(1 + \xi_i \xi)$$

where $\xi \in [-1, +1]$. In a similar vane we can define

$$h_i(\eta) = \frac{1}{2}(1 + \eta_i \eta)$$

where $\eta \in [-1, +1]$. In these two expressions the index is defined as follows: $i = 0, ..., N$. If we wish to use the 1D basis functions to construct 2D functions, then we take the tensor-product which amounts to just defining $i = 1, ..., (N + 1)^2$ where we now have to define ξ_i and η_i not just at two grid points as in 1D but now at four points. These four points are the vertices (corners) of the quadrilateral reference element defined as $(\xi_i, \eta_i) = [(-1, -1), (+1, -1), (-1, +1), (+1, +1)]$. At this point we can now define the 2D basis functions as

$$\psi_i(\xi, \eta) = \frac{1}{2}(1 + \xi_i \xi) \frac{1}{2}(1 + \eta_i \eta)$$

or, more compactly as

$$\psi_i(\xi, \eta) = \frac{1}{4}(1 + \xi_i \xi)(1 + \eta_i \eta).$$

From this expression we can define the derivatives as follows

$$\frac{\partial \psi_i}{\partial \xi}(\xi, \eta) = \frac{1}{4}\xi_i(1 + \eta_i \eta)$$

and

$$\frac{\partial \psi_i}{\partial \eta}(\xi, \eta) = \frac{1}{4}\eta_i(1 + \xi_i \xi).$$

12.11.2 Metric Terms

To compute the metric terms we follow Sec. 12.7 and expand the coordinates as follows

$$\mathbf{x}_N^{(e)}(\xi, \eta) = \sum_{j=1}^{M_N} \psi_j(\xi, \eta) \mathbf{x}_j^{(e)}.$$

Taking the derivative yields

$$\frac{\partial \mathbf{x}_N^{(e)}}{\partial \xi}(\xi, \eta) = \sum_{j=1}^{M_N} \frac{\partial \psi_j}{\partial \xi}(\xi, \eta) \mathbf{x}_j^{(e)}$$

where, for linear elements, yields

$$\frac{\partial \mathbf{x}_N^{(e)}}{\partial \xi}(\xi, \eta) = \sum_{j=1}^{4} \frac{1}{4} \xi_j \left(1 + \eta_j \eta\right) \mathbf{x}_j^{(e)}.$$

We can now expand the summation to obtain

$$\frac{\partial \mathbf{x}_N^{(e)}}{\partial \xi}(\xi, \eta) = \frac{1}{4} \left(\xi_1 \left(1 + \eta_1 \eta\right) \mathbf{x}_1^{(e)} + \xi_2 \left(1 + \eta_2 \eta\right) \mathbf{x}_2^{(e)} \right)$$

$$+ \frac{1}{4} \left(\xi_3 \left(1 + \eta_3 \eta\right) \mathbf{x}_3^{(e)} + \xi_4 \left(1 + \eta_4 \eta\right) \mathbf{x}_4^{(e)} \right).$$

Let us assume that x is along the ξ direction and y is along η. This means that since $\xi_i = (-1, +1, -1, +1)$ and $\eta_i = (-1, -1, +1, +1)$ then $x_1^{(e)} = x_3^{(e)}$ and $x_2^{(e)} = x_4^{(e)}$ and similarly $y_1^{(e)} = y_2^{(e)}$ and $y_3^{(e)} = y_4^{(e)}$. Let us call $x_1^{(e)} = x_o^{(e)}$ and $x_2^{(e)} = x_o^{(e)} + \Delta x^{(e)}$. We can now show that $\frac{\partial x_N^{(e)}}{\partial \xi}$ is

$$\frac{\partial x_N^{(e)}}{\partial \xi}(\xi, \eta) = \frac{1}{4} \left(-(1 - \eta) x_o^{(e)} + (1 - \eta) \left(x_o^{(e)} + \Delta x^{(e)} \right) \right)$$

$$+ \frac{1}{4} \left(-(1 + \eta) x_o^{(e)} + (1 + \eta) \left(x_o^{(e)} + \Delta x^{(e)} \right) \right)$$

which simplifies to

$$\frac{\partial x_N^{(e)}}{\partial \xi} = \frac{\Delta x^{(e)}}{2}.$$

If we did this for the other terms we would find that

$$\frac{\partial x_N^{(e)}}{\partial \eta} = \frac{\partial y_N^{(e)}}{\partial \xi} = 0$$

and

$$\frac{\partial y_N^{(e)}}{\partial \eta} = \frac{\Delta y^{(e)}}{2}.$$

With these terms known, we can now evaluate the determinant of the Jacobian to yield

$$J^{(e)} \equiv |J| = \frac{\partial x_N^{(e)}}{\partial \xi} \frac{\partial y_N^{(e)}}{\partial \eta} - \frac{\partial x_N^{(e)}}{\partial \eta} \frac{\partial y_N^{(e)}}{\partial \xi} = \frac{\Delta x^{(e)} \Delta y^{(e)}}{4}.$$

12.11.3 Derivatives in Physical Space

The derivatives in physical coordinates are defined as

$$\frac{\partial \psi_i}{\partial x} = \frac{\partial \psi_i}{\partial \xi} \frac{\partial \xi}{\partial x} + \frac{\partial \psi_i}{\partial \eta} \frac{\partial \eta}{\partial x}$$

and

$$\frac{\partial \psi_i}{\partial y} = \frac{\partial \psi_i}{\partial \xi} \frac{\partial \xi}{\partial y} + \frac{\partial \psi_i}{\partial \eta} \frac{\partial \eta}{\partial y}$$

where, when we substitute the metric terms (using Eq. (12.16)) and the derivatives of the basis functions in the reference element yield

$$\frac{\partial \psi_i}{\partial x} = \frac{1}{4} \xi_i (1 + \eta_i \eta) \frac{2}{\Delta x^{(e)}}$$

and

$$\frac{\partial \psi_i}{\partial y} = \frac{1}{4} \eta_i (1 + \xi_i \xi) \frac{2}{\Delta y^{(e)}}.$$

12.11.4 Laplacian Matrix

With these definitions in place we can now define the Laplacian matrix as follows, since

$$L_{ij}^{(e)} \equiv \int_{\Omega_e} \nabla \psi_i \cdot \nabla \psi_j \, d\Omega_e = \int_{\hat{\Omega}} \nabla \psi_i \cdot \nabla \psi_j \, J^{(e)}(\xi) d\hat{\Omega}$$

we can now substitute for these to get

$$L_{ij}^{(e)} = \frac{1}{16} \int_{-1}^{+1} \int_{-1}^{+1} \left(\xi_i \xi_j (1 + \eta_i \eta)(1 + \eta_j \eta) \frac{4}{(\Delta x^{(e)})^2} \right) \frac{\Delta x^{(e)} \Delta y^{(e)}}{4} \, d\xi \, d\eta$$

$$+ \frac{1}{16} \int_{-1}^{+1} \int_{-1}^{+1} \left(\eta_i \eta_j (1 + \xi_i \xi)(1 + \xi_j \xi) \frac{4}{(\Delta y^{(e)})^2} \right) \frac{\Delta x^{(e)} \Delta y^{(e)}}{4} \, d\xi \, d\eta.$$

Integrating yields the following expression

$$
\begin{aligned}
L_{ij}^{(e)} &= \frac{\Delta y^{(e)}}{16\Delta x^{(e)}} \left(\xi_i \xi_j \xi\right)_{-1}^{+1} \left(\eta + \frac{1}{2}(\eta_i + \eta_j)\eta^2 + \frac{1}{3}\eta_i \eta_j \eta^3\right)_{-1}^{+1} \\
&+ \frac{\Delta x^{(e)}}{16\Delta y^{(e)}} \left(\eta_i \eta_j \eta\right)_{-1}^{+1} \left(\xi + \frac{1}{2}(\xi_i + \xi_j)\xi^2 + \frac{1}{3}\xi_i \xi_j \xi^3\right)_{-1}^{+1}
\end{aligned}
$$

where upon evaluating yields

$$
L_{ij}^{(e)} = \frac{\Delta x^{(e)} \Delta y^{(e)}}{12} \left(\frac{1}{\left(\Delta x^{(e)}\right)^2} \xi_i \xi_j (3 + \eta_i \eta_j) + \frac{1}{\left(\Delta y^{(e)}\right)^2} \eta_i \eta_j (3 + \xi_i \xi_j)\right).
$$

Using the fact that the reference element coordinates are defined as follows $(\xi, \eta)_1 = (-1, -1)$, $(\xi, \eta)_2 = (+1, -1)$, $(\xi, \eta)_3 = (-1, +1)$, $(\xi, \eta)_4 = (+1, +1)$ yields

$$
L_{ij}^{(e)} = \frac{\Delta y^{(e)}}{6\Delta x^{(e)}} \begin{pmatrix} 2 & -2 & 1 & -1 \\ -2 & 2 & -1 & 1 \\ 1 & -1 & 2 & -2 \\ -1 & 1 & -2 & 2 \end{pmatrix} + \frac{\Delta x^{(e)}}{6\Delta y^{(e)}} \begin{pmatrix} 2 & 1 & -2 & -1 \\ 1 & 2 & -1 & -2 \\ -2 & -1 & 2 & 1 \\ -1 & -2 & 1 & 2 \end{pmatrix}
$$

where $\Delta x^{(e)}$ and $\Delta y^{(e)}$ are the grid spacings along each direction for each element. Assuming that $\Delta x^{(e)} = \Delta y^{(e)}$ yields the simplified expression

$$
L_{ij}^{(e)} = \frac{1}{6} \begin{pmatrix} 4 & -1 & -1 & -2 \\ -1 & 4 & -2 & -1 \\ -1 & -2 & 4 & -1 \\ -2 & -1 & -1 & 4 \end{pmatrix}. \tag{12.45}
$$

Note that the row-sum and column-sum of this matrix add up to zero.

Exercise Why does the row-sum of the Laplacian matrix result in zero? Why does the column-sum add to zero? [hint: think Taylor series and what these matrices represent.] ☐

12.11.5 Mass Matrix

The mass matrix is defined as

$$
M_{ij}^{(e)} \equiv \int_{\Omega_e} \psi_i \psi_j \, d\Omega_e = \int_{\hat{\Omega}} \psi_i \psi_j J^{(e)}(\xi) \, d\hat{\Omega}
$$

which, after using the definition of the basis functions, we can write it as

$$
M_{ij}^{(e)} = \frac{1}{16} \int_{-1}^{+1} \int_{-1}^{+1} (1 + \xi_i \xi)(1 + \eta_i \eta)(1 + \xi_j \xi)(1 + \eta_j \eta) \frac{\Delta x^{(e)} \Delta y^{(e)}}{4} \, d\xi \, d\eta.
$$

Integrating yields the following expression

$$M_{ij}^{(e)} = \frac{\Delta x^{(e)} \Delta y^{(e)}}{64} \left(\xi + \frac{1}{2}(\xi_i + \xi_j)\xi^2 + \frac{1}{3}\xi_i\xi_j\xi^3 \right)_{-1}^{+1} \left(\eta + \frac{1}{2}(\eta_i + \eta_j)\eta^2 + \frac{1}{3}\eta_i\eta_j\eta^3 \right)_{-1}^{+1}.$$

Evaluating the integrals yields

$$M_{ij}^{(e)} = \frac{\Delta x^{(e)} \Delta y^{(e)}}{64} \left(2 + \frac{2}{3}\xi_i\xi_j \right) \left(2 + \frac{2}{3}\eta_i\eta_j \right).$$

Introducing the values of the reference coordinates $(\xi, \eta)_1 = (-1, -1)$, $(\xi, \eta)_2 = (+1, -1)$, $(\xi, \eta)_3 = (-1, +1)$, $(\xi, \eta)_4 = (+1, +1)$ yields

$$M_{ij}^{(e)} = \frac{\Delta x^{(e)} \Delta y^{(e)}}{36} \begin{pmatrix} 4 & 2 & 2 & 1 \\ 2 & 4 & 1 & 2 \\ 2 & 1 & 4 & 2 \\ 1 & 2 & 2 & 4 \end{pmatrix}. \tag{12.46}$$

As a sanity check, diagonalizing Eq. (12.46) results in

$$M_{ij}^{(e)} = \frac{\Delta x^{(e)} \Delta y^{(e)}}{4} \begin{pmatrix} 1 & 0 & 0 & 0 \\ 0 & 1 & 0 & 0 \\ 0 & 0 & 1 & 0 \\ 0 & 0 & 0 & 1 \end{pmatrix}. \tag{12.47}$$

Exercise Why does Eq. (12.47) give us confidence that we have computed the mass matrix correctly?   □

12.11.6 Matrix Equations on the Reference Element

Let us now put the element matrices together into one matrix equation for the reference element. The element matrix problem

$$-L_{ij}^{(e)} q_j = M_{ij}^{(e)} f_j$$

now reads

$$-\left[\frac{\Delta y^{(e)}}{6\Delta x^{(e)}} \begin{pmatrix} 2 & -2 & 1 & -1 \\ -2 & 2 & -1 & 1 \\ 1 & -1 & 2 & -2 \\ -1 & 1 & -2 & 2 \end{pmatrix} + \frac{\Delta x^{(e)}}{6\Delta y^{(e)}} \begin{pmatrix} 2 & 1 & -2 & -1 \\ 1 & 2 & -1 & -2 \\ -2 & -1 & 2 & 1 \\ -1 & -2 & 1 & 2 \end{pmatrix} \right] \begin{pmatrix} q_1 \\ q_2 \\ q_3 \\ q_4 \end{pmatrix}$$

$$= \frac{\Delta x^{(e)} \Delta y^{(e)}}{36} \begin{pmatrix} 4 & 2 & 2 & 1 \\ 2 & 4 & 1 & 2 \\ 2 & 1 & 4 & 2 \\ 1 & 2 & 2 & 4 \end{pmatrix} \begin{pmatrix} f_1 \\ f_2 \\ f_3 \\ f_4 \end{pmatrix}. \tag{12.48}$$

Unfortunately, continuing with this equation will reveal little about the structure of the resulting differencing stencil.

12.11.7 Difference Equation for the Laplacian Operator

Looking at the left-hand side term of Eq. (12.48) we immediately note that the first term represents the approximation to $\frac{\partial^2 q}{\partial x^2}$ whereas the second term represents $\frac{\partial^2 q}{\partial y^2}$. Therefore, to get a better sense of what the differencing stencils for CG look like, let us replace the coefficients f_j on the right-hand side with $\nabla^2 q_j$. This then gives the following two matrix problems

$$-\frac{\Delta y^{(e)}}{6\Delta x^{(e)}} \begin{pmatrix} 2 & -2 & 1 & -1 \\ -2 & 2 & -1 & 1 \\ 1 & -1 & 2 & -2 \\ -1 & 1 & -2 & 2 \end{pmatrix} \begin{pmatrix} q_1 \\ q_2 \\ q_3 \\ q_4 \end{pmatrix} = \frac{\Delta x^{(e)} \Delta y^{(e)}}{36} \begin{pmatrix} 4 & 2 & 2 & 1 \\ 2 & 4 & 1 & 2 \\ 2 & 1 & 4 & 2 \\ 1 & 2 & 2 & 4 \end{pmatrix} \begin{pmatrix} q_{xx,1} \\ q_{xx,2} \\ q_{xx,3} \\ q_{xx,4} \end{pmatrix} \quad (12.49a)$$

$$-\frac{\Delta x^{(e)}}{6\Delta y^{(e)}} \begin{pmatrix} 2 & 1 & -2 & -1 \\ 1 & 2 & -1 & -2 \\ -2 & -1 & 2 & 1 \\ -1 & -2 & 1 & 2 \end{pmatrix} \begin{pmatrix} q_1 \\ q_2 \\ q_3 \\ q_4 \end{pmatrix} = \frac{\Delta x^{(e)} \Delta y^{(e)}}{36} \begin{pmatrix} 4 & 2 & 2 & 1 \\ 2 & 4 & 1 & 2 \\ 2 & 1 & 4 & 2 \\ 1 & 2 & 2 & 4 \end{pmatrix} \begin{pmatrix} q_{yy,1} \\ q_{yy,2} \\ q_{yy,3} \\ q_{yy,4} \end{pmatrix}. \quad (12.49b)$$

12.11.7.1 Solution at the Global GridPoint (i, j)

Although Galerkin methods do not use a structured grid stencil but rather a 1D indexing common to all unstructured grid stencils (using i instead of i, j in 2D), let us use a structured grid stencil for illustration purposes only. Let us discuss the above Laplacian matrix equations in the context of 4 elements sharing the global gridpoint (i, j) which in its unstructured grid stencil is referred to as the grid point 5 in Fig. 12.1. To get the solution of the Laplacian problem at this gridpoint requires applying DSS to the element equations given by Eq. (12.49). This means that we have to visit the four elements claiming the global gridpoint (i, j) to get each of their contributions to this point; to simplify the discussion we assume that the elements all have the same Δx and Δy spacings and let us diagonalize the mass matrix as well. Doing so results in the following global matrix problem

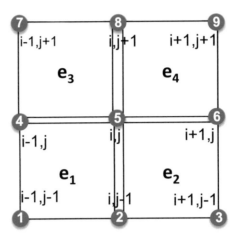

Fig. 12.1: The contribution of the four elements (e_1, e_2, e_3, and e_4) to the global gridpoint 5 (i, j).

$$\frac{\Delta x \Delta y}{4}\begin{pmatrix} q_{xx,1} \\ 2q_{xx,2} \\ q_{xx,3} \\ 2q_{xx,4} \\ 4q_{xx,5} \\ 2q_{xx,6} \\ q_{xx,7} \\ 2q_{xx,8} \\ q_{xx,9} \end{pmatrix} = -\frac{\Delta y}{6\Delta x}\begin{pmatrix} 2 & -2 & 0 & 1 & -1 & 0 & 0 & 0 & 0 \\ -2 & 4 & -2 & -1 & 2 & -1 & 0 & 0 & 0 \\ 0 & -2 & 2 & 0 & -1 & 1 & 0 & 0 & 0 \\ 1 & -1 & 0 & 4 & -4 & 0 & 1 & -1 & 0 \\ -1 & 2 & -1 & -4 & 8 & -4 & -1 & 2 & -1 \\ 0 & -1 & 1 & 0 & -4 & 4 & 0 & -1 & 1 \\ 0 & 0 & 0 & 1 & -1 & 0 & 2 & -2 & 0 \\ 0 & 0 & 0 & -1 & 2 & -1 & -2 & 4 & -2 \\ 0 & 0 & 0 & 0 & -1 & 1 & 0 & -2 & 2 \end{pmatrix}\begin{pmatrix} q_1 \\ q_2 \\ q_3 \\ q_4 \\ q_5 \\ q_6 \\ q_7 \\ q_8 \\ q_9 \end{pmatrix} \qquad (12.50)$$

where Table 12.1 shows the global gridpoint numbers of the grid shown in Fig. 12.1 (in the blue circles). Figure 12.1 also gives the relation between the global gridpoint (GGP) numbers and their associated finite difference indices (i, j).

Table 12.1: intma(i,e) array for $N = 1$ associated with Fig. 12.1.

i	e=1	e=2	e=3	e=4
1	1	2	4	5
2	2	3	5	6
3	4	5	7	8
4	5	6	8	9

Total Contribution to the GridPoint (i, j)

From row 5 of Eq. (12.50) we see that the global solution at the gridpoint 5 is

$$q_{xx,(5)} = -\frac{1}{6\Delta x^2}(-q_1 + 2q_2 - q_3)$$
$$-\frac{1}{6\Delta x^2}(-4q_4 + 8q_5 - 4q_6)$$
$$-\frac{1}{6\Delta x^2}(-q_7 + 2q_8 - q_9)$$

and rewriting it in terms of finite difference indices we get

$$q_{xx,(i,j)} = -\frac{1}{6\Delta x^2}\left(-q_{i-1,j-1} + 2q_{i,j-1} - q_{i+1,j-1}\right)$$
$$-\frac{1}{6\Delta x^2}\left(-4q_{i-1,j} + 8q_{i,j} - 4q_{i+1,j}\right)$$
$$-\frac{1}{6\Delta x^2}\left(-q_{i-1,j+1} + 2q_{i,j+1} - q_{i+1,j+1}\right). \tag{12.51}$$

Consistency of the Laplacian Operator

It is not readily obvious that the expression that we have derived for the Laplacian operator along the x-direction is a consistent operator. To verify this we now expand it using Taylor series at the point (i,j). This results in the following expressions:

$$q_{i-1,j-1} = q_{i,j} - \Delta y q_{i,j}^{(0,1)} + \frac{\Delta y^2}{2}q_{i,j}^{(0,2)} - \frac{\Delta y^3}{6}q_{i,j}^{(0,3)}$$
$$-\Delta x q_{i,j}^{(1,0)} + \Delta x \Delta y q_{i,j}^{(1,1)} - \frac{\Delta x \Delta y^2}{2}q_{i,j}^{(1,2)}$$
$$+\frac{\Delta x^2}{2}q_{i,j}^{(2,0)} - \frac{\Delta x^2 \Delta y}{2}q_{i,j}^{(2,1)} - \frac{\Delta x^3}{6}q_{i,j}^{(3,0)} + O\left(\Delta^4\right)$$

where the superscripts $q^{(m,n)}$ refer to the mth and n^{th} derivatives with respect to x and y, respectively, and $O\left(\Delta^4\right)$ refers to the remaining fourth order terms such as $O(\Delta x^4, \Delta y^4, \Delta x^2 \Delta y^2, \Delta x^3 \Delta y, \Delta y \Delta x^3)$ that we leave behind in our expansion. For the remainder of the terms we get

$$q_{i,j-1} = q_{i,j} - \Delta y q_{i,j}^{(0,1)} + \frac{\Delta y^2}{2}q_{i,j}^{(0,2)} - \frac{\Delta y^3}{6}q_{i,j}^{(0,3)} + O\left(\Delta^4\right),$$

$$q_{i+1,j-1} = q_{i,j} - \Delta y q_{i,j}^{(0,1)} + \frac{\Delta y^2}{2} q_{i,j}^{(0,2)} - \frac{\Delta y^3}{6} q_{i,j}^{(0,3)}$$
$$+ \Delta x q_{i,j}^{(1,0)} - \Delta x \Delta y q_{i,j}^{(1,1)} + \frac{\Delta x \Delta y^2}{2} q_{i,j}^{(1,2)}$$
$$+ \frac{\Delta x^2}{2} q_{i,j}^{(2,0)} - \frac{\Delta x^2 \Delta y}{2} q_{i,j}^{(2,1)} + \frac{\Delta x^3}{6} q_{i,j}^{(3,0)} + O\left(\Delta^4\right),$$

$$q_{i-1,j} = q_{i,j} - \Delta x q_{i,j}^{(1,0)} + \frac{\Delta x^2}{2} q_{i,j}^{(2,0)} - \frac{\Delta x^3}{6} q_{i,j}^{(3,0)} + O\left(\Delta^4\right),$$

$$q_{i+1,j} = q_{i,j} + \Delta x q_{i,j}^{(1,0)} + \frac{\Delta x^2}{2} q_{i,j}^{(2,0)} + \frac{\Delta x^3}{6} q_{i,j}^{(3,0)} + O\left(\Delta^4\right),$$

$$q_{i-1,j+1} = q_{i,j} - \Delta x q_{i,j}^{(1,0)} + \frac{\Delta x^2}{2} q_{i,j}^{(2,0)} - \frac{\Delta x^3}{6} q_{i,j}^{(3,0)}$$
$$+ \Delta y q_{i,j}^{(0,1)} - \Delta x \Delta y q_{i,j}^{(1,1)} + \frac{\Delta x^2 \Delta y}{2} q_{i,j}^{(2,1)}$$
$$+ \frac{\Delta y^2}{2} q_{i,j}^{(0,2)} - \frac{\Delta x \Delta y^2}{2} q_{i,j}^{(1,2)} + \frac{\Delta y^3}{6} q_{i,j}^{(0,3)} + O\left(\Delta^4\right),$$

$$q_{i,j+1} = q_{i,j} + \Delta y q_{i,j}^{(0,1)} + + \frac{\Delta y^2}{2} q_{i,j}^{(0,2)} + \frac{\Delta y^3}{6} q_{i,j}^{(0,3)} + O\left(\Delta^4\right),$$

$$q_{i+1,j+1} = q_{i,j} + \Delta x q_{i,j}^{(1,0)} + \frac{\Delta x^2}{2} q_{i,j}^{(2,0)} - \frac{\Delta x^3}{6} q_{i,j}^{(3,0)}$$
$$+ \Delta y q_{i,j}^{(0,1)} + \Delta x \Delta y q_{i,j}^{(1,1)} + \frac{\Delta x^2 \Delta y}{2} q_{i,j}^{(2,1)}$$
$$+ \frac{\Delta y^2}{2} q_{i,j}^{(0,2)} + \frac{\Delta x \Delta y^2}{2} q_{i,j}^{(1,2)} + \frac{\Delta y^3}{6} q_{i,j}^{(0,3)} + O\left(\Delta^4\right).$$

Summing the first three terms of Eq. (12.51) yields

$$-q_{i-1,j-1} + 2q_{i,j-1} - q_{i+1,j-1} = -\Delta x^2 q_{xx} + \Delta x^2 \Delta y q_{xxy} + O\left(\Delta^4\right).$$

The next three terms yield

$$-4q_{i-1,j} + 8q_{i,j} - 4q_{i+1,j} = -4\Delta x^2 q_{xx} + O\left(\Delta^4\right)$$

and for the last three terms

$$-q_{i-1,j+1} + 2q_{i,j+1} - q_{i+1,j+1} = -\Delta x^2 q_{xx} - \Delta x^2 \Delta y q_{xxy} + O\left(\Delta^4\right).$$

Summing these three relations yields $-6\Delta x^2 q_{xx} + O\left(\Delta^4\right)$ which then means that
Eq. (12.51) becomes

$$q_{xx} = q_{xx} + O\left(\Delta^2\right).$$

Doing the same kind of analysis for $\frac{\partial^2 q}{\partial y^2}$ shows that

$$q_{yy} = q_{yy} + O\left(\Delta^2\right).$$

The value of doing this analysis is that we have shown the consistency of the numerical method which tells us that we are indeed solving the correct differential equation and that we are doing so (for the case of $N = 1$ degree basis functions) with second order accuracy. These results can be extended to higher order polynomials N where we would find that the order of the scheme is $O(\Delta^{N+1})$.

12.12 2D Elliptic Equation

Now that we know how to construct a second derivative with the CG method we can use this information to solve elliptic equations. Let us put everything together that we have learned so far and summarize the solution procedure. Starting with the element-wise Poisson problem we get: find $q_N^{(e)} \in H^1(\Omega)$ such that

$$\int_{\Omega_e} \psi_i \nabla^2 q_N^{(e)} \, d\Omega_e = \int_{\Omega_e} \psi_i f_N^{(e)} \, d\Omega_e \ \ \forall \ \psi \in H^1(\Omega). \qquad (12.52)$$

Using integration by parts and the divergence theorem, we write

$$\int_{\Gamma_e} \psi_i \hat{\mathbf{n}} \cdot \nabla q_N^{(e)} \, d\Gamma_e - \int_{\Omega_e} \nabla \psi_i \cdot \nabla q_N^{(e)} \, d\Omega_e = \int_{\Omega_e} \psi_i f_N^{(e)} \, d\Omega_e \qquad (12.53)$$

where $i = 1, \ldots, M_N$ and $e = 1, \ldots, N_e$. Assuming that we defined the problem statement of Eq. (12.52) as having homogeneous Dirichlet boundary conditions then we can write the function space of our solution as $q_N^{(e)} \in H_0^1(\Omega)$. Regardless of the type of Dirichlet boundary condition, Eq. (12.53) becomes

$$-\int_{\Omega_e} \nabla \psi_i \cdot \nabla q_N^{(e)} \, d\Omega_e = \int_{\Omega_e} \psi_i f_N^{(e)} \, d\Omega_e \qquad (12.54)$$

with the Dirichlet boundary condition

$$q|_{\Gamma_{\mathcal{D}}} = g(\mathbf{x})$$

imposed in either strong form (as we showed in Alg. 12.17) or using the lift operator presented previously. Introducing the basis function expansion

$$q_N^{(e)}(x) = \sum_{j=1}^{M_N} \psi_j(x) q_j^{(e)}$$

and

$$f_N^{(e)}(x) = \sum_{j=1}^{M_N} \psi_j(x) f_j^{(e)}$$

we can write Eq. (12.54) as follows

$$-\int_{\Omega_e} \boldsymbol{\nabla}\psi_i \cdot \boldsymbol{\nabla}\psi_j \, d\Omega_e \, q_j^{(e)} = \int_{\Omega_e} \psi_i(x)\psi_j(x) \, d\Omega_e \, f_j^{(e)} \tag{12.55}$$

with $i, j = 1, \ldots, M_N$ and $e = 1, \ldots, N_e$. Introducing the mapping from the physical element Ω_e to the reference element $\hat{\Omega}$ gives

$$-\sum_{k=1}^{M_Q} w_k J_k^{(e)} \boldsymbol{\nabla}\psi_{ik} \cdot \boldsymbol{\nabla}\psi_{jk} \, q_j^{(e)} = \sum_{k=1}^{M_Q} w_k J_k^{(e)} \psi_{ik}\psi_{jk} f_j^{(e)} \tag{12.56}$$

where the gradient terms are obtained from the expression

$$\frac{\partial\psi}{\partial\mathbf{x}} = \frac{\partial\psi}{\partial\xi}\frac{\partial\xi}{\partial\mathbf{x}} + \frac{\partial\psi}{\partial\eta}\frac{\partial\eta}{\partial\mathbf{x}}$$

which requires the metric terms of the mapping from Ω_e to $\hat{\Omega}$ (see Alg. 12.1). Equation (12.56) has the following corresponding element matrix form

$$-L_{ij}^{(e)} q_j^{(e)} = M_{ij}^{(e)} f_j^{(e)}$$

where

$$L_{ij}^{(e)} = \sum_{k=1}^{M_Q} w_k J_k^{(e)} \boldsymbol{\nabla}\psi_{ik} \cdot \boldsymbol{\nabla}\psi_{jk}$$

and

$$M_{ij}^{(e)} = \sum_{k=1}^{M_Q} w_k J_k^{(e)} \psi_{ik}\psi_{jk}.$$

Invoking the *direct stiffness summation* (DSS) operator yields the following global matrix problem

$$-L_{IJ} \, q_J = M_{IJ} f_J$$

where q_J, f_J are the values of q and f at the global gridpoints $J = 1, \ldots, N_P$ and

$$M_{IJ} = \bigwedge_{e=1}^{N_e} M_{ij}^{(e)}, \qquad L_{IJ} = \bigwedge_{e=1}^{N_e} L_{ij}^{(e)},$$

where the DSS operator performs the summation via the mapping $(i, e) \to (I)$ where $i = 0, \ldots, N_\xi$, $j = 0, \ldots, N_\eta$, $e = 1, \ldots, N_e$, $(I, J) = 1, \ldots, N_p$, and N_e, N_p are the total number of elements and gridpoints (see Alg. 12.11).

12.12.1 Algorithm for the 2D Elliptic Equation

The entire solution procedure for solving elliptic equations in two-dimensions is summarized in Alg. 12.18. Note that the first thing we need to do is to construct the element matrices which are then used to form the global matrix problem by virtue of the DSS operator. At this point, we can impose the Dirichlet boundary conditions in a strong sense by modifying the global matrix problem. The final step is to construct the linear algebra problem $Ax = b$ and solve for x. This can be accomplished by standard linear algebra techniques such as LU factorization (direct solver) or by iterative solvers (e.g., conjugate gradient or GMRES). It should become clear at

---

Algorithm 12.18 Recipe for solving the 2D elliptic equation with CG.

---

function CG(q)

    Construct $M_{ij}^{(e)}$, $L_{ij}^{(e)}$, and $B_i^{(e)}$ ▷ use Algs. 12.6, 12.9

    Construct M_{IJ}, L_{IJ}, and B_I via DSS   ▷ use Algs. 12.11, 12.12

    Construct $R_I = M_{IJ} f_J$

    Modify L_{IJ} and $R_I - B_I$ to satisfy Dirichlet boundary conditions   ▷ use Alg. 12.17

    Solve $-L_{IJ} q_J = R_I - B_I$ ▷ $Ax = b$ linear system

end function

---

this point that Alg. 12.18 can be used to solve elliptic equations in any dimension (one, two, or three) as long as we construct the basis functions and corresponding element matrices in monolithic form (i.e., not writing the basis functions and element matrices in terms of tensor-products). For time-independent problems, such as this one, there is little to be gained by tensor-product forms so the approach presented in Alg. 12.18 would be the recommended approach for solving this class of problems with the CG method.

12.13 Analysis of the Matrix Properties of the Spatial Operators

Let us now analyze the sparsity patterns of both M and L, that is, the mass and Laplacian matrices. In addition, let us compute the eigenvalues of the Laplacian matrix to confirm that L is well-conditioned. For the analysis given below, we assume homogeneous Dirichlet boundary conditions. In addition, the matrices shown are for a grid with $N_e = 4 \times 4$ elements using $N = 4$ polynomial order; an example of this grid is shown in Fig. 12.2.

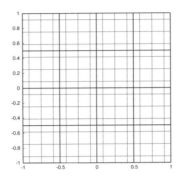

Fig. 12.2: The quadrilateral grid with $N_e = 4 \times 4$ elements with $N = 4$ polynomial order used for the matrix analysis. The thick (blue) lines denote the element boundaries and the thin (red) lines the interpolation points (Lobatto points in this case).

12.13.1 Sparsity Pattern of the Mass Matrix

Figure 12.3 shows the sparsity pattern for the mass matrix M using both exact $(Q = N + 1)$ and inexact $(Q = N)$ integration. Figure 12.3 shows that the mass

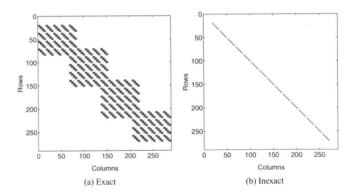

(a) Exact   (b) Inexact

Fig. 12.3: Sparsity patterns for the CG mass matrix M with $N_e = 4 \times 4$, $N = 4$ for (a) exact and (b) inexact integration.

matrix is not full but rather sparse; one can see the overlap where the four elements are touching. This figure shows that the mass matrix is tightly banded, that is, it has a small bandwidth (subdiagonals) and that the matrix becomes diagonal when inexact integration is used.

12.13.2 Sparsity Pattern of the Laplacian Matrix

Figure 12.4 shows the sparsity pattern for the Laplacian matrix L for exact and inexact integration. Similarly to the mass matrix, Fig. 12.4 shows that the Laplacian

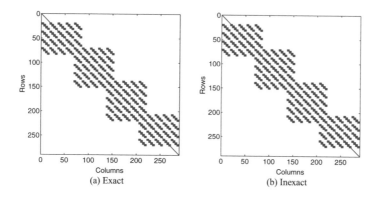

(a) Exact   (b) Inexact

Fig. 12.4: Sparsity patterns for the CG Laplacian matrix L with $N_e = 4 \times 4$, $N = 4$ for (a) exact and (b) inexact integration.

matrix is not completely full, that is, it has a small bandwidth ($O(2N + 1)^2$). For the Laplacian matrix, there is no difference in the sparsity pattern between exact and inexact integration; however, this does not imply that the matrices are identical (in fact, they are not).

12.13.3 Eigenvalue Analysis of Spatial Operator

Figure 12.5 shows the eigenvalues of the Laplacian ($-L$) are all on the real axis and, although we have normalized the values, we can state that the condition number of this matrix is rather small (i.e., it is well-conditioned). This is good news because in order to construct the numerical solution requires the inversion of L which can be achieved with Gaussian elimination; for large systems, this matrix problem is best solved with matrix-free methods using either the conjugate gradient or Generalized Minimal Residual (GMRES) method (see [326]).

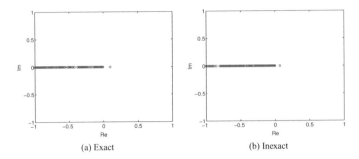

(a) Exact   (b) Inexact

Fig. 12.5: Eigenvalues of the CG Laplacian matrix $-L$ with $N_e = 4 \times 4$, $N = 4$ for (a) exact and (b) inexact integration.

12.14 Example of 2D Poisson Equation

Suppose we wish to solve the continuous second order differential equation

$$\nabla^2 q(x, y) = f(x, y) \qquad \forall (x, y) \in [-1, 1]^2$$

where the exact solution is given by

$$q(x, y) = \sin(c\pi x) \sin(c\pi y) \tag{12.57}$$

and

$$f(x, y) = -2(c\pi)^2 \sin(c\pi x) \sin(c\pi y)$$

where c is any constant. For integer values of c, the Dirichlet boundary conditions are homogeneous. For all values of c the Dirichlet boundary conditions are defined as

$$g(-1, y) = \sin(-c\pi) \sin(c\pi y)$$
$$g(+1, y) = \sin(+c\pi) \sin(c\pi y)$$
$$g(x, -1) = \sin(c\pi x) \sin(-c\pi)$$
$$g(x, +1) = \sin(c\pi x) \sin(c\pi y);$$

however, we shall use $c = 1$ for this example (we obtain a wave number 2 because the domain is defined as $(x, y) \in [-1, 1]^2$).

12.14.1 Error Norm

To be more precise, when elements of different sizes are used, it is more accurate to write the norm in integral form as follows

$$L^2 = \sqrt{\frac{\int_\Omega \left(q^{(num)} - q^{(exact)}\right)^2 d\Omega}{\int_\Omega \left(q^{(exact)}\right)^2 d\Omega}} \tag{12.58}$$

where we use quadrature for the integrals as follows: let

$$\int_\Omega f \, d\Omega = \sum_{e=1}^{N_e} \sum_{k=1}^{M_Q} w_k J_k^{(e)} f_k^{(e)}$$

for any function f, with volume Jacobian $J^{(e)}$ and quadrature weights w. The errors below use the integral form of the error norm.

12.14.2 Solution Accuracy

Figure 12.6 shows the analytic and CG numerical solutions using $N = 16$ order polynomials and $N_e = 4 \times 4$ elements for a total of $N_p = 4225$ gridpoints. The grid

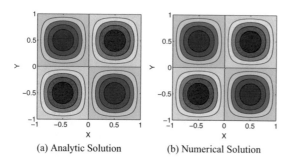

(a) Analytic Solution   (b) Numerical Solution

Fig. 12.6: CG solution contours for the 2D Poisson equation with $c = 1$ for a $N_e = 4 \times 4$ grid with polynomial order $N = 16$ for the (a) analytic and (b) numerical solutions.

used for this computation is shown in Fig. 12.7. Figure 12.6 shows that there is no difference between the analytic and numerical solutions. Let us now see how well our model behaves by varying the number of elements and the polynomial order of the basis functions.

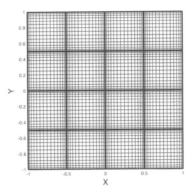

Fig. 12.7: A quadrilateral grid with $N_e = 4 \times 4$ elements with $N = 16$ polynomial order.

Figure 12.8 shows the convergence rates for various polynomial orders, N, for a total number of gridpoints N_p where, for d-dimesional space, $N_p = \left(\sqrt{N_e} N + 1 \right)^d$; for convenience we maintain the number of elements, polynomial order, and number of points constant along all coordinate directions. Figure 12.8 shows that there is little

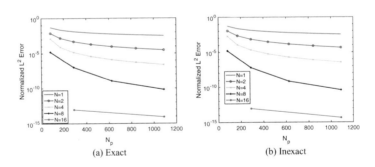

Fig. 12.8: Convergence rates of CG for the 2D Poisson equation with $c = 1$ for polynomial orders $N = 1$, $N = 2$, $N = 4$, $N = 8$, and $N = 16$ using a total number of gridpoints N_p for (a) exact integration ($Q = N + 1$) and (b) inexact integration ($Q = N$).

difference between exact and inexact integration at least in the rates of convergence. The main differences occur for low values of N (≤ 4), however, as N increases the difference between exact and inexact integration all but disappear. Furthermore, the accuracy of the CG method, especially for the high-order polynomials, approaches machine double precision (1×10^{-16}).

12.15 Computational Cost of High-Order

To verify whether high-order is indeed worth the cost requires not just looking at the convergence rates but also at the computational time required by the methods. Let us now make the elliptic problem a bit more challenging by modifying the analytic solution by using $c = 2$ in Eq. (12.57).

12.15.1 Solution Accuracy

Figure 12.9 shows the analytic solution in both contour and surface plot form. Figure 12.10 shows the convergence rates and computational cost (wallclock time in

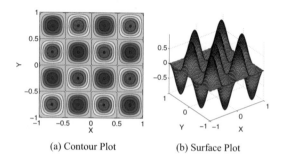

(a) Contour Plot   (b) Surface Plot

Fig. 12.9: The analytic solution for the 2D Poisson equation with $c = 2$ in (a) contour and (b) surface plot form.

seconds) for various polynomial orders N. Because this problem is more difficult than for $c = 1$ (more variation in the analytic solution) Fig. 12.10a shows that accuracy gains are indeed realized for high-order polynomials including $N = 16$. However, from Fig. 12.10b the cost of high-order is not at all clear. To see this better, let us now zoom into the plot and only show small values of the wallclock times (5 seconds total time). Figure 12.11 shows that to achieve, for example, an accuracy of 10^{-5}, it is far more efficient to use high-order. In this specific case, this accuracy is achieved most efficiently by $N = 8$. In fact, if we were to do this type of study for various problems, we would find that polynomial orders near $N = 8$ are in fact the most efficient for a certain level of accuracy (near machine single precision). However, if more precision is required, say 10^{-10}, then it is best to further increase the polynomial order; for this level of accuracy, Fig. 12.11 shows that the most efficient is $N = 16$.

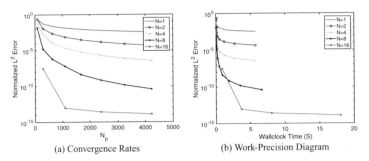

(a) Convergence Rates   (b) Work-Precision Diagram

Fig. 12.10: Solution accuracy of CG with inexact integration for the 2D Poisson equation with $c = 2$ for polynomial orders $N = 1$, $N = 2$, $N = 4$, $N = 8$, and $N = 16$ using a total number of gridpoints N_p. The plots show the following results: (a) convergence rates and (b) work-precision (cost).

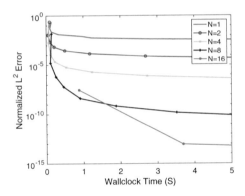

Fig. 12.11: Work-Precision: Error versus wallclock time for CG with inexact integration for the 2D Poisson equation with $c = 2$ for polynomial orders $N = 1$, $N = 2$, $N = 4$, $N = 8$, and $N = 16$.

Chapter 13
2D Discontinuous Galerkin Methods for Elliptic Equations

13.1 Introduction

In Ch. 12 we discussed the solution of the continuous Galerkin method for solving multi-dimensional elliptic problems. Let us now describe the solution of two-dimensional elliptic problems with the discontinuous Galerkin method. Although the extension to multi-dimensions is straightforward, to simplify the discussion we focus on two dimensions. This chapter focuses on the discretization of scalar elliptic problems (e.g., Poisson problem) using the local discontinuous Galerkin (LDG) method. We reserve the symmetric interior penalty Galerkin (SIPG) method for Ch. 14.

This chapter is organized as follows. In Sec. 13.2 we describe the continuous form of the elliptic equation used for introducing the LDG method and in Sec. 13.3 we discuss its integral form. In Sec. 13.4 we describe the construction of the basis functions which leads to the element equations defined in Sec. 13.5. In Sec. 13.6 two different solution strategies for building the Laplacian operator with DG are discussed. This then leads to Sec. 13.7 where the LDG algorithm is described. In Sec. 13.8 we analyze the sparsity and eigenvalues of the global matrices. We end this chapter with an example for solving a Poisson problem using the LDG method in Sec. 13.9.

13.2 2D Elliptic Equation

Let us begin by writing the two-dimensional elliptic equation in the following form

$$\nabla^2 q = f(\mathbf{x}) \tag{13.1}$$

© The Editor(s) (if applicable) and The Author(s), under exclusive license
to Springer Nature Switzerland AG 2020
F. X. Giraldo, *An Introduction to Element-Based Galerkin Methods on Tensor-Product Bases*, Texts in Computational Science and Engineering 24,
https://doi.org/10.1007/978-3-030-55069-1_13

where $q = q(\mathbf{x})$ is our scalar solution variable, $\nabla = \left(\frac{\partial}{\partial x}\hat{\mathbf{i}} + \frac{\partial}{\partial y}\hat{\mathbf{j}} \right)$ is the 2D gradient operator, and $\nabla^2 = \nabla \cdot \nabla$ is the Laplacian operator, and f is a known forcing function.

13.3 Weak Integral Form

In Ch. 9 we saw that in order to define a consistent numerical method requires first decomposing the second order operator into a series of two first order operators: this is called the *local discontinuous Galerkin* method and is written in continuous form as follows

$$\mathbf{Q}_N^{(e)} = \nabla q_N^{(e)}$$
$$\nabla \cdot \mathbf{Q}_N^{(e)} = f_N^{(e)}(\mathbf{x})$$

which makes use of the so-called *flux form*.

Next, we expand the variables as follows

$$q_N^{(e)}(\mathbf{x}) = \sum_{j=1}^{M_N} \psi_j(\mathbf{x}) q_j^{(e)},$$

$$f_N^{(e)}(\mathbf{x}) = \sum_{j=1}^{M_N} \psi_j(\mathbf{x}) f_j^{(e)},$$

and

$$\mathbf{Q}_N^{(e)}(\mathbf{x}) = \sum_{j=1}^{M_N} \psi_j(\mathbf{x}) \mathbf{Q}_j^{(e)}.$$

The integer M_N is a function of N which determines the number of points inside each element Ω_e. For the case of tensor-product methods $M_N = (N+1)^d$ for d-dimensions, where for convenience we assume that the polynomial order in all spatial directions is constant. Next, we substitute $q_N^{(e)}$ and $f_N^{(e)}$, into the PDE, multiply by a test function and integrate within the local domain yielding the weak integral form: find $q_N^{(e)} \in L^2$ such that

$$\int_{\Omega_e} \mathbf{Q}_N^{(e)} \cdot \boldsymbol{\tau}_i \, d\Omega_e = \int_{\Omega_e} \nabla q_N^{(e)} \cdot \boldsymbol{\tau}_i \, d\Omega_e \qquad \forall \boldsymbol{\tau} \in L^2$$

$$\int_{\Omega_e} \psi_i \nabla \cdot \mathbf{Q}_N^{(e)} \, d\Omega_e = \int_{\Omega_e} \psi_i f_N^{(e)} \, d\Omega_e \qquad \forall \psi \in L^2$$

with $i = 1, \dots, M_N$, where ψ and τ are the locally defined basis functions in each element Ω_e such that

$$\Omega = \bigcup_{e=1}^{N_e} \Omega_e$$

with N_e denoting the number of elements that, when summed, comprise the global domain Ω. We define τ in Sec. 13.4.

13.4 Basis Functions and the Reference Element

Recall that in Ch. 12 we constructed 2D basis functions on the reference element $(\xi, \eta) \in [-1, +1]^2$ using the tensor-product notation

$$\psi_i(\xi, \eta) = h_j(\xi) \otimes h_k(\eta)$$

where h are the 1D basis functions, $\otimes$ is the tensor-product operator, and the 1D indices vary as follows $j, k = 0, ..., N$ with the 2D index varying as $i = 1, ..., M_N$ where $M_N = (N + 1)^2$ in two dimensions. The map from the 1D local indices (j, k) to the 2D local index i is defined as follows: $i = j + 1 + k(N + 1)$.

With this definition in place, we can now expand the solution variable as follows

$$q_N^{(e)}(\mathbf{x}) = \sum_{j=1}^{M_N} \psi_j(\mathbf{x}) q_j^{(e)}$$

which implies the approximation of the gradient operator to be

$$\nabla q_N^{(e)}(\mathbf{x}) = \sum_{j=1}^{M_N} \nabla \psi_j(\mathbf{x}) q_j^{(e)}$$

where the partial derivatives are defined as follows

$$\frac{\partial q_N^{(e)}(\mathbf{x})}{\partial x} = \sum_{j=1}^{M_N} \frac{\partial \psi_j(\mathbf{x})}{\partial x} q_j^{(e)} \quad \text{and} \quad \frac{\partial q_N^{(e)}(\mathbf{x})}{\partial y} = \sum_{j=1}^{M_N} \frac{\partial \psi_j(\mathbf{x})}{\partial y} q_j^{(e)}.$$

To compute the metric terms for the reference element, we use the mappings defined in Sec. 12.7.

The vector function $\boldsymbol{\tau}$ is a tensor composed of the function ψ. In other words

$$\boldsymbol{\tau} = \psi \mathbf{I}_2 \quad \text{with} \quad \mathbf{I}_2 = \begin{pmatrix} 1 & 0 \\ 0 & 1 \end{pmatrix}$$

where $\mathbf{I}_2$ denotes the rank-2 identity matrix. Let us now review how to construct the element-wise problem using the local discontinuous Galerkin method using what we learned in Ch. 9.

13.5 Element Equations on a Single Element

In DG recall that we construct the Laplacian operator

$$\nabla^2 q_N^{(e)}$$

by the two-step process: let

$$\mathbf{Q}_N^{(e)} = \nabla q_N^{(e)}$$

be an auxiliary variable that is a placeholder for the gradient of q defined in a consistent DG sense and is a vector defined as $\mathbf{Q} = Q^{(x)}\,\hat{\mathbf{i}} + Q^{(y)}\,\hat{\mathbf{j}}$ where $\hat{\mathbf{i}}$ and $\hat{\mathbf{j}}$ are the directional unit vectors of the coordinate system, with corresponding components $Q^{(x)}$ and $Q^{(y)}$. Next, we compute the divergence of this new variable to construct the Laplacian as follows

$$\nabla \cdot \nabla q_N^{(e)} = \nabla \cdot \mathbf{Q}_N^{(e)}.$$

Therefore, in order to solve the Poisson problem

$$\nabla^2 q = f(\mathbf{x})$$

requires the construction of the auxiliary (vector-valued) variable $\mathbf{Q}$ such that

$$\mathbf{Q}_N^{(e)} = \nabla q_N^{(e)}.$$

After this has been done, we then solve the Poisson problem

$$\nabla \cdot \mathbf{Q}_N^{(e)} = f_N^{(e)}.$$

13.5.1 First Step: Evaluating the Auxiliary Variable

Let us begin our discussion by evaluating the auxiliary variable $\mathbf{Q}$. Beginning with

$$\mathbf{Q}_N^{(e)} = \nabla q_N^{(e)},$$

we multiply by the tensor test function $\boldsymbol{\tau}$ and integrate within Ω_e to get

$$\int_{\Omega_e} \mathbf{Q}_N^{(e)} \cdot \boldsymbol{\tau}_i \, d\Omega_e = \int_{\Omega_e} \nabla q_N^{(e)} \cdot \boldsymbol{\tau}_i \, d\Omega_e. \tag{13.2}$$

Next, we need to approximate $q_N^{(e)}$ and $\mathbf{Q}_N^{(e)}$ as follows: let

$$q_N^{(e)}(x, y) = \sum_{j=1}^{M_N} \psi_j(x, y) q_j^{(e)}$$

and

$$\mathbf{Q}_N^{(e)}(x, y) = \sum_{j=1}^{M_N} \psi_j(x, y) \mathbf{Q}_j^{(e)}.$$

Using the product rule

$$\nabla q_N^{(e)} \cdot \boldsymbol{\tau}_i = \nabla \cdot \left(\boldsymbol{\tau}_i q_N^{(e)}\right) - (\nabla \cdot \boldsymbol{\tau}_i) q_N^{(e)}$$

and substituting into the right-hand side of Eq. (13.2) yields

$$\int_{\Omega_e} \mathbf{Q}_N^{(e)} \cdot \boldsymbol{\tau}_i \, d\Omega_e = \int_{\Gamma_e} (\hat{\mathbf{n}} \cdot \boldsymbol{\tau}_i) q_N^{(e)} \, d\Gamma_e - \int_{\Omega_e} (\boldsymbol{\nabla} \cdot \boldsymbol{\tau}_i) q_N^{(e)} \, d\Omega_e \qquad (13.3)$$

where we have invoked the divergence theorem on the first term on the right-hand side.

13.5.1.1 Matrix-Vector Problem

We can now recast Eq. (13.3) in matrix form as follows

$$M_{ij}^{(e)} \left(\mathbf{Q}_j^{(e)} \cdot \mathbf{I}_2 \right) = \left(\mathbf{F}_{ij}^{(e)} \right)^T \left(q_j^{(*,e)} \mathbf{I}_2 \right) - \tilde{\mathbf{D}}_{ij}^{(e)} q_j^{(e)} \qquad (13.4)$$

where we have included the numerical flux in the first term on the right and where the element matrices are

$$M_{ij}^{(e)} = \int_{\Omega_e} \psi_i \psi_j \, d\Omega_e,$$

$$\mathbf{F}_{ij}^{(e)} = \int_{\Gamma_e} \psi_i \psi_j \hat{\mathbf{n}}(\mathbf{x}) \, d\Gamma_e,$$

and

$$\tilde{\mathbf{D}}_{ij}^{(e)} = \int_{\Omega_e} \boldsymbol{\nabla} \psi_i \psi_j \, d\Omega_e$$

where $i, j = 1, \ldots, M_N$ and $e = 1, \ldots, N_e$. For the flux matrix $\mathbf{F}$ the numerical flux is defined as $q^{(*,e)} = \alpha q^{(e)} + (1 - \alpha) q^{(k)}$ where k is the face neighbor of e and $\hat{\mathbf{n}}$ is the unit normal vector of the face shared by the elements e and k; we write it as $\hat{\mathbf{n}}(\mathbf{x})$ to remind the reader that the normal vector (unlike in 1D where it was a constant) is now a function of the coordinate along the face defined by Γ_e.

13.5.2 Second Step: Evaluate the Poisson Problem

Now that we know the auxiliary variable $\mathbf{Q}$, which is the placeholder for the gradient of q, we can now take the divergence of this vector function to compute the Laplacian. We now solve the Poisson problem

$$\boldsymbol{\nabla} \cdot \mathbf{Q}_N^{(e)} = f_N^{(e)}.$$

To solve this problem using the usual Galerkin machinery requires us to multiply by a test function and integrate within Ω_e to get

$$\int_{\Omega_e} \psi_i \boldsymbol{\nabla} \cdot \mathbf{Q}_N^{(e)} \, d\Omega_e = \int_{\Omega_e} \psi_i f_N^{(e)} \, d\Omega_e. \qquad (13.5)$$

Using the product rule

$$\psi_i \nabla \cdot \mathbf{Q}_N^{(e)} = \nabla \cdot \left(\psi_i \mathbf{Q}_N^{(e)} \right) - \nabla \psi_i \cdot \mathbf{Q}_N^{(e)}$$

and substituting into the left-hand side of Eq. (13.5) yields

$$\int_{\Gamma_e} \psi_i \hat{\mathbf{n}} \cdot \mathbf{Q}_N^{(*,e)} \, d\Omega_e - \int_{\Omega_e} \nabla \psi_i \cdot \mathbf{Q}_N^{(e)} \, d\Omega_e = \int_{\Omega_e} \psi_i f_N^{(e)} \, d\Omega_e \qquad (13.6)$$

where we have included the numerical flux in the first term on the left-hand side after invoking the divergence theorem.

13.5.2.1 Matrix-Vector Problem

Substituting a basis function expansion for $f_N^{(e)}$ and $\mathbf{Q}_N^{(e)}$ as follows

$$f_N^{(e)}(x, y) = \sum_{j=1}^{M_N} \psi_j(x, y) f_j^{(e)}$$

and

$$\mathbf{Q}_N^{(e)}(x, y) = \sum_{j=1}^{M_N} \psi_j(x, y) \mathbf{Q}_j^{(e)},$$

into Eq. (13.6) yields the matrix-vector problem

$$\left(\mathbf{F}_{ij}^{(e)} \right)^T \mathbf{Q}_j^{(*,e)} - \left(\tilde{\mathbf{D}}_{ij}^{(e)} \right)^T \mathbf{Q}_j^{(e)} = M_{ij}^{(e)} f_j^{(e)} \qquad (13.7)$$

where the element matrices are the same ones we used for evaluating the auxiliary variable $\mathbf{Q}$ (see Eq. (13.4)). Here we use a similar numerical flux as for q (but the order is reversed)

$$\mathbf{Q}^{(*,e)} = (1 - \alpha)\mathbf{Q}^{(e)} + \alpha \mathbf{Q}^{(k)}$$

where, once again, k is the face neighbor of e.

13.6 Solution Strategy

13.6.1 Approach I

For either parabolic or elliptic-hyperbolic equations that are both initial and boundary value problems, the solution strategy for solving the Laplacian operators is much more straightforward because if we already have an initial solution for q then all we need to do is construct an approximation for $\mathbf{Q} = \nabla q$. Once this approximation has been constructed we can then use it to approximate the Laplacian.

For a purely elliptic equation, such as the Poisson problem, there is no such initial solution and so this represents a boundary value problem. In this case, the solution of the Laplacian operator is more difficult because we cannot use an initial value of q to compute $\mathbf{Q} = \nabla q$ which can then be used to solve $\nabla \cdot \mathbf{Q} = f$. For a purely elliptic problem we have to solve for q and $\mathbf{Q}$ simultaneously. To see what this means, we must first rewrite Eq. (13.4) in its component form as follows

$$M_{ij}^{(e)} Q_j^{(x,e)} = F_{ij}^{(x,e)} q_j^{(*,e)} - \tilde{D}_{ij}^{(x,e)} q_j^{(e)} \tag{13.8a}$$

$$M_{ij}^{(e)} Q_j^{(y,e)} = F_{ij}^{(y,e)} q_j^{(*,e)} - \tilde{D}_{ij}^{(y,e)} q_j^{(e)} \tag{13.8b}$$

where

$$F_{ij}^{(x,e)} = \int_{\Gamma_e} \psi_i \psi_j \hat{n}^{(x)} \, d\Gamma_e, \qquad F_{ij}^{(y,e)} = \int_{\Gamma_e} \psi_i \psi_j \hat{n}^{(y)} \, d\Gamma_e,$$

and

$$\tilde{D}_{ij}^{(x,e)} = \int_{\Omega_e} \frac{\partial \psi_i}{\partial x} \psi_j \, d\Omega_e, \qquad \tilde{D}_{ij}^{(y,e)} = \int_{\Omega_e} \frac{\partial \psi_i}{\partial y} \psi_j \, d\Omega_e$$

with $i, j = 1, \ldots, M_N$ and $e = 1, \ldots, N_e$. Writing this system in compact vector form and multiplying by the inverse mass matrix yields for Eq. (13.4)

$$\left(\mathbf{Q}_i^{(e)} \cdot \mathbf{I}_2 \right) = \left(\hat{\mathbf{F}}_{ij}^{(e)} \right)^T \left(q_j^{(*,e)} \mathbf{I}_2 \right) - \hat{\mathbf{D}}_{ij}^{(e)} q_j^{(e)} \tag{13.9}$$

where

$$\hat{\mathbf{F}}_{ij}^{(e)} = \left(M_{ik}^{(e)} \right)^{-1} \mathbf{F}_{kj}^{(e)}$$

and

$$\hat{\mathbf{D}}_{ij}^{(e)} = \left(M_{ik}^{(e)} \right)^{-1} \mathbf{D}_{kj}^{(e)}.$$

Substituting Eq. (13.9) into Eq. (13.7) yields

$$\left(\mathbf{F}_{ik}^{(e)} \right)^T \left[\left(\hat{\mathbf{F}}_{kj}^{(e)} \right)^T \left(q_j^{(*,e)} \mathbf{I}_2 \right) - \hat{\mathbf{D}}_{kj}^{(e)} q_j^{(e)} \right]^{(*,e)}$$

$$- \left(\tilde{\mathbf{D}}_{ik}^{(e)} \right)^T \left[\left(\hat{\mathbf{F}}_{kj}^{(e)} \right)^T \left(q_j^{(*,e)} \mathbf{I}_2 \right) - \hat{\mathbf{D}}_{kj}^{(e)} q_j^{(e)} \right]^{(e)} = M_{ij}^{(e)} f_j^{(e)} \tag{13.10}$$

where the first term on the left side spans not just the neighbors of e but also the neighbors of the neighbors which are denoted by $(*, e)$ operating on $(*, e)$. Although Eq. (13.10) is technically correct and uses element matrices that we have defined already, it is complicated to translate into a numerical algorithm due to the challenge posed by the need to represent the neighbors of neighbors. Let us try a different approach and is in fact the extension of the approach presented in Ch. 9 for the construction of elliptic operators in one dimension.

13.6.2 Approach II

Let us begin by rewriting Eq. (13.9) as follows

$$
\begin{pmatrix} MQ^{(x)} \\ MQ^{(y)} \end{pmatrix} = \begin{pmatrix} \left(F_q^{(x)} - \tilde{D}^{(x)} \right) q \\ \left(F_q^{(y)} - \tilde{D}^{(y)} \right) q \end{pmatrix} \tag{13.11}
$$

where, in a similar fashion to the presentation of LDG in Ch. 9, the flux matrix F contains the numerical flux while $\tilde{D}$ is a block diagonal matrix and so the right-hand side denotes a global matrix-vector problem, where we shall define the matrices shortly. Next, we let $\hat{D}_v = F_v - \tilde{D}$ where the subscript v denotes either q or Q. The reason why we need to distinguish between the q and Q flux matrices is because they may include boundary conditions which are not necessarily the same for both variables. Note that the differentiation matrix is the same for both q and Q. We can now rewrite Eq. (13.11) as follows

$$
\begin{pmatrix} MQ^{(x)} \\ MQ^{(y)} \end{pmatrix} = \begin{pmatrix} \hat{D}_q^{(x)} q \\ \hat{D}_q^{(y)} q \end{pmatrix}. \tag{13.12}
$$

Next, let us now rewrite Eq. (13.7) as follows

$$
\mathbf{F}_Q^T \mathbf{Q} - \tilde{\mathbf{D}}^T \mathbf{Q} = M f \tag{13.13}
$$

which can be further simplified to

$$
\hat{\mathbf{D}}_Q^T \mathbf{Q} = M f \tag{13.14}
$$

where $\hat{\mathbf{D}}_Q = \mathbf{F}_Q - \tilde{\mathbf{D}}$. Expanding Eq. (13.14) gives

$$
\hat{D}_Q^{(x)} Q^{(x)} + \hat{D}_Q^{(y)} Q^{(y)} = M f. \tag{13.15}
$$

Substituting Eq. (13.12) for $\mathbf{Q}$ into Eq. (13.15) yields

$$
\hat{D}_Q^{(x)} M^{-1} \hat{D}_q^{(x)} q + \hat{D}_Q^{(y)} M^{-1} \hat{D}_q^{(y)} q \equiv \left(\hat{D}_Q^{(x)} M^{-1} \hat{D}_q^{(x)} + \hat{D}_Q^{(y)} M^{-1} \hat{D}_q^{(y)} \right) q = M f \tag{13.16}
$$

which reveals that the Laplacian operator in the LDG method is in fact

$$
L = \hat{D}_Q^{(x)} M^{-1} \hat{D}_q^{(x)} + \hat{D}_Q^{(y)} M^{-1} \hat{D}_q^{(y)} \equiv \hat{\mathbf{D}}_Q^T M^{-1} \hat{\mathbf{D}}_q \tag{13.17}
$$

which is constructable because M is block diagonal and non-singular[1]. Furthermore, we can see that the LDG construction of the Poisson problem can be generalized to three dimensions quite easily.

Exercise Derive the three-dimensional version of Eq. (13.17). □

13.7 Algorithm for LDG

Let us now describe the algorithms required for constructing the element and global matrices required by the LDG method. The construction of the element mass matrix is identical to that for the CG method and has already been outlined in Alg. 12.6, where we also need the DSS operator described in Alg. 12.11. Recall from Eq. (13.17) that the LDG Laplacian matrix L is constructed as the product of the global differentiation $\tilde{D}$, flux F, and mass M matrices. Let us now describe the algorithm for constructing the element differentiation matrix.

13.7.1 Element Differentiation Matrix

Algorithm 13.1 shows the steps required for constructing the weak form differentiation matrix $\tilde{D}^{(e)}$. Note that this matrix requires the gradient of the basis function $\nabla\psi$, which is obtained using Eqs. (12.7) and (12.8). In order to build the gradient of the basis functions requires the metric terms. For the metric terms we use Alg. 12.1. The term $J^{(e)}$ represents the determinant of the volume Jacobian. Once $\tilde{D}^{(e)}$ is constructed, we can then build the global matrix $\tilde{D}$ using the DSS operator from Alg. 12.11.

Alternatively, we can build the weak form differentiation matrix as a collection of vector calls. Algorithm 13.2 shows such an approach which is more useful in the sense that for non-elliptic equations (e.g., time-dependent parabolic systems) we typically do not need to build the entire matrix but rather only need the action of the matrix on the solution vector and thereby usually only require a vector. Using this vector approach, we can build the global matrix $\tilde{D}$ by calling algorithm Alg. 13.2 to construct the vector R by feeding the vector q as the identity matrix as shown in Alg. 13.3. Note that the global differentiation matrix $\tilde{D}$ has an entry for each spatial dimension but, for simplicity, in Alg. 13.3 we do not explicitly show this.

Remark 13.1 The construction of the action of a matrix on a vector is an important strategy in order to develop code that is useful for constructing discrete operators for various types of partial differential equations. We highly encourage code developers to use this approach.

---

[1] We know it is non-singular since it is a mass matrix which is defined as the inner product of the basis functions.

Algorithm 13.1 Construction of matrix $\tilde{\mathbf{D}}^{(e)}$ with exact integration.

---

function D_MATRIX_WEAK

$\quad \tilde{\mathbf{D}}^{(e)} = \mathbf{0}$

$\quad$ **for** $e = 1 : N_e$ **do** $\hfill \triangleright$ loop over elements

$\quad\quad$ **for** $k = 1 : M_Q$ **do** $\hfill \triangleright$ loop over integration points

$\quad\quad\quad$ **for** $j = 1 : M_N$ **do** $\hfill \triangleright$ loop over columns of $\tilde{\mathbf{D}}^{(e)}$

$\quad\quad\quad\quad$ **for** $i = 1 : M_N$ **do** $\hfill \triangleright$ loop over rows of $\tilde{\mathbf{D}}^{(e)}$

$\quad\quad\quad\quad\quad \tilde{\mathbf{D}}_{ij}^{(e)} \mathrel{+}= w_k J_k^{(e)} \nabla \psi_{ik} \psi_{jk}$

$\quad\quad\quad\quad$ **end for**

$\quad\quad\quad$ **end for**

$\quad\quad$ **end for**

$\quad$ **end for**

end function

Algorithm 13.2 Construction of differentiation vector $\mathbf{R}$ with exact integration.

---

function D_VECTOR_WEAK(q)

$\quad \mathbf{R} = \mathbf{0}$

$\quad$ **for** $e = 1 : N_e$ **do** $\hfill \triangleright$ loop over elements

$\quad\quad$ **for** $k = 1 : M_Q$ **do** $\hfill \triangleright$ loop over integration points

$\quad\quad\quad q_k = 0$

$\quad\quad\quad$ **for** $j = 1 : M_N$ **do**

$\quad\quad\quad\quad J = intma(j, e)$ $\hfill \triangleright$ get global gridpoint index

$\quad\quad\quad\quad q_k \mathrel{+}= \psi_{j,k} q_J$ $\hfill \triangleright$ interpolate q-vector onto integration points

$\quad\quad\quad$ **end for**

$\quad\quad\quad$ **for** $i = 1 : M_N$ **do** $\hfill \triangleright$ loop over interpolation points

$\quad\quad\quad\quad I = intma(i, e)$ $\hfill \triangleright$ get global gridpoint index

$\quad\quad\quad\quad \mathbf{R}_I \mathrel{+}= w_k J_k^{(e)} \nabla \psi_{ik} q_k$

$\quad\quad\quad$ **end for**

$\quad\quad$ **end for**

$\quad$ **end for**

end function

Algorithm 13.3 Construction of a matrix from vector form.

---

function VECTOR_TO_MATRIX(q)

$\quad \mathbf{M} = \mathbf{0}$

$\quad$ **for** $I = 1 : N_p$ **do** $\hfill \triangleright$ loop over global gridpoints

$\quad\quad q = \mathbf{0}$

$\quad\quad q(I) = 1$

$\quad\quad \mathbf{R}$=D_VECTOR_WEAK(q) $\hfill \triangleright$ form R-vector with q input

$\quad\quad \mathbf{M}(:, I) = \mathbf{R}(:)$

$\quad$ **end for**

end function

13.7.2 Global Flux Matrix

Let us now turn to the algorithm for constructing the global flux matrix $\mathbf{F}$. Algorithm 13.4 shows the algorithm for constructing the vector $\mathbf{R}$ that can then be used in Alg. 13.3 to construct the global matrix $\mathbf{F}$. There are a few data structures that need explanation[2]. The $face$ data structure contains the information for each face. The first two locations $face(1:2, f)$ contain the local position of the face f on the left and right elements. For a quadrilateral it would be either 1, 2, 3, or 4 where 1, 2 refer to the ξ direction with $\eta = -1, +1$ and 3, 4 refer to the η direction with $\xi = -1, +1$. The last two locations store the element of the left and right neighbor. The data structures $mapL$ and $mapR$ tell us the local index of the point in a tensor-product grid in order to determine the global gridpoint number of the points along the face f. The array $normal$ stores the normal vectors of each face in the grid and $J^{(f)}$ stores the face Jacobian - in 2D, the face Jacobian is the length of the face and in 3D it represents the surface area (see Eq. (12.35)). The vector w stores the one-dimensional quadrature weights, while h is the one-dimensional basis functions. Finally, α and β in the numerical flux computation are weights whereby $\alpha \in [0, 1]$ and $\beta = 1 - \alpha$.

At this point, we have all the information to construct the Laplacian matrix L for the LDG method from Eq. (13.17) where $\hat{\mathbf{D}}_q = \mathbf{F}_q - \tilde{\mathbf{D}}$ and $\hat{\mathbf{D}}_Q = \mathbf{F}_Q - \tilde{\mathbf{D}}$.

13.8 Analysis of the Matrix Properties of the Spatial Operators

Let us now analyze the sparsity patterns of both the mass M and Laplacian L matrices as well as analyze the eigenvalues of the Laplacian matrix, just as we did in Ch. 12 for CG. For the analysis given below, we assume homogeneous Dirichlet boundary conditions. In addition, the matrices analyzed are for a grid with $N_e = 4 \times 4$ elements using $N = 4$ polynomial order; an example of this grid is shown in Fig. 12.2.

13.8.1 Sparsity Pattern of the Mass Matrix

Figure 13.1 shows the sparsity pattern for the mass matrix M using both exact $(Q = N + 1)$ and inexact $(Q = N)$ integration. Figure 13.1 shows that the mass matrix is block diagonal for exact integration and diagonal for inexact integration.

---

[2] code for constructing these data structures can be found a
https://github.com/fxgiraldo/Element-based-Galerkin-Methods/tree/master/Chapters/Chapter_13

Algorithm 13.4 Construction of flux vector R with exact integration.

---

function F_VECTOR_WEAK(q)
  $\mathbf{R} = \mathbf{0}$
  for $f = 1 : N_f$ **do**   ▷ loop over faces
                                                                                      ▷ get left element information
    $p_L = face(1, f); e_L = face(3, f)$
    for $i = 1 : N + 1$ **do**   ▷ loop over interpolation points
      $i_L = mapL(1, i, p_L); j_L = mapL(2, i, p_R)$
      $I_L(i) = intma(i_L, j_L, e_L); q_L(i) = q(I_L(i))$
    end for
                                                                                      ▷ get right element information
    $p_R = face(2, f); e_R = face(4, f)$
    if $e_R > 0$ **then**
      for $i = 1 : N + 1$ **do**   ▷ loop over interpolation points
        $i_R = mapR(1, i, p_R); j_R = mapR(2, i, p_R)$
        $I_R(i) = intma(i_R, j_R, e_R); q_R(i) = q(I_R(i))$
      end for
    else
      for $i = 1 : N + 1$ **do**
        $q_R(i) = q_D(I_L(i))$ ▷ impose Dirichlet boundary condition
      end for
    end if
                                                                                      ▷ loop over integration points
    for $k = 1 : Q + 1$ **do**
      $w_k = J^{(f)}(k, f)$
      $n_L^{(x)} = normal(1, k, f); n_L^{(y)} = normal(2, k, f); n_R^{(x)} = -n_L^{(x)}; n_R^{(y)} = -n_L^{(y)}$
      $qL = 0; qR = 0$
      for $i = 1 : N + 1$ **do**   ▷ interpolate onto integration points
        $qL \mathrel{+}= h(i, k) \cdot q_L(i); qR \mathrel{+}= h(i, k) \cdot q_R(i)$
      end for
      $q^{(*)} = \alpha \cdot qL + \beta \cdot qR$ ▷ construct numerical flux
                                                                                      ▷ loop over interpolation points of left element
      for $i = 1 : N + 1$ **do**
        $\mathbf{R}(I_L(i), 1) \mathrel{+}= w_k \cdot n_L^{(x)} \cdot h(i, k) \cdot q^{(*)}$
        $\mathbf{R}(I_L(i), 2) \mathrel{+}= w_k \cdot n_L^{(y)} \cdot h(i, k) \cdot q^{(*)}$
      end for
                                                                                      ▷ loop over interpolation points of right element
      if $e_R > 0$ **then**
        for $i = 1 : N + 1$ **do**
          $\mathbf{R}(I_R(i), 1) \mathrel{+}= w_k \cdot n_R^{(x)} \cdot h(i, k) \cdot q^{(*)}$
          $\mathbf{R}(I_R(i), 2) \mathrel{+}= w_k \cdot n_R^{(y)} \cdot h(i, k) \cdot q^{(*)}$
        end for
      end if
    end for
  end for
end function

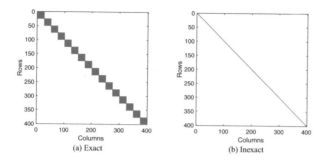

Fig. 13.1: Sparsity patterns for the DG mass matrix M with $N_e = 4 \times 4$, $N = 4$ for (a) exact and (b) inexact integration.

13.8.2 Sparsity Pattern of the Laplacian Matrix

Figure 13.2 shows the sparsity pattern for the Laplacian matrix L for exact and inexact integration with $\alpha = 1$. Figure 13.2 shows that the Laplacian matrix is not

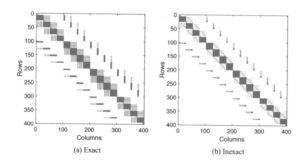

Fig. 13.2: Sparsity patterns for the LDG Laplacian matrix L with $N_e = 4 \times 4$, $N = 4$ for (a) exact and (b) inexact integration.

full, that is, it has a small bandwidth $(O(2N)^2)$. For the Laplacian matrix, there is no difference in the sparsity pattern between exact and inexact integration although the matrices are not identical.

13.8.3 Eigenvalue Analysis of Spatial Operator

Figure 13.3 shows the eigenvalues of the Laplacian $(-L)$ are all on the real axis and, although the eigenvalues have been normalized, the spectrum is rather similar to that obtained for the CG method (for $\alpha = \frac{1}{2}$, the spectrum is very similar and becomes

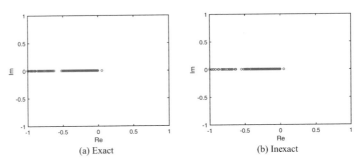

(a) Exact   (b) Inexact

Fig. 13.3: Eigenvalues of the LDG Laplacian matrix $-L$ with $N_e = 4 \times 4$, $N = 4$ for (a) exact and (b) inexact integration.

less so for $\alpha \neq \frac{1}{2}$). The fact that the eigenspectra of the DG method is similar to that of CG is important because it means that any iterative method for solving the the Poisson problem for CG is equally applicable to the DG method with LDG.

13.9 Example of 2D Poisson Equation

Let us consider the solution of the same Poisson problem we used in Ch. 12, that is,

$$\nabla^2 q(x, y) = f(x, y) \qquad \forall (x, y) \in [-1, 1]^2$$

with the exact solution

$$q(x, y) = \sin(c\pi x) \sin(c\pi y)$$

and source function

$$f(x, y) = -2(c\pi)^2 \sin(c\pi x) \sin(c\pi y).$$

Recall that for integer values of c, the Dirichlet boundary conditions are homogeneous. For all values of c the Dirichlet boundary conditions are defined as

$$g(-1, y) = \sin(-c\pi) \sin(c\pi y)$$
$$g(+1, y) = \sin(+c\pi) \sin(c\pi y)$$
$$g(x, -1) = \sin(c\pi x) \sin(-c\pi)$$
$$g(x, +1) = \sin(c\pi x) \sin(c\pi y);$$

where we use $c = 1$ in what follows.

13.9.1 Solution Accuracy

In Fig. 12.6 we showed the analytic and numerical solutions for CG using $N = 16$ order polynomials and $N_e = 4 \times 4$ elements for a total of $N_p = 4225$ gridpoints. Let us now see how well the LDG method behaves by varying the number of elements and the polynomial order of the basis functions. We use the integral form of the L^2 norm given by Eq. (12.58) for computing error norms.

Figure 13.4 shows the convergence rates for various polynomial orders, N, for a total number of gridpoints N_p. Figure 13.4 shows that there is little difference in the

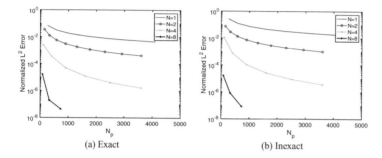

Fig. 13.4: Convergence rates of LDG for the 2D Poisson equation with $c = 1$ for polynomial orders $N = 1$, $N = 2$, $N = 4$, and $N = 8$ using a total number of gridpoints N_p for (a) exact integration ($Q = N + 1$) and (b) inexact integration ($Q = N$).

convergence rates between exact and inexact integration.

To discern the cost of various polynomial orders we show *work-precision* diagrams in Fig. 13.5. Regardless of exact or inexact integration, we conclude that for a given level of accuracy ($< 10^{-4}$), high-order ($N \geq 4$) is more efficient. The fact that the solution is smooth means that high-order will be more efficient but even when the solution is not necessarily smooth everywhere, high-order can offer increased precision in smooth regions.

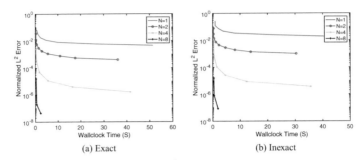

(a) Exact   (b) Inexact

Fig. 13.5: Work-Precision: Error versus wallclock time for LDG for the 2D Poisson equation with $c = 1$ for polynomial orders $N = 1$, $N = 2$, $N = 4$, and $N = 8$ with (a) exact integration ($Q = N + 1$) and (b) inexact integration ($Q = N$).

Chapter 14
2D Unified Continuous and Discontinuous Galerkin Methods for Elliptic Equations

14.1 Introduction

In Ch. 12 we saw that the CG method can be applied in a relatively straightforward manner to solve 2D elliptic equations. All that is required is to know how to compute first derivatives. As in the 1D case, a judicious use of Green's first identity allows for a simple discretization of the Laplacian operator using the so-called *primal formulation*. In Chs. 9 and 13 for the DG method we did not use the simpler primal formulation but rather wrote the second order operator as a system of first order operators and then used the local discontinuous Galerkin (LDG) method.

In Ch. 12 we used a primal formulation for CG but in Ch. 13 we used a flux formulation for DG. Since it would be advantageous to unite both the CG and DG methods within a single source code, then we need to find a method that can work equally well for both CG and DG discretizations of elliptic operators. Let us derive a general form for such an approach that falls under the category of interior penalty methods. Of course, the simplicity and elegance of the LDG method makes it a very attractive approach that should also be considered when selecting a method for second order operators. We leave this decision up to the reader and will now introduce the interior penalty method which allows us to unite CG and DG within the same code base.

14.2 Elliptic Equation

Let us define the Poisson problem to be

$$\nabla^2 q(\mathbf{x}) = f(\mathbf{x}) \tag{14.1}$$

F. X. Giraldo, *An Introduction to Element-Based Galerkin Methods on Tensor-Product Bases*, Texts in Computational Science and Engineering 24, https://doi.org/10.1007/978-3-030-55069-1_14

where $\mathbf{x} \in \Omega$ with the Dirichlet boundary condition

$$q|_{\Gamma_D} = g(\mathbf{x})$$

and Neumann boundary condition

$$(\hat{\mathbf{n}} \cdot \nabla q)|_{\Gamma_N} = h(\mathbf{x})$$

where Γ_D and Γ_N denote the Dirichlet and Neumann boundaries of the domain Ω. Let us now describe the flux formulation for solving this problem.

14.3 Flux Formulation

A natural way to derive discretization methods for elliptic equations is to exploit the fact that an elliptic operator is defined as the divergence of the gradient of a function, this is known as the *flux formulation*. In other words, we rewrite the original Poisson equation as follows

$$\mathbf{Q}(\mathbf{x}) = \nabla q(\mathbf{x})$$

and

$$\nabla \cdot \mathbf{Q}(\mathbf{x}) = f(\mathbf{x})$$

with the Dirichlet and Neumann boundary conditions defined as above. Let us now apply the usual Galerkin machinery to this problem to derive the following discrete form

$$\int_{\Omega_e} \boldsymbol{\tau}_i \cdot \mathbf{Q}_N^{(e)} \, d\Omega_e = \int_{\Omega_e} \boldsymbol{\tau}_i \cdot \nabla q_N^{(e)} \, d\Omega_e \qquad (14.2a)$$

$$\int_{\Omega_e} \psi_i \nabla \cdot \mathbf{Q}_N^{(e)} \, d\Omega_e = \int_{\Omega_e} \psi_i f_N^{(e)} \, d\Omega_e \qquad (14.2b)$$

where $\boldsymbol{\tau}$ and ψ are the test functions, $\Omega = \bigcup_{e=1}^{N_e} \Omega_e$ defines the global domain, and $\Gamma_i \bigcup \Gamma_b = \bigcup_{e=1}^{N_e} \Gamma_e$ defines the boundary, where Γ_e are all element faces, Γ_i are all interior faces, and Γ_b are all boundary faces, with $i = 1, \ldots, M_N$ and $e = 1, \ldots, N_e$. For brevity, we have omitted writing the test functions and solution q as functions of $\mathbf{x}$, although this should be understood.

Integrating Eqs. (14.2) by parts yields

$$\int_{\Omega_e} \boldsymbol{\tau}_i \cdot \mathbf{Q}_N^{(e)} \, d\Omega_e = \int_{\Gamma_e} \hat{\mathbf{n}} \cdot (\boldsymbol{\tau}_i q_N)^{(*,e)} \, d\Gamma_e - \int_{\Omega_e} \nabla \cdot \boldsymbol{\tau}_i q_N^{(e)} \, d\Omega_e \qquad (14.3a)$$

$$\int_{\Gamma_e} \hat{\mathbf{n}} \cdot (\psi_i \mathbf{Q}_N)^{(*,e)} \, d\Gamma_e - \int_{\Omega_e} \nabla \psi_i \cdot \mathbf{Q}_N^{(e)} \, d\Omega_e = \int_{\Omega_e} \psi_i f_N^{(e)} \, d\Omega_e \qquad (14.3b)$$

where $\hat{\mathbf{n}}$ is the unit normal vector pointing outward from the element Ω_e to its neighbor.

14.4 Primal Formulation

If we let $\tau = \psi \mathbf{I}_2$, where $\mathbf{I}_2$ is the rank-2 identity matrix, we recover the LDG method, with the proper selection of numerical fluxes. For example, the simplest numerical flux that yields an LDG solution is $w^{(*,e)} = \{w\}$ where the operator $\{\cdot\}$ denotes the typical averaging operator for the function w. The difficulty with LDG, as we saw in Chs. 9 and 13 is that it is complicated to derive a primal formulation. In order to derive a primal formulation, we need to choose the test function τ differently. Looking at Eq. (14.3) we can see that if we choose $\tau = \boldsymbol{\nabla}\psi$ then we are able to substitute Eq. (14.3a) into (14.3b). Replacing τ with $\boldsymbol{\nabla}\psi$ in Eq. (14.3) yields

$$\int_{\Omega_e} \boldsymbol{\nabla}\psi_i \cdot \mathbf{Q}_N^{(e)} \, d\Omega_e = \int_{\Gamma_e} \hat{\mathbf{n}} \cdot (\boldsymbol{\nabla}\psi_i q_N)^{(*,e)} \, d\Gamma_e - \int_{\Omega_e} \boldsymbol{\nabla}^2 \psi_i q_N^{(e)} \, d\Omega_e \quad (14.4a)$$

$$\int_{\Gamma_e} \hat{\mathbf{n}} \cdot (\psi_i \mathbf{Q}_N)^{(*,e)} \, d\Gamma_e - \int_{\Omega_e} \boldsymbol{\nabla}\psi_i \cdot \mathbf{Q}_N^{(e)} \, d\Omega_e = \int_{\Omega_e} \psi_i f_N^{(e)} \, d\Omega_e. \quad (14.4b)$$

Substituting Eq. (14.4a) for the first term into Eq. (14.4b) for the second term yields the *primal formulation*

$$\int_{\Gamma_e} \hat{\mathbf{n}} \cdot (\psi_i \mathbf{Q}_N)^{(*,e)} \, d\Gamma_e - \int_{\Gamma_e} \hat{\mathbf{n}} \cdot (\boldsymbol{\nabla}\psi_i q_N)^{(*,e)} \, d\Gamma_e$$
$$+ \int_{\Omega_e} \boldsymbol{\nabla}^2 \psi_i q_N^{(e)} \, d\Omega_e = \int_{\Omega_e} \psi_i f_N^{(e)} \, d\Omega_e. \quad (14.5)$$

However, Eq. (14.5) tells us that ψ must live in H^2 which is far too restrictive, although perfectly acceptable. To make the functional space of ψ less restrictive we apply integration by parts one more time using Eq. (14.4a) rearranged in the following way

$$\int_{\Omega_e} \boldsymbol{\nabla}^2 \psi_i q_N^{(e)} \, d\Omega_e = -\int_{\Omega_e} \boldsymbol{\nabla}\psi_i \cdot \mathbf{Q}_N^{(e)} \, d\Omega_e + \int_{\Gamma_e} \hat{\mathbf{n}} \cdot (\boldsymbol{\nabla}\psi_i q_N)^{(e)} \, d\Gamma_e \quad (14.6)$$

where, here, we do not replace the boundary integral with a numerical flux. Substituting this relation into Eq. (14.5) yields

$$\int_{\Gamma_e} \hat{\mathbf{n}} \cdot (\psi_i \mathbf{Q}_N)^{(*,e)} \, d\Gamma_e + \int_{\Gamma_e} \hat{\mathbf{n}} \cdot \left[(\boldsymbol{\nabla}\psi_i q_N)^{(e)} - (\boldsymbol{\nabla}\psi_i q_N)^{(*,e)} \right] d\Gamma_e$$
$$- \int_{\Omega_e} \boldsymbol{\nabla}\psi_i \cdot \mathbf{Q}_N^{(e)} \, d\Omega_e = \int_{\Omega_e} \psi_i f_N^{(e)} \, d\Omega_e. \quad (14.7)$$

Next, if we make the assumption that $\psi^{(*,e)} = \psi$ and $\nabla\psi^{(*,e)} = \nabla\psi$ we can simplify Eq. (14.7) as follows

$$
\int_{\Gamma_e} \psi_i \hat{\mathbf{n}} \cdot \mathbf{Q}_N^{(*,e)} \, d\Gamma_e + \int_{\Gamma_e} \hat{\mathbf{n}} \cdot \nabla\psi_i \left[q_N^{(e)} - q_N^{(*,e)} \right] d\Gamma_e
$$
$$
- \int_{\Omega_e} \nabla\psi_i \cdot \mathbf{Q}_N^{(e)} \, d\Omega_e = \int_{\Omega_e} \psi_i f_N^{(e)} \, d\Omega_e. \qquad (14.8)
$$

Remark 14.1 Let us now discuss the assumption on the basis functions that we just made. Clearly, if the basis functions ψ are symmetric then the value of ψ on the left and right of any face Γ_e will be identical so the assumption $\psi^{(*,e)} = \psi$ is valid for any (CG or DG) method that uses Lagrange polynomials, provided that the integration points are symmetric in the element, which is true for Legendre and Lobatto points. The second assumption that $\nabla\psi^{(*,e)} = \nabla\psi$ is only true for elements that are of the same length (i.e., conforming elements). For non-conforming elements, we will have to compute $\nabla\psi^{(*,e)}$ differently.

Next, we replace $\mathbf{Q}$ with ∇q in Eq. (14.8) to arrive at the final form

$$
\int_{\Gamma_e} \psi_i \hat{\mathbf{n}} \cdot \nabla q_N^{(*,e)} \, d\Gamma_e + \int_{\Gamma_e} \hat{\mathbf{n}} \cdot \nabla\psi_i \left[q_N^{(e)} - q_N^{(*,e)} \right] d\Gamma_e
$$
$$
- \int_{\Omega_e} \nabla\psi_i \cdot \nabla q_N^{(e)} \, d\Omega_e = \int_{\Omega_e} \psi_i f_N^{(e)} \, d\Omega_e. \qquad (14.9)
$$

With proper numerical fluxes, Eq. (14.9) represents the symmetric interior penalty Galerkin (SIPG) method.

Exercise Equation (14.9) contains all of the terms that we found in the weak form Laplacian for the CG method described in Eq. (12.36) with an additional flux integral (the second term on the left-hand side). Under what conditions will the second term in Eq. (14.9) vanish? Under these conditions, the Laplacian operator for CG and DG are exactly the same. □

14.5 Symmetric Interior Penalty Galerkin (SIPG) Method

To derive the SIPG method from Eq. (14.9), we must first define numerical fluxes. Note that we need to define fluxes for both ∇q and q. Assuming that q is a scalar, we see that, as in the LDG method, there is no need to use anything else except the centered flux defined as:

$$
q_N^{(*,e)} = \left\{ q^{(e,k)} \right\} \qquad (14.10)
$$

which denotes the mean flux of the element given as $q^{(e)}$ and that for its k neighbor defined as $q^{(k)}$. On the other hand, since ∇q is a vector, we may want to use some form of upwinding for the numerical flux. In the LDG method, we had some options and we chose to use a centered flux. However, to derive the SIPG in a more general form let us introduce the upwind flux

$$
\nabla q_N^{(*,e)} = \left\{ \nabla q^{(e,k)} \right\} - \hat{\mathbf{n}} \mu \llbracket q^{(e,k)} \rrbracket \qquad (14.11)
$$

where $\mu > 0$ is a penalty term that, as of yet, is undefined. If $\mu = 0$ then we recover a centered flux; otherwise, the flux is similar to an upwind flux. In Eq. (14.11) μ plays the role of the scaling of the maximum wave speed (as in the Rusanov flux for hyperbolic equations) and has dimensions similar to $1/\Delta x$. Substituting the fluxes in Eqs. (14.10) and (14.11) into Eq. (14.9) yields

$$- \int_{\Omega_e} \nabla \psi_i \cdot \nabla q_N^{(e)} \, d\Omega_e + \int_{\Gamma_e} \psi_i \hat{\mathbf{n}} \cdot \left[\left\{ \nabla q_N^{(e,k)} \right\} - \hat{\mathbf{n}} \mu \left(q_N^{(k)} - q_N^{(e)} \right) \right] d\Gamma_e$$

$$+ \int_{\Gamma_e} \hat{\mathbf{n}} \cdot \nabla \psi_i \left[q_N^{(e)} - \left\{ q_N^{(e,k)} \right\} \right] d\Gamma_e = \int_{\Omega_e} \psi_i f_N^{(e)} \, d\Omega_e \qquad (14.12)$$

where $i = 1, \ldots, M_N$ and $e = 1, \ldots, N_e$, which is the final form of the SIPG method and, in fact, is similar to the approach we described in Eq. (9.2) in Ch. 9[1].

14.6 Algorithm for the SIPG Method

Equation (14.12) shows that there are three integrals required on the left-hand side for the SIPG method. The first integral is a volume integral that is in fact identical to the weak form Laplacian matrix required by the CG method. Next, there are two flux integrals. Let us now describe the algorithms required for constructing the global matrix required by the SIPG method. In what follows, we use the approach introduced in Ch. 13 whereby we construct only the action of the operators on the solution vector. With this approach, we are able to construct either global matrices (required in time-independent problems) or matrix-vector products (useful for constructing efficient methods for time-dependent problems where the construction of the global matrix may not strictly be required).

To construct the weak form Laplacian matrix (arising from integration by parts) we use Alg. 12.9 to construct the element matrix followed by the DSS operator given in Alg. 12.11. This yields the first term (volume integral) of Eq. (14.12). However, here, we are interested in the vector approach which we now describe. Algorithm 14.1 looks very similar to Alg. 12.9 used to construct the global weak form Laplacian matrix. The difference is that we now use the solution vector q to construct the action of the Laplacian on q. Once the volume integral is stored in the vector $\mathbf{R}$ we then augment it with the flux terms. Let us now describe the algorithm for constructing the flux integrals which are the second and third terms on the left-hand side of Eq. (14.12).

Algorithm 14.2 describes the steps required for constructing the flux terms in the SIPG Laplacian matrix, where we reuse the map (I_L, I_R) from Alg. 13.4 for the flux matrix of the LDG method. The description of a few terms in Alg. 14.2 is in order. The vector $\hat{\mathbf{n}}$ denotes the outward pointing normal vector from the left element L to the right element R. A good choice for computing the jump weight μ can be found

---

[1] The difference is in the negative of the operator and in the fact that we have explicitly defined the numerical fluxes.

Algorithm 14.1 SIPG Laplacian volume integral contribution to the vector **R**.

---

function VOLUME_INTEGRAL_LAPLACIAN_SIPG(q)
   $\mathbf{R} = 0$
   for $e = 1 : N_e$ **do**   ▷ loop over elements
      for $k = 1 : M_Q$ **do**   ▷ loop over integration points
         $\nabla q_k = 0$
         for $j = 1 : M_N$ **do**
            $J = intma(j, e)$
            $\nabla q_k \mathrel{+}= \nabla \psi_{jk} q_J$ ▷ compute derivative on integration points
         end for
         for $i = 1 : M_N$ **do**   ▷ loop over interpolation points
            $I = intma(i, e)$
            $\mathbf{R}_I \mathrel{+}= w_k J_k^{(e)} \nabla \psi_{ik} \cdot \nabla q_k$
         end for
      end for
   end for
end function

---

in Shahbazi [340] as follows $\mu = \max\left(\mu^{(L)}, \mu^{(R)}\right)$ where

$$\mu_k^{(e)} = \mu_C \frac{N(N+1)}{2} \frac{J_k^{(f)}}{J_k^{(e)}}$$

where k is the local gridpoint index, e is the element index, N is the polynomial order, $J^{(f)}$ and $J^{(e)}$ are the face and volume Jacobians, and μ_C is a constant (for the results in Sec. 14.8 we use $\mu_C = 1 \times 10^5$); large values of μ_C are essential for enforcing Dirichlet boundary conditions. Note that the contributions to the vector **R** from the left and right elements are essentially the same but reversed in sign. If q is continuous across the faces, then we can see that this is indeed the case.

Once Algs. 14.1 and 14.2 have been called consecutively, we can then use Alg. 13.3 to construct the global representation of the SIPG Laplacian matrix. At this point, we may also impose strong form boundary conditions as described in Alg. 12.17.

14.7 Analysis of the Matrix Properties of the Spatial Operators

Let us now analyze the sparsity pattern and eigenvalues of the Laplacian matrix L. For the analysis given below, we assume homogeneous Dirichlet boundary conditions. In addition, the matrices analyzed are for a grid with $N_e = 4 \times 4$ elements using $N = 4$ polynomial order; an example of this grid is shown in Fig. 12.2.

Algorithm 14.2 SIPG flux integral contribution to the vector **R**.

---

function FLUX_INTEGRAL_LAPLACIAN_SIPG(q)
  for $f = 1 : N_f$ **do**   ▷ loop over faces
    for $k = 1 : Q + 1$ **do**   ▷ loop over integration points
      for $i = 1 : N + 1$ **do**
        Construct $\nabla q_k^{(L)}$ and $\nabla q_k^{(R)}$ ▷ if external boundary, use BCs for $q^{(R)}$

$$\nabla q_k^{(*,e)} = \tfrac{1}{2}\left[\nabla q_k^{(L)} + \nabla q_k^{(R)} - \mu_k \hat{\mathbf{n}}_k \left(q_k^{(R)} - q_k^{(L)}\right)\right]$$

$$q_k^{(*,e)} = \tfrac{1}{2}\left[q_k^{(L)} + q_k^{(R)}\right]$$

$$L = I_L(i)$$

$$\mathbf{R}_L \mathrel{+}= w_k J_k^{(f)} \psi_{i,k}\left(\hat{\mathbf{n}}_k \cdot \nabla q_k^{(*,e)}\right)$$

$$\mathbf{R}_L \mathrel{+}= w_k J_k^{(f)}\left(\hat{\mathbf{n}}_k \cdot \nabla \psi_{i,k}^{(L)}\right)\left(q_k^{(L)} - q_k^{(*,e)}\right)$$

      end for

                                 ▷ loop over interpolation points of right element
    if $e_R > 0$ **then**
      for $i = 1 : N + 1$ **do**
        $R = I_R(i)$

$$\mathbf{R}_R \mathrel{-}= w_k J_k^{(f)} \psi_{i,k}\left(\hat{\mathbf{n}}_k \cdot \nabla q_k^{(*,e)}\right)$$

$$\mathbf{R}_R \mathrel{-}= w_k J_k^{(f)}\left(\hat{\mathbf{n}}_k \cdot \nabla \psi_{i,k}^{(R)}\right)\left(q_k^{(R)} - q_k^{(*,e)}\right)$$

      end for
    end if
    end for
  end for
end function

14.7.1 Sparsity Pattern of the Laplacian Matrix

Figure 14.1 shows the sparsity pattern for the Laplacian matrix L for exact and inexact integration. Figure 14.1 shows that the Laplacian matrix is not full, i.e., it has

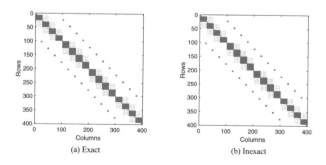

(a) Exact   (b) Inexact

Fig. 14.1: Sparsity patterns for the SIPG Laplacian matrix L with $N_e = 4 \times 4$, $N = 4$ for (a) exact and (b) inexact integration.

a small bandwidth (similar to LDG). For the Laplacian matrix, there is no difference in the sparsity pattern between exact and inexact integration although the matrices are not identical.

14.7.2 Eigenvalue Analysis of Spatial Operator

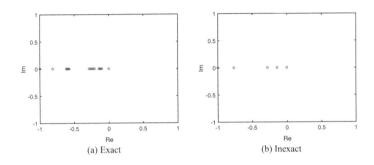

(a) Exact   (b) Inexact

Fig. 14.2: Eigenvalues of the SIPG Laplacian matrix $-L$ with $N_e = 4 \times 4$, $N = 4$ for (a) exact and (b) inexact integration.

Figure 14.2 shows the eigenvalues of the Laplacian $(-L)$ are all on the real axis and, although the eigenvalues have been normalized, the spectrum is rather similar to that obtained for the CG method (Ch. 12) and that for the DG method with LDG (Ch. 13).

14.8 Example of 2D Poisson Equation

Let us consider the solution of the same Poisson problem we used in Chs. 12 and 13, i.e,

$$\nabla^2 q(x, y) = f(x, y) \qquad \forall (x, y) \in [-1, 1]^2$$

with the exact solution

$$q(x, y) = \sin(c\pi x) \sin(c\pi y)$$

and source function

$$f(x, y) = -2(c\pi)^2 \sin(c\pi x) \sin(c\pi y)$$

where we use $c = 1$ in what follows.

14.8.1 Solution Accuracy

In Fig. 12.6 we showed the analytic and numerical solutions for CG using $N = 16$ order polynomials and $N_e = 4 \times 4$ elements for a total of $N_p = 4225$ gridpoints. Let us now see how well the DG method with SIPG behaves by varying the number of elements and the polynomial order of the basis functions. To compute error norms, we use the integral form of the L^2 norm given by Eq. (12.58).

Figure 14.3 shows the convergence rates for various polynomial orders, N, for a total number of gridpoints N_p. Figure 14.3 shows that there is little difference in the

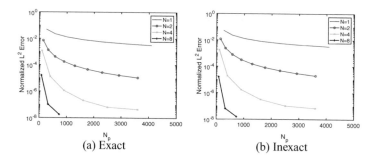

Fig. 14.3: Convergence rates of SIPG for the 2D Poisson equation with $c = 1$ for polynomial orders $N = 1$, $N = 2$, $N = 4$, and $N = 8$ using a total number of gridpoints N_p for (a) exact integration $(Q = N + 1)$ and (b) inexact integration $(Q = N)$.

convergence rates between exact and inexact integration.

To discern the cost of various polynomial orders we show work-precision diagrams in Fig. 14.4. Regardless of exact or inexact integration, we conclude that for a given level of accuracy $(< 10^{-4})$, high-order $(N \geq 4)$ is clearly more efficient. The fact that the solution is smooth means that high-order will be more efficient but even when the solution is not necessarily smooth everywhere, high-order can offer increased precision in smooth regions.

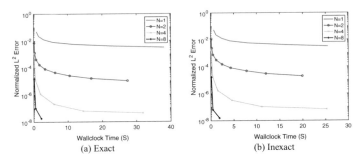

Fig. 14.4: Work-Precision: Error versus wallclock time for SIPG for the 2D Poisson equation with $c = 1$ for polynomial orders $N = 1$, $N = 2$, $N = 4$, and $N = 8$ with (a) exact integration ($Q = N + 1$) and (b) inexact integration ($Q = N$).

Chapter 15
2D Continuous Galerkin Methods for Hyperbolic Equations

15.1 Introduction

In Chs. 5 and 8 we discussed the implementation of continuous Galerkin (CG) methods in one dimension for hyperbolic and elliptic equations, respectively. In Ch. 12 we introduced the extension of the CG method to two dimensions by describing its implementation for elliptic partial differential equations (PDEs). In this chapter we extend the CG method for the application of hyperbolic equations in two dimensions. We also discuss the addition of diffusion operators.

This chapter is organized as follows. In Sec. 15.2 we introduce the continuous PDE and in Sec. 15.3 describe its integral form. In Sec. 15.4 we write the element equations on the reference element. We use this result in Sec. 15.5 to construct the global problem required for CG. To clarify the construction of the algorithms, in Sec. 15.6 we derive the advection and Laplacian matrices, and then write the resulting matrix problem on the reference element for linear polynomials ($N = 1$). In Sec. 15.7 we discuss algorithms for constructing the CG solution, and advection matrix. We discuss in detail the algorithms using non-tensor-product versus tensor-product approaches and introduce sum factorization. Finally, in Sec. 15.8 we show numerical results for the two-dimensional advection equation including convergence rates and computational efficiency measures (i.e., work-precision diagrams).

15.2 2D Advection-Diffusion Equation

Let us begin our discussion with the two-dimensional advection-diffusion equation which can be written in the following form

$$\frac{\partial q}{\partial t} + \mathbf{u} \cdot \nabla q = \nu \nabla^2 q \tag{15.1}$$

© The Editor(s) (if applicable) and The Author(s), under exclusive license
to Springer Nature Switzerland AG 2020
F. X. Giraldo, *An Introduction to Element-Based Galerkin Methods on Tensor-Product Bases*, Texts in Computational Science and Engineering 24,
https://doi.org/10.1007/978-3-030-55069-1_15

where $q = q(\mathbf{x}, t)$ is the scalar solution variable, $\nabla = \left(\frac{\partial}{\partial x} \hat{\mathbf{i}} + \frac{\partial}{\partial y} \hat{\mathbf{j}} \right)$ is the 2D gradient operator, $\hat{\mathbf{i}}$ and $\hat{\mathbf{j}}$ are the directional unit vectors in 2D Cartesian space, and ν is the viscosity coefficient which we assume, for the moment, to be a constant.

15.3 Integral Form

To construct the integral form, we begin by expanding the solution variable as follows

$$q_N^{(e)}(\mathbf{x}, t) = \sum_{j=1}^{M_N} \psi_j(\mathbf{x}) q_j^{(e)}(t)$$

with

$$\mathbf{u}_N^{(e)}(\mathbf{x}) = \sum_{j=1}^{M_N} \psi_j(\mathbf{x}) \mathbf{u}_j^{(e)}$$

where $\mathbf{u}^{(e)}(\mathbf{x}) = u^{(e)}(\mathbf{x})\hat{\mathbf{i}} + v^{(e)}(\mathbf{x})\hat{\mathbf{j}}$ is the velocity vector with u and v components along $\hat{\mathbf{i}}$ (x-direction) and $\hat{\mathbf{j}}$ (y-direction). The integer M_N is a function of N which determines the number of points inside each element Ω_e. For the case of quadrilateral elements $M_N = (N + 1)^2$, where for simplicity we assume that the polynomial order is constant along both spatial dimensions (although they need not be).

Next, we substitute $q_N^{(e)}$ and $\mathbf{u}_N^{(e)}$, into the PDE, multiply by a test function and integrate within each element yielding the problem statement: find $q_N^{(e)} \in H^1$ such that

$$\int_{\Omega_e} \psi_i \frac{\partial q_N^{(e)}}{\partial t} \, d\Omega_e + \int_{\Omega_e} \psi_i \mathbf{u}_N^{(e)} \cdot \nabla q_N^{(e)} \, d\Omega_e = \nu \int_{\Omega_e} \psi_i \nabla^2 q_N^{(e)} \, d\Omega_e \qquad \forall \psi \in H^1$$

$$(15.2)$$

where $i = 1, \ldots, M_N$ and

$$\Omega = \bigcup_{e=1}^{N_e} \Omega_e$$

with N_e denoting the number of elements that, when summed, comprise the global domain Ω. In what follows, we use the basis function and metric terms described in Sec. 12.4.

15.4 Element Equations on the Reference Element

We need to employ integration by parts in Eq. (15.2), in order to bring down the second order operator (∇^2) to a product of first order operators, since we are using the space H^1.

15.4.1 Integration by Parts for the Diffusion Operator

To maintain the solution q in the vector space H^1 requires using the product rule

$$\nabla \cdot \left(\psi_i \nabla q_N^{(e)} \right) = \nabla \psi_i \cdot \nabla q_N^{(e)} + \psi_i \nabla^2 q_N^{(e)}$$

which we rearrange as follows

$$\psi_i \nabla^2 q_N^{(e)} = \nabla \cdot \left(\psi_i \nabla q_N^{(e)} \right) - \nabla \psi_i \cdot \nabla q_N^{(e)}.$$

Substituting this identity into the integral form (15.2) and using the divergence theorem leads to the following element integral equation

$$\int_{\Omega_e} \psi_i \frac{\partial q_N^{(e)}}{\partial t} \, d\Omega_e + \int_{\Omega_e} \psi_i \mathbf{u}_N^{(e)} \cdot \nabla q_N^{(e)} \, d\Omega_e = \nu \int_{\Gamma_e} \psi_i g(\mathbf{x}) \, d\Gamma_e \qquad (15.3)$$
$$- \nu \int_{\Omega_e} \nabla \psi_i \cdot \nabla q_N^{(e)} \, d\Omega_e$$

where, as in Ch. 8, the known function $g(\mathbf{x})$ is the Neumann boundary condition defined as $\hat{\mathbf{n}} \cdot \nabla q_N |_{\Gamma_N} = g(\mathbf{x})$ where Γ_N denotes the part of Γ that satisfies this boundary condition. Substituting the expansion for $q_N^{(e)}$ yields

$$\int_{\Omega_e} \psi_i \left(\sum_{j=1}^{M_N} \psi_j \frac{\partial q_j^{(e)}}{\partial t} \right) d\Omega_e + \int_{\Omega_e} \psi_i \left(\sum_{k=1}^{M_N} \psi_k \mathbf{u}_k^{(e)} \right) \cdot \left(\sum_{j=1}^{M_N} \nabla \psi_j q_j^{(e)} \right) d\Omega_e =$$
$$+ \nu \int_{\Gamma_e} \psi_i g(\mathbf{x}) \, d\Gamma_e - \nu \int_{\Omega_e} \nabla \psi_i \cdot \left(\sum_{j=1}^{M_N} \nabla \psi_j q_j^{(e)} \right) d\Omega_e.$$

In practice, we only compute the boundary integral at the global domain boundary Γ since, by the C^0 condition of CG, this integral cancels at the interior boundaries between elements. However, we saw in Ch. 8 that we can define this term at all element edges as long as $g(\mathbf{x})$ is defined continuously in the interior.

Remark 15.1 Making $g(\mathbf{x})$ continuous leads the flux term to vanish at interior element faces. However, at the exterior boundary, this term will not vanish and will help satisfy Neumann boundary conditions. Making $g(\mathbf{x})$ continuous (e.g., $g(\mathbf{x}) = 0$) in the interior allows for a way to unite the CG and DG methods which we discuss in Ch. 17.

Remark 15.2 Equation (15.4) is defined for each physical element. To convert from the physical to the reference element equations requires introducing the metric terms. We introduce these in the next section.

15.4.2 Matrix-Vector Problem Resulting from Exact Integration

At this point we have to decide whether we are going to evaluate the integrals using co-located quadrature (which results in inexact integration) or not (that, for a specific number of quadrature points, results in exact integration). Let us first see what happens when we use exact integration.

Let $M_Q = (Q + 1)^2$ where $Q = \frac{3}{2}N + \frac{1}{2}$ Lobatto points integrate $3N$ polynomials exactly (since the advection term is a 3N degree polynomial)[1]. This quadrature rule yields the following matrix-vector problem

$$M_{ij}^{(e)} \frac{dq_j^{(e)}}{dt} + A_{ij}^{(e)}(\mathbf{u})q_j^{(e)} = \nu B_i^{(e)} - \nu L_{ij}^{(e)} q_j^{(e)}$$

where the above matrices are defined as follows

$$M_{ij}^{(e)} \equiv \int_{\Omega_e} \psi_i \psi_j \, d\Omega_e = \sum_{l=1}^{M_Q} w_l J_l^{(e)} \psi_{il} \psi_{jl}$$

is the mass matrix where w_l and $J_l^{(e)}$ are the quadrature weights and determinant of the volume Jacobian evaluated at the quadrature points $\boldsymbol{\xi}_l$,

$$A_{ij}^{(e)}(\mathbf{u}^{(e)}) \equiv \int_{\Omega_e} \psi_i \left(\sum_{k=1}^{M_N} \psi_k \mathbf{u}_k^{(e)} \right) \cdot \nabla \psi_j \, d\Omega_e = \sum_{l=1}^{M_Q} w_l J_l^{(e)} \psi_{il} \left(\sum_{k=1}^{M_N} \psi_{kl} \mathbf{u}_k^{(e)} \right) \cdot \nabla \psi_{jl}$$

$$(15.4)$$

is the advection matrix, which is nothing more than the strong form differentiation matrix with a velocity approximation embedded,

$$B_i^{(e)} \equiv \int_{\Gamma_e} \psi_i g(\mathbf{x}) \, d\Gamma_e = \sum_{l=0}^{Q} w_l^{(f)} J_l^{(f)} \psi_{il} g(\mathbf{x}_l)$$

is the boundary vector resulting from the integration by parts of the Laplacian operator where the superscript (f) denotes the face lying on Γ_e, and

$$L_{ij}^{(e)} \equiv \int_{\Omega_e} \nabla \psi_i \cdot \nabla \psi_j \, d\Omega_e = \sum_{l=1}^{M_Q} w_l J_l^{(e)} \nabla \psi_{il} \cdot \nabla \psi_{jl}$$

is the Laplacian matrix that we have already seen in Ch. 12. To compute terms such as $\nabla \psi$ on the reference element, we follow the approach described in Sec. 12.7. There is no problem integrating the above matrices exactly, except for the issue of having to deal with a non-diagonal mass matrix. To get around this complexity, we can use

---

[1] For Legendre points, we could use $Q = \frac{3}{2}N$ points to achieve exact integration of a $3N$ polynomial but these points are not used for CG.

mass lumping. However, let us look at the other choice we have at our disposal, namely, inexact integration which should only be done for tensor-product bases[2].

15.4.3 Matrix-Vector Problem Resulting from Inexact Integration

Looking at the mass, advection, boundary, and Laplacian matrices we see that they represent integrals of polynomials of degree: 2N for the mass and Laplacian matrices, 3N for the advection matrix, and 2N-1 for the boundary vector. Thus if we use inexact integration, i.e., 2N-1 integration, we only commit a small numerical error for the mass and Laplacian matrices. For the advection matrix, the error is a bit larger, but note that for the boundary vector we commit no such error and, in fact, obtain this term exactly! Let us now use inexact integration and see what the resulting matrix-vector problem looks like. In this case, $M_Q = (Q + 1)^2 = M_N$ where $Q = N$ which leads to the following matrix-vector problem

$$M_{ij}^{(e)} \frac{dq_j^{(e)}}{dt} + \left(\mathbf{u}_i^{(e)} \right)^T D_{ij}^{(e)} q_j^{(e)} = v B_i^{(e)} - v L_{ij}^{(e)} q_j^{(e)}$$

where changes in our matrix-vector problem from exact to inexact integration occur in all the matrices. Let us begin with the mass matrix and replace the integral by quadrature as follows

$$M_{ij}^{(e)} = \sum_{l=1}^{M_N} w_l J_l^{(e)} \psi_{il} \psi_{jl}.$$

By virtue of cardinality, ψ_{il} is non-zero only when $i = l$ which means that we can now replace l by i and remove the summation to yield

$$M_{ij}^{(e)} = w_i J_i^{(e)} \delta_{ij},$$

where we have applied cardinality to $\psi_{jl} = \psi_{ji}$ and replaced it by δ_{ij}. Note that the mass matrix has now simplified to a vector which we can write as

$$M_i^{(e)} = w_i J_i^{(e)}.$$

Using this idea, we can also simplify the boundary vector as follows

$$B_i^{(e)} = w_i^{(f)} J_i^{(f)} g(\mathbf{x}_i)$$

---

[2] For inexact integration on triangles see, e.g., [87, 276, 167, 191, 277]. We recommend not using inexact integration on triangles or tetrahedra because no set of points have been found that act as good interpolation and integration points.

and the Laplacian matrix

$$L_{ij}^{(e)} = \sum_{l=1}^{M_N} w_l J_l^{(e)} \nabla \psi_{il} \cdot \nabla \psi_{jl}$$

does not simplify at all.

Let us now take a closer look at the advection matrix

$$A_{ij}^{(e)}(\mathbf{u}) = \int_{\Omega_e} \psi_i \left(\sum_{k=1}^{M_N} \psi_k \mathbf{u}_k^{(e)} \right) \cdot \nabla \psi_j \, d\Omega_e.$$

Replacing the integral with quadrature yields

$$A_{ij}^{(e)}(\mathbf{u}) = \sum_{l=1}^{M_N} w_l J_l^{(e)} \psi_{il} \left(\sum_{k=1}^{M_N} \psi_{kl} \mathbf{u}_k^{(e)} \right) \cdot \nabla \psi_{jl}.$$

By the cardinality of the Lagrange polynomial basis function ψ we see that in the summation involving $\mathbf{u}^{(e)}$ we get a non-zero value for ψ_{kl} when $k = l$ and also for ψ_{il} we get a non-zero value for $i = l$. Taking these simplifications into account yields

$$A_{ij}^{(e)}(\mathbf{u}) = w_i J_i^{(e)} \mathbf{u}_i^{(e)} \cdot \nabla \psi_{ji}$$

where we have replaced the quadrature sum l by i since $i = l$ for a non-zero advection term. Rearranging yields

$$A_{ij}^{(e)}(\mathbf{u}) = \mathbf{u}_i^{(e)} \cdot w_i J_i^{(e)} \nabla \psi_{ji}.$$

We can further simplify the right-hand side to its final form

$$A_{ij}^{(e)}(\mathbf{u}) = \left(\mathbf{u}_i^{(e)} \right)^T \mathbf{D}_{ij}^{(e)} \tag{15.5}$$

where $\mathbf{D}$ is the strong form differentiation matrix defined as follows

$$\mathbf{D}_{ij}^{(e)} \equiv \int_{\Omega_e} \psi_i \nabla \psi_j \, d\Omega_e = \sum_{i=1}^{M_N} w_i J_i^{(e)} \nabla \psi_{ji}$$

and is in fact similar to the 1D differentiation matrix we saw in Ch. 5.

15.5 Global Matrix-Vector Problem

Once all of the element matrices have been defined, we can then use the *direct stiffness summation* operation to construct the global matrices. That is, we construct

$$M_{IJ} = \bigwedge_{e=1}^{N_e} M_{ij}^{(e)}, \qquad A_{IJ}(\mathbf{u}) = \bigwedge_{e=1}^{N_e} A_{ij}^{(e)}(\mathbf{u}), \qquad B_I = \bigwedge_{e=1}^{N_e} B_i^{(e)}, \qquad L_{IJ} = \bigwedge_{e=1}^{N_e} L_{ij}^{(e)}$$

such that $I, J = 1, \ldots, N_p$ are the total number of gridpoints in the mesh[3] and $i, j = 1, \ldots, M_N$ are the total number of degrees of freedom per element with N_e elements, which then allows us to write the global matrix problem for exact integration as

$$M_{IJ} \frac{dq_J}{dt} + A_{IJ}(\mathbf{u})q_J = \nu B_I - \nu L_{IJ}q_J$$

and for inexact integration as

$$M_I \frac{dq_I}{dt} + \mathbf{u}_I^T \mathbf{D}_{IJ}q_J = \nu B_I - \nu L_{IJ}q_J$$

since M is diagonal for inexact integration.

Left multiplying by the inverse mass matrix yields

$$\frac{dq_I}{dt} + M_{IK}^{-1} A_{KJ}(\mathbf{u})q_J = \nu M_{IK}^{-1} B_K - \nu M_{IK}^{-1} L_{KJ}q_J$$

for exact integration, and

$$\frac{dq_I}{dt} + \mathbf{u}_I^T M_I^{-1} \mathbf{D}_{IJ}q_J = \nu M_I^{-1} B_I - \nu M_I^{-1} L_{IJ}q_J$$

for inexact integration. We can write the final semi-discrete form for exact integration as

$$\frac{dq_I}{dt} + \hat{A}_{IJ}(\mathbf{u})q_J = \nu \hat{B}_I - \nu \hat{L}_{IJ}q_J \tag{15.6}$$

with the matrices now defined as

$$\hat{A}_{IJ}(\mathbf{u}) = M_{IK}^{-1} A_{KJ}(\mathbf{u}), \qquad \hat{B}_I = M_{IK}^{-1} B_K, \qquad \text{and} \qquad \hat{L}_{IJ} = M_{IK}^{-1} L_{KJ}.$$

Remark 15.3 For exact integration we cannot solve the problem in the form given in Eq. (15.6) since it is impractical to construct the inverse of a matrix. Instead, we merely write it in this form to reveal the structure of the difference equations at the grid points. In practice, we would solve this equation using Gaussian elimination or some Krylov subspace iterative method to solve the equations without actually computing the inverse of M.

For inexact integration we get the final form

$$\frac{dq_I}{dt} + \mathbf{u}_I^T \hat{\mathbf{D}}_{IJ}q_J = \nu \hat{B}_I - \nu \hat{L}_{IJ}q_J \tag{15.7}$$

with the matrices defined as follows

$$\hat{\mathbf{D}}_{IJ}(\mathbf{u}) = M_I^{-1} \mathbf{D}_{IJ}(\mathbf{u}), \qquad \hat{B}_I = M_I^{-1} B_I, \qquad \text{and} \qquad \hat{L}_{IJ} = M_I^{-1} L_{IJ}.$$

---

[3] For quadrilateral elements with equal order polynomial N and an equal number of elements in the x and y directions yields $N_p = (\sqrt{N_e} \cdot N + 1)^2$.

Remark 15.4 For inexact integration we can actually solve the problem in the form given in Eq. (15.7). Since the mass matrix M_I is diagonal, then using the inverse mass matrix is practical since it just represents division by the diagonal elements. The inexact integration form is extremely efficient to execute because no matrix inversion is required for explicit time-integrators, and can be exploited to construct very efficient implicit and implicit-explicit time-integrators (e.g., see [161, 156, 158, 165, 160]). In spite of the integration error committed, we strongly recommend the inexact integration approach.

15.6 Example of 2D CG for Linear Elements

Let us construct the matrix problem for the 2D advection-diffusion equation using linear elements $N = 1$ as we did for the 2D elliptic equation in Sec. 12.11. We go through this analysis because learning how to construct the matrix-vector problem by hand is an invaluable tool for verifying the accuracy of your algorithms and checking the linear ($N = 1$) case should be the first step in this process. To simplify the discussion let us assume that we are far away from boundaries such that the global matrix problem reads

$$M_{IJ} \frac{dq_J}{dt} + A_{IJ}(\mathbf{u})q_J = -\nu L_{IJ}q_J$$

where we now have to define the element matrices $M_{ij}^{(e)}$, $A_{ij}^{(e)}(\mathbf{u})$, and $L_{ij}^{(e)}$ which we then use to construct the global matrices M_{IJ}, $A_{IJ}(\mathbf{u})$, and L_{IJ} via direct stiffness summation (DSS).

15.6.1 2D Basis Functions

Recall that for linear elements, the 1D basis function defined at the element grid point i is

$$h_i(\xi) = \frac{1}{2}(1 + \xi_i\xi)$$

where $\xi \in [-1, +1]$ and $\xi_i = (-1, +1)$ for $i = 1, 2$. In a similar vein we can define

$$h_i(\eta) = \frac{1}{2}(1 + \eta_i\eta)$$

where $\eta \in [-1, +1]$ and, again, $\eta_i = (-1, +1)$ for $i = 1, 2$. In these two expressions the index is defined as follows: $i = 1, \ldots, N + 1$. If we wish to use the 1D basis functions to construct 2D functions, then we take the tensor-product which amounts to just defining $i = 1, \ldots, (N + 1)^2$ where we now have to define ξ_i and η_i not just at two grid points as in 1D but now four points in 2D. These four points are the vertices of the quadrilateral reference element. At this point we can now define the 2D basis functions as

$$\psi_i(\xi, \eta) = \frac{1}{2}(1 + \xi_i \xi) \frac{1}{2}(1 + \eta_i \eta)$$

or, more compactly as

$$\psi_i(\xi, \eta) = \frac{1}{4}(1 + \xi_i \xi)(1 + \eta_i \eta)$$

where we now define ξ_i and η_i as follows

$$\xi_i = (-1, +1, -1, +1) \quad \text{and} \quad \eta_i = (-1, -1, +1, +1)$$

for $i = 1, \ldots, 4$ where i is numbered from bottom to top and left to right. From this expression we can now define the derivatives as follows

$$\frac{\partial \psi_i}{\partial \xi}(\xi, \eta) = \frac{1}{4}\xi_i (1 + \eta_i \eta)$$

and

$$\frac{\partial \psi_i}{\partial \eta}(\xi, \eta) = \frac{1}{4}\eta_i (1 + \xi_i \xi).$$

15.6.2 Metric Terms

To compute the metric terms, we follow the approach described in Sec. 12.7 and, specifically for linear elements we follow Sec. 12.11.2 to arrive at the following

$$\frac{\partial x_N^{(e)}}{\partial \xi} = \frac{\Delta x^{(e)}}{2}, \quad \frac{\partial x_N^{(e)}}{\partial \eta} = \frac{\partial y_N^{(e)}}{\partial \xi} = 0, \quad \frac{\partial y_N^{(e)}}{\partial \eta} = \frac{\Delta y^{(e)}}{2},$$

and

$$J^{(e)} = \frac{\Delta x^{(e)} \Delta y^{(e)}}{4}.$$

15.6.3 Derivatives in Physical Space

From Sec. 12.11.3, the derivatives of the basis functions in the reference element are given as

$$\frac{\partial \psi_i}{\partial x} = \frac{1}{4}\xi_i (1 + \eta_i \eta) \frac{2}{\Delta x^{(e)}}$$

and

$$\frac{\partial \psi_i}{\partial y} = \frac{1}{4}\eta_i (1 + \xi_i \xi) \frac{2}{\Delta y^{(e)}}.$$

15.6.4 Mass and Laplacian Matrices

From Ch. 12 we found that the mass and Laplacian matrices are

$$M_{ij}^{(e)} = \frac{\Delta x^{(e)} \Delta y^{(e)}}{36} \begin{pmatrix} 4 & 2 & 2 & 1 \\ 2 & 4 & 1 & 2 \\ 2 & 1 & 4 & 2 \\ 1 & 2 & 2 & 4 \end{pmatrix}$$

and

$$L_{ij}^{(e)} = \frac{\Delta y^{(e)}}{6\Delta x^{(e)}} \begin{pmatrix} 2 & -2 & 1 & -1 \\ -2 & 2 & -1 & 1 \\ 1 & -1 & 2 & -2 \\ -1 & 1 & -2 & 2 \end{pmatrix} + \frac{\Delta x^{(e)}}{6\Delta y^{(e)}} \begin{pmatrix} 2 & 1 & -2 & -1 \\ 1 & 2 & -1 & -2 \\ -2 & -1 & 2 & 1 \\ -1 & -2 & 1 & 2 \end{pmatrix}.$$

Therefore, the only new matrix that we need to construct in order to be able to solve the 2D advection-diffusion equation is the advection matrix. Let us now discuss the construction of this matrix.

15.6.5 Advection Matrix

Recall from Eq. (15.4) that the integral form of the advection matrix is written as follows

$$A_{ij}^{(e)}(\mathbf{u}) = \int_{\Omega_e} \psi_i \left(\sum_{k=1}^{M_N} \psi_k \mathbf{u}_k^{(e)} \right) \cdot \nabla \psi_j \, d\Omega_e$$

which we now simplify by assuming that $\mathbf{u}$ is constant. In this case, the advection matrix reduces to the simple form

$$A_{ij}^{(e)}(\mathbf{u}) = u \int_{\Omega_e} \psi_i \frac{\partial \psi_j}{\partial x} \, d\Omega_e + v \int_{\Omega_e} \psi_i \frac{\partial \psi_j}{\partial y} \, d\Omega_e$$

which we write in matrix form as

$$A_{ij}^{(e)}(\mathbf{u}) = u D_{ij}^{(e,x)} + v D_{ij}^{(e,y)},$$

where

$$D_{ij}^{(e,x)} = \int_{\Omega_e} \psi_i \frac{\partial \psi_j}{\partial x} \, d\Omega_e$$

and

$$D_{ij}^{(e,y)} = \int_{\Omega_e} \psi_i \frac{\partial \psi_j}{\partial y} \, d\Omega_e$$

are the 2D versions of the 1D differentiation matrices that we have already seen in Ch. 5. Let us now evaluate these differentiation matrices.

Using Eq. (12.16), we define the metric terms as follows

$$\frac{\partial \xi}{\partial x} = \frac{2}{\Delta x^{(e)}}, \quad \frac{\partial \eta}{\partial y} = \frac{2}{\Delta y^{(e)}}, \quad \frac{\partial \xi}{\partial y} = \frac{\partial \eta}{\partial x} = 0, \quad J^{(e)} = \frac{\Delta x^{(e)} \Delta y^{(e)}}{4}$$

since, for simplicity, we assume that the physical coordinate system is aligned with the reference element coordinate system. Using the metric terms in the derivatives of the basis functions

$$\frac{\partial \psi_j}{\partial x} = \frac{\partial \psi_j}{\partial \xi} \frac{\partial \xi}{\partial x} + \frac{\partial \psi_j}{\partial \eta} \frac{\partial \eta}{\partial x},$$

$$\frac{\partial \psi_j}{\partial y} = \frac{\partial \psi_j}{\partial \xi} \frac{\partial \xi}{\partial y} + \frac{\partial \psi_j}{\partial \eta} \frac{\partial \eta}{\partial y},$$

allows us to write the differentiation matrices written in terms of the physical coordinates

$$D_{ij}^{(e,x)} = \int_{\Omega_e} \psi_i \frac{\partial \psi_j}{\partial x} \, d\Omega_e, \quad D_{ij}^{(e,y)} = \int_{\Omega_e} \psi_i \frac{\partial \psi_j}{\partial y} \, d\Omega_e$$

to be written in terms of the reference element coordinates as follows

$$D_{ij}^{(e,x)} = \int_{-1}^{+1} \int_{-1}^{+1} \frac{1}{4} (1 + \xi_i \xi)(1 + \eta_i \eta) \frac{1}{4} \xi_j (1 + \eta_j \eta) \frac{2}{\Delta x^{(e)}} \frac{\Delta x^{(e)} \Delta y^{(e)}}{4} \, d\xi \, d\eta$$

$$D_{ij}^{(e,y)} = \int_{-1}^{+1} \int_{-1}^{+1} \frac{1}{4} (1 + \xi_i \xi)(1 + \eta_i \eta) \frac{1}{4} \eta_j (1 + \xi_j \xi) \frac{2}{\Delta y^{(e)}} \frac{\Delta x^{(e)} \Delta y^{(e)}}{4} \, d\xi \, d\eta.$$

Simplifying and integrating yields

$$D_{ij}^{(e,x)} = \frac{\Delta y^{(e)}}{32} \left[\xi_j \xi + \frac{1}{2} \xi_i \xi_j \xi^2 \right]_{-1}^{+1} \left[\eta + \frac{1}{2} (\eta_i + \eta_j) \eta^2 + \frac{1}{3} \eta_i \eta_j \eta^3 \right]_{-1}^{+1}$$

$$D_{ij}^{(e,y)} = \frac{\Delta x^{(e)}}{32} \left[\eta_j \eta + \frac{1}{2} \eta_i \eta_j \eta^2 \right]_{-1}^{+1} \left[\xi + \frac{1}{2} (\xi_i + \xi_j) \xi^2 + \frac{1}{3} \xi_i \xi_j \xi^3 \right]_{-1}^{+1}.$$

Evaluating the matrices at the bounds of integration and simplifying yields the compact form

$$D_{ij}^{(e,x)} = \frac{\Delta y^{(e)}}{24} \xi_j (3 + \eta_i \eta_j), \quad D_{ij}^{(e,y)} = \frac{\Delta x^{(e)}}{24} \eta_j (3 + \xi_i \xi_j).$$

Substituting for the values of the reference element coordinates (ξ, η) yields the following matrices

$$D_{ij}^{(e,x)} = \frac{\Delta y^{(e)}}{12} \begin{pmatrix} -2 & 2 & -1 & 1 \\ -2 & 2 & -1 & 1 \\ -1 & 1 & -2 & 2 \\ -1 & 1 & -2 & 2 \end{pmatrix}$$

and

$$D_{ij}^{(e,y)} = \frac{\Delta x^{(e)}}{12} \begin{pmatrix} -2 & -1 & 2 & 1 \\ -1 & -2 & 1 & 2 \\ -2 & -1 & 2 & 1 \\ -1 & -2 & 1 & 2 \end{pmatrix}.$$

Since the advection matrix is defined as

$$A_{ij}^{(e)}(\mathbf{u}) = uD_{ij}^{(e,x)} + vD_{ij}^{(e,y)},$$

then we can write it as

$$A_{ij}^{(e)}(\mathbf{u}) = u\frac{\Delta y^{(e)}}{12} \begin{pmatrix} -2 & 2 & -1 & 1 \\ -2 & 2 & -1 & 1 \\ -1 & 1 & -2 & 2 \\ -1 & 1 & -2 & 2 \end{pmatrix} + v\frac{\Delta x^{(e)}}{12} \begin{pmatrix} -2 & -1 & 2 & 1 \\ -1 & -2 & 1 & 2 \\ -2 & -1 & 2 & 1 \\ -1 & -2 & 1 & 2 \end{pmatrix}.$$

15.6.6 Matrix Equations on the Reference Element

Let us now put the element matrices together into one matrix equation for the reference element. The element matrix problem

$$M_{ij}^{(e)}\frac{dq_j^{(e)}}{dt} + A_{ij}^{(e)}(\mathbf{u})q_j^{(e)} = -vL_{ij}^{(e)}q_j^{(e)}$$

now reads

$$\frac{\Delta x^{(e)}\Delta y^{(e)}}{36} \begin{pmatrix} 4 & 2 & 2 & 1 \\ 2 & 4 & 1 & 2 \\ 2 & 1 & 4 & 2 \\ 1 & 2 & 2 & 4 \end{pmatrix} \frac{d}{dt} \begin{pmatrix} q_1^{(e)} \\ q_2^{(e)} \\ q_3^{(e)} \\ q_4^{(e)} \end{pmatrix}$$

$$+ \left[u\frac{\Delta y^{(e)}}{12} \begin{pmatrix} -2 & 2 & -1 & 1 \\ -2 & 2 & -1 & 1 \\ -1 & 1 & -2 & 2 \\ -1 & 1 & -2 & 2 \end{pmatrix} + v\frac{\Delta x^{(e)}}{12} \begin{pmatrix} -2 & -1 & 2 & 1 \\ -1 & -2 & 1 & 2 \\ -2 & -1 & 2 & 1 \\ -1 & -2 & 1 & 2 \end{pmatrix} \right] \begin{pmatrix} q_1^{(e)} \\ q_2^{(e)} \\ q_3^{(e)} \\ q_4^{(e)} \end{pmatrix} =$$

$$- v\left[\frac{\Delta y^{(e)}}{6\Delta x^{(e)}} \begin{pmatrix} 2 & -2 & 1 & -1 \\ -2 & 2 & -1 & 1 \\ 1 & -1 & 2 & -2 \\ -1 & 1 & -2 & 2 \end{pmatrix} + \frac{\Delta x^{(e)}}{6\Delta y^{(e)}} \begin{pmatrix} 2 & 1 & -2 & -1 \\ 1 & 2 & -1 & -2 \\ -2 & -1 & 2 & 1 \\ -1 & -2 & 1 & 2 \end{pmatrix} \right] \begin{pmatrix} q_1^{(e)} \\ q_2^{(e)} \\ q_3^{(e)} \\ q_4^{(e)} \end{pmatrix}.$$

In the next section we show how to use the resulting element equations to construct the global matrix-vector problem for CG.

15.7 Algorithms for the CG Global Matrix-Vector Problem

15.7.1 Non-Tensor-Product Approach

Let us now describe the construction of the RHS operator for the advection equation since the matrix $A(\mathbf{u})$ is the most challenging operator to deal with in this equation. Let us begin with the recipe for constructing the matrix A with exact integration as defined in Eq. (15.4) where the key steps are described in Alg. 15.1. Algorithm

---

Algorithm 15.1 Recipe 1: construction of matrix $A^{(e)}$ with exact integration.

function A_MATRIX(**u**)
   $A^{(e)} = \mathbf{0}$
   for $e = 1 : N_e$ **do**   ▷ loop over elements
      for $l = 1 : M_Q$ **do**   ▷ loop over integration points
       $\mathbf{u}_l = 0$
       for $k = 1 : M_N$ **do**
          $\mathbf{u}_l += \psi_{kl}\mathbf{u}_k^{(e)}$ ▷ interpolate to the integration points
       end for
       for $j = 1 : M_N$ **do**   ▷ loop over columns of $A^{(e)}$
        for $i = 1 : M_N$ **do**   ▷ loop over rows of $A^{(e)}$
          $A_{ij}^{(e)} += w_l J_l^{(e)} \psi_{il} \left(\mathbf{u}_l \cdot \nabla \psi_{jl} \right)$
        end for
       end for
      end for
   end for
end function

15.1 shows that the construction of the matrix A is similar to the construction of other matrices we have already seen, such as the differentiation matrix. In fact, Eq. (15.4) shows that there is clearly a differentiation matrix embedded in A except that the gradient operator is multiplied (via an inner product) by the velocity vector. The key step to recognize is that for each element and for each integration point, we first interpolate the velocity vector to the integration point l and then loop over the elements of the local element matrix (i and j loops). The derivatives in physical space, of course, have to be mapped to those in terms of the reference element space but otherwise the construction is straightforward. Once we construct $A^{(e)}$ via Alg. 15.1 we then use Alg. 12.11 to construct the global matrix A via DSS.

Exercise Compute the complexity for building the matrix $A^{(e)}$ in Alg. 15.1.   □

To see how inexact integration changes the construction of the matrix $A^{(e)}$, let us now look at Alg. 15.2 which describes the $A^{(e)}$ matrix defined in Eq. (15.5). We can see that Alg. 15.2 has one fewer loop than Alg. 15.1 because, with inexact integration, we are able to exploit cardinality and thereby replace the l loop by i.

Exercise Compute the complexity for building the matrix $A^{(e)}$ in Alg. 15.2.   □

Algorithm 15.2 Recipe 2: construction of matrix $A^{(e)}$ with inexact integration.

function A_MATRIX($A^{(e)}(\mathbf{u})$)
   $A^{(e)} = \mathbf{0}$
   for $e = 1 : N_e$ **do**   ▷ loop over elements
      for $j = 1 : M_N$ **do**   ▷ loop over columns of $A^{(e)}$
         for $i = 1 : M_N$ **do**   ▷ loop over rows of $A^{(e)}$
$$A_{ij}^{(e)} + = w_i J_i^{(e)} \left(\mathbf{u}_i^{(e)} \cdot \boldsymbol{\nabla}\psi_{ji} \right)$$
        end for
      end for
   end for
end function

Once we have constructed the global matrix A we can now solve the advection equation as given in Alg. 15.3. Algorithm 15.3 shows that once we have constructed

---

Algorithm 15.3 CG 2D wave equation using A matrix approach.

function CG(q)
   Construct $M^{(e)}$ and $A^{(e)}$ ▷ using Alg. 12.6 and Alg. 15.1 or 15.2
   Construct M and A using DSS   ▷ using Alg. 12.11
   for $n = 0 : N_t$ **do**   ▷ time loop
$$R_I(q^n) = -A_{IJ}q_J^n$$ ▷ build global RHS vector
$$\hat{R}_I(q^n) = M_{IJ}^{-1} R_J(q^n)$$ ▷ apply boundary conditions
$$\tfrac{d}{dt}q_I = \hat{R}_I(q^n)$$ ▷ evolve equations forward in time to get q^{n+1}
   end for
end function

---

the global matrices (M and A in this case) that the solution algorithm becomes quite simple. In fact, we only need to march the equations forward in time where we use the global matrices to construct a RHS vector, which is a function of the solution vector q. Once we construct $\hat{R}$ we then enforce the boundary conditions here. At this point, the problem has simplified to a system of ordinary differential equations (ODEs) which can be solved by any ODE solver such as the RK3 method presented in Ch. 5 or by other time-integrators discussed in Ch. 20.

Algorithms 15.1 and 15.2 are a good way to approach the solution of the advection diffusion equation provided that the velocity vector $\mathbf{u}$ is time-independent. However, if the velocity also changes in time then this approach is not so useful because we need to rebuild $A(\mathbf{u})$ at every time-step in Alg. 15.3. Therefore, let us propose another approach whereby we only build and store the global mass matrix (since it never changes with time) and instead of building the matrix A we build the RHS vector R directly.

Algorithm 15.4 shows the steps for building $R^{(e)}$. Note that this function needs to be called whenever the solution vector $q^{(e)}$ or $\mathbf{u}^{(e)}$ changes which, for a time-dependent problem, is at every time-step and at every stage of a multi-stage ODE solver.

Algorithm 15.4 Recipe 1: construction of vector $R^{(e)}$ with exact integration.

---

function R_VECTOR(q, **u**)
  $R^{(e)} = \mathbf{0}$
  for $e = 1 : N_e$ **do**   ▷ loop over elements
    for $l = 1 : M_Q$ **do**   ▷ loop over integration points
      $\mathbf{u}_l = 0; \nabla q_l = 0$
      for $k = 1 : M_N$ **do**   ▷ interpolate to the integration points
        $\mathbf{u}_l += \psi_{kl}\mathbf{u}_k^{(e)}$
        $\nabla q_l += \nabla\psi_{kl}q_k^{(e)}$
      end for
      for $i = 1 : M_N$ **do**   ▷ loop over rows of $R^{(e)}$
        $R_i^{(e)} += w_l J_l^{(e)}\psi_{il}(\mathbf{u}_l \cdot \nabla q_l)$
      end for
    end for
  end for
end function

If we wish to use inexact integration, then the algorithm simplifies a bit as shown in Alg. 15.5. For inexact integration, the l loop becomes an i loop and, via cardinality

---

Algorithm 15.5 Recipe 2: construction of vector $R^{(e)}$ with inexact integration.

---

function R_VECTOR(q, **u**)
  $R^{(e)} = \mathbf{0}$
  for $e = 1 : N_e$ **do**   ▷ loop over elements
    for $i = 1 : M_N$ **do**   ▷ loop over integration points
      $\nabla q_i = 0$
      for $k = 1 : M_N$ **do**   ▷ compute gradient on integration points
        $\nabla q_i += \nabla\psi_{ik}q_k^{(e)}$
      end for
      $R_i^{(e)} += w_i J_i^{(e)}\left(\mathbf{u}_i^{(e)} \cdot \nabla q_i\right)$
    end for
  end for
end function

---

we can remove the basis function ψ_{il} but otherwise the algorithm looks quite similar to the exact integration version.

Note that if we construct $R^{(e)}(q, \mathbf{u})$ instead of $A^{(e)}(\mathbf{u})$ then we cannot use Alg. 15.3 to construct the CG solution. Instead we modify this algorithm to suit our needs which we now describe in Alg. 15.6. Algorithm 15.6 is rather similar to Alg. 15.3 except that we never construct the matrix A. For more complex problems (such as nonlinear problems) this is the preferred way to solve the problem because it will become increasingly costly to construct global matrices such as A and because it is impractical to build such a matrix when it needs to be recalculated whenever **u** changes, which is the case for most problems of interest.

Algorithm 15.6 CG 2D wave equation using R vector approach.

---

function CG(q)
    Construct $M^{(e)}$ ▹ using Alg. 12.6
    Construct M using DSS   ▹ using Alg. 12.11
    for $n = 0 : N_t$ **do**   ▹ time loop
        Construct $R^{(e)}(q, \mathbf{u})$ ▹ use Alg. 15.4 or 15.5
        Construct R_I using DSS   ▹ using Alg. 12.12
        $\hat{R}_I = M_{IJ}^{-1} R_J$ ▹ apply boundary conditions
        $\frac{d}{dt} q_I = \hat{R}_I$ ▹ evolve equations forward in time to get q^{n+1}
    end for
end function

15.7.2 Tensor-Product Approach

We end this section by showing the advantage of using the tensor-product approach for decreasing the number of operations - this approach is the *sum factorization* method introduced by Orszag [291]. To simplify the discussion we only consider inexact integration (although a similar strategy can be applied to exact integration although it is more complicated to explain). To describe this approach let us revisit Eq. (15.5)

$$A_{ij}^{(e)}(\mathbf{u}) = \left(\mathbf{u}_i^{(e)}\right)^T w_i J_i^{(e)} \nabla \psi_{ji} \equiv \left(\mathbf{u}_i^{(e)}\right)^T \mathbf{D}_{ij}^{(e)}$$

where we recognize that to make this approach less costly and hence more efficient requires only optimizing the differentiation matrix $\mathbf{D}^{(e)}$. Let us rewrite the differentiation matrix as

$$\mathbf{D}_{I,J}^{(e)} = w_I J_I^{(e)} \nabla \psi_{J,I} \equiv w_I J_I^{(e)} \left(\frac{\partial \psi_{J,I}}{\partial x} \hat{\mathbf{i}} + \frac{\partial \psi_{J,I}}{\partial y} \hat{\mathbf{j}} \right)$$

where we replaced (i, j) by (I, J) and will redefine the two-dimensional basis functions ψ as a product of one-dimensional basis functions h in the following manner. Let

$$\psi_{J,I} = h_{k,i} h_{l,j}$$

where $I = (i + 1) + j(N + 1)$ and $J = (k + 1) + l(N + 1)$ with $i, j, k, l = 0, \dots, N$. To explain the subscripts, let us rewrite the 2D basis functions in the following way

$$\psi_{J,I} = \psi_J(\xi_I, \eta_I)$$

where the connection to the 1D functions is as follows

$$\psi_J(\xi_I, \eta_I) = h_k(\xi_i) h_l(\eta_j) \equiv h_{k,i} h_{l,j}.$$

Using this information and replacing ψ in the differentiation matrix with the tensor-product of the h functions allows us to rewrite the differentiation matrix as

$$\mathbf{D}_{I,J}^{(e)} = w_i w_j J_{i,j}^{(e)} \left(\frac{\partial}{\partial x} \left(h_k(\xi_i) h_l(\eta_j) \right) \hat{\mathbf{i}} + \frac{\partial}{\partial y} \left(h_k(\xi_i) h_l(\eta_j) \right) \hat{\mathbf{j}} \right) \tag{15.8}$$

where now we replace the physical coordinates (x, y) with the reference element coordinates (ξ, η). We can immediately see that in the differentiation matrix above, the x-derivative term can be expanded as follows

$$\frac{\partial}{\partial x} \left(h_k(\xi_i) h_l(\eta_j) \right) = \frac{\partial}{\partial \xi} \left(h_k(\xi_i) h_l(\eta_j) \right) \frac{\partial \xi_{i,j}}{\partial x} + \frac{\partial}{\partial \eta} \left(h_k(\xi_i) h_l(\eta_j) \right) \frac{\partial \eta_{i,j}}{\partial x}$$

which can be rewritten as

$$\frac{\partial}{\partial x} \left(h_k(\xi_i) h_l(\eta_j) \right) = \left(\frac{dh_k(\xi_i)}{d\xi} h_l(\eta_j) \right) \frac{\partial \xi_{i,j}}{\partial x} + \left(h_k(\xi_i) \frac{dh_l(\eta_j)}{d\eta} \right) \frac{\partial \eta_{i,j}}{\partial x}$$

where we have written the reference element derivatives as total derivatives to denote that the function h is either purely a function of ξ (i.e., $h = h(\xi)$) or purely a function of η (i.e., $h = h(\eta)$) and the metric terms $\frac{\partial \xi}{\partial x}$ are computed using Alg. 12.1.

Using cardinality, we can simplify this expression as follows

$$\frac{\partial}{\partial x} \left(h_k(\xi_i) h_l(\eta_j) \right) = \frac{dh_k(\xi_i)}{d\xi} \frac{\partial \xi_{i,j}}{\partial x} + \frac{dh_l(\eta_j)}{d\eta} \frac{\partial \eta_{i,j}}{\partial x}$$

with a similar expression for the y-derivative

$$\frac{\partial}{\partial y} \left(h_k(\xi_i) h_l(\eta_j) \right) = \frac{dh_k(\xi_i)}{d\xi} \frac{\partial \xi_{i,j}}{\partial y} + \frac{dh_l(\eta_j)}{d\eta} \frac{\partial \eta_{i,j}}{\partial y}.$$

This is the key to constructing an efficient algorithm and is what we will use below to construct efficient derivatives of $\frac{\partial q}{\partial \xi}$ and $\frac{\partial q}{\partial \eta}$. Before we describe the algorithm using the sum factorization approach let us first see how this concept allows us to simplify an integral of the type

$$\int_{\Omega_e} \psi_I \frac{\partial q_N^{(e)}}{\partial x} d\Omega_e = \sum_{K=1}^{M_Q} w_K J_K^{(e)} \psi_{I,K} \sum_{J=1}^{M_N} \frac{\partial \psi_{J,K}}{\partial x} q_J^{(e)}. \tag{15.9}$$

Assuming inexact integration (Q=N), we can invoke cardinality to remove the K-loop as follows

$$\int_{\Omega_e} \psi_I \frac{\partial q_N^{(e)}}{\partial x} d\Omega_e = w_I J_I^{(e)} \sum_{J=1}^{M_N} \frac{\partial \psi_{J,I}}{\partial x} q_J^{(e)}. \tag{15.10}$$

Introducing tensor-product basis functions, we can now write

$$\int_{\Omega_e} \psi_I \frac{\partial q_N^{(e)}}{\partial x} d\Omega_e = w_{i,j} J_{i,j}^{(e)} \sum_{k=0}^{N} \sum_{l=0}^{N} \left(\frac{dh_{k,i}}{d\xi} h_{l,j} \frac{\partial \xi_{i,j}}{\partial x} q_{k,l}^{(e)} + h_{k,i} \frac{dh_{l,j}}{d\eta} \frac{\partial \eta_{i,j}}{\partial x} q_{k,l}^{(e)} \right). \tag{15.11}$$

Invoking cardinality, we can eliminate the l-loop in the first sum and replace l with j and the k-loop in the second sum and replace k with i to get

$$\int_{\Omega_e} \psi_I \frac{\partial q_N^{(e)}}{\partial x} d\Omega_e = w_{i,j} J_{i,j}^{(e)} \left(\sum_{k=0}^{N} \frac{dh_{k,i}}{d\xi} \frac{\partial \xi_{i,j}}{\partial x} q_{k,j}^{(e)} + \sum_{l=0}^{N} \frac{dh_{l,j}}{d\eta} \frac{\partial \eta_{i,j}}{\partial x} q_{i,l}^{(e)} \right). \quad (15.12)$$

The final form is obtained by substituting l in the second loop with k to get

$$\int_{\Omega_e} \psi_I \frac{\partial q_N^{(e)}}{\partial x} d\Omega_e = w_{i,j} J_{i,j}^{(e)} \sum_{k=0}^{N} \left(\frac{dh_{k,i}}{d\xi} \frac{\partial \xi_{i,j}}{\partial x} q_{k,j}^{(e)} + \frac{dh_{k,j}}{d\eta} \frac{\partial \eta_{i,j}}{\partial x} q_{i,k}^{(e)} \right) \quad (15.13)$$

which is the form presented in Alg. 15.7 which uses the *sum factorization* approach to construct the RHS vector $R^{(e)}$.

Algorithm 15.7 highlights the main steps of this approach. What is important to note about Algorithm 15.7 is that there are only three loops (of dimension N) for each element. This tells us that for each element, the cost of constructing the $R^{(e)}$ vector is $O(N^3)$. Recall the complexity of Algs. 15.4 and 15.5 to see how the new algorithm differs. Once $R^{(e)}$ is constructed using Alg. 15.7 it can be used directly in

---

Algorithm 15.7 Sum Factorization: construction of vector $R^{(e)} = \left(\mathbf{u}^{(e)} \right)^T \mathbf{D}^{(e)} q^{(e)}$ with inexact integration and tensor-product bases.

---

function R_VECTOR(q,**u**)
  $R^{(e)} = \mathbf{0}$
  for $e = 1 : N_e$ **do**   ▷ loop over elements
    for $i = 0 : N$ **do**   ▷ loop over integration points
      for $j = 0 : N$ **do**
        $I = (i+1) + j(N+1)$ ▷ tensor-product to non-tensor-product map
        $\frac{\partial q}{\partial \xi} = 0;\ \frac{\partial q}{\partial \eta} = 0$
        for $k = 0 : N$ **do**   ▷ build derivatives in (ξ, η) space
          $\frac{\partial q}{\partial \xi} + = \frac{dh_{k,i}}{d\xi} q_{k,j}^{(e)}$
          $\frac{\partial q}{\partial \eta} + = \frac{dh_{k,j}}{d\xi} q_{i,k}^{(e)}$
        end for
        $\frac{\partial q}{\partial x} = \frac{\partial q}{\partial \xi} \frac{\partial \xi_{i,j}}{\partial x} + \frac{\partial q}{\partial \eta} \frac{\partial \eta_{i,j}}{\partial x}$ ▷ build derivatives in (x, y) space
        $\frac{\partial q}{\partial y} = \frac{\partial q}{\partial \xi} \frac{\partial \xi_{i,j}}{\partial y} + \frac{\partial q}{\partial \eta} \frac{\partial \eta_{i,j}}{\partial y}$
        $R_I^{(e)} = w_i w_j J_{i,j}^{(e)} \left(u_i^{(e)} \frac{\partial q}{\partial x} + v_i^{(e)} \frac{\partial q}{\partial y} \right)$
      end for
    end for
  end for
end function

Alg. 15.6.

Exercise Compare the complexity of Algs. 15.4 and 15.5 with 15.7.   □

Exercise Write the *sum factorization* algorithm for computing $\mathbf{D}^{(e)}$ for exact integration and compute the complexity of the algorithm. $\square$

15.8 Example of 2D Hyperbolic Equation Problem

Suppose we wish to solve the continuous first order differential equation (i.e., the *advection equation*)

$$\frac{\partial q}{\partial t} + \mathbf{u} \cdot \nabla q = 0 \qquad \forall (x, y) \in [-1, 1]^2$$

where $q = q(x, y, t)$ and $\mathbf{u} = \mathbf{u}(x, y)$ with $\mathbf{u} = (u, v)^T$. Let the velocity field be

$$u(x, y) = y \qquad \text{and} \qquad v(x, y) = -x$$

which forms a velocity field that rotates fluid particles in a clockwise direction. Note that this velocity field is divergence-free, i.e., the following condition is satisfied

$$\nabla \cdot \mathbf{u} = 0.$$

Clearly, this problem represents a 2D wave equation which is a hyperbolic (initial value) problem that requires an initial condition. Let that initial condition be the Gaussian

$$q(x, y, 0) = \exp(-\sigma \left[(x - x_c)^2 + (y - y_c)^2 \right]) \qquad (15.14)$$

where $(x_c, y_c) = (-0.5, 0)$ is the initial center of the Gaussian and $\sigma = 32$ controls its steepness. The analytic solution is given as

$$q(x, y, t) = q(x - ut, y - vt, 0)$$

where periodicity is enforced at all four boundaries.

Figure 15.1 shows a sample grid and solution after one revolution for polynomial order $N = 16$ with $N_e = 2 \times 2$ total elements. The thick blue lines in Fig. 15.1a outline the element grid boundaries (2×2) while the thin red lines show the grid formed by the internal degrees of freedom $((N + 1) \times (N + 1))$.

15.8.1 Solution Accuracy

Figure 15.2 shows the convergence rates for various polynomial orders, N, for a total number of gridpoints N_p where, for quadrilaterals, $N_p = (N_e^{(x)} N + 1)(N_e^{(y)} N + 1)$, where $N_e^{(s)}$ denotes the number of elements along the coordinate direction (s). To

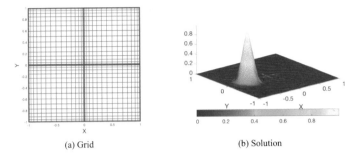

(a) Grid   (b) Solution

Fig. 15.1: CG solution for the 2D hyperbolic equation for polynomial order $N = 16$ and $N_e = 2 \times 2$ showing the (a) grid and (b) solution after one revolution.

measure the order of accuracy, we use the integral form of the L^2 norm given by Eq. (12.58) and evolve the solution in time using the SSP-RK3 time-integrator presented in Ch. 5 using a Courant number of $C = 0.25$ for all simulations.

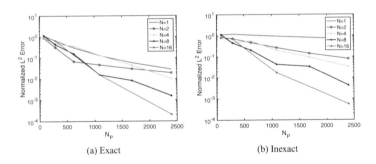

(a) Exact   (b) Inexact

Fig. 15.2: Convergence rates of CG for the 2D hyperbolic equation for polynomial orders $N = 1$, $N = 2$, $N = 4$, $N = 8$, and $N = 16$ using a total number of gridpoints N_p for (a) exact integration ($Q = N + 1$) and (b) inexact integration ($Q = N$) with Lobatto points.

Figure 15.2 shows that there is little difference in the rates of convergence between exact and inexact integration. The biggest differences occur for values of $N \leq 4$, however, as N increases the differences between exact and inexact integration diminish.

15.8.2 Computational Cost of High-Order

Let us now measure the wallclock time required by each simulation to see the relationship between accuracy and time-to-solution. Figure 15.3 shows the computational cost (wallclock time in seconds) for various polynomial orders N. Figure

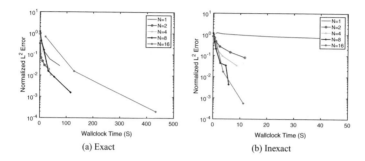

(a) Exact   (b) Inexact

Fig. 15.3: Work-Precision: Convergence rates as a function of wallclock time (in seconds) of CG for the 2D hyperbolic equation for polynomial orders $N = 1$, $N = 2$, $N = 4$, $N = 8$, and $N = 16$ for (a) exact integration ($Q = N + 1$) and (b) inexact integration ($Q = N$) with Lobatto points.

15.3 shows that to achieve, for example, an accuracy of 10^{-3} it is more efficient to use high-order ($N = 8$ for exact integration and $N = 16$ for inexact). Furthermore, Fig. 15.3 shows that the cost of using inexact integration is far less than that for exact integration; for inexact integration we used sum factorization. Although Fig. 15.2 does show that exact integration is more accurate than inexact, one needs to weigh the cost of the algorithm as well. For some applications we seek the most accurate solution possible whereas for other types of problems (industrial applications) we are willing to accept a less accurate solution for a gain in computational efficiency.

.

Chapter 16
2D Discontinuous Galerkin Methods for Hyperbolic Equations

16.1 Introduction

In Chs. 6 and 9 we discussed the implementation of discontinuous Galerkin methods in one dimension for hyperbolic and elliptic equations, respectively. In Ch. 15 we described the CG discretization for hyperbolic equations in two-dimensions and discussed the addition of second order operators to these equations. This chapter focuses on the DG method for hyperbolic equations in two-dimensions although we also discuss the addition of elliptic operators. In order to construct a DG discretization of these equations, we make extensive use of the results presented in Ch. 6 for the DG discretization of one-dimensional hyperbolic systems and the results from Ch. 13 for the DG discretization of two-dimensional elliptic systems.

This chapter is organized as follows. In Sec. 16.2 we introduce the continuous hyperbolic PDE that we discretize by the DG method. In Secs. 16.5 and 16.6 we describe the weak and strong form DG discretizations, respectively. The heart of the chapter is Sec. 16.8 where the construction of the algorithms are described in detail including both non-tensor-products and tensor-product bases. We close this chapter with an example problem for the 2D wave equation.

16.2 2D Advection-Diffusion Equation

Let us begin the discussion with the two-dimensional advection-diffusion equation which we wrote previously in Ch. 15 in the following form

$$\frac{\partial q}{\partial t} + \mathbf{u} \cdot \nabla q = \nabla \cdot (\nu \nabla q) \qquad (16.1)$$

© The Editor(s) (if applicable) and The Author(s), under exclusive license
to Springer Nature Switzerland AG 2020
F. X. Giraldo, *An Introduction to Element-Based Galerkin Methods on
Tensor-Product Bases*, Texts in Computational Science and Engineering 24,
https://doi.org/10.1007/978-3-030-55069-1_16

where $q = q(\mathbf{x}, t)$ is the scalar solution variable, $\nabla = \left(\frac{\partial}{\partial x}\hat{\mathbf{i}} + \frac{\partial}{\partial y}\hat{\mathbf{j}}\right)$ is the 2D gradient operator, $\hat{\mathbf{i}}$ and $\hat{\mathbf{j}}$ are the directional unit vectors in 2D Cartesian space, and v is the viscosity coefficient. The form described in Eq. (16.1) is known as the advection or non-conservation form of the equation. The alternate form

$$\frac{\partial q}{\partial t} + \nabla \cdot (q\mathbf{u} - v\nabla q) = q\nabla \cdot \mathbf{u} \qquad (16.2)$$

is known as the conservation form. Note that we can now recast this equation in the general balance law form

$$\frac{\partial q}{\partial t} + \nabla \cdot \mathbf{f} = S(q)$$

where $\mathbf{f} = (q\mathbf{u} - v\nabla q)$ and $S(q) = q\nabla \cdot \mathbf{u}$. For the case that the velocity field is divergence-free ($\nabla \cdot \mathbf{u} = 0$) we obtain the conservation law

$$\frac{\partial q}{\partial t} + \nabla \cdot \mathbf{f} = 0.$$

To simplify the discussion of the DG method we assume a divergence-free velocity field in the discretization of the advection-diffusion equation in conservation form. The reason why we use the conservation form and not the advection form is that DG requires the equations in strict conservation form. The equations must be written in conservation (also known as flux form) because only in this form do the fluxes make physical sense[1]. In other words, only in this form will the physically significant quantities (such as mass, momentum, and energy fluxes) enter the numerical solution. Another reason why it is important to discuss DG with the conservation form of the equations is because only in this form can one expect the numerical solution to converge to the physically correct solution when discontinuities (such as shocks) are present. Note that this is true for either CG or DG and for any other numerical method. It is has been shown that, in the presence of discontinuities, not using the conservation form of the equations may not yield the correct solution [249,250,205].

16.3 Integral Form

To construct the integral form, we begin by expanding the solution variable q as follows

$$q_N^{(e)}(\mathbf{x}, t) = \sum_{i=1}^{M_N} \psi_i(\mathbf{x}) q_i^{(e)}(t)$$

with

$$\mathbf{u}_N^{(e)}(\mathbf{x}) = \sum_{i=1}^{M_N} \psi_i(\mathbf{x}) \mathbf{u}_i^{(e)}$$

---

[1] The DG method can be, and has been, used on equations in non-conservation form but one must be particularly careful to ensure that the fluxes make physical sense - if not then the method may not be consistent.

and $\mathbf{f}_N^{(e)} = q_N^{(e)} \mathbf{u}_N^{(e)}$ where $\mathbf{u}(\mathbf{x}) = u(\mathbf{x})\hat{\mathbf{i}} + v(\mathbf{x})\hat{\mathbf{j}}$ is the velocity vector with u and v components along $\hat{\mathbf{i}}$ (x-direction) and $\hat{\mathbf{j}}$ (y-direction). For simplicity we define $M_N = (N+1)^2$ which denotes the number of points inside each element Ω_e. Next, we substitute $q_N^{(e)}$, $\mathbf{u}_N^{(e)}$, and $\mathbf{f}_N^{(e)}$ into the PDE, multiply by a test function and integrate within the local domain Ω_e yielding the integral form: find $q \in L^2$ such that

$$\int_{\Omega_e} \psi_i \frac{\partial q_N^{(e)}}{\partial t} \, d\Omega_e + \int_{\Gamma_e} \psi_i \hat{\mathbf{n}}^{(e,l)} \cdot \mathbf{f}_N^{(*,l)} \, d\Gamma_e - \int_{\Omega_e} \nabla \psi_i \cdot \mathbf{f}_N^{(e)} \, d\Omega_e = 0 \qquad \forall \psi \in L^2$$

(16.3)

with $i = 1, \ldots, M_N$ and $e = 1, \ldots, N_e$, where $\mathbf{f}^{(*,l)} = \mathbf{f}_{inv}^{(*,l)} + \mathbf{f}_{visc}^{(*,l)}$ is the numerical flux function which we assume to be the Rusanov flux for the inviscid terms

$$\mathbf{f}_{inv}^{(*,l)} = \frac{1}{2} \left[\mathbf{f}_{inv}^{(l)} + \mathbf{f}_{inv}^{(e)} - |\lambda| \hat{\mathbf{n}}^{(e,l)} \left(q^{(l)} - q^{(e)} \right) \right]$$

and the average flux [22] for the viscous terms

$$\mathbf{f}_{visc}^{(*,l)} = \frac{1}{2} \left[\mathbf{f}_{visc}^{(l)} + \mathbf{f}_{visc}^{(e)} \right]$$

where the superscript (l) denotes the face neighbor of (e), $\hat{\mathbf{n}}^{(e,l)}$ is the unit normal vector of the face shared by elements e and l, and λ is the maximum wave speed of the system; for the advection-diffusion equation λ is just the maximum normal velocity $\hat{\mathbf{n}} \cdot \mathbf{u}|_{\Gamma_e}$.

16.4 Basis Functions and the Reference Element

As we did in Chs. 12 - 15 we construct the 2D basis functions on the reference element as a tensor-product of the 1D basis functions as follows

$$\psi_i(\xi, \eta) = h_j(\xi) \otimes h_k(\eta)$$

where h are the 1D basis functions we have discussed already, $\otimes$ is the tensor-product, and the 1D indices vary as follows $j, k = 0, \ldots, N$ with the 2D index varying as $i = 1, \ldots, M_N$. To get from the 1D indices (j, k) to the 2D index i requires the mapping $i = (j+1) + k(N+1)$.

With this definition in place, we can now expand the solution variable q as follows

$$q_N^{(e)}(\mathbf{x}, t) = \sum_{i=1}^{M_N} \psi_i(\mathbf{x}) q_i^{(e)}(t)$$

which implies the approximation of the gradient operator to be

$$\nabla q_N^{(e)}(\mathbf{x}, t) = \sum_{i=1}^{M_N} \nabla \psi_i(\mathbf{x}) q_i^{(e)}(t)$$

where the partial derivatives are defined as follows

$$\frac{\partial q_N^{(e)}(\mathbf{x}, t)}{\partial x} = \sum_{i=1}^{M_N} \frac{\partial \psi_i(\mathbf{x})}{\partial x} q_i^{(e)}(t)$$

and

$$\frac{\partial q_N^{(e)}(\mathbf{x}, t)}{\partial y} = \sum_{i=1}^{M_N} \frac{\partial \psi_i(\mathbf{x})}{\partial y} q_i^{(e)}(t).$$

Since we perform all of our computations in the reference element with coordinates (ξ, η) then we must transform the derivatives from (x, y) to (ξ, η).

Using the chain rule, we write the derivatives of the basis functions as

$$\frac{\partial \psi(x(\xi, \eta), y(\xi, \eta))}{\partial x} = \frac{\partial \psi(\xi, \eta)}{\partial \xi} \frac{\partial \xi(x, y)}{\partial x} + \frac{\partial \psi(\xi, \eta)}{\partial \eta} \frac{\partial \eta(x, y)}{\partial x}$$

and

$$\frac{\partial \psi(x(\xi, \eta), y(\xi, \eta))}{\partial y} = \frac{\partial \psi(\xi, \eta)}{\partial \xi} \frac{\partial \xi(x, y)}{\partial y} + \frac{\partial \psi(\xi, \eta)}{\partial \eta} \frac{\partial \eta(x, y)}{\partial y}$$

where we use the derivatives $\frac{\partial \psi}{\partial \xi}$ and metric terms $\frac{\partial \xi}{\partial x}$ from Ch. 12. Let us now describe the weak and strong form DG methods. Let us begin with the weak form which is the original DG method.

16.5 Element Equations on a Single Element: Weak Form

Equation (16.3) is the element equation that we need to solve for the *weak form* DG representation. For the purely inviscid case ($v = 0$) the solution of this problem is rather straightforward. However, when viscosity is present, we must consider how to handle the Laplacian operator. Recall from Ch. 13 that one possibility for handling the Laplacian operator is by using the *local discontinuous Galerkin* method. Let us revisit how to incorporate this method.

16.5.1 Local Discontinuous Galerkin Method for the Diffusion Operator

To ensure that the DG discrete form of the Laplacian operator is consistent we mentioned in Chs. 9 and 13 that it is best to approach the discretization via the *flux formulation*

$$\mathbf{Q} = \nabla q \quad \text{and} \quad \nabla \cdot \mathbf{Q} = \nabla^2 q.$$

For time-dependent problems, the construction of the LDG operators is much more straightforward because we can compute the auxiliary variable $\mathbf{Q}$ as follows

$$M_{ij}^{(e)} \left(Q_j^{(e)} \cdot \mathbf{I}_2 \right) = \sum_{l=1}^{N_{FN}} \left(\mathbf{F}_{ij}^{(e,l)} \right)^T \left(q_j^{(*,l)} \mathbf{I}_2 \right) - \tilde{\mathbf{D}}_{ij}^{(e)} q_j^{(e)}$$

where, e.g., $q^{(*,l)} = \frac{1}{2} \left[q^{(l)} + q^{(e)} \right]$, the values of q used are those at the current time t^n, and N_{FN} denotes the number of face neighbors that the element e has. For example, for quadrilaterals $N_{FN} = 4$ and for tensor-product bases we have $N_{FN} = 2d$ where d is the spatial dimension. The variable $\mathbf{Q}$ obtained here is now used in the construction of the flux variable $\mathbf{f}$ in the element equations.

16.5.2 Matrix-Vector Problem Resulting from Exact Integration

At this point we have to decide whether we are going to evaluate the integrals using co-located quadrature (which results in inexact integration) or non-colocated quadrature (that, for a specific number of quadrature points, results in exact integration). Let us first see what happens when we use exact integration. For quadrilateral elements let $M_Q = (Q + 1)^2$ where $Q = \frac{3}{2}N + \frac{1}{2}$ Lobatto points integrate $3N$ polynomials exactly (recall that the advection term is a 3N degree polynomial). This quadrature rule yields the following matrix-vector problem

$$M_{ij}^{(e)} \frac{dq_j^{(e)}}{dt} + \sum_{l=1}^{N_{FN}} \left(\mathbf{F}_{ij}^{(e,l)} \right)^T \mathbf{f}_j^{(*,l)} - \left(\tilde{\mathbf{D}}_{ij}^{(e)} \right)^T \mathbf{f}_j^{(e)} = 0$$

with $i, j = 1, \ldots, M_N$ and $e = 1, \ldots, N_e$, where we have replaced the partial time derivative with a total derivative to emphasize that $q_j^{(e)}$ is only a function of time. Further note that the above matrices are defined as follows

$$M_{ij}^{(e)} \equiv \int_{\Omega_e} \psi_i \psi_j \, d\Omega_e = \sum_{k=1}^{M_Q} w_k J_k^{(e)} \psi_{ik} \psi_{jk}$$

is the mass matrix where w_k and $J_k^{(e)}$ are the quadrature weights and determinant of the volume Jacobian evaluated at the quadrature point ξ_k,

$$\mathbf{F}_{ij}^{(e,l)} \equiv \int_{\Gamma_e} \psi_i \psi_j \hat{\mathbf{n}}^{(e,l)} \, d\Gamma_e = \sum_{k=0}^{Q} w_k^{(l)} J_k^{(l)} \psi_{ik} \psi_{jk} \hat{\mathbf{n}}_k^{(e,l)}$$

is the flux matrix where $w_k^{(l)}$ and $J_k^{(l)}$ are the quadrature weights and determinant of the face Jacobian (see Eq. (12.35)) of the face formed by the interface of elements e

and l, and

$$\tilde{\mathbf{D}}_{ij}^{(e)} \equiv \int_{\Omega_e} \nabla \psi_i \psi_j \, d\Omega_e = \sum_{k=1}^{M_Q} w_k J_k^{(e)} \nabla \psi_{ik} \psi_{jk}$$

is the weak form differentiation matrix that we have already seen in Ch. 13. There
is no difficulty in integrating the above matrices exactly, even for a non-diagonal
mass matrix. In the DG method having a non-diagonal mass matrix poses very little
difficulty since the global representation of this matrix is block-diagonal. Inverting
the mass matrix yields the final matrix problem

$$\frac{dq_i^{(e)}}{dt} + \sum_{l=1}^{N_{FN}} \left(\hat{\mathbf{F}}_{ij}^{(e,l)} \right)^T \mathbf{f}_j^{(*,l)} - \left(\hat{\mathbf{D}}_{ij}^{(e)} \right)^T \mathbf{f}_j^{(e)} = 0$$

where

$$\hat{\mathbf{F}}_{ij}^{(e,l)} = \left(M_{ik}^{(e)} \right)^{-1} \mathbf{F}_{kj}^{(e,l)}$$

and

$$\hat{\mathbf{D}}_{ij}^{(e)} = \left(M_{ik}^{(e)} \right)^{-1} \tilde{\mathbf{D}}_{kj}^{(e)}$$

are the flux and differentiation matrices premultiplied by the inverse mass matrix.

16.5.3 Matrix-Vector Problem Resulting from Inexact Integration

Looking at the mass, flux, and differentiation matrices we see that they represent
integrals of polynomials of degree: $2N$ for the mass and flux matrices and $3N$ for the
differentiation matrix. Note that the differentiation matrix is written above as a $2N$
polynomial but in reality, due to the nonlinear nature of the flux $\mathbf{f}$ we obtain a $3N$
polynomial matrix; however, this is of little importance in the following discussion
as we concentrate on the effects of inexact integration on the mass and flux matrices.

As we saw in Ch. 15, by using inexact integration, i.e., $2N$-1 integration, we only
commit a small numerical error for the mass and flux matrices; for the differentiation
matrix, this error is a bit larger. Let us now use inexact integration and see what the
resulting matrix-vector problem looks like. In this case, we let $M_Q = (Q+1)^2 = M_N$
where $Q = N$ to get the following matrix-vector problem

$$M_i^{(e)} \frac{dq_i^{(e)}}{dt} + \sum_{l=1}^{N_{FN}} \left(\mathbf{F}_i^{(e,l)} \right)^T \mathbf{f}_i^{(*,l)} - \left(\tilde{\mathbf{D}}_{ij}^{(e)} \right)^T \mathbf{f}_j^{(e)} = 0 \qquad (16.4)$$

with $i, j = 1, \ldots, M_N$ and $e = 1, \ldots, N_e$, where changes in our matrix-vector problem
from exact to inexact integration occur in all the matrices. The mass matrix simplifies
to

$$M_{ij}^{(e)} = w_i J_i^{(e)} \delta_{ij},$$

which can now be stored simply as the vector

$$M_i^{(e)} = w_i J_i^{(e)}.$$

The flux matrix becomes

$$\mathbf{F}_{ij}^{(e,l)} = w_i^{(l)} J_i^{(l)} \hat{\mathbf{n}}_i^{(e,l)} \delta_{ij}$$

where the flux matrix only affects the boundary of the element and can be stored as the vector

$$\mathbf{F}_i^{(e,l)} = w_i^{(l)} J_i^{(l)} \hat{\mathbf{n}}_i^{(e,l)}.$$

Finally, the differentiation matrix is

$$\tilde{\mathbf{D}}_{ij}^{(e)} = w_j J_j^{(e)} \nabla \psi_{ij}$$

where by cardinality, we replaced k with j and removed the summation.

16.6 Element Equations on a Single Element: Strong Form

To construct the strong form with DG, we begin with the product rule

$$\nabla \cdot \left(\psi_i \mathbf{f}_N^{(e)} \right) = \nabla \psi_i \cdot \mathbf{f}_N^{(e)} + \psi_i \nabla \cdot \mathbf{f}_N^{(e)}$$

and, after integration, yields

$$\int_{\Gamma_e} \psi_i \left(\hat{\mathbf{n}}^{(e,l)} \cdot \mathbf{f}_N^{(e)} \right) d\Gamma_e = \int_{\Omega_e} \nabla \psi_i \cdot \mathbf{f}_N^{(e)} d\Omega_e + \int_{\Omega_e} \psi_i \nabla \cdot \mathbf{f}_N^{(e)} d\Omega_e. \qquad (16.5)$$

Rearranging gives

$$\int_{\Omega_e} \psi_i \nabla \cdot \mathbf{f}_N^{(e)} d\Omega_e = \int_{\Gamma_e} \psi_i \left(\hat{\mathbf{n}}^{(e,l)} \cdot \mathbf{f}_N^{(*,l)} \right) d\Gamma_e - \int_{\Omega_e} \nabla \psi_i \cdot \mathbf{f}_N^{(e)} d\Omega_e \qquad (16.6)$$

where in the first term on the right-hand side we replaced the discontinuous flux by a continuous numerical flux function. Using Eq. (16.5) we can rewrite the last term in Eq. (16.6) as follows

$$- \int_{\Omega_e} \nabla \psi_i \cdot \mathbf{f}_N^{(e)} d\Omega_e = - \int_{\Gamma_e} \psi_i \left(\hat{\mathbf{n}}^{(e,l)} \cdot \mathbf{f}_N^{(e)} \right) d\Gamma_e + \int_{\Omega_e} \psi_i \nabla \cdot \mathbf{f}_N^{(e)} d\Omega_e$$

which can now be substituted into Eq. (16.6) to obtain

$$\int_{\Omega_e} \psi_i \nabla \cdot \mathbf{f}_N^{(e)} d\Omega_e = \int_{\Gamma_e} \psi_i \hat{\mathbf{n}}^{(e,l)} \cdot \left(\mathbf{f}_N^{(*,l)} - \mathbf{f}_N^{(e)} \right) d\Gamma_e + \int_{\Omega_e} \psi_i \nabla \cdot \mathbf{f}_N^{(e)} d\Omega_e.$$

Substituting this relation into the element equation gives

$$\int_{\Omega_e} \psi_i \frac{\partial q_N^{(e)}}{\partial t} \, d\Omega_e + \int_{\Gamma_e} \psi_i \hat{\mathbf{n}}^{(e,l)} \cdot \left(\mathbf{f}_N^{(*,l)} - \mathbf{f}_N^{(e)} \right) \, d\Gamma_e + \int_{\Omega_e} \psi_i \nabla \cdot \mathbf{f}_N^{(e)} \, d\Omega_e = 0 \quad (16.7)$$

$\forall \psi \in L^2$, $i = 1, \dots, M_N$, and $e = 1, \dots, N_e$, which we refer to as the *strong form* DG representation.

16.6.1 Matrix-Vector Problem Resulting from Exact Integration

Once again, let us consider the differences in the matrix-vector representation when we evaluate the integrals using exact and inexact integration. Let us first explore the exact integration case. This quadrature rule yields the following matrix-vector problem

$$M_{ij}^{(e)} \frac{dq_j^{(e)}}{dt} + \sum_{l=1}^{N_{FN}} \left(\mathbf{F}_{ij}^{(e,l)} \right)^T \left(\mathbf{f}^{(*,l)} - \mathbf{f}^{(e)} \right)_j + \left(\mathbf{D}_{ij}^{(e)} \right)^T \mathbf{f}_j^{(e)} = 0$$

where the above matrices are defined as follows

$$M_{ij}^{(e)} \equiv \int_{\Omega_e} \psi_i \psi_j \, d\Omega_e = \sum_{k=1}^{M_Q} w_k J_k^{(e)} \psi_{ik} \psi_{jk}$$

is the mass matrix,

$$\mathbf{F}_{ij}^{(e,l)} \equiv \int_{\Gamma_e} \psi_i \psi_j \hat{\mathbf{n}}^{(e,l)} \, d\Gamma_e = \sum_{k=0}^{Q} w_k^{(l)} J_k^{(l)} \psi_{ik} \psi_{jk} \hat{\mathbf{n}}_k^{(e,l)}$$

is the flux matrix, and

$$\mathbf{D}_{ij}^{(e)} \equiv \int_{\Omega_e} \psi_i \nabla \psi_j \, d\Omega_e = \sum_{k=1}^{M_Q} w_k J_k^{(e)} \psi_{ik} \nabla \psi_{jk}$$

is the strong form differentiation matrix.
Inverting the mass matrix yields the final matrix problem

$$\frac{dq_i^{(e)}}{dt} + \sum_{l=1}^{N_{FN}} \left(\hat{\mathbf{F}}_{ij}^{(e,l)} \right)^T \left(\mathbf{f}^{(*,l)} - \mathbf{f}^{(e)} \right)_j + \left(\hat{\mathbf{D}}_{ij}^{(e)} \right)^T \mathbf{f}_j^{(e)} = 0$$

where

$$\hat{\mathbf{F}}_{ij}^{(e,l)} = \left(M_{ik}^{(e)} \right)^{-1} \mathbf{F}_{kj}^{(e,l)}$$

and

$$\hat{\mathbf{D}}_{ij}^{(e)} = \left(M_{ik}^{(e)} \right)^{-1} \mathbf{D}_{kj}^{(e)}$$

are the flux and differentiation matrices premultiplied by the inverse mass matrix.

16.6.2 Matrix-Vector Problem Resulting from Inexact Integration

Let us now consider the matrix-vector problem for the strong form DG with inexact integration. For inexact integration we let $M_Q = (Q+1)^2 = M_N$ where $Q = N$ which yields the following matrix-vector problem

$$M_i^{(e)} \frac{dq_i^{(e)}}{dt} + \sum_{l=1}^{N_{FN}} \left(\mathbf{F}_i^{(e,l)} \right)^T \left(\mathbf{f}^{(*,l)} - \mathbf{f}^{(e)} \right)_i + \left(\mathbf{D}_{ij}^{(e)} \right)^T \mathbf{f}_j^{(e)} = 0 \qquad (16.8)$$

where changes in the matrix-vector problem from exact to inexact integration occur in all the matrices. The mass matrix simplifies to

$$M_i^{(e)} = w_i J_i^{(e)},$$

the flux matrix becomes

$$\mathbf{F}_i^{(e,l)} = w_i^{(l)} J_i^{(l)} \hat{\mathbf{n}}_i^{(e,l)},$$

and the strong form differentiation matrix is

$$\mathbf{D}_{ij}^{(e)} = w_i J_i^{(e)} \nabla \psi_{ji}.$$

Substituting these matrices into the element equation (16.8) yields

$$w_i J_i^{(e)} \frac{dq_i^{(e)}}{dt} + \sum_{l=1}^{N_{FN}} w_i^{(l)} J_i^{(l)} \left(\hat{\mathbf{n}}_i^{(e,l)} \right)^T \left(\mathbf{f}^{(*,l)} - \mathbf{f}^{(e)} \right)_i + w_i J_i^{(e)} \left(\nabla \psi_{ji} \right)^T \mathbf{f}_j^{(e)} = 0. \quad (16.9)$$

Dividing by $w_i J_i^{(e)}$ yields

$$\frac{dq_i^{(e)}}{dt} + \sum_{l=1}^{N_{FN}} \frac{w_i^{(l)} J_i^{(l)}}{w_i J_i^{(e)}} \left(\hat{\mathbf{n}}_i^{(e,l)} \right)^T \left(\mathbf{f}^{(*,l)} - \mathbf{f}^{(e)} \right)_i + \left(\nabla \psi_{ji} \right)^T \mathbf{f}_j^{(e)} = 0 \qquad (16.10)$$

which looks like a collocation method with a penalty term to handle discontinuous fluxes. In fact, for this particular form of the DG method and with very specific choices of penalty terms both the DG method and the spectral multi-domain penalty method (see, e.g., [232, 198, 234, 196, 197, 239, 173]) can be shown to be equivalent [121]. Another value of this form of the DG method is that it now looks a lot like a high-order finite difference method [97] which then allows many of the improvements to these methods to be extended directly to the DG method (such as entropy-stability; see, e.g., [1]).

16.7 Example of 2D DG for Linear Elements

Let us discuss the explicit construction of the matrix problem for the 2D advection equation using linear elements $N = 1$. As a reminder, this analysis is carried out to give the reader a way to verify that the codes they write are correct, at least for the simplest case. We discuss the advection equation rather than the advection-diffusion equation in order to simplify the discussion of the element matrix problem

$$M_{ij}^{(e)} \frac{dq_j^{(e)}}{dt} + \sum_{l=1}^{N_{FN}} \left(\mathbf{F}_{ij}^{(e,l)} \right)^T \mathbf{f}_j^{(*,l)} - \left(\tilde{\mathbf{D}}_{ij}^{(e)} \right)^T \mathbf{f}_j^{(e)} = 0$$

where we now define the flux as $\mathbf{f} = q\mathbf{u}$, instead of $\mathbf{f} = q\mathbf{u} - \nu\nabla q$, which vastly simplifies the solution strategy by obviating the need to discuss the viscous terms[2].

16.7.1 2D Basis Functions

In Ch. 15 we wrote the 2D linear basis functions as

$$\psi_i(\xi, \eta) = \frac{1}{2}(1 + \xi_i \xi)\frac{1}{2}(1 + \eta_i \eta)$$

or, more compactly as

$$\psi_i(\xi, \eta) = \frac{1}{4}(1 + \xi_i \xi)(1 + \eta_i \eta).$$

From this expression we defined the derivatives as follows

$$\frac{\partial \psi_i}{\partial \xi}(\xi, \eta) = \frac{1}{4}\xi_i(1 + \eta_i \eta), \quad \frac{\partial \psi_i}{\partial \eta}(\xi, \eta) = \frac{1}{4}\eta_i(1 + \xi_i \xi).$$

16.7.2 Metric Terms

In Ch. 15 we also defined the metric terms as follows

$$\frac{\partial x_N}{\partial \xi} = \frac{\Delta x}{2}, \quad \frac{\partial x_N}{\partial \eta} = \frac{\partial y_N}{\partial \xi} = 0, \quad \frac{\partial y_N}{\partial \eta} = \frac{\Delta y}{2}, \quad J^{(e)} = \frac{\Delta x^{(e)} \Delta y^{(e)}}{4}.$$

---

[2] We ignore the viscous terms because they have already been discussed in detail in Chs. 13 and 14.

16.7.3 Derivatives in Physical Space

We defined the derivatives in physical coordinates in Ch. 15 as

$$\frac{\partial \psi_i}{\partial x} = \frac{1}{4}\xi_i\,(1 + \eta_i\eta)\,\frac{2}{\Delta x}$$

and

$$\frac{\partial \psi_i}{\partial y} = \frac{1}{4}\eta_i\,(1 + \xi_i\xi)\,\frac{2}{\Delta y}.$$

16.7.4 Mass Matrix

From Ch. 13 we found that the mass matrix is

$$M_{ij}^{(e)} = \frac{\Delta x^{(e)}\Delta y^{(e)}}{36}\begin{pmatrix} 4\ 2\ 2\ 1 \\ 2\ 4\ 1\ 2 \\ 2\ 1\ 4\ 2 \\ 1\ 2\ 2\ 4 \end{pmatrix}.$$

We still need to construct the flux matrix $\mathbf{F}$ and the weak form differentiation matrix $\tilde{\mathbf{D}}^{(e)}$.

16.7.5 Differentiation Matrix

The weak form differentiation matrix is written as

$$\tilde{\mathbf{D}}_{ij}^{(e)} = \int_{\Omega_e} \nabla\psi_i\psi_j\,d\Omega_e.$$

Beginning with

$$\tilde{D}_{ij}^{(e,x)} = \int_{\Omega_e} \frac{\partial \psi_i}{\partial x}\psi_j\,d\Omega_e \qquad D_{ij}^{(e,y)} = \int_{\Omega_e} \frac{\partial \psi_i}{\partial y}\psi_j\,d\Omega_e$$

we find that

$$\frac{\partial \psi_i}{\partial x} = \frac{\partial \psi_i}{\partial \xi}\frac{\partial \xi}{\partial x} + \frac{\partial \psi_i}{\partial \eta}\frac{\partial \eta}{\partial x},$$

$$\frac{\partial \psi_i}{\partial y} = \frac{\partial \psi_i}{\partial \xi}\frac{\partial \xi}{\partial y} + \frac{\partial \psi_i}{\partial \eta}\frac{\partial \eta}{\partial y},$$

where, since we are assuming that x is along ξ and y is along η, we get the metric terms

$$\frac{\partial \xi}{\partial x} = \frac{2}{\Delta x} \qquad \frac{\partial \eta}{\partial y} = \frac{2}{\Delta y} \qquad \frac{\partial \xi}{\partial y} = \frac{\partial \eta}{\partial x} = 0.$$

This now allows us to write the differentiation matrices in terms of the reference element coordinates as follows

$$\tilde{D}_{ij}^{(e,x)} = \int_{-1}^{+1} \int_{-1}^{+1} \frac{1}{4}\xi_i (1 + \eta_i\eta) \frac{1}{4} (1 + \xi_j\xi)(1 + \eta_j\eta) \frac{2}{\Delta x^{(e)}} \frac{\Delta x^{(e)}\Delta y^{(e)}}{4} \, d\xi \, d\eta$$

$$\tilde{D}_{ij}^{(e,y)} = \int_{-1}^{+1} \int_{-1}^{+1} \frac{1}{4}\eta_i (1 + \xi_i\xi) \frac{1}{4} (1 + \xi_i\xi)(1 + \eta_i\eta) \frac{2}{\Delta y^{(e)}} \frac{\Delta x^{(e)}\Delta y^{(e)}}{4} \, d\xi \, d\eta.$$

Simplifying and integrating yields

$$\tilde{D}_{ij}^{(e,x)} = \frac{\Delta y^{(e)}}{32} \left[\xi_i\xi + \frac{1}{2}\xi_i\xi_j\xi^2\right]_{-1}^{+1} \left[\eta + \frac{1}{2}(\eta_i + \eta_j)\eta^2 + \frac{1}{3}\eta_i\eta_j\eta^3\right]_{-1}^{+1}$$

$$\tilde{D}_{ij}^{(e,y)} = \frac{\Delta x^{(e)}}{32} \left[\eta_i\eta + \frac{1}{2}\eta_i\eta_j\eta^2\right]_{-1}^{+1} \left[\xi + \frac{1}{2}(\xi_i + \xi_j)\xi^2 + \frac{1}{3}\xi_i\xi_j\xi^3\right]_{-1}^{+1}.$$

Evaluating the terms above at the bounds of integration and simplifying yields the compact form

$$\tilde{D}_{ij}^{(e,x)} = \frac{\Delta y^{(e)}}{24}\xi_i(3 + \eta_i\eta_j) \qquad \tilde{D}_{ij}^{(e,y)} = \frac{\Delta x^{(e)}}{24}\eta_i(3 + \xi_i\xi_j).$$

Substituting for the values of the reference element coordinates, $\xi_i = (-1, +1, -1, +1)$ and $\eta_i = (-1, -1, +1, +1)$, yields the matrix form

$$\tilde{D}_{ij}^{(e,x)} = \frac{\Delta y^{(e)}}{12} \begin{pmatrix} -2 & -2 & -1 & -1 \\ 2 & 2 & -1 & 1 \\ -1 & -1 & -2 & -2 \\ 1 & 1 & 2 & 2 \end{pmatrix}$$

and

$$\tilde{D}_{ij}^{(e,y)} = \frac{\Delta x^{(e)}}{12} \begin{pmatrix} -2 & -1 & -2 & -1 \\ -1 & -2 & -1 & -2 \\ 2 & -1 & 2 & 1 \\ 1 & 2 & 1 & 2 \end{pmatrix}.$$

16.7.6 Flux Matrix

Recall that the flux matrix is defined as

$$\mathbf{F}_{ij}^{(e,l)} = \int_{\Gamma_e} \psi_i\psi_j \hat{\mathbf{n}}^{(e,l)} \, d\Gamma_e$$

where l is the face neighbor of e and $\hat{\mathbf{n}}^{(e,l)}$ is the outward pointing unit normal vector of the face shared by the elements e and l, pointing from e to l. For linear elements the basis functions are $\psi_i(\xi, \eta) = \frac{1}{4}(1 + \xi_i\xi)(1 + \eta_i\eta)$ where $i, j = 1, \ldots, M_N$ that, for linear quadrilateral elements, is $M_N = (N + 1)^2 = 4$. Substituting this definition into the flux matrix yields

$$F_{ij}^{(e,l)} = \int_{-1}^{+1} \frac{1}{16}(1 + \xi_i\xi)(1 + \eta_i\eta)(1 + \xi_j\xi)(1 + \eta_j\eta) \; J^{(l)}(\xi, \eta) \; \hat{\mathbf{n}}^{(e,l)} \; ds$$

where ds is either $d\xi$ or $d\eta$, depending on which of the four faces of the element e we are evaluating and $\xi_i, \xi_j, \eta_i, \eta_j$ are the vertices of the face $f_k = (e, l_k)$. Let us use Fig. 16.1 to discuss the face components of the flux.

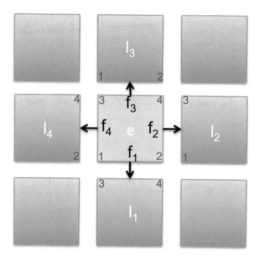

Fig. 16.1: The element e, its four vertices (red font), its four faces (f_1, f_2, f_3, f_4) and its four face neighbors (l_1, l_2, l_3, l_4).

Looking at Fig. 16.1 we see that for face 1 (f_1), which is shared by elements e and l_1, the integral is along the ξ direction where $\eta = -1$ so that we get the following integral

$$F_{ij}^{(e,l_1)} = \int_{-1}^{+1} \frac{1}{16}\left[1 + (\xi_i + \xi_j)\xi + \xi_i\xi_j\xi^2\right](1 - \eta_i)(1 - \eta_j) \frac{\Delta x^{(1)}}{2} \hat{\mathbf{n}}^{(e,l_1)} \; d\xi.$$

Integrating yields

$$F_{ij}^{(e,l_1)} = \frac{\Delta x^{(1)} \; \hat{\mathbf{n}}^{(e,l_1)}}{32}\left[\xi + \frac{1}{2}(\xi_i + \xi_j)\xi^2 + \frac{1}{3}\xi_i\xi_j\xi^3\right]_{-1}^{+1}(1 - \eta_i)(1 - \eta_j)$$

which, upon evaluating at the bounds of integration, yields

$$F_{ij}^{(e,l_1)} = \frac{\Delta x^{(1)} \, \hat{\mathbf{n}}^{(e,l_1)}}{48} \left(3 + \xi_i \xi_j \right) \left(1 - \eta_i \right) \left(1 - \eta_j \right).$$

Substituting the values of $(\xi, \eta)_i$ and $(\xi, \eta)_j$ yields the matrix

$$F_{ij}^{(e,l_1)} = \frac{\Delta x^{(1)} \, \hat{\mathbf{n}}^{(e,l_1)}}{48} \begin{pmatrix} 16 & 8 & 0 & 0 \\ 8 & 16 & 0 & 0 \\ 0 & 0 & 0 & 0 \\ 0 & 0 & 0 & 0 \end{pmatrix}$$

and simplifying to its final form yields

$$F_{ij}^{(e,l_1)} = \frac{\Delta x^{(1)} \, \hat{\mathbf{n}}^{(e,l_1)}}{6} \begin{pmatrix} 2 & 1 & 0 & 0 \\ 1 & 2 & 0 & 0 \\ 0 & 0 & 0 & 0 \\ 0 & 0 & 0 & 0 \end{pmatrix}$$

which is just the 1D mass matrix for the face $f_1 = (e, l_1)$. For this face, its normal vector is $\hat{\mathbf{n}}^{(e,l_1)} = 0\hat{\mathbf{i}} - 1\hat{\mathbf{j}}$ where $\hat{\mathbf{i}}$ and $\hat{\mathbf{j}}$ are the Cartesian coordinate directional vectors. Following this approach for the remaining three faces yields the three matrices defined below. For face 2 we get

$$F_{ij}^{(e,l_2)} = \frac{\Delta y^{(2)} \, \hat{\mathbf{n}}^{(e,l_2)}}{6} \begin{pmatrix} 0 & 0 & 0 & 0 \\ 0 & 2 & 0 & 1 \\ 0 & 0 & 0 & 0 \\ 0 & 1 & 0 & 2 \end{pmatrix}$$

where the normal vector here is $\hat{\mathbf{n}}^{(e,l_2)} = 1\hat{\mathbf{i}} + 0\hat{\mathbf{j}}$. For face 3 we get

$$F_{ij}^{(e,l_3)} = \frac{\Delta x^{(3)} \, \hat{\mathbf{n}}^{(e,l_3)}}{6} \begin{pmatrix} 0 & 0 & 0 & 0 \\ 0 & 0 & 0 & 0 \\ 0 & 0 & 2 & 1 \\ 0 & 0 & 1 & 2 \end{pmatrix}$$

where the normal vector is $\hat{\mathbf{n}}^{(e,l_3)} = 0\hat{\mathbf{i}} + 1\hat{\mathbf{j}}$. Finally, for face 4 we get

$$F_{ij}^{(e,l_4)} = \frac{\Delta y^{(4)} \, \hat{\mathbf{n}}^{(e,l_4)}}{6} \begin{pmatrix} 2 & 0 & 1 & 0 \\ 0 & 0 & 0 & 0 \\ 1 & 0 & 2 & 0 \\ 0 & 0 & 0 & 0 \end{pmatrix}$$

where the normal vector is $\hat{\mathbf{n}}^{(e,l_4)} = -1\hat{\mathbf{i}} + 0\hat{\mathbf{j}}$.

16.7.7 Numerical Flux Function

In order to solve the weak form element matrix problem

$$M_{ij}^{(e)} \frac{dq_j^{(e)}}{dt} + \sum_{l=1}^{N_{FN}} \left(\mathbf{F}_{ij}^{(e,l)} \right)^T \mathbf{f}_j^{(*,l)} - \left(\tilde{\mathbf{D}}_{ij}^{(e)} \right)^T \mathbf{f}_j^{(e)} = 0$$

requires us to know the exact form of the numerical flux $\mathbf{f}^{(*,l)}$. For simplicity, let us use the Rusanov flux[3] which is defined as

$$\mathbf{f}_j^{(*,l)} = \frac{1}{2} \left[\mathbf{f}_j^{(l)} + \mathbf{f}_j^{(e)} - |\lambda| \hat{\mathbf{n}}^{(e,l)} \left(q_j^{(l)} - q_j^{(e)} \right) \right].$$

To see the structure of the product

$$\left(\mathbf{F}_{ij}^{(e,l)} \right)^T \mathbf{f}_j^{(*,l)}$$

let us first extract the normal vector from the flux matrix and write this product as follows

$$F_{ij}^{(e,l)} \left(\hat{\mathbf{n}}^{(e,l)} \cdot \mathbf{f}_j^{(*,l)} \right)$$

where

$$F_{ij}^{(e,l)} = \int_{\Gamma_e} \psi_i \psi_j \, d\Gamma_e.$$

For face 1 (see Fig. 16.1), we then have the following numerical flux function

$$\hat{\mathbf{n}}^{(e,l_1)} \cdot \mathbf{f}_j^{(*,l_1)} = \frac{1}{2} \left[\hat{\mathbf{n}}^{(e,l_1)} \cdot \left(\mathbf{f}_j^{(l_1)} + \mathbf{f}_j^{(e)} \right) - |\lambda| \left(q_j^{(l_1)} - q_j^{(e)} \right) \right].$$

More specifically, for $j = 1, 2$ we get

$$\hat{\mathbf{n}}^{(e,l_1)} \cdot \mathbf{f}_1^{(*,l_1)} = \frac{1}{2} \left[\hat{\mathbf{n}}^{(e,l_1)} \cdot \left(\mathbf{f}_3^{(l_1)} + \mathbf{f}_1^{(e)} \right) - |\lambda|_{1,3}^{(e,l_1)} \left(q_3^{(l_1)} - q_1^{(e)} \right) \right],$$

$$\hat{\mathbf{n}}^{(e,l_1)} \cdot \mathbf{f}_2^{(*,l_1)} = \frac{1}{2} \left[\hat{\mathbf{n}}^{(e,l_1)} \cdot \left(\mathbf{f}_4^{(l_1)} + \mathbf{f}_2^{(e)} \right) - |\lambda|_{2,4}^{(e,l_1)} \left(q_4^{(l_1)} - q_2^{(e)} \right) \right],$$

and for $j = 3, 4$ we do not need to define the flux function since it will be multiplied by rows 3 and 4 of $F^{(e,l_1)}$ which are zero.

Substituting the values for $\hat{\mathbf{n}}^{(e,l_1)}$ and $\mathbf{f}$ yields

$$\hat{\mathbf{n}}^{(e,l_1)} \cdot \mathbf{f}_1^{(*,l_1)} = \frac{1}{2} \left[-\left((qv)_3^{(l_1)} + (qv)_1^{(e)} \right) - v_{1,3}^{(e,l_1)} \left(q_3^{(l_1)} - q_1^{(e)} \right) \right],$$

---

[3] The choice of the Riemann solver does not affect this discussion. Other Riemann solvers, such as those discussed in Ch. 18, can be substituted for Rusanov here with few modifications.

which, after simplifying, yields

$$\hat{\mathbf{n}}^{(e,l_1)} \cdot \mathbf{f}_1^{(*,l_1)} = -(qv)_3^{(l_1)},$$

and

$$\hat{\mathbf{n}}^{(e,l_1)} \cdot \mathbf{f}_2^{(*,l_1)} = -(qv)_4^{(l_1)}$$

where we have taken

$$|\lambda|_{1,3}^{(e,l_1)} = |\hat{\mathbf{n}}^{(e,l_1)} \cdot \mathbf{u}^{(e,l_1)}|_{1,3} = v_{1,3}^{(e,l_1)}$$

and

$$|\lambda|_{2,4}^{(e,l_1)} = |\hat{\mathbf{n}}^{(e,l_1)} \cdot \mathbf{u}^{(e,l_1)}|_{2,4} = v_{2,4}^{(e,l_1)}$$

and have assumed v to be continuous. For face 2, we get

$$\hat{\mathbf{n}}^{(e,l_2)} \cdot \mathbf{f}_2^{(*,l_2)} = (qu)_2^{(e)},$$

and

$$\hat{\mathbf{n}}^{(e,l_2)} \cdot \mathbf{f}_4^{(*,l_2)} = (qu)_4^{(e)}$$

where we have taken

$$|\lambda|_{2,1}^{(e,l_2)} = |\hat{\mathbf{n}}^{(e,l_2)} \cdot \mathbf{u}^{(e,l_2)}|_{2,1} = u_{2,1}^{(e,l_2)}$$

and

$$|\lambda|_{4,3}^{(e,l_2)} = |\hat{\mathbf{n}}^{(e,l_2)} \cdot \mathbf{u}^{(e,l_2)}|_{4,3} = u_{4,3}^{(e,l_2)}$$

and have assumed u to be continuous. For face 3, we get

$$\hat{\mathbf{n}}^{(e,l_3)} \cdot \mathbf{f}_3^{(*,l_3)} = (qv)_3^{(e)},$$

and

$$\hat{\mathbf{n}}^{(e,l_3)} \cdot \mathbf{f}_4^{(*,l_3)} = (qv)_4^{(e)}$$

where we have taken

$$|\lambda|_{3,1}^{(e,l_3)} = |\hat{\mathbf{n}}^{(e,l_3)} \cdot \mathbf{u}^{(e,l_3)}|_{3,1} = v_{3,1}^{(e,l_3)}$$

and

$$|\lambda|_{4,2}^{(e,l_3)} = |\hat{\mathbf{n}}^{(e,l_3)} \cdot \mathbf{u}^{(e,l_3)}|_{4,2} = v_{4,2}^{(e,l_3)}$$

and have assumed v to be continuous. Finally, for face 4 we get

$$\hat{\mathbf{n}}^{(e,l_4)} \cdot \mathbf{f}_1^{(*,l_4)} = -(qu)_2^{(l_4)},$$

and

$$\hat{\mathbf{n}}^{(e,l_4)} \cdot \mathbf{f}_3^{(*,l_4)} = -(qu)_4^{(l_4)}$$

where we have taken

$$|\lambda|_{1,2}^{(e,l_4)} = |\hat{\mathbf{n}}^{(e,l_4)} \cdot \mathbf{u}^{(e,l_4)}|_{1,2} = u_{1,2}^{(e,l_4)}$$

and

$$|\lambda|_{3,4}^{(e,l_4)} = |\hat{\mathbf{n}}^{(e,l_4)} \cdot \mathbf{u}^{(e,l_4)}|_{3,4} = u_{3,4}^{(e,l_4)}$$

and have assumed u to be continuous.

Remark 16.1 By going through the exercise of what the numerical flux function is doing, it should become obvious that this function is just favoring the upwind direction. For example, in Fig. 16.1, if we assume that the flow field in u is moving from left to right and in v from bottom to top we see that the Rusanov flux function is merely favoring the upwind direction. This is the idea behind numerical flux functions because, as we saw in Ch. 1, for hyperbolic equations using the upwind direction results in a stable numerical method.

16.8 Algorithms for the DG Matrix-Vector Problem

Let us now describe the algorithms for constructing the matrix-vector problem for DG. First we describe the construction of the differentiation and flux matrices for non tensor-products and then the tensor-product approach.

16.8.1 Non-Tensor-Product Approach

In this section, we describe the construction of the strong and weak form differentiation matrices as well as the flux matrix. We begin with the strong form differentiation matrix.

16.8.1.1 Strong Form Differentiation Matrix

Algorithm 16.1 Construction of the matrix $\mathbf{D}^{(e)}$ with exact integration.

```
function D_MATRIX_STRONG
    D(e) = 0
    for e = 1 : Ne do                                    ▷ loop over elements
        for k = 1 : MQ do                                ▷ loop over integration points
            for j = 1 : MN do                            ▷ loop over columns of D(e)
                for i = 1 : MN do                        ▷ loop over rows of D(e)
                    D(e)ij + = wk J(e)k ψik ∇ψjk
                end for
            end for
        end for
    end for
end function
```

Algorithm 16.1 shows the construction of the matrix $\mathbf{D}^{(e)}$. The structure of the algorithm is straightforward. We always require a loop through all the elements; note that for each element the operations are completely independent from the other elements in the grid. Then for each element, we require looping over all the integration points (k-loop). For each integration point we must construct the derivative of the basis functions in terms of physical variables. This operation is in fact executed using the chain rule as follows

$$\frac{\partial \psi_{jk}}{\partial x} = \frac{\partial \psi_{jk}}{\partial \xi} \frac{\partial \xi_k}{\partial x} + \frac{\partial \psi_{jk}}{\partial \eta} \frac{\partial \eta_k}{\partial x}$$

and

$$\frac{\partial \psi_{jk}}{\partial y} = \frac{\partial \psi_{jk}}{\partial \xi} \frac{\partial \xi_k}{\partial y} + \frac{\partial \psi_{jk}}{\partial \eta} \frac{\partial \eta_k}{\partial y}$$

where we only compute the metric terms and the derivatives of the basis functions once at the beginning of the computation or if the grid changes.

Exercise Compute the complexity for building the matrix $\mathbf{D}^{(e)}$ in Alg. 16.1 taking into account the computation of the derivatives in physical space as a function of the derivatives in the reference element and the metric terms.   □

To see how inexact integration changes the construction of the matrix $\mathbf{D}^{(e)}$, let us now look at Alg. 16.2. We can see that Alg. 16.2 has one fewer loop than Alg.

Algorithm 16.2 Construction of the matrix $\mathbf{D}^{(e)}$ with inexact integration.

---

function D_MATRIX_STRONG
  for $e = 1 : N_e$ **do**   ▷ loop over elements
    for $j = 1 : M_N$ **do**   ▷ loop over columns of $\mathbf{D}^{(e)}$
      for $i = 1 : M_N$ **do**   ▷ loop over rows of $\mathbf{D}^{(e)}$
        $\mathbf{D}^{(e)}_{ij} = w_i J^{(e)}_i \nabla \psi_{ji}$
      end for
    end for
  end for
end function

16.1 because, with inexact integration, we are able to exploit cardinality and thereby replace the k loop by i. Another difference between these two algorithms is that in the inexact case, the values of $\mathbf{D}^{(e)}$ for the i and j loops are only visited once. This means that we do not require initializing the matrix and we do not need to worry about possible conflicts regarding which *thread*[4] is accessing this specific location since no conflicts in memory access can occur.

Exercise Compute the complexity for building the matrix $\mathbf{D}^{(e)}$ in Alg. 16.2.   □

---

[4] The classical example is in the OpenMP paradigm where an *atomic* pragma needs to be used to tell the threads that only one thread can access this memory at a time. This is obviously not a good idea if vectorization of a loop is expected.

16.8.1.4 Weak Form Differentiation Matrix

The weak form differentiation matrix was outlined in Alg. 13.1 for exact integration. In fact, comparing Alg. 13.1 and 16.2 shows that the construction of the weak and strong form differentiation matrices are rather similar.

Exercise Compute the complexity for building the matrix $\tilde{\mathbf{D}}^{(e)}$ in Alg. 13.1. □

To see how inexact integration changes the construction of the matrix $\tilde{\mathbf{D}}^{(e)}$, let us now look at Alg. 16.3. We can see that Alg. 16.3 has one fewer loop than Alg.

Algorithm 16.3 Construction of the matrix $\tilde{\mathbf{D}}^{(e)}$ with inexact integration.

function D_MATRIX_WEAK
  for $e = 1 : N_e$ **do**   ▷ loop over elements
    for $j = 1 : M_N$ **do**   ▷ loop over columns of $\tilde{\mathbf{D}}^{(e)}$
      for $i = 1 : M_N$ **do**   ▷ loop over rows of $\tilde{\mathbf{D}}^{(e)}$
        $\tilde{\mathbf{D}}_{ij}^{(e)} = w_j J_j^{(e)} \nabla \psi_{ij}$
      end for
    end for
  end for
end function

13.1 because, with inexact integration, we are able to exploit cardinality and thereby replace the k loop by j.

Exercise Compute the complexity for building the matrix $\tilde{\mathbf{D}}^{(e)}$ in Alg. 16.3. □

16.8.1.7 Flux Matrix

Let us now turn to the flux matrix. Recall that the flux matrix looks very much like a mass matrix with a normal unit vector embedded. Algorithm 16.4 describes the steps required in constructing the flux matrix with exact integration.

Algorithm 16.4 requires a careful explanation. The need for the element loop is clear, i.e., we form the flux matrix for each element. However, the l-loop is different from any other loop we have seen so far. Since each element has N_{FN} *face neighbors*, then we need a different matrix for each face as we saw in Sec. 16.7.6. For this reason, the k-loop below the l-loop has dimension $M_Q^{(l)}$ because we do not need all of the integration points but only those that are on the face l of the element e. Note that the quadrature weights w and the corresponding face Jacobian J have a superscript (e, l). This is to remind the reader that the weight and Jacobian are associated with the face l of the element e. In fact, we saw in Sec. 16.7.6 that in a two-dimensional problem, these integrals are line (one-dimensional) integrals [5]. For

---

[5] In three-dimensions the flux integrals will be area (two-dimensional) integrals.

Algorithm 16.4 Construction of the matrix $\mathbf{F}^{(e,l)}$ with exact integration.

function F_MATRIX
  $\mathbf{F}^{(e,l)} = 0$
  for $e = 1 : N_e$ **do**   ▷ loop over elements
    for $l = 1 : N_{FN}$ **do**   ▷ loop over neighboring faces of the element e
      for $k = 1 : M_Q^{(l)}$ **do**   ▷ integration points along the face between (e, l)
        for $j = 1 : M_N$ **do**   ▷ loop over columns of $\mathbf{F}^{(e,l)}$
          for $i = 1 : M_N$ **do**   ▷ loop over rows of $\mathbf{F}^{(e,l)}$
$$\mathbf{F}_{ij}^{(e,l)} + = w_k^{(e,l)} J_k^{(e,l)} \psi_{ik}\psi_{jk}\hat{\mathbf{n}}_k^{(e,l)}$$
         end for
        end for
      end for
    end for
  end for
end function

---

this particular two-dimensional example, the weights $w^{(e,l)}$ are the one-dimensional quadrature weights we described in Ch. 4. For straight-sided elements, the Jacobian is $J^{(e,l)} = \frac{\Delta s}{2}$ where Δs is the length (in physical coordinates) of the face defined by (e, l). The vector $\hat{\mathbf{n}}^{(e,l)}$ in Alg. 16.4 is the unit normal vector pointing from the element e to the neighboring element on the other side of the face l.

If we now used inexact integration, then we can replace the integration loop k by i to derive the algorithm presented in Alg. 16.5. Algorithm 16.5 is vastly simpler than

---

Algorithm 16.5 Construction of the matrix $\mathbf{F}^{(e,l)}$ with inexact integration.

function F_MATRIX
  for $e = 1 : N_e$ **do**   ▷ loop over elements
    for $l = 1 : N_{FN}$ **do**   ▷ loop over neighboring faces of the element e
      for $i = 1 : M_N^{(l)}$ **do**   ▷ loop over rows of $\mathbf{F}^{(e,l)}$
$$\mathbf{F}_i^{(e,l)} = w_i^{(e,l)} J_i^{(e,l)} \hat{\mathbf{n}}_i^{(e,l)}$$
      end for
    end for
  end for
end function

---

its exact integration counterpart. In fact, we can see that the flux matrix has simplified to a collection of vectors. That is for each element e we have $l = 1, \ldots, N_{FN}$ flux vectors.

16.8.1.8 Solution Algorithm

Once we have constructed all of the element matrices, we are ready to solve the PDE. Algorithm 16.6 shows the main steps in constructing a DG solution for the advection equation. Algorithm 16.6 shows that once we have constructed the element

---

Algorithm 16.6 DG solution algorithm for the 2D wave equation using element matrices.

---

function DG(q,f)

    Construct $M^{(e)}$ ▷ using Alg. 12.6

    Construct $\mathbf{D}^{(e)}$ or $\tilde{\mathbf{D}}^{(e)}$ ▷ using Alg. 16.1 or Alg. 13.1

    Construct $\mathbf{F}^{(e,l)}$ ▷ using Alg. 16.4

    for $n = 1 : N_t$ **do**   ▷ time loop

$$R_i^{(e)} = \left(\tilde{\mathbf{D}}_{ij}^{(e)}\right)^T \mathbf{f}_j^n - \sum_{l=1}^{N_{FN}} F_{ij}^{(e,l)} \left(\mathbf{f}_j^n\right)^{(*,l)}$$ ▷ build RHS vector

$$\hat{R}_i^{(e)} = \left(M_{ij}^{(e)}\right)^{-1} R_j^{(e)}$$

$$\frac{d}{dt} q_i^{(e)} = \hat{R}_i^{(e)}$$ ▷ evolve equations forward in time to get q^{n+1}

    end for

end function

---

matrices (M, $\tilde{\mathbf{D}}^{(e)}$, and $\mathbf{F}^{(e,l)}$ in this case) that the solution algorithm becomes quite simple. Upon constructing the element matrices, we then loop through the time-steps (n-loop) where, at each time-step, we construct the RHS vector $R^{(e)}$ which is formed by taking the action of $\tilde{\mathbf{D}}^{(e)}$ and $\mathbf{F}^{(e,l)}$ on the flux vector $\mathbf{f}$. Once $R^{(e)}$ is constructed, we then left-multiply it by the inverse mass matrix and denote it as $\hat{R}^{(e)}$. At this point, we use this vector to evolve the equations forward in time to get q^{n+1}.

As we discussed in Ch. 15 the difficulty with this approach where we construct matrices is that this only works for linear problems and for static grids (grids that never change). Whenever the grid changes (even for a linear problem) requires the re-construction of all the element matrices. To avoid this problem we must construct the RHS vector $R^{(e)}$ at every time-step without explicitly building the element matrices. The optimal approach for this strategy is to recognize that the differentiation matrix represents a so-called *volume* integral (area integral in 2D and a volume integral in 3D) while the flux matrix represents a so-called *flux* (or boundary) integral. Let us now describe the contribution of each of these types of integrals to the RHS vector.

Algorithm 16.7 describes the weak form volume integral. One can see that it resembles the construction of the weak form differentiation matrix; in fact, this operation has the weak form differentiation embedded since this term represents the action of the weak form differentiation matrix on the vector $\mathbf{f}$. The element loop is required since we build all operators in an element-based approach. The k-loop is the integration loop. For each integration point, we need to interpolate the flux vector to the integration points - this is the role of the j-loop. The i-loop then loops over all the rows of the vector $R^{(e)}$ and performs the volume integration. The quadrature weights w_k are the two-dimensional quadrature weights with the corresponding volume Jacobian $J_k^{(e)}$. This completes the contribution of the volume integrals.

Algorithm 16.8 describes the flux integral. The steps of this algorithm also require a careful explanation. The l-loop goes over all the faces in the grid. Each face is associated with a left and right element (the superscripts (L) and (R)) with quadrature points $M_Q^{(l)}$ along the face (the k-loop).

Algorithm 16.7 Weak form volume integral contribution to $R^{(e)}$ with exact integration.

function R_VECTOR($\mathbf{f}$)
   $R^{(e)} = 0$
   for $e = 1 : N_e$ **do**   ▷ loop over elements
      for $k = 1 : M_Q$ **do**   ▷ loop over integration points
         $\mathbf{f}_k = 0$
         for $j = 1 : M_N$ **do**   ▷ interpolate to the integration points
            $\mathbf{f}_k + = \psi_{kj}\mathbf{f}_j$
         end for
         for $i = 1 : M_N$ **do**   ▷ loop over rows of $R^{(e)}$
            $R_i^{(e)} + = w_k J_k^{(e)} \boldsymbol{\nabla}\psi_{ik} \cdot \mathbf{f}_k^{(e)}$
         end for
      end for
   end for
end function

Algorithm 16.8 Flux integral contribution to $R^{(e)}$ with exact integration.

function R_VECTOR($\mathbf{f}$)
   for $l = 1 : N_F$ **do**   ▷ loop over faces
      for $k = 1 : M_Q^{(l)}$ **do**   ▷ loop over integration points along face l
         $\mathbf{f}_k^{(L)} = 0$
         $\mathbf{f}_k^{(R)} = 0$
         for $j = 1 : M_N$ **do**   ▷ interpolate to the integration points
            $\mathbf{f}_k^{(L)} + = \psi_{kj}\mathbf{f}_j^{(L)}$ ▷ left flux
            $\mathbf{f}_k^{(R)} + = \psi_{kj}\mathbf{f}_j^{(R)}$ ▷ right flux
         end for
         $\mathbf{f}_k^{(*)} = \frac{1}{2}\left[\mathbf{f}_k^{(L)} + \mathbf{f}_k^{(R)} - |\lambda|\hat{\mathbf{n}}_k^{(l)}\left(q_k^{(R)} - q_k^{(L)} \right) \right]$ ▷ build continuous flux
         for $i = 1 : M_N$ **do**   ▷ update $R^{(e)}$
            $R_i^{(L)} - = w_k^{(l)} J_k^{(l)}\psi_{ik}\left(\hat{\mathbf{n}}_k^{(l)} \cdot \mathbf{f}_k^{(*)} \right)$
            $R_i^{(R)} + = w_k^{(l)} J_k^{(l)}\psi_{ik}\left(\hat{\mathbf{n}}_k^{(l)} \cdot \mathbf{f}_k^{(*)} \right)$
         end for
      end for
   end for
end function

The j-loop is then used to interpolate the left and right element fluxes along the quadrature points $M_Q^{(l)}$ of the face l. Once these fluxes are computed they are then used to compute a numerical flux that is now continuous at the face l and contains a flux definition that is physical in some sense. In the example above we use the Rusanov flux although other fluxes can be used (see Ch. 18). Once this continuous numerical flux $\mathbf{f}^{(*)}$ is constructed it is then used to augment the value of the RHS vector for both the left and right elements. Note that the sign of the flux is opposite for both the left and right elements. This represents the fact that the flux leaving one element (negative sign) must be equal to the flux entering the neighboring element (positive sign). This concludes the contribution of the flux integral to the RHS vector.

Imposing Boundary Conditions

Note that a special concession needs to be considered in the last step of the algorithm in the case that the face l represents a physical boundary. In that case, there is only a contribution to the left element since there is no right element. In fact, if we need to impose boundary conditions we do so by introducing them via $\mathbf{f}^{(R)}$; e.g., Dirichlet boundary conditions or no-flux (hard wall) boundary conditions can be introduced this way. Neumann boundary conditions can also be introduced but this is done for second order operators where $\mathbf{f}$ represents a gradient operator.

Exercise Compute the complexity of Algs. 16.7 and 16.8. □

Exercise Algorithms 16.7 and 16.8 represent the solution algorithm for the weak form DG. Construct the analogous algorithms for the strong form DG. □

Let us now explore the simplification of the solution algorithms when using inexact integration. In Alg. 16.9 we see that via cardinality we can remove the k-

---

Algorithm 16.9 Weak form volume integral contribution to $R^{(e)}$ with inexact integration.

```
function R_VECTOR(f)
    R^(e) = 0
    for e = 1 : N_e do                                    ▷ loop over elements
        for j = 1 : M_N do                                ▷ loop over integration points
            for i = 1 : M_N do                            ▷ loop over rows of R^(e)
                R_i^(e) + = w_j J_j^(e) ∇ψ_{ij} · f_j^(e)
            end for
        end for
    end for
end function
```

---

loop and replace k by j everywhere to collapse the algorithm to the two M_N loops for each element e. The savings here is in not having to interpolate the flux vector $\mathbf{f}$ to the integration points since they are the same as the interpolation points.

Algorithm 16.10 shows the simplification to the flux terms which, in this case, is rather significant. Again, due to cardinality, we no longer need to interpolate $\mathbf{f}$ to the integration points (k-loop is replaced by j) and the rows of the RHS vector (i-loop) and the j-loop are the same resulting in only one loop (i-loop) for each face l of the grid. In Alg. 16.10 the term $\tau_i^{(e,l)}$ is one that we have not seen before. Recall that in Alg. 16.8 we used $M_Q^{(l)}$ quadrature points. These quadrature points and corresponding weights are non-zero only for those points that lie on the face l. Now that we have replaced the quadrature by M_N we must account for the fact that we are not using all M_N interpolation points of the element. This is so because the flux integral is a line integral and not an area integral (or what we have been calling volume integrals). Therefore the integer array τ is a *trace* array that is zero for all $i = 1, \ldots, M_N$ except for those points that lie on the face l. For example, consider

---

Algorithm 16.10 Flux integral contribution to $R^{(e)}$ with inexact integration.

---

function R_VECTOR($\mathbf{f}$)
   for $l = 1 : N_F$ **do**   ▷ loop over faces
      for $i = 1 : M_N$ **do**   ▷ loop over rows of $R^{(e)}$

$$\mathbf{f}_i^{(*)} = \tfrac{1}{2} \left[\mathbf{f}_i^{(L)} + \mathbf{f}_i^{(R)} - |\lambda| \hat{\mathbf{n}}_i^{(l)} \left(q_i^{(R)} - q_i^{(L)} \right) \right] \qquad \text{▷ build continuous flux}$$

$$R_i^{(L)} - = w_i^{(l)} J_i^{(l)} \left(\hat{\mathbf{n}}_i^{(l)} \cdot \mathbf{f}_i^{(*)} \right) \tau_i^{(e,l)}$$

$$R_i^{(R)} + = w_i^{(l)} J_i^{(l)} \left(\hat{\mathbf{n}}_i^{(l)} \cdot \mathbf{f}_i^{(*)} \right) \tau_i^{(e,l)}$$

      end for
   end for
end function

Fig. 16.1 for the element e. The trace array for a specific element e has four faces $l = 1, \ldots, 4$. For the linear polynomial case presented in Fig. 16.1 it also has four gridpoints $i = 1, \ldots, 4$. The values of the trace array are as follows: for the bottom face defined by the gridpoints $(1, 2)$ they are $\tau^{(e, l_1)} = (1, 1, 0, 0)$ while those for the right face defined by the gridpoints $(2, 4)$ are $\tau^{(e, l_2)} = (0, 1, 0, 1)$ and so on.

One final comment on the solution algorithm is in order. Recall that in Alg. 16.6 we first construct the element matrices and then use them to construct the RHS vector. In contrast, in Algs. 16.7 and 16.8 or 16.9 and 16.10 we construct the RHS vector directly without the need to build and store element matrices. We can identify a few advantages in NOT building and storing the element matrices. First, we can save on memory by not constructing matrices for each element. Second, we can use this approach for both linear and nonlinear problems. Third, if the grid changes (as in adaptive mesh refinement, see Ch. 19) this will not pose any additional complication to our solution approach. A final, although perhaps less obvious advantage, is that when we build the action of a matrix on a vector, it is equivalent to performing the matrix-vector operation in a banded matrix storage form. That is, we never visit elements of the matrix that are zero which means that the algorithm using this approach will have optimal complexity.

16.8.2 Tensor-Product Approach

We end this section by showing the advantage of using the tensor-product approach for decreasing the number of operations. We use the *sum factorization* method already described in Ch. 15. Furthermore, to make it simpler we only consider inexact integration.

Recall that the volume integral contribution comes from the term

$$\int_{\Omega_e} \nabla \psi_I \cdot \mathbf{f}_N^{(e)} d\Omega_e$$

which, in quadrature form becomes

$$\sum_{K=1}^{M_N} w_K^{(e)} J_K^{(e)} \nabla \psi_{I,K} \cdot \mathbf{f}_K^{(e)} \qquad (16.11)$$

where $I, K = 1, \ldots, M_N$. Let us now describe how to write the 2D basis function $\psi_{I,K}$ as the tensor-product of 1D basis function. Recall the following tensor-product definition

$$\psi_{I,K} = h_{i,k} \, h_{j,l}$$

where $i, j, k, l = 0, \ldots, N$. By this statement we mean the following

$$\psi_I(\boldsymbol{\xi}_K) = h_i(\xi_k) \, h_j(\eta_l).$$

where the indices are related as follows $I = (i+1)+j(N+1)$ and $K = (k+1)+l(N+1)$. Next, we compute $\frac{\partial \psi}{\partial x}$ and use the chain rule to rewrite the derivatives in terms of (ξ, η) as follows

$$\frac{\partial}{\partial x}\psi_{I,K} \equiv \frac{\partial}{\partial x}\left(h_i(\xi_k)h_j(\eta_l)\right) = \frac{\partial}{\partial \xi}\left(h_i(\xi_k)h_j(\eta_l)\right)\frac{\partial \xi_{k,l}}{\partial x} + \frac{\partial}{\partial \eta}\left(h_i(\xi_k)h_j(\eta_l)\right)\frac{\partial \eta_{k,l}}{\partial x}.$$

Using the chain rule results in the following x-derivative

$$\frac{\partial}{\partial x}\psi_{I,K} = \frac{dh_{i,k}}{d\xi}h_{j,l}\frac{\partial \xi_{k,l}}{\partial x} + h_{i,k}\frac{dh_{j,l}}{d\eta}\frac{\partial \eta_{k,l}}{\partial x}$$

with a similar expression for the y-derivative

$$\frac{\partial}{\partial y}\psi_{I,K} = \frac{dh_{i,k}}{d\xi}h_{j,l}\frac{\partial \xi_{k,l}}{\partial y} + h_{i,k}\frac{dh_{j,l}}{d\eta}\frac{\partial \eta_{k,l}}{\partial y}.$$

Using this information we can now rewrite Eq. (16.11) in tensor-product form as follows

$$\sum_{K=1}^{M_N} w_K^{(e)} J_K^{(e)} \nabla \psi_{I,K} \cdot \mathbf{f}_K^{(e)} = \sum_{k=0}^{N}\sum_{l=0}^{N} w_{k,l}^{(e)} J_{k,l}^{(e)} \left(\frac{dh_{i,k}}{d\xi}h_{j,l}\frac{\partial \xi_{k,l}}{\partial x} + h_{i,k}\frac{dh_{j,l}}{d\eta}\frac{\partial \eta_{k,l}}{\partial x} \right) f_{k,l}^{(e,x)}$$

$$+ \sum_{k=0}^{N}\sum_{l=0}^{N} w_{k,l}^{(e)} J_{k,l}^{(e)} \left(\frac{dh_{i,k}}{d\xi}h_{j,l}\frac{\partial \xi_{k,l}}{\partial y} + h_{i,k}\frac{dh_{j,l}}{d\eta}\frac{\partial \eta_{k,l}}{\partial y} \right) f_{k,l}^{(e,y)} \qquad (16.12)$$

where $\mathbf{f}^{(e)} = f^{(e,x)}\hat{\mathbf{i}} + f^{(e,y)}\hat{\mathbf{j}}$. Factoring the terms $\frac{dh_{i,k}}{d\xi} h_{j,l}$ and $h_{i,k}\frac{dh_{j,l}}{d\eta}$ we can rewrite Eq. (16.12) as follows

$$
\sum_{K=1}^{M_N} w_K^{(e)} J_K^{(e)} \nabla \psi_{I,K} \cdot \mathbf{f}_K^{(e)} = \sum_{k=0}^{N} \sum_{l=0}^{N} w_{k,l}^{(e)} J_{k,l}^{(e)} \frac{dh_{i,k}}{d\xi} h_{j,l} \left(\frac{\partial \xi_{k,l}}{\partial x} f_{k,l}^{(e,x)} + \frac{\partial \xi_{k,l}}{\partial y} f_{k,l}^{(e,y)} \right)
$$

$$
+ \sum_{k=0}^{N} \sum_{l=0}^{N} w_{k,l}^{(e)} J_{k,l}^{(e)} h_{i,k} \frac{dh_{j,l}}{d\eta} \left(\frac{\partial \eta_{k,l}}{\partial x} f_{k,l}^{(e,x)} + \frac{\partial \eta_{k,l}}{\partial y} f_{k,l}^{(e,y)} \right). \tag{16.13}
$$

Using cardinality we see that $h_{j,l}$ in the first term on the right-hand side of Eq. (16.13) is non-zero if and only if $j = l$ and in the second term $h_{i,k}$ is non-zero if and only if $i = k$ which means we can remove the l-loop and replace l with j from the first term and remove the k-loop from the second term and replace k with i to get

$$
\sum_{i=0}^{N} \sum_{j=0}^{N} \sum_{k=0}^{N} w_{k,j}^{(e)} J_{k,j}^{(e)} \frac{dh_{i,k}}{d\xi} h_{j,j} \left(\frac{\partial \xi_{k,j}}{\partial x} f_{k,j}^{(e,x)} + \frac{\partial \xi_{k,j}}{\partial y} f_{k,j}^{(e,y)} \right)
$$

$$
+ \sum_{i=0}^{N} \sum_{j=0}^{N} \sum_{l=0}^{N} w_{i,l}^{(e)} J_{i,l}^{(e)} h_{i,i} \frac{dh_{j,l}}{d\eta} \left(\frac{\partial \eta_{i,l}}{\partial x} f_{i,l}^{(e,x)} + \frac{\partial \eta_{i,l}}{\partial y} f_{i,l}^{(e,y)} \right). \tag{16.14}
$$

Swapping the indices $i \leftrightarrow k$ in the first term and $j \leftrightarrow l$ in the second allows us to rewrite Eq. (16.14) as follows

$$
\sum_{i=0}^{N} \sum_{j=0}^{N} \sum_{k=0}^{N} w_{i,j}^{(e)} J_{i,j}^{(e)} \frac{dh_{k,i}}{d\xi} h_{j,j} \left(\frac{\partial \xi_{i,j}}{\partial x} f_{i,j}^{(e,x)} + \frac{\partial \xi_{i,j}}{\partial y} f_{i,j}^{(e,y)} \right)
$$

$$
+ \sum_{i=0}^{N} \sum_{j=0}^{N} \sum_{l=0}^{N} w_{i,j}^{(e)} J_{i,j}^{(e)} h_{i,i} \frac{dh_{l,j}}{d\eta} \left(\frac{\partial \eta_{i,j}}{\partial x} f_{i,j}^{(e,x)} + \frac{\partial \eta_{i,j}}{\partial y} f_{i,j}^{(e,y)} \right). \tag{16.15}
$$

To write a streamlined algorithm we simply rename the variable l with k in the second term in Eq. (16.15) which results in

$$
\sum_{i=0}^{N} \sum_{j=0}^{N} \sum_{k=0}^{N} w_{i,j}^{(e)} J_{i,j}^{(e)} \frac{dh_{k,i}}{d\xi} h_{j,j} \left(\frac{\partial \xi_{i,j}}{\partial x} f_{i,j}^{(e,x)} + \frac{\partial \xi_{i,j}}{\partial y} f_{i,j}^{(e,y)} \right)
$$

$$
+ \sum_{i=0}^{N} \sum_{j=0}^{N} \sum_{k=0}^{N} w_{i,j}^{(e)} J_{i,j}^{(e)} h_{i,i} \frac{dh_{k,j}}{d\eta} \left(\frac{\partial \eta_{i,j}}{\partial x} f_{i,j}^{(e,x)} + \frac{\partial \eta_{i,j}}{\partial y} f_{i,j}^{(e,y)} \right). \tag{16.16}
$$

Equation (16.16) can now be used to construct the algorithm which we list in Alg. 16.11. Using cardinality, we can drop the terms $h_{j,j}$ in the first term and $h_{i,i}$ in the second. However, we leave them in Alg. 16.11 in order to make it clear why we are storing each of the sums in a separate location of the vector $R^{(e)}$.

Exercise Write the analog of Alg. 16.11 for the strong form differentiation.   □

Exercise Write the exact integration version of Alg. 16.11.   □

Algorithm 16.11 Weak form volume integral contribution to $R^{(e)}$ with inexact integration, tensor-products and sum factorization.

function R_VECTOR($\mathbf{f}$)
  $R^{(e)} = 0$
  for $e = 1 : N_e$ **do**
    for $i = 0 : N$ **do**
      for $j = 0 : N$ **do**
        for $k = 0 : N$ **do**
$$R^{(e)}_{kj} + = w_i w_j J^{(e)}_{ij} \frac{dh_{ki}}{d\xi} h_{j,j} \left(\frac{\partial \xi_{ij}}{\partial x} f^{(x)}_{ij} + \frac{\partial \xi_{ij}}{\partial y} f^{(y)}_{ij} \right) \qquad \triangleright \text{ derivative along } \xi$$
$$R^{(e)}_{ik} + = w_i w_j J^{(e)}_{ij} h_{i,i} \frac{dh_{kj}}{d\eta} \left(\frac{\partial \eta_{ij}}{\partial x} f^{(x)}_{ij} + \frac{\partial \eta_{ij}}{\partial y} f^{(y)}_{ij} \right) \qquad \triangleright \text{ derivative along } \eta$$
        end for
      end for
    end for
  end for
end function

Exercise Write the exact integration and strong form version of Alg. 16.11.   □

Let us now write the inexact integration tensor-product form of the flux integral. Algorithm 16.12 shows that the algorithm for the flux integral contribution is simple

Algorithm 16.12 Flux integral contribution to $R^{(e)}$ with inexact integration and tensor-products.

function R_VECTOR($\mathbf{f}$)
  for $l = 1 : N_f$ **do** $\triangleright$ loop over faces
    $p_L = face(1, l); p_R = face(2, l)$
    $L = face(3, l); R = face(4, l)$
    for $i = 0 : N$ **do** $\triangleright$ loop over certain rows of $R^{(e)}$
      $i_L = mapL(1, i, p_L); j_L = mapL(2, i, p_L)$
      $i_R = mapR(1, i, p_R); j_R = mapR(2, i, p_R)$
$$\mathbf{f}^{(*)}_i = \tfrac{1}{2} \left[\mathbf{f}^{(L)}_{i_L,j_L} + \mathbf{f}^{(R)}_{i_R,j_R} - |\lambda| \hat{\mathbf{n}}^{(l)}_i \left(q^{(R)}_{i_R,j_R} - q^{(L)}_{i_L,j_L} \right) \right] \qquad \triangleright \text{ Rusanov flux}$$
$$R^{(L)}_{i_L,j_L} - = w^{(l)}_i J^{(l)}_i \left(\hat{\mathbf{n}}^{(l)}_i \cdot \mathbf{f}^{(*)}_i \right)$$
$$R^{(R)}_{i_R,j_R} + = w^{(l)}_i J^{(l)}_i \left(\hat{\mathbf{n}}^{(l)}_i \cdot \mathbf{f}^{(*)}_i \right)$$
    end for
  end for
end function

- the only difficulty is in determining which value of $R^{(e)}$ is augmented by the flux values. This is the role of the integer arrays $face$, $mapL$ and $mapR$ [6]. The $face$ data structure has dimensions $face(1 : 4, 1 : N_f)$ where N_f are the number of faces. The $face$ data structure stores the following information: $p_L = face(1, l)$ stores the position of the face l within the element on the left side while $p_R = face(2, l)$ stores

---

[6] Routines to construct these data structures can be found in the folder "2d_data_structures" in the github site for this text.

this information for the right side. We can see such a configuration in Fig. 16.1 where the local gridpoint numbers $(1, 2)$ define the bottom face of e while $(3, 4)$ define the top face of e, etc. The face data structure also contains the following information $L = face(3, l)$ and $R = face(4, l)$ which are the indices of the left (L) and right (R) elements. The role of $mapL(1 : 2, l, p_L)$ is to point to the tensor-product indices of the points on the face in relation to the left element. It plays the role of the trace vector $\tau^{(l)}$ that we described previously. The data structure $mapR(1 : 2, l, p_R)$ performs the same role as $mapL$ but for the right element.

16.9 Example of 2D Hyperbolic Equation Problem

Let us now apply the DG algorithms we have constructed to solve the continuous first order differential equation

$$\frac{\partial q}{\partial t} + \mathbf{u} \cdot \nabla q = 0 \qquad \forall (x, y) \in [-1, 1]^2$$

where $q = q(x, y, t)$ and $\mathbf{u} = \mathbf{u}(x, y)$ with $\mathbf{u} = (u, v)^T$. Let the velocity field be

$$u(x, y) = y \qquad \text{and} \qquad v(x, y) = -x$$

which forms a velocity field that rotates fluid particles in a clockwise direction, which was solved with the CG method in Ch. 15. Recall that this velocity field satisfies the divergence-free condition

$$\nabla \cdot \mathbf{u} = 0.$$

In order to make this problem well-posed requires an initial condition which, here, we define as follows

$$q(x, y, 0) = \exp\left(-\sigma \left[(x - x_c)^2 + (y - y_c)^2\right]\right) \qquad (16.17)$$

where $(x_c, y_c) = (-0.5, 0)$ is the initial center of the Gaussian and the parameter $\sigma = 32$ controls its steepness. The analytic solution to this problem is given as

$$q(x, y, t) = q(x - ut, y - vt, 0)$$

where periodicity is enforced at all four boundaries. Also a Courant number of $C = 0.25$ is used for all the simulations with the SSP-RK3 time-integrator presented in Ch. 5.

16.9.1 Error Norm

We use the integral form of the L^2 norm given by Eq. (12.58).

16.9.2 Lobatto Points

In this section, we use Nth order Lobatto points for interpolation and the Lobatto points $Q = N$ (inexact) and $Q = N + 1$ (exact) for integration. In Sec. 16.9.3 we present results for Legendre points.

16.9.2.1 Solution Accuracy

Figure 16.2 shows the convergence rates for various polynomial orders, N, for a total number of gridpoints N_p where, for quadrilaterals, $N_p = (N_e^{(x)}(N+1))(N_e^{(y)}(N+1))$, where $N_e^{(s)}$ denotes the number of elements along the coordinate direction (s). Figure

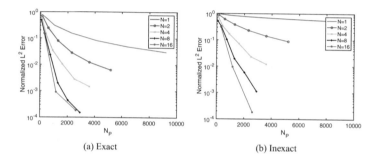

(a) Exact   (b) Inexact

Fig. 16.2: Convergence rates of DG (Lobatto Points) for the 2D hyperbolic equation for polynomial orders $N = 1, N = 2, N = 4, N = 8$, and $N = 16$ using a total number of gridpoints N_p for (a) exact integration ($Q = N + 1$) and (b) inexact integration ($Q = N$).

16.2 shows that there is little difference between exact and inexact integration in terms of their rates of convergence. The main differences occur for low values of N (≤ 4), however, as N increases (beyond $N = 8$) the difference between exact and inexact integration disappears.

16.9.2.2 Computational Cost of High-Order

Figure 16.3 shows the computational cost (wallclock time in seconds) for various polynomial orders. The exact integration results are obtained using the non-tensor-product form whereas the inexact integration results use the tensor-product form as well as sum factorization. In other words, this work-precision diagram represents the difference in performance in applying the DG method in a naive way (exact integration with non-tensor-products) compared to the most efficient way (inexact integration with tensor-products and sum factorization). Figure 16.3 shows that for

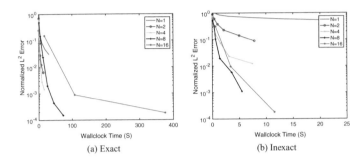

(a) Exact   (b) Inexact

Fig. 16.3: Convergence rates as a function of wallclock time (in seconds) of DG (Lobatto Points) for the 2D hyperbolic equation for polynomial orders $N = 1$, $N = 2$, $N = 4$, $N = 8$, and $N = 16$ using a total number of gridpoints N_p for (a) exact integration ($Q = N + 1$) and (b) inexact integration ($Q = N$).

exact integration, the highest order $N = 16$ is far too expensive. For exact integration, $N = 8$ achieves an error accuracy of 10^{-3} most efficiently, with $N = 4$ a close second. However, if we insist on more accuracy, then $N = 8$ easily wins. This conclusion is more obvious in the case of inexact integration where $N = 8$ and $N = 16$ win even at accuracies near 10^{-2}.

16.9.3 Legendre Points

Figure 16.4 shows the convergence rates with Lagrange polynomials using Legendre points for both interpolation and integration. Figure 16.4 shows that the Legendre points give level of accuracies somewhere between those for exact and inexact Lobatto points. However, the cost of the Legendre points is similar to that for inexact Lobatto points.

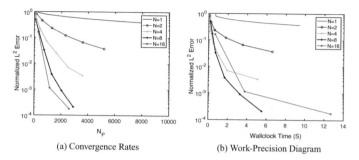

(a) Convergence Rates   (b) Work-Precision Diagram

Fig. 16.4: Convergence rates of DG (Legendre Points) for the 2D hyperbolic equation for polynomial orders $N = 1$, $N = 2$, $N = 4$, $N = 8$, and $N = 16$ using a total number of gridpoints N_p for (a) error versus number of points and (b) error versus wallclock time.

Chapter 17
2D Continuous/Discontinuous Galerkin Methods for Hyperbolic Equations

17.1 Introduction

In Chs. 5 and 6 we described the implementation of the CG and DG methods, respectively, for the solution of scalar hyperbolic equations in one dimension. Then in Ch. 7 we showed how to unify these two methods. In a similar fashion, we presented the CG and DG method for two-dimensional conservation laws in Chs. 15 and 16. In this chapter we describe how to unify the CG and DG methods for the solution of two-dimensional scalar balance laws using the *local element-wise storage* described in Ch. 7. This chapter is organized as follows: in Sec. 17.2 we describe the 2D advection-diffusion equations that we discretize in this chapter. Then in Sec. 17.3 we derive the weak integral form of the equation. This is followed by Sec. 17.4 where we describe the basis functions. This leads to Secs. 17.5 and 17.6 where we describe the discretized equations on a single element using both the weak and strong forms, respectively. The heart of the chapter is Sec. 17.7 where we describe the algorithms for constructing a unified CG/DG discretization of the 2D advection-diffusion equation. We end the chapter in Sec. 17.8 by solving a 2D wave equation with both the CG and DG methods in a unified way.

17.2 2D Advection-Diffusion Equation

Let us begin our discussion with the two-dimensional advection-diffusion equation written as follows

$$\frac{\partial q}{\partial t} + \nabla \cdot \mathbf{f} = \nabla \cdot (\nu \nabla q) \tag{17.1}$$

© The Editor(s) (if applicable) and The Author(s), under exclusive license
to Springer Nature Switzerland AG 2020
F. X. Giraldo, *An Introduction to Element-Based Galerkin Methods on
Tensor-Product Bases*, Texts in Computational Science and Engineering 24,
https://doi.org/10.1007/978-3-030-55069-1_17

where $\mathbf{f} = q\mathbf{u}$ and, for simplicity, assume that the velocity is divergence-free and that v is constant. Although Eq. (17.1) appears like a balance law, it can also be written as a conservation law if we let

$$\mathbf{F} = \mathbf{f} - v\nabla q$$

which then allows us to write

$$\frac{\partial q}{\partial t} + \nabla \cdot \mathbf{F} = 0.$$

In what follows, we construct unified CG/DG methods for the advection-diffusion equation in conservation form; the reasons for using the conservation form are discussed in Ch. 16. Let us now construct the integral form of the continuous partial differential equation.

17.3 Weak Integral Form

To construct the weak integral form, we begin by expanding the solution variable q as follows

$$q_N^{(e)}(\mathbf{x}, t) = \sum_{j=1}^{M_N} \psi_j(\mathbf{x}) q_j^{(e)}(t)$$

with

$$\mathbf{u}_N^{(e)}(\mathbf{x}) = \sum_{j=1}^{M_N} \psi_j(\mathbf{x}) \mathbf{u}_j^{(e)}$$

and $\mathbf{f}_N^{(e)} = q_N^{(e)} \mathbf{u}_N^{(e)}$ where $\mathbf{u}(\mathbf{x}) = u(\mathbf{x})\hat{\mathbf{i}} + v(\mathbf{x})\hat{\mathbf{j}}$ is the velocity vector with u and v components along $\hat{\mathbf{i}}$ (x-direction) and $\hat{\mathbf{j}}$ (y-direction). The integer M_N is a function of N which determines the number of points inside each element Ω_e. Recall that we have used this idea in all the chapters describing 2D problems such as Chs. 12 through 14 on elliptic problems and Chs. 15 and 16 on hyperbolic problems.

Next, we substitute $q_N^{(e)}$, $\mathbf{u}_N^{(e)}$, and $\mathbf{f}_N^{(e)}$ into the PDE, multiply by a test function and integrate within the local domain Ω_e yielding the weak integral form: find $q \in S$ $\forall \psi \in S$ such that

$$\int_{\Omega_e} \psi_i \frac{\partial q_N^{(e)}}{\partial t} d\Omega_e + \int_{\Gamma_e} \psi_i \hat{\mathbf{n}} \cdot \mathbf{f}_N^{(*)} d\Gamma_e - \int_{\Omega_e} \nabla\psi_i \cdot \mathbf{f}_N^{(e)} d\Omega_e = \qquad (17.2)$$

$$v \int_{\Gamma_e} \psi_i \hat{\mathbf{n}} \cdot \nabla q_N^{(*)} d\Gamma_e + v \int_{\Gamma_e} \hat{\mathbf{n}} \cdot \nabla\psi_i \left[q_N^{(e)} - q_N^{(*)} \right] d\Gamma_e - v \int_{\Omega_e} \nabla\psi_i \cdot \nabla q_N^{(e)} d\Omega_e$$

with $i = 1, \dots, M_N$ and $e = 1, \dots, N_e$, where $\mathbf{f}^{(*)}$ is the numerical flux function which we take as the Rusanov flux

$$\mathbf{f}^{(*)} = \left\{ \mathbf{f}^{(e,l)} \right\} - \frac{1}{2} \hat{\mathbf{n}} |\lambda| [\![q^{(e,l)}]\!]$$

and, using the SIPG method from Ch. 14 we write

$$q_N^{(*)} = \left\{ q^{(e,l)} \right\}$$

and

$$\nabla q_N^{(*)} = \left\{ \nabla q^{(e,l)} \right\} - \hat{\mathbf{n}} \mu [\![q^{(e,l)}]\!]$$

where the superscript (l) denotes the face neighbor of (e), $\hat{\mathbf{n}}$ is the unit normal vector of the face shared by elements e and l, λ is the maximum wave speed of the system and μ is the penalty term defined in Ch. 14. The function space S is a placeholder and can be either H^1 for the CG method or L^2 for DG as described in Ch. 2.

17.4 2D Basis Functions and the Reference Element

Recall from Chs. 15 and 16 that the 2D basis functions on the reference element are constructed from the tensor-product of the 1D basis functions as follows

$$\psi_i(\xi, \eta) = h_j(\xi) \otimes h_k(\eta)$$

where h are the 1D basis functions, $\otimes$ is the tensor-product operator, and the 1D indices vary as follows $j, k = 0, ..., N$ with the 2D index varying as $i = 1, ..., M_N$ where $M_N = (N + 1)^2$. To get from the 1D local indices (j, k) to the 2D local index i requires the mapping $i = j + 1 + k(N + 1)$. In order to construct the derivatives of the solution vector and basis functions, we follow the discussion in Sec. 16.4.

17.5 Element Equations on a Single Element: Weak Form

Equation (17.2) represents the element equation that we need to solve. At this point we must decide whether to evaluate the integrals exactly or not, and whether we will construct element matrices or just form the RHS vector. Let us begin with exact integration and address the other choices momentarily.

17.5.1 Matrix-Vector Problem Resulting from Exact Integration

Let us now consider the differences in the resulting matrices when we evaluate the integrals using co-located quadrature (i.e., inexact integration) or non-co-located quadrature (that, for a specific number of quadrature points, results in exact integration). Let us first consider exact integration. Substituting the SIPG numerical fluxes from Ch. 14 into Eq. (17.2) yields the following integral form

$$
\int_{\Omega_e} \psi_i \frac{\partial q_N^{(e)}}{\partial t} \, d\Omega_e + \int_{\Gamma_e} \psi_i \hat{\mathbf{n}} \cdot \mathbf{f}_N^{(*)} \, d\Gamma_e - \int_{\Omega_e} \boldsymbol{\nabla} \psi_i \cdot \mathbf{f}_N^{(e)} \, d\Omega_e = \qquad (17.3)
$$

$$
\nu \int_{\Gamma_e} \psi_i \hat{\mathbf{n}} \cdot \boldsymbol{\nabla} \psi_j \, d\Gamma_e \left\{ q^{(e,l)} \right\}_j - \nu \int_{\Gamma_e} \mu \psi_i \psi_j \, d\Gamma_e [\![q^{(e,l)}]\!]_j
$$

$$
+ \; \nu \int_{\Gamma_e} \hat{\mathbf{n}} \cdot \boldsymbol{\nabla} \psi_i \psi_j \, d\Gamma_e \left(q^{(e)} - \left\{ q^{(e,l)} \right\} \right)_j - \nu \int_{\Omega_e} \boldsymbol{\nabla} \psi_i \cdot \boldsymbol{\nabla} q_N^{(e)} \, d\Omega_e.
$$

From Eq. (17.3) we can now write the matrix-vector problem as follows

$$
M_{ij}^{(e)} \frac{dq_j^{(e)}}{dt} + \sum_{l=1}^{N_{FN}} \left(\mathbf{F}_{ij}^{(e,l)} \right)^T \mathbf{f}_j^{(*,l)} - \left(\tilde{\mathbf{D}}_{ij}^{(e)} \right)^T \mathbf{f}_j^{(e)} =
$$

$$
\nu \sum_{l=1}^{N_{FN}} \left[\bar{F}_{ij}^{(e,l)} \left\{ q^{(e,l)} \right\}_j - \bar{\bar{F}}_{ij}^{(e,l)} [\![q^{(e,l)}]\!]_j + \bar{\bar{\bar{F}}}_{ij}^{(e,l)} \left(q^{(e)} - \left\{ q^{(e,l)} \right\} \right)_j \right]
$$

$$
- \; \nu L_{ij}^{(e)} q_j^{(e)} \qquad\qquad\qquad\qquad\qquad\qquad\qquad\qquad\qquad\qquad (17.4)
$$

with $i, j = 1, \ldots, M_N$ and $e = 1, \ldots, N_e$, where the mass matrix is

$$
M_{ij}^{(e)} \equiv \int_{\Omega_e} \psi_i \psi_j \, d\Omega_e = \sum_{k=1}^{M_Q} w_k J_k^{(e)} \psi_{ik} \psi_{jk},
$$

the flux matrix for the first order operator is

$$
\mathbf{F}_{ij}^{(e,l)} \equiv \int_{\Gamma_e} \psi_i \psi_j \hat{\mathbf{n}}^{(e,l)} \, d\Gamma_e = \sum_{k=0}^{Q} w_k^{(l)} J_k^{(l)} \psi_{ik} \psi_{jk} \hat{\mathbf{n}}_k^{(e,l)},
$$

the weak form differentiation matrix is

$$
\tilde{\mathbf{D}}_{ij}^{(e)} \equiv \int_{\Omega_e} \boldsymbol{\nabla} \psi_i \psi_j \, d\Omega_e = \sum_{k=1}^{M_Q} w_k J_k^{(e)} \boldsymbol{\nabla} \psi_{ik} \psi_{jk},
$$

the first SIPG flux matrix is

$$
\bar{F}_{ij}^{(e,l)} \equiv \int_{\Gamma_e} \psi_i \hat{\mathbf{n}}^{(e,l)} \cdot \boldsymbol{\nabla} \psi_j \, d\Gamma_e = \sum_{k=0}^{Q} w_k^{(l)} J_k^{(l)} \psi_{ik} \hat{\mathbf{n}}_k^{(e,l)} \cdot \boldsymbol{\nabla} \psi_{jk},
$$

the second SIPG flux matrix is

$$
\bar{\bar{F}}_{ij}^{(e,l)} \equiv \int_{\Gamma_e} \mu \psi_i \psi_j \, d\Gamma_e = \sum_{k=0}^{Q} w_k^{(l)} J_k^{(l)} \mu_k \psi_{ik} \psi_{jk},
$$

the third SIPG flux matrix is

$$\bar{\bar{F}}_{ij}^{(e,l)} \equiv \int_{\Gamma_e} \hat{\mathbf{n}}^{(e,l)} \cdot \boldsymbol{\nabla}\psi_i \psi_j \, d\Gamma_e = \sum_{k=0}^{Q} w_k^{(l)} J_k^{(l)} \hat{\mathbf{n}}_k^{(e,l)} \cdot \boldsymbol{\nabla}\psi_{ik} \psi_{jk},$$

and the Laplacian matrix is

$$L_{ij}^{(e)} \equiv \int_{\Omega_e} \boldsymbol{\nabla}\psi_i \cdot \boldsymbol{\nabla}\psi_j \, d\Omega_e = \sum_{k=1}^{M_Q} w_k J_k^{(e)} \boldsymbol{\nabla}\psi_{ik} \cdot \boldsymbol{\nabla}\psi_{jk},$$

which are described in Chs. 12 and 14. In fact, we have seen all of these matrices before, in one form or another[1].

17.5.2 Matrix-Vector Problem Resulting from Inexact Integration

Let us now use inexact integration and see what the resulting matrix-vector problem looks like. In this case, we let $M_Q \equiv (Q+1)^2 = M_N$ where $Q = N$ to get the following element matrices. The mass matrix simplifies to

$$M_{ij}^{(e)} = w_i J_i^{(e)} \delta_{ij},$$

the flux matrix for the first order operator becomes

$$\mathbf{F}_{ij}^{(e,l)} = w_i^{(l)} J_i^{(l)} \hat{\mathbf{n}}_i^{(e,l)} \delta_{ij},$$

the weak form differentiation matrix is

$$\tilde{\mathbf{D}}_{ij}^{(e)} = \sum_{k=1}^{M_N} w_k J_k^{(e)} \boldsymbol{\nabla}\psi_{ik} \psi_{jk},$$

and the SIPG flux matrices become

$$\bar{F}_{ij}^{(e,l)} = w_i^{(l)} J_i^{(l)} \hat{\mathbf{n}}_i^{(e,l)} \cdot \boldsymbol{\nabla}\psi_{ji},$$

$$\bar{\bar{F}}_{ij}^{(e,l)} = w_i^{(l)} J_i^{(l)} \mu_i \delta_{ij},$$

$$\bar{\bar{\bar{F}}}_{ij}^{(e,l)} = \sum_{j=0}^{N} w_j^{(l)} J_j^{(l)} \hat{\mathbf{n}}_j^{(e,l)} \cdot \boldsymbol{\nabla}\psi_{ij},$$

and the Laplacian matrix remains unchanged

$$L_{ij}^{(e)} = \sum_{k=1}^{M_N} w_k J_k^{(e)} \boldsymbol{\nabla}\psi_{ik} \cdot \boldsymbol{\nabla}\psi_{jk}.$$

---

[1] The only exceptions are the three SIPG flux matrices which we did not write out explicitly in this form in Ch. 14 but, instead, presented it in algorithmic form.

17.6 Element Equations on a Single Element: Strong Form

Let us now explore the strong form of the unified CG/DG method. To construct the strong form, we need to apply integration by parts twice, but only to the first order operator, which results in

$$\int_{\Omega_e} \psi_i \nabla \cdot \mathbf{f}_N^{(e)} \, d\Omega_e = \int_{\Gamma_e} \psi_i \mathbf{n} \cdot \left(\mathbf{f}_N^{(*,l)} - \mathbf{f}_N^{(e)} \right) d\Gamma_e + \int_{\Omega_e} \psi_i \nabla \cdot \mathbf{f}_N^{(e)} \, d\Omega_e.$$

Substituting this expression into Eq. (17.3) yields the strong integral form: find $q \in S$ $\forall \psi \in S$ such that

$$\int_{\Omega_e} \psi_i \frac{\partial q_N^{(e)}}{\partial t} \, d\Omega_e + \int_{\Gamma_e} \psi_i \mathbf{n} \cdot \left(\mathbf{f}_N^{(*,l)} - \mathbf{f}_N^{(e)} \right) d\Gamma_e + \int_{\Omega_e} \psi_i \nabla \cdot \mathbf{f}_N^{(e)} \, d\Omega_e =$$

$$\nu \int_{\Gamma_e} \psi_i \hat{\mathbf{n}} \cdot \nabla \psi_j \, d\Gamma_e \left\{ q^{(e,l)} \right\}_j - \nu \int_{\Gamma_e} \mu \psi_i \psi_j \, d\Gamma_e [\![q^{(e,l)}]\!]_j \qquad (17.5)$$

$$+ \nu \int_{\Gamma_e} \hat{\mathbf{n}} \cdot \nabla \psi_i \psi_j \, d\Gamma_e \left(q^{(e)} - \left\{ q^{(e,l)} \right\} \right)_j - \nu \int_{\Omega_e} \nabla \psi_i \cdot \nabla q_N^{(e)} \, d\Omega_e$$

where the second order terms remain unchanged.

17.6.1 Matrix-Vector Problem Resulting from Exact Integration

Exact quadrature yields the following matrix-vector problem

$$M_{ij}^{(e)} \frac{dq_j^{(e)}}{dt} + \sum_{l=1}^{N_{FN}} \left(\mathbf{F}_{ij}^{(e,l)} \right)^T \left(\mathbf{f}^{(*,l)} - \mathbf{f}^{(e)} \right)_j + \left(\mathbf{D}_{ij}^{(e)} \right)^T \mathbf{f}_j^{(e)} =$$

$$\nu \sum_{l=1}^{N_{FN}} \left[\bar{F}_{ij}^{(e,l)} \left\{ q^{(e,l)} \right\}_j - \bar{\bar{F}}_{ij}^{(e,l)} [\![q^{(e,l)}]\!]_j + \bar{\bar{\bar{F}}}_{ij}^{(e,l)} \left(q^{(e)} - \left\{ q^{(e,l)} \right\} \right)_j \right]$$

$$- \nu L_{ij}^{(e)} q_j^{(e)} \qquad (17.6)$$

where the only matrix that changes when going from weak to strong form is the differentiation matrix

$$\mathbf{D}_{ij}^{(e)} = \int_{\Omega_e} \psi_i \nabla \psi_j \, d\Omega_e = \sum_{k=1}^{M_Q} w_k J_k^{(e)} \psi_{ik} \nabla \psi_{jk},$$

which is now the strong form differentiation matrix.

17.6.2 Matrix-Vector Problem Resulting from Inexact Integration

Using inexact (co-located) Lobatto integration ($Q = N$) simplifies the matrices in Eq. (17.6) as follows: the mass matrix simplifies to

$$M_{ij}^{(e)} = w_i J_i^{(e)} \delta_{ij},$$

the first order flux matrix becomes

$$\mathbf{F}_{ij}^{(e,l)} = w_i^{(l)} J_i^{(l)} \hat{\mathbf{n}}_i^{(e,l)} \delta_{ij},$$

and the differentiation matrix is

$$\mathbf{D}_{ij}^{(e)} = w_i J_i^{(e)} \boldsymbol{\nabla} \psi_{ji},$$

while the matrices corresponding to the second order term remain identical to the weak form inexact integration which we write below for completeness

$$\bar{F}_{ij}^{(e,l)} = w_i^{(l)} J_i^{(l)} \hat{\mathbf{n}}_i^{(e,l)} \cdot \boldsymbol{\nabla} \psi_{ji},$$

$$\bar{\bar{F}}_{ij}^{(e,l)} = w_i^{(l)} J_i^{(l)} \mu_i \delta_{ij},$$

$$\bar{\bar{\bar{F}}}_{ij}^{(e,l)} = \sum_{j=0}^{N} w_j^{(l)} J_j^{(l)} \hat{\mathbf{n}}_j^{(e,l)} \cdot \boldsymbol{\nabla} \psi_{ij},$$

and

$$L_{ij}^{(e)} = \sum_{k=1}^{M_N} w_k J_k^{(e)} \boldsymbol{\nabla} \psi_{ik} \cdot \boldsymbol{\nabla} \psi_{jk}.$$

Substituting these matrices into the element equation (17.6) yields

$$w_i J_i^{(e)} \frac{dq_i^{(e)}}{dt} + \sum_{l=1}^{N_{FN}} w_i^{(l)} J_i^{(l)} \hat{\mathbf{n}}_i^{(e,l)} \cdot \left(\mathbf{f}^{(*,l)} - \mathbf{f}^{(e)} \right)_i + w_i J_i^{(e)} \boldsymbol{\nabla} \psi_{ji} \cdot \mathbf{f}_j^{(e)} =$$

$$\nu \sum_{l=1}^{N_{FN}} \left[w_i^{(l)} J_i^{(l)} \hat{\mathbf{n}}_i^{(e,l)} \cdot \boldsymbol{\nabla} \psi_{ji} \left\{ q^{(e,l)} \right\}_j - w_i^{(l)} J_i^{(l)} \mu_i [\![q^{(e,l)}]\!]_i \right.$$

$$\left. + \sum_{j=0}^{N} w_j^{(l)} J_j^{(l)} \hat{\mathbf{n}}_j^{(e,l)} \cdot \boldsymbol{\nabla} \psi_{ij} \left(q^{(e)} - \left\{ q^{(e,l)} \right\} \right)_j \right]$$

$$- \nu \sum_{k=1}^{M_N} w_k J_k^{(e)} \boldsymbol{\nabla} \psi_{ik} \cdot \boldsymbol{\nabla} \psi_{jk} q_j^{(e)} \qquad (17.7)$$

which, as in Ch. 16, looks like a high-order finite difference method and hence is rather simple to code.

17.7 Algorithms for a Unified CG/DG Matrix-Vector Problem

Let us now describe the algorithms for constructing the matrix-vector problem given by Eq. (17.7). We only focus on the tensor-product approach which we introduced in Chs. 15 and 16 and only consider the strong form with inexact integration on Lobatto points. The reader should be readily able to extend the algorithms below to use either weak form or non-tensor-products. We only consider nodal functions with Lobatto points because, although DG can use a variety of basis functions and interpolation points, CG can only use nodal functions with Lobatto points since it needs the points along the boundary in order to enforce continuity via DSS as described in Ch. 12.

17.7.1 First Order Operator

We begin with the description of the algorithms for constructing the first order operator. First we discuss the volume integral term, followed by the flux integral term.

17.7.1.1 Volume Integral Term

Using the sum factorization approach for tensor-product basis functions described in Secs. 15.7.2 and 16.8.2, we can write the algorithm for the strong form volume integral contribution. Let us begin with the volume integral term written as follows

$$\int_{\Omega_e} \psi_I \nabla \cdot \mathbf{f}_N \, d\Omega_e \approx \sum_{K=1}^{M_N} w_K J_K^{(e)} \psi_{I,K} (\nabla \cdot \mathbf{f})_K = w_I J_I^{(e)} \psi_{I,I} (\nabla \cdot \mathbf{f})_I$$

where $I, K = 1, \ldots, M_N$ and have used cardinality to remove the sum. Note that in general we may write

$$(\nabla \cdot \mathbf{f})_K = \sum_{i=1}^{M_N} \nabla \psi_{I,K} \cdot \mathbf{f}_I.$$

Replacing the monolithic basis functions with tensor-products as follows

$$\psi_{I,K} = h_{i,k}(\xi) h_{j,l}(\eta)$$

where $i, j, k, l = 0, \ldots, N$, and writing the flux vector $\mathbf{f} = f^{(x)}\hat{\mathbf{i}} + f^{(y)}\hat{\mathbf{j}}$ allows us to write the following

$$w_I J_I^{(e)} \psi_{I,I} (\nabla \cdot \mathbf{f})_I = w_i w_j J_{i,j}^{(e)} \left(\frac{\partial \xi_{i,j}}{\partial x} \frac{dh_{k,i}}{d\xi} h_{l,j}(\eta) + \frac{\partial \eta_{i,j}}{\partial x} h_{k,i}(\xi) \frac{dh_{l,j}}{d\eta} \right) f_{k,l}^{(x)}$$

$$+ w_i w_j J_{i,j}^{(e)} \left(\frac{\partial \xi_{i,j}}{\partial x} \frac{dh_{k,i}}{d\xi} h_{l,j}(\eta) + \frac{\partial \eta_{i,j}}{\partial x} h_{k,i}(\xi) \frac{dh_{l,j}}{d\eta} \right) f_{k,l}^{(x)}.$$

Using cardinality we can simplify the volume integral as follows

$$
w_I J_I^{(e)} \psi_{I,I} \left(\nabla \cdot \mathbf{f} \right)_I = w_i w_j J_{i,j}^{(e)} \left(\frac{\partial \xi_{i,j}}{\partial x} \frac{dh_{k,i}}{d\xi} f_{k,j}^{(x)} + \frac{\partial \eta_{i,j}}{\partial x} \frac{dh_{l,j}}{d\eta} f_{i,l}^{(x)} \right)
$$
$$
+ w_i w_j J_{i,j}^{(e)} \left(\frac{\partial \xi_{i,j}}{\partial x} \frac{dh_{k,i}}{d\xi} f_{k,j}^{(y)} + \frac{\partial \eta_{i,j}}{\partial x} \frac{dh_{l,j}}{d\eta} f_{i,l}^{(x)} \right).
$$

Swapping l for k yields the final form

$$
w_I J_I^{(e)} \psi_{I,I} \left(\nabla \cdot \mathbf{f} \right)_I = w_i w_j J_{i,j}^{(e)} \left(\frac{\partial \xi_{i,j}}{\partial x} \frac{dh_{k,i}}{d\xi} f_{k,j}^{(x)} + \frac{\partial \eta_{i,j}}{\partial x} \frac{dh_{k,j}}{d\eta} f_{i,k}^{(x)} \right)
$$
$$
+ w_i w_j J_{i,j}^{(e)} \left(\frac{\partial \xi_{i,j}}{\partial x} \frac{dh_{k,i}}{d\xi} f_{k,j}^{(y)} + \frac{\partial \eta_{i,j}}{\partial x} \frac{dh_{k,j}}{d\eta} f_{i,k}^{(x)} \right)
$$

which we present in Alg. 17.1.

Algorithm 17.1 First order operator volume integral contribution to $R^{(e)}$ with strong form, inexact integration, tensor-products, and sum factorization.

---

function R_VECTOR($\mathbf{f}$)
  $R^{(e)} = 0$
  for $e = 1 : N_e$ **do**   ▷ loop over elements
    for $i = 0 : N$ **do**   ▷ loop over integration points
      for $j = 0 : N$ **do**
        $I = (i + 1) + j(N + 1)$ ▷ tensor-product to non-tensor-product map
        $\frac{\partial \mathbf{f}}{\partial \xi} = 0; \frac{\partial \mathbf{f}}{\partial \eta} = 0$
        for $k = 0 : N$ **do**   ▷ build derivatives in (ξ, η) space
          $\frac{\partial \mathbf{f}}{\partial \xi} += \frac{dh_{k,i}}{d\xi} \mathbf{f}_{k,j}$
          $\frac{\partial \mathbf{f}}{\partial \eta} += \frac{dh_{k,j}}{d\xi} \mathbf{f}_{i,k}$
        end for
        $\frac{\partial f^{(x)}}{\partial x} = \frac{\partial f^{(x)}}{\partial \xi} \frac{\partial \xi_{i,j}}{\partial x} + \frac{\partial f^{(x)}}{\partial \eta} \frac{\partial \eta_{i,j}}{\partial x}$ ▷ build derivatives in (x, y) space
        $\frac{\partial f^{(y)}}{\partial y} = \frac{\partial f^{(y)}}{\partial \xi} \frac{\partial \xi_{i,j}}{\partial y} + \frac{\partial f^{(y)}}{\partial \eta} \frac{\partial \eta_{i,j}}{\partial y}$
        $R_I^{(e)} = -w_i w_j J_{i,j}^{(e)} \left(\frac{\partial f^{(x)}}{\partial x} + \frac{\partial f^{(y)}}{\partial y} \right)$
      end for
    end for
  end for
end function

Exercise Write the analog of Alg. 17.1 for the weak form. You can check your answer with the result presented in Alg. 16.11.   □

Exercise Write the exact integration version of Alg. 17.1.   □

17.7.1.4 Flux Integral Term

Let us now write the tensor-product form of the flux integral with inexact integration using Lobatto points. Algorithm 17.2 highlights the main steps in constructing the strong form flux integral contribution. Note that this algorithm is almost identical to the weak form flux integral presented in Alg. 16.12. In fact, comparing these two algorithms we note that we can write a unified code for the flux integral contributions for both the weak and strong forms. This is the role of the switch δ_{strong} which yields the weak form for $\delta_{\text{strong}} = 0$ and the strong form for $\delta_{\text{strong}} = 1$. Note that Alg. 17.2 makes use of the face data structures $face$, $mapL$ and $mapR$ described in Ch. 13. This concludes the description of the algorithms for the first row of Eq. (17.7).

Algorithm 17.2 First order operator flux integral contribution to $R^{(e)}$ with strong form, inexact integration, and tensor-products.

---

function R_VECTOR($\mathbf{f}$)
    for $l = 1 : N_f$ **do**   ▷ loop over faces
        $p_L = face(1, l)$; $p_R = face(2, l)$
        $L = face(3, l)$; $R = face(4, l)$
        for $i = 0 : N$ **do**   ▷ loop over certain rows of $R^{(e)}$
            $i_L = mapL(1, i, p_L)$; $j_L = mapL(2, i, p_L)$; $I_L = (i_L + 1) + j_L(N + 1)$
            $i_R = mapR(1, i, p_R)$; $j_R = mapR(2, i, p_R)$; $I_R = (i_R + 1) + j_R(N + 1)$
            $\mathbf{f}_i^{(*)} = \frac{1}{2}\left[\mathbf{f}_{i_L,j_L}^{(L)} + \mathbf{f}_{i_R,j_R}^{(R)} - |\lambda|\hat{\mathbf{n}}_i^{(l)}\left(q_{i_R,j_R}^{(R)} - q_{i_L,j_L}^{(L)}\right)\right]$ ▷ Rusanov flux
            $R_{I_L}^{(L)} - = w_i^{(l)} J_i^{(l)}\left[\hat{\mathbf{n}}_i^{(l)} \cdot \left(\mathbf{f}_i^{(*)} - \delta_{\text{strong}}\mathbf{f}_i^{(L)}\right)\right]$
            $R_{I_R}^{(R)} + = w_i^{(l)} J_i^{(l)}\left[\hat{\mathbf{n}}_i^{(l)} \cdot \left(\mathbf{f}_i^{(*)} - \delta_{\text{strong}}\mathbf{f}_i^{(R)}\right)\right]$
        end for
    end for
end function

17.7.2 Second Order Operator

Let us now describe the algorithms required for constructing the second order operator. As before, we first discuss the volume integral term and then the flux integral term.

17.7.2.1 Volume Integral Term

The volume integral for the second order term involves the third row of Eq. (17.7). We already discussed this algorithm in Ch. 14 but only in the context of the monolithic (non-tensor-product) form. This term is the Laplacian operator

$$\int_{\Omega_e} \nabla \psi_I \cdot \nabla q_N \, d\Omega_e \approx \sum_{K=1}^{M_N} w_K J_K^{(e)} \nabla \psi_{I,K} \cdot \nabla q_K$$

where

$$\nabla \psi_{I,K} = \frac{\partial \psi_{I,K}}{\partial x} \hat{\mathbf{i}} + \frac{\partial \psi_{I,K}}{\partial y} \hat{\mathbf{j}},$$

with

$$\frac{\partial \psi_{I,K}}{\partial x} = \frac{\partial \psi_{I,K}}{\partial \xi} \frac{\partial \xi_K}{\partial x} + \frac{\partial \psi_{I,K}}{\partial \eta} \frac{\partial \eta_K}{\partial x}$$

and

$$\frac{\partial \psi_{I,K}}{\partial y} = \frac{\partial \psi_{I,K}}{\partial \xi} \frac{\partial \xi_K}{\partial y} + \frac{\partial \psi_{I,K}}{\partial \eta} \frac{\partial \eta_K}{\partial y}.$$

Using tensor-product bases we note that

$$\frac{\partial \psi_{I,K}}{\partial \xi} = \frac{\partial}{\partial \xi} \left(h_{i,k}(\xi) h_{j,l}(\eta) \right)$$

$$= \frac{dh_{i,k}(\xi)}{d\xi} h_{j,l}(\eta)$$

and

$$\frac{\partial \psi_{I,K}}{\partial \eta} = \frac{\partial}{\partial \eta} \left(h_{i,k}(\xi) h_{j,l}(\eta) \right)$$

$$= h_{i,k}(\xi) \frac{dh_{j,l}(\eta)}{d\eta}.$$

These expressions now allow us to write

$$\frac{\partial \psi_{I,K}}{\partial x} = \frac{dh_{i,k}(\xi)}{d\xi} h_{j,l}(\eta) \frac{\partial \xi_{k,l}}{\partial x} + h_{i,k}(\xi) \frac{dh_{j,l}(\eta)}{d\eta} \frac{\partial \eta_{k,l}}{\partial x}$$

and, using cardinality, we write

$$\frac{\partial \psi_{I,K}}{\partial x} = \frac{dh_{i,k}(\xi)}{d\xi} \frac{\partial \xi_{k,l}}{\partial x} + \frac{dh_{j,l}(\eta)}{d\eta} \frac{\partial \eta_{k,l}}{\partial x}.$$

Replacing j with i in the second term yields

$$\frac{\partial \psi_{I,K}}{\partial x} = \frac{dh_{i,k}(\xi)}{d\xi} \frac{\partial \xi_{k,l}}{\partial x} + \frac{dh_{i,l}(\eta)}{d\eta} \frac{\partial \eta_{k,l}}{\partial x}.$$

A similar expression can be found for the y-derivative

$$\frac{\partial \psi_{I,K}}{\partial y} = \frac{dh_{i,k}(\xi)}{d\xi} \frac{\partial \xi_{k,l}}{\partial y} + \frac{dh_{i,l}(\eta)}{d\eta} \frac{\partial \eta_{k,l}}{\partial y}$$

which completes the definition of $\nabla \psi_{I,K}$. We now need to define ∇q_K.

Let us write

$$\nabla q_K = \frac{\partial q_K}{\partial x}\hat{\mathbf{i}} + \frac{\partial q_K}{\partial y}\hat{\mathbf{j}}$$

where

$$\frac{\partial q_K}{\partial x} = \sum_{i=0}^{N}\sum_{j=0}^{N}\left(\frac{dh_{i,k}(\xi)}{d\xi}h_{j,l}(\eta)\frac{\partial \xi_{k,l}}{\partial x} + h_{i,k}(\xi)\frac{dh_{j,l}(\eta)}{d\eta}\frac{\partial \eta_{k,l}}{\partial x}\right)q_{i,j}$$

$$= \sum_{i=0}^{N}\frac{dh_{i,k}(\xi)}{d\xi}\frac{\partial \xi_{k,l}}{\partial x}q_{i,l} + \sum_{j=0}^{N}\frac{dh_{j,l}(\eta)}{d\eta}\frac{\partial \eta_{k,l}}{\partial x}q_{k,j}$$

where we used cardinality in the last step in order to remove one of the sums.
Replacing j with i in the second term yields

$$\frac{\partial q_K}{\partial x} \equiv \frac{\partial q_{k,l}}{\partial x} = \sum_{i=0}^{N}\left[\frac{dh_{i,k}(\xi)}{d\xi}\frac{\partial \xi_{k,l}}{\partial x}q_{i,l} + \frac{dh_{i,l}(\eta)}{d\eta}\frac{\partial \eta_{k,l}}{\partial x}q_{k,i}\right]$$

$$= \frac{\partial q_{k,l}}{\partial \xi}\frac{\partial \xi_{k,l}}{\partial x} + \frac{\partial q_{k,l}}{\partial \eta}\frac{\partial \eta_{k,l}}{\partial x} \tag{17.8}$$

with the following y-derivative

$$\frac{\partial q_K}{\partial y} \equiv \frac{\partial q_{k,l}}{\partial y} = \sum_{i=0}^{N}\left[\frac{dh_{i,k}(\xi)}{d\xi}\frac{\partial \xi_{k,l}}{\partial y}q_{i,l} + \frac{dh_{i,l}(\eta)}{d\eta}\frac{\partial \eta_{k,l}}{\partial y}q_{k,i}\right]$$

$$= \frac{\partial q_{k,l}}{\partial \xi}\frac{\partial \xi_{k,l}}{\partial y} + \frac{\partial q_{k,l}}{\partial \eta}\frac{\partial \eta_{k,l}}{\partial y}. \tag{17.9}$$

Now we are ready to construct the term $\nabla \psi_{I,K} \cdot \nabla q_K$.
 Let us write

$$\nabla \psi_{I,K} \cdot \nabla q_K = \frac{\partial \psi_{I,K}}{\partial x}\frac{\partial q_K}{\partial x} + \frac{\partial \psi_{I,K}}{\partial y}\frac{\partial q_K}{\partial y}$$

where

$$\frac{\partial \psi_{I,K}}{\partial x}\frac{\partial q_K}{\partial x} = \left(\frac{dh_{i,k}(\xi)}{d\xi}\frac{\partial \xi_{k,l}}{\partial x} + \frac{dh_{i,l}(\eta)}{d\eta}\frac{\partial \eta_{k,l}}{\partial x}\right)\frac{\partial q_{k,l}}{\partial x}$$

and

$$\frac{\partial \psi_{I,K}}{\partial y}\frac{\partial q_K}{\partial y} = \left(\frac{dh_{i,k}(\xi)}{d\xi}\frac{\partial \xi_{k,l}}{\partial y} + \frac{dh_{i,l}(\eta)}{d\eta}\frac{\partial \eta_{k,l}}{\partial y}\right)\frac{\partial q_{k,l}}{\partial y}.$$

Putting all this together allows us to write the volume integral contribution as follows

$$\int_{\Omega_e}\nabla \psi_I \cdot \nabla q_N\, d\Omega_e \approx \sum_{k=0}^{N}\sum_{l=0}^{N}w_k w_l J_{k,l}^{(e)}\left(\frac{dh_{i,k}(\xi)}{d\xi}\frac{\partial \xi_{k,l}}{\partial x} + \frac{dh_{i,l}(\eta)}{d\eta}\frac{\partial \eta_{k,l}}{\partial x}\right)\frac{\partial q_{k,l}}{\partial x}$$

$$+ \sum_{k=0}^{N}\sum_{l=0}^{N}w_k w_l J_{k,l}^{(e)}\left(\frac{dh_{i,k}(\xi)}{d\xi}\frac{\partial \xi_{k,l}}{\partial y} + \frac{dh_{i,l}(\eta)}{d\eta}\frac{\partial \eta_{k,l}}{\partial y}\right)\frac{\partial q_{k,l}}{\partial y}.$$

Rearranging gives us the final form

$$\int_{\Omega_e} \nabla \psi_I \cdot \nabla q_N \, d\Omega_e \approx \sum_{k=0}^{N} \sum_{l=0}^{N} w_k w_l J_{k,l}^{(e)} \frac{dh_{i,k}(\xi)}{d\xi} h_{l,l}(\eta) \left(\frac{\partial \xi_{k,l}}{\partial x} \frac{\partial q_{k,l}}{\partial x} + \frac{\partial \xi_{k,l}}{\partial y} \frac{\partial q_{k,l}}{\partial y} \right)$$

$$+ \sum_{k=0}^{N} \sum_{l=0}^{N} w_k w_l J_{k,l}^{(e)} h_{k,k}(\xi) \frac{dh_{i,l}(\eta)}{d\eta} \left(\frac{\partial \eta_{k,l}}{\partial x} \frac{\partial q_{k,l}}{\partial x} + \frac{\partial \eta_{k,l}}{\partial y} \frac{\partial q_{k,l}}{\partial y} \right) \qquad (17.10)$$

where we have reintroduced the terms omitted due to cardinality.

Algorithm 17.3 highlights the procedure for constructing Eq. (17.10). First we construct ∇q_K using Eqs. (17.8) and (17.9). Then we use these derivatives to construct the remaining part of Eq. (17.10). Note that the cardinality terms ($h_{l,l}(\eta)$ and $h_{k,k}(\xi)$ in red font) clarify which locations of the vector $R^{(e)}$ to update. It is interesting to compare and contrast this algorithm which uses tensor-products with Alg. 14.1 which uses the monolithic form of the basis functions.

Algorithm 17.3 Second order operator volume integral contribution to $R^{(e)}$ with strong form, inexact integration, and tensor-products.

---

function R_VECTOR(q)
  for $e = 1 : N_e$ **do**   ▷ loop over elements
    for $k = 0 : N$ **do**   ▷ loop over integration points
      for $l = 0 : N$ **do**
        $\frac{\partial q_{k,l}}{\partial \xi} = 0; \frac{\partial q_{k,l}}{\partial \eta} = 0$
        for $i = 0 : N$ **do**
          $\frac{\partial q_{k,l}}{\partial \xi} + = \frac{dh_{i,k}}{d\xi} q_{i,l}$
          $\frac{\partial q_{k,l}}{\partial \eta} + = \frac{dh_{i,l}}{d\xi} q_{k,i}$
        end for
        $\frac{\partial q_{k,l}}{\partial x} = \frac{\partial q_{k,l}}{\partial \xi} \frac{\partial \xi_{k,l}}{\partial x} + \frac{\partial q}{\partial \eta} \frac{\partial \eta_{k,l}}{\partial x}$ ▷ build ∇q_K
        $\frac{\partial q_{k,l}}{\partial y} = \frac{\partial q_{k,l}}{\partial \xi} \frac{\partial \xi_{k,l}}{\partial y} + \frac{\partial q}{\partial \eta} \frac{\partial \eta_{k,l}}{\partial y}$
        $(\nabla \xi \cdot \nabla q)_{k,l} = \frac{\partial \xi_{k,l}}{\partial x} \frac{\partial q_{k,l}}{\partial x} + \frac{\partial \xi_{k,l}}{\partial y} \frac{\partial q_{k,l}}{\partial y}$
        $(\nabla \eta \cdot \nabla q)_{k,l} = \frac{\partial \eta_{k,l}}{\partial x} \frac{\partial q_{k,l}}{\partial x} + \frac{\partial \eta_{k,l}}{\partial y} \frac{\partial q_{k,l}}{\partial y}$
        for $i = 0 : N$ **do**
          $I_\xi = (i + 1) + l(N + 1); I_\eta = (k + 1) + i(N + 1)$
          $R_{I_\xi}^{(e)} + = w_k w_l J_{k,l}^{(e)} \frac{dh_{i,k}(\xi)}{d\xi} h_{l,l}(\eta)(\nabla \xi \cdot \nabla q)_{k,l}$
          $R_{I_\eta}^{(e)} + = w_k w_l J_{k,l}^{(e)} h_{k,k}(\xi) \frac{dh_{i,l}(\eta)}{d\eta}(\nabla \eta \cdot \nabla q)_{k,l}$
        end for
      end for
    end for
  end for
end function

Exercise Compare the complexity of Alg. 17.3 with Alg. 14.1.   □

17.7.2.3 Flux Integral Term

The contribution of the flux integral arising from the second order operator involves the second row of Eq. (17.7) which we write below for convenience

$$v \sum_{l=1}^{N_{FN}} \left[w_i^{(l)} J_i^{(l)} \hat{\mathbf{n}}_i^{(e,l)} \cdot \nabla \psi_{ji} \left\{ q^{(e,l)} \right\}_j - w_i^{(l)} J_i^{(l)} \mu_i [\![q^{(e,l)}]\!]_i \right]$$

$$+ v \sum_{l=1}^{N_{FN}} \sum_{j=0}^{N} w_j^{(l)} J_j^{(l)} \hat{\mathbf{n}}_j^{(e,l)} \cdot \nabla \psi_{ij} \left(q^{(e)} - \left\{ q^{(e,l)} \right\} \right)_j$$

which we now discretize using inexact integration and tensor-products. Let us rewrite this expression as follows

$$v \sum_{l=1}^{N_{FN}} \left[w_i^{(l)} J_i^{(l)} \hat{\mathbf{n}}_i^{(e,l)} \cdot \nabla q_i^{(*,l)} + \sum_{k=0}^{N} w_k^{(l)} J_k^{(l)} \hat{\mathbf{n}}_k^{(e,l)} \cdot \nabla \psi_{i,k} \left(q^{(e)} - q^{(*,l)} \right)_k \right]$$

where we have combined two terms and modified the index from j to k in the last term. Algorithm 17.4 shows the steps required to form the contribution of the second order flux terms to the vector $R^{(e)}$. The only steps that still need further explanation are the gradient terms (red font) which can be computed efficiently using tensor-products. Algorithms 17.5 and 17.6 describe the construction of these gradient terms. Note that we need to know which face of the element we are working on (the indices p_L and p_R). This completes the contribution of the second order operator terms to the vector $R^{(e)}$.

17.7.3 Construction of the Global Matrix Problem

Once the right-hand side vector $R^{(e)}$ has been constructed, we then need to build its global representation using the DSS operator defined in Algs. 12.11 and 12.12 for the mass matrix and right-hand side vector, respectively. Upon the completion of this step, we now have a global matrix problem that we can solve.

17.8 Example of 2D Hyperbolic Equation Problem

Let us now apply the unified CG/DG algorithms we have constructed to solve the hyperbolic equation we solved in Eq. (15.14) in Sec. 15.8 and Eq. (16.17) in Sec. 16.9 where we use the L^2 norm given in Eq. (12.58). Because the algorithms in this chapter are unified for both CG and DG, we can only use Lobatto points. We do not include the second order operators because they have already been discussed at length in Ch. 14.

Algorithm 17.4 Second order operator flux integral contribution to $R^{(e)}$ with strong form, inexact integration, and tensor-products.

---

function R_VECTOR(**f**)
  for $l = 1 : N_f$ **do**   ▷ loop over faces
    $p_L = face(1, l); p_R = face(2, l)$
    $L = face(3, l); R = face(4, l)$
    for $i = 0 : N$ **do**   ▷ loop over certain rows of $R^{(e)}$
        $i_L = mapL(1, i, p_L); j_L = mapL(2, i, p_L); I_L = (i_L + 1) + j_L(N + 1)$
        $i_R = mapR(1, i, p_R); j_R = mapR(2, i, p_R); I_R = (i_R + 1) + j_R(N + 1)$
        $\nabla q_i^{(L)} = \text{GRAD\_Q}(q, L, p_L, i); \quad \nabla q_i^{(R)} = \text{GRAD\_Q}(q, R, p_R, i)$
        $\nabla q_i^{(*,l)} = \frac{1}{2}\left[\nabla q_i^{(L)} + \nabla q_i^{(R)} - \hat{\mathbf{n}}_i^{(l)}\mu_i\left(q_i^{(R)} - q_i^{(L)}\right)\right]$

        $R_{I_L}^{(L)} \mathrel{+}= w_i^{(l)}J_i^{(l)}\hat{\mathbf{n}}_i^{(l)} \cdot \nabla q_i^{(*,l)}$
        $R_{I_R}^{(R)} \mathrel{-}= w_i^{(l)}J_i^{(l)}\hat{\mathbf{n}}_i^{(l)} \cdot \nabla q_i^{(*,l)}$
    end for

    for $k = 0 : N$ **do**   ▷ loop over integration points
        $q_k^{(*,l)} = \frac{1}{2}\left[q_k^{(L)} + q_k^{(R)}\right]$
        for $i = 0 : N$ **do**   ▷ compute gradient on integration points
            $i_L = mapL(1, i, p_L); j_L = mapL(2, i, p_L); I_L = (i_L + 1) + j_L(N + 1)$
            $i_R = mapR(1, i, p_R); j_R = mapR(2, i, p_R); I_R = (i_R + 1) + j_R(N + 1)$
            $\nabla\psi_{i,k}^{(L)} = \text{GRAD\_}\psi(h, L, p_L, i, k); \quad \nabla\psi_{i,k}^{(R)} = \text{GRAD\_}\psi(h, R, p_R, i, k)$
            $R_{I_L}^{(L)} \mathrel{+}= w_i^{(l)}J_i^{(l)}\hat{\mathbf{n}}_i^{(l)} \cdot \nabla\psi_{i,k}^{(L)}\left(q_k^{(L)} - q_k^{(*,l)}\right)$
            $R_{I_R}^{(R)} \mathrel{-}= w_i^{(l)}J_i^{(l)}\hat{\mathbf{n}}_i^{(l)} \cdot \nabla\psi_{i,k}^{(R)}\left(q_k^{(R)} - q_k^{(*,l)}\right)$
        end for
    end for
  end for
end function

17.8.0.1 Solution Accuracy

Figure 17.1 shows the convergence rates for both CG and DG with inexact integration using Lobatto points. We use the SSP-RK3 time-integrator from Ch. 5 with a Courant number of $C = 0.25$. Both CG and DG yield similar results. The difference in the number of total gridpoints N_p is due to the fact that for DG the total number of points is $N_p^{DG} = (N_e^{(x)}(N + 1))(N_e^{(y)}(N + 1))$, where $N_e^{(s)}$ denotes the number of elements along the coordinate direction (s), while for CG it is $N_p^{CG} = (N_e^{(x)} \cdot N + 1)(N_e^{(y)} \cdot N + 1)$. The larger number of gridpoints for DG as compared to CG is due to the fact that DG uses duplicate gridpoints at the faces of the elements. We need to remind the reader that we can use the same gridpoint storage with CG (which we call local element-wise or LEW storage) but here we use the global gridpoint storage (GGP) to show the difference in the resulting degrees of freedom. We show in Ch. 19 that LEW storage has some advantages, e.g., when adaptive mesh refinement is of interest.

Figure 17.2 shows the work-precision diagrams for both CG and DG. Both CG and DG yield similar results; however, for more complex problems we often find

---

Algorithm 17.5 Tensor-product construction of $\nabla q_k^{(e)}$.

---

function GRAD_Q(q, e, p, k)
  if $p = 1$ **then**   ▷ Face 1: $\eta = -1$
    for $i = 0 : N$ **do**
$$\frac{\partial \psi}{\partial \xi} = \frac{\partial h_{i,k}}{\partial \xi} h_{0,0}; \quad \frac{\partial \psi}{\partial \eta} = h_{i,k} \frac{\partial h_{0,0}}{\partial \eta}$$
$$\frac{\partial q_{k,0}}{\partial \xi} + = \frac{\partial \psi_{k,0}}{\partial \xi} q_i; \quad \frac{\partial q}{\partial \eta} + = \frac{\partial \psi}{\partial \eta} q_i$$
    end for
$$\frac{\partial q}{\partial x} = \frac{\partial q_{k,0}}{\partial \xi} \frac{\partial \xi_{k,0}^{(e)}}{\partial x} + \frac{\partial q_{k,0}}{\partial \eta} \frac{\partial \eta_{k,0}^{(e)}}{\partial x}$$
$$\frac{\partial q}{\partial y} = \frac{\partial q_{k,0}}{\partial \xi} \frac{\partial \xi_{k,0}^{(e)}}{\partial y} + \frac{\partial q_{k,0}}{\partial \eta} \frac{\partial \eta_{k,0}^{(e)}}{\partial y}$$
  else if $p = 2$ **then**   ▷ Face 2: $\xi = +1$
    for $i = 0 : N$ **do**
$$\frac{\partial \psi}{\partial \xi} = \frac{\partial h_{N,N}}{\partial \xi} h_{i,k}; \quad \frac{\partial \psi}{\partial \eta} = h_{N,N} \frac{\partial h_{i,k}}{\partial \eta}$$
$$\frac{\partial q_{N,k}}{\partial \xi} + = \frac{\partial \psi}{\partial \xi} q_i; \quad \frac{\partial q_{N,k}}{\partial \eta} + = \frac{\partial \psi}{\partial \eta} q_i$$
    end for
$$\frac{\partial q}{\partial x} = \frac{\partial q_{N,k}}{\partial \xi} \frac{\partial \xi_{N,k}^{(e)}}{\partial x} + \frac{\partial q_{N,k}}{\partial \eta} \frac{\partial \eta_{N,k}^{(e)}}{\partial x}$$
$$\frac{\partial q}{\partial y} = \frac{\partial q_{N,k}}{\partial \xi} \frac{\partial \xi_{N,k}^{(e)}}{\partial y} + \frac{\partial q_{N,k}}{\partial \eta} \frac{\partial \eta_{N,k}^{(e)}}{\partial y}$$
  else if $p = 3$ **then**   ▷ Face 3: $\eta = +1$
    for $i = 0 : N$ **do**
$$\frac{\partial \psi}{\partial \xi} = \frac{\partial h_{i,k}}{\partial \xi} h_{N,N}; \quad \frac{\partial \psi}{\partial \eta} = h_{i,k} \frac{\partial h_{N,N}}{\partial \eta}$$
$$\frac{\partial q_{k,N}}{\partial \xi} + = \frac{\partial \psi}{\partial \xi} q_i; \quad \frac{\partial q_{k,N}}{\partial \eta} + = \frac{\partial \psi}{\partial \eta} q_i$$
    end for
$$\frac{\partial q}{\partial x} = \frac{\partial q_{k,N}}{\partial \xi} \frac{\partial \xi_{k,N}^{(e)}}{\partial x} + \frac{\partial q_{k,N}}{\partial \eta} \frac{\partial \eta_{k,N}^{(e)}}{\partial x}$$
$$\frac{\partial q}{\partial y} = \frac{\partial q_{k,N}}{\partial \xi} \frac{\partial \xi_{k,N}^{(e)}}{\partial y} + \frac{\partial q_{k,N}}{\partial \eta} \frac{\partial \eta_{k,N}^{(e)}}{\partial y}$$
  else if $p = 4$ **then**   ▷ Face 4: $\xi = -1$
    for $i = 0 : N$ **do**
$$\frac{\partial \psi}{\partial \xi} = \frac{\partial h_{0,0}}{\partial \xi} h_{i,k}; \quad \frac{\partial \psi}{\partial \eta} = h_{0,0} \frac{\partial h_{i,k}}{\partial \eta}$$
$$\frac{\partial q_{0,k}}{\partial \xi} + = \frac{\partial \psi}{\partial \xi} q_i; \quad \frac{\partial q_{0,k}}{\partial \eta} + = \frac{\partial \psi}{\partial \eta} q_i$$
    end for
$$\frac{\partial q}{\partial x} = \frac{\partial q_{0,k}}{\partial \xi} \frac{\partial \xi_{0,k}^{(e)}}{\partial x} + \frac{\partial q_{0,k}}{\partial \eta} \frac{\partial \eta_{0,k}^{(e)}}{\partial x}$$
$$\frac{\partial q}{\partial y} = \frac{\partial q_{0,k}}{\partial \xi} \frac{\partial \xi_{0,k}^{(e)}}{\partial y} + \frac{\partial q_{0,k}}{\partial \eta} \frac{\partial \eta_{0,k}^{(e)}}{\partial y}$$
  end if
end function

---

that CG is faster particularly due to the smaller degrees of freedom. Figure 17.2 shows that $N = 8$ and $N = 16$ yield the best compromise between accuracy and computational cost for this particular case.

Algorithm 17.6 Tensor-product construction of $\nabla \psi_k^{(e)}$.

---

function GRAD_$\psi(h, e, p, i, k)$
 if $p = 1$ **then** $\triangleright$ Face 1: $\eta = -1$
$$\frac{\partial \psi}{\partial \xi} = \frac{\partial h_{i,k}}{\partial \xi} h_{0,0}; \quad \frac{\partial \psi}{\partial \eta} = h_{i,k} \frac{\partial h_{0,0}}{\partial \eta}$$
$$\frac{\partial \psi}{\partial x} = \frac{\partial \psi}{\partial \xi} \frac{\partial \xi_{k,0}^{(e)}}{\partial x} + \frac{\partial \psi}{\partial \eta} \frac{\partial \eta_{k,0}^{(e)}}{\partial x}$$
$$\frac{\partial \psi}{\partial y} = \frac{\partial \psi}{\partial \xi} \frac{\partial \xi_{k,0}^{(e)}}{\partial y} + \frac{\partial \psi}{\partial \eta} \frac{\partial \eta_{k,0}^{(e)}}{\partial y}$$
 else if $p = 2$ **then** $\triangleright$ Face 2: $\xi = +1$
$$\frac{\partial \psi}{\partial \xi} = \frac{\partial h_{N,N}}{\partial \xi} h_{i,k}; \quad \frac{\partial \psi}{\partial \eta} = h_{N,N} \frac{\partial h_{i,k}}{\partial \eta}$$
$$\frac{\partial \psi}{\partial x} = \frac{\partial \psi}{\partial \xi} \frac{\partial \xi_{N,k}^{(e)}}{\partial x} + \frac{\partial \psi}{\partial \eta} \frac{\partial \eta_{N,k}^{(e)}}{\partial x}$$
$$\frac{\partial \psi}{\partial y} = \frac{\partial \psi}{\partial \xi} \frac{\partial \xi_{N,k}^{(e)}}{\partial y} + \frac{\partial \psi}{\partial \eta} \frac{\partial \eta_{N,k}^{(e)}}{\partial y}$$
 else if $p = 3$ **then** $\triangleright$ Face 3: $\eta = +1$
$$\frac{\partial \psi}{\partial \xi} = \frac{\partial h_{i,k}}{\partial \xi} h_{N,N}; \quad \frac{\partial \psi}{\partial \eta} = h_{i,k} \frac{\partial h_{N,N}}{\partial \eta}$$
$$\frac{\partial \psi}{\partial x} = \frac{\partial \psi}{\partial \xi} \frac{\partial \xi_{k,N}^{(e)}}{\partial x} + \frac{\partial \psi}{\partial \eta} \frac{\partial \eta_{k,N}^{(e)}}{\partial x}$$
$$\frac{\partial \psi}{\partial y} = \frac{\partial \psi}{\partial \xi} \frac{\partial \xi_{k,N}^{(e)}}{\partial y} + \frac{\partial \psi}{\partial \eta} \frac{\partial \eta_{k,N}^{(e)}}{\partial y}$$
 else if $p = 4$ **then** $\triangleright$ Face 4: $\xi = -1$
$$\frac{\partial \psi}{\partial \xi} = \frac{\partial h_{0,0}}{\partial \xi} h_{i,k}; \quad \frac{\partial \psi}{\partial \eta} = h_{0,0} \frac{\partial h_{i,k}}{\partial \eta}$$
$$\frac{\partial \psi}{\partial x} = \frac{\partial \psi}{\partial \xi} \frac{\partial \xi_{0,k}^{(e)}}{\partial x} + \frac{\partial \psi}{\partial \eta} \frac{\partial \eta_{0,k}^{(e)}}{\partial x}$$
$$\frac{\partial \psi}{\partial y} = \frac{\partial \psi}{\partial \xi} \frac{\partial \xi_{0,k}^{(e)}}{\partial y} + \frac{\partial \psi}{\partial \eta} \frac{\partial \eta_{0,k}^{(e)}}{\partial y}$$
 end if
end function

Algorithm 17.7 Unified CG/DG solution algorithm for the 2D hyperbolic-elliptic equation using R vector approach.

---

function CGDG($\mathbf{f}, q$)
 Construct $M^{(e)}$ $\triangleright$ using Alg. 12.6
 Construct M using DSS $\triangleright$ using Alg. 12.11
 for $n = 1 : N_t$ **do** $\triangleright$ time loop
  Construct $R^{(e)}(\mathbf{f}, q)$ $\triangleright$ use Algs. 17.1, 17.2, 17.3 and 17.4
  Construct R_I using DSS $\triangleright$ using Alg. 12.12
  $\hat{R}_I = M_{IJ}^{-1} R_J$ $\triangleright$ apply boundary conditions
  $\frac{d}{dt} q_I = \hat{R}_I$ $\triangleright$ evolve equations forward in time to get q^{n+1}
 end for
end function

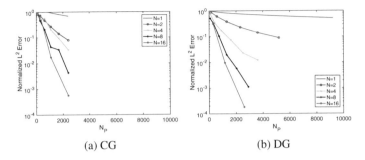

Fig. 17.1: Convergence rates for the 2D hyperbolic equation for polynomial orders $N = 1$, $N = 2$, $N = 4$, $N = 8$, and $N = 16$ using a total number of gridpoints N_p for (a) CG and (b) DG with inexact integration on Lobatto points.

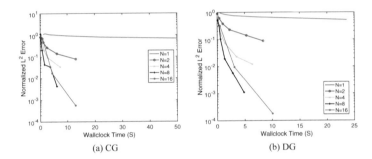

Fig. 17.2: Work-Precision: Convergence rates as a function of wallclock time for the 2D hyperbolic equation for polynomial orders $N = 1$, $N = 2$, $N = 4$, $N = 8$, and $N = 16$ using a total number of gridpoints N_p for (a) CG and (b) DG with inexact integration on Lobatto points.

Part IV
Advanced Topics

Part IV treats advanced topics related to element-based Galerkin (EBG) methods but that are not strictly necessary to understand the fundamentals of EBG methods. Chapter 18 is the most relevant chapter in Part IV to the development of EBG methods because it describes strategies for stabilizing EBG methods. In this chapter, we describe limiters, artificial diffusion, filters, and entropy-stable methods. Then in Ch. 19, we discuss the implementation of adaptive mesh refinement in the context of EBG methods. Here, we discuss the three classes of mesh refinement including r-, h-, and p-refinement. In Ch. 20 we briefly outline various categories of time-integration methods including explicit, implicit, implicit-explicit, semi-Lagrangian, and multirate methods. Finally, we end the textbook with Ch. 21 which describes a promising Galerkin method known as the hybridized discontinuous Galerkin method that seeks to combine the advantages of the discontinuous Galerkin method with the efficiency of the continuous Galerkin method.

Chapter 18
Stabilization of High-Order Methods

18.1 Introduction

In this chapter we address the question of how to stabilize high-order methods, focusing on the application of these methods to hyperbolic equations, particularly for wave propagation. All numerical methods suffer from aliasing errors, particularly methods that have little to no inherent dissipation such as high-order methods; we saw this in the dissipation-dispersion analysis of CG and DG methods in Ch. 7. In order to avoid numerical instabilities caused by such errors, high-order methods must use some form of stabilization mechanism, whereby this mechanism usually takes the form of dissipation. Note that this dissipation can be viewed in numerous ways, some not so favorably. However, another way of viewing these mechanisms is to acknowledge that energy must be allowed to cascade from the higher frequencies to the lower frequencies (Kolmogorov scales). This is in fact a very physical argument that confirms the need for such dissipation. For element-based Galerkin methods, the options available to us for introducing dissipation include: diffusion or hyper-diffusion (also called hyper-viscosity) operators that are tailored to only target high-frequency noise, low-pass filters which curtail the solution in the frequency domain, element-based limiters, and upwinding methods. In this chapter, we refer to diffusion-based stabilization as *hyper-diffusion* and only discuss upwinding methods for the discontinuous Galerkin method because it is essential to the general construction of DG methods since all DG methods must use a numerical flux and this flux, we have already seen in Chs. 6 and 7, performs best when it is based on upwinding methods. It should be noted that upwinding methods are also available for CG methods such as in the Petrov-Galerkin method [107] and in the relatively new discontinuous Petrov-Galerkin method [44]. Unlike the Bubnov-Galerkin method where the test and trial functions are always the same, the Petrov-Galerkin method uses different test and trial functions. We do not discuss Petrov-Galerkin methods in detail but cover

F. X. Giraldo, *An Introduction to Element-Based Galerkin Methods on Tensor-Product Bases*, Texts in Computational Science and Engineering 24, https://doi.org/10.1007/978-3-030-55069-1_18

other related upwinding-type methods for both CG and DG, such as the *streamline upwinding Petrov-Galerkin* (SUPG) method [48], the *variational multi-scale* (VMS) method [209], and a new stabilization known as the *dynamic sub-grid scale* (DSGS) method [268]. We also discuss a relatively new approach to stabilize numerical methods using an approach referred to as *provably stable* methods.

The goal of this chapter is, by no means, to give an exhaustive coverage of stabilization methods as this area of research is quite vast and ever changing. Instead, the idea we take is to give the reader sufficient information to understand the basic mechanisms of stabilization, under simple examples, so that the reader can then extend them to more difficult problems. As an example, our discussion on limiters will not attempt to cover all the limiters available (there are just too many) but instead we select a few limiters and describe them in detail - particularly those that allow the reader to stabilize their numerical methods for the class of problems described in this text.

The remainder of this chapter is organized as follows. In Sec. 18.2 we describe *aliasing* and the errors associated with this phenomenon. Then in Sec. 18.3 we describe the approach to dealiasing using low-pass filters that curtail the high-frequency waves responsible for causing aliasing errors. In Sec. 18.4 we discuss the construction of Riemann solvers, which is the natural way to include dissipation within Godunov methods such as the discontinuous Galerkin method. Then in Sec. 18.5 we address the use of limiters to stabilize the solution and satisfy positivity preservation of variables (such as tracers) that must remain positive definite. In Sec. 18.6 we cover methods based on diffusion and hyper-diffusion operators whereby the viscosity coefficients are computed adaptively, i.e., the viscosity coefficient is based on some characteristic of the local element-wise solution. We refer to this class of methods as *local adaptive viscosity* (LAV) methods; we begin this section with a discussion on the general construction of hyper-diffusion operators. We end this chapter in Sec. 18.7 with provably stable methods, with a discussion on energy-stable and entropy-stable methods.

18.2 Aliasing Error

Like any high order method, both CG and DG methods are prone to suffer from *aliasing errors*. Aliasing errors occur when the grid cannot resolve the high frequency waves and, as a consequence, interprets them as truncated lower frequency waves. This can cause severe oscillations in the solution and can eventually make the numerical method unstable.

An example of this is shown in Fig. 18.1 for the one-dimensional wave equation

$$\frac{\partial q}{\partial t} + u \frac{\partial q}{\partial x} = 0 \tag{18.1}$$

using CG with 4th order polynomials. For this simulation, we use periodic boundary conditions and the steep Gaussian initial condition defined as follows

$$q(x,0) = \exp\left(-60x^2\right) \tag{18.2}$$

where $x \in [-1, +1]$ with $u = 2$. We use this initial condition throughout this chapter when we show dispersion plots. Figure 18.1a shows the numerical solution after a

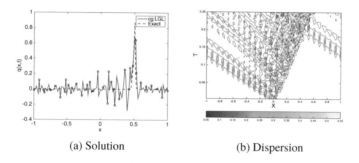

(a) Solution   (b) Dispersion

Fig. 18.1: The (a) solution and (b) dispersion plots for the 1D wave equation with no stabilization for CG using 25 elements and 4th order polynomials.

certain time and Fig. 18.1b shows the error as a function of space and time (x-t diagram). Figure 18.1b shows the dispersion experienced by the numerical solution. Since the velocity field is fixed moving from left to right, we should only see errors with a positive slope (moving to the right in the x-t diagram). The negative slopes (waves moving to the left) that we see in this plot demonstrate the dispersion errors in the solution. It should be understood that the dispersion errors we see in Fig. 18.1b are due to the fact that the CG method is inherently non-dissipative. Comparing the behavior of the DG method we see that with no dissipation (using a centered numerical flux) we get an identical result to CG shown in Fig. 18.2. In Fig. 18.3 we

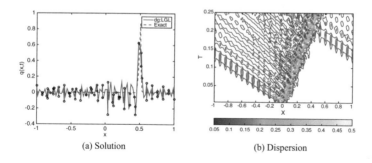

(a) Solution   (b) Dispersion

Fig. 18.2: The (a) solution and (b) dispersion plots for the 1D wave equation with a centered flux for DG using 25 elements and 4th order polynomials.

show the removal of much of the dispersion error by using an upwind flux (Rusanov).
Let us now give a simple example to explain the production of aliasing errors.

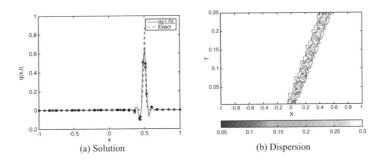

(a) Solution   (b) Dispersion

Fig. 18.3: The (a) solution and (b) dispersion plots for the 1D wave equation with an
upwind flux (Rusanov) for DG using 25 elements and 4th order polynomials.

Example 18.1 Taking the nonlinear term

$$\left(q_N \frac{\partial q_N}{\partial x} \right)_j$$

at the jth gridpoint and introducing Fourier modes

$$q_N(x_j) = \sum_{k=0}^{N} v_k \exp\left[ij\left(k\Delta x\right)\right], \quad \frac{\partial q_N}{\partial x}(x_j) = \sum_{l=0}^{N} v_l \exp\left[ij\left(l\Delta x\right)\right], \quad i = \sqrt{-1},$$

yields

$$\left(q_N \frac{\partial q_N}{\partial x} \right)_j = \sum_{k=0}^{N} \sum_{l=0}^{N} v_k \exp\left[ij\left(k\Delta x\right)\right] v_l \exp\left[ij\left(l\Delta x\right)\right]$$

$$= \sum_{k=0}^{N} \sum_{l=0}^{N} v_k v_l \exp\left[ij\left(k+l\right)\Delta x\right]$$

where $(k + l) \Delta x$ can become larger than the maximum value π associated with the
minimum wavelength resolved by the grid - in other words $\lambda_{min} = 2\Delta x$ (where
$k, l = \frac{2\pi}{\lambda}$). Therefore, since the grid cannot resolve the frequency $(k + l) \Delta x$ the
signal will be interpretted as $2\pi - (k + l) \Delta x$.

Example 18.1 explains how aliasing errors may arise in the evaluation of nonlinear
terms. However, it is also possible to produce such errors for linear problems. In
Ex. 18.1 even if the velocity is known but is expanded via basis functions, aliasing

may occur. This occurs, for example, when using exact integration with nodal bases or when using modal bases. In the case of nodal bases with inexact integration, the resulting errors are not due to nonlinear aliasing (since the velocity is sampled at co-located interpolation and integration points) but rather due purely to linear dispersion errors.

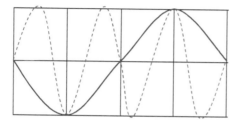

Fig. 18.4: Aliasing Error: When the grid is not sufficiently refined, high frequency waves (dashed red curve) may be interpreted as low frequency waves (solid blue curve) by the numerical scheme. This is known as aliasing.

The situation of high frequency waves being interpreted as low frequency waves is illustrated schematically in Fig. 18.4. Note that the high frequency wave (red dashed line) is interpreted by the grid and, consequently, the numerical method as the thick (blue) line.

To better explain *aliasing* let us consider the following domain: $x \in [-0.05, +0.05]$ with the high-frequency wave defined as $q(x) = \cos(2\pi f x)$ where f is the frequency of the original signal. Figure 18.5 illustrates various sampling rates (solid blue line) of the original high-frequency signal (dashed red line) for a $f = 60$ Hz signal. The *Nyquist frequency* $f_N = \frac{1}{2}f$ is the cross-over frequency at which we can represent the original signal faithfully (Fig. 18.5b). If we go below this value (see Fig. 18.5a) we do not recover the correct frequency - this is an aliasing error which will give rise to *Gibbs phenomena*; Gibbs phenomenon is defined as the overshoots and undershoots exhibited by Fourier series and other eigenfunctions near discontinuities (see [147, 148]). If we go above the Nyquist frequency, then we are guaranteed to represent the correct frequency of the signal (see Fig. 18.5c).

Low order methods do not suffer as severely from aliasing errors because they are already dissipative and thereby either damp or ignore the high frequency waves. In essence, low order schemes have implicit stabilization mechanisms. However, this implicit stabilization is not scale selective and thus can damp most of the waves including those containing pertinent information. One clear advantage of high-order methods is that they offer the possibility to choose which signals to retain and which to discard, albeit with a bit of worthwhile effort.

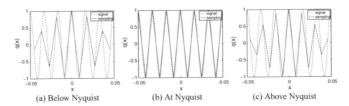

Fig. 18.5: Aliasing: the original signal of 60 Hz (red) is sampled at various rates (a) $f_s = \frac{60}{110} f$ (below Nyquist frequency), (b) $f_s = \frac{60}{120} f$ (at Nyquist frequency), and (c) $f_s = \frac{60}{130} f$ (above Nyquist frequency) .

18.3 Spectral Filters

One of the advantages of high-order Galerkin methods is that they are constructed upon the eigenfunctions of the Sturm-Liouville operator which means that these methods can be considered to be representations of the solution in the frequency-domain (i.e., modal or spectral space). Note that if modal functions are used, then this representation is immediately obvious. However, although nodal functions represent the solution in the time-domain (i.e., physical space), the solution using this approach can also be recast in the frequency-domain by using the nodal-to-modal transform that we define below.

In this text, we have mainly emphasized the construction of numerical solutions using the nodal basis functions as such

$$q_N(\xi) = \sum_{n=0}^{N} q_n \psi_n(\xi) \qquad (18.3)$$

because the expansion coefficients q_n have a physical significance, i.e., they represent the solution of q in the time-domain (nodal space) at the gridpoint "n". For filtering (removing waves that may corrupt the quality of the solution) it only makes sense to work in the frequency-domain (modal space) which can be written as follows:

$$q_N(\xi) = \sum_{m=0}^{N} \tilde{q}_m \psi_m(\xi) \qquad (18.4)$$

where $\tilde{q}, \psi$ are the expansion coefficients and the modal basis functions, respectively. The objective of filtering is to damp the high frequency waves that may be aliased and, as a result, cause spurious oscillations. In nodal space the coefficients q_{N-1} and q_N correspond to the solution values of the two right most collocation points of the element, whereas in modal space these represent the two highest order modes. Thus by mapping to modal space we can then isolate and remove these high order modes.

Example 18.2 Figure 18.6a shows the approximation functions for a $N = 8$ nodal expansion. Note that all of the functions are high order and so each of the 9 functions contain the high order modes (high frequency waves). In modal space the basis

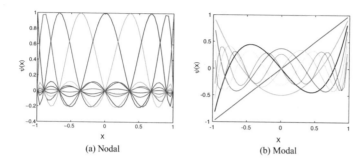

(a) Nodal   (b) Modal

Fig. 18.6: Basis function expansion for $N = 8$ for (a) nodal and (b) modal polynomials.

functions ψ_m are the wave/frequency modes (and $\tilde{q}_m$ are their amplitudes) Figure 18.6b shows the approximation functions for a $N = 8$ modal expansion. Note that the functions ψ_{N-1} and ψ_N correspond to the two highest frequency modes (the dotted lines).

To simplify the exposition, let us rewrite the modal basis functions as $\phi(\xi)$ and rewrite Eq. (18.4) as

$$q_N(\xi) = \sum_{m=0}^{N} \tilde{q}_m \phi_m(\xi). \tag{18.5}$$

Defining the inner product (with respect to the left-hand side of Eq. (18.5)) via numerical quadrature yields

$$(q_N(\xi), \phi_k) = \sum_{j=0}^{N} w_j q_N(\xi_j)\phi_k(\xi_j)$$

and, replacing $q_N(\xi_j)$ by its modal expansion (Eq. (18.5)) yields

$$\sum_{j=0}^{N} w_j q_N(\xi_j)\phi_k(\xi_j) = \sum_{j=0}^{N} w_j \left(\sum_{m=0}^{N} \tilde{q}_m \phi_m(\xi_j) \right) \phi_k(\xi_j)$$

$$= \sum_{j=0}^{N} w_j \tilde{q}_k \phi_k^2(\xi_j)$$

where we have used orthogonality to simplify the right-hand side. Letting

$$\gamma_k = \sum_{j=0}^{N} w_j \phi_k^2(\xi_j)$$

we can write

$$\widetilde{q}_k = \frac{1}{\gamma_k} \sum_{j=0}^{N} w_j q_N(\xi_j) \phi_k(\xi_j). \tag{18.6}$$

This equation defines the *Legendre transform*. Once we have the modal coefficients, we can filter the function as follows

$$q_N^F(\xi) = (1 - \mu) q_N(\xi) + \mu \sum_{k=0}^{N} \omega_k \widetilde{q}_k \phi_k(\xi) \tag{18.7}$$

where ω_k is the low-pass filter

$$\omega_k = \begin{cases} 1 & k < s \\ \sigma_k \left(\frac{k-s}{N-s} \right) & s \le k \le N \end{cases}$$

where $\{\sigma_k(x) \mid x(k) = \frac{k-s}{N-s}\}$ can be any smoothly decaying function (see Vandeven's theorem discussed below) and $\mu \in [0, 1]$ is a weighting coefficient which determines how much filtering is applied. For $\mu = 0$ no filter is applied and for $\mu = 1$ a full-strength filter is applied. For example, $\sigma(x)$ can be the polynomial function

$$\sigma_k(x) = 1 - \alpha_F x(k)^F \quad s \le k \le N$$

where α_F is some weight, k are the wavenumbers, $k = s, \dots, N$ are the wavenumbers of the modes to be filtered, and F is the order of the polynomial. Typical values used are $\alpha_F = 1$, $F = 2$, and $s = \frac{2}{3}N$ [128] that yields the *quadratic filter*. For $s = \frac{2}{3}N$ only the last third of the spectrum is modified, as is typical of so-called triangular truncation used in spectral methods.

Another possible choice is the exponential filter [200]

$$\sigma_k(x) = \exp \left[-\alpha_F x(k)^F \right] \quad s \le k \le N$$

with the three chosen constants α_F, s, and F, where α_F scales the exponential, s controls which modes to curtail, while F determines the steepness of the filter. Typical constants chosen are $\alpha_F = 17$, $s = 0$, and $F = 18$. Note that $s = 0$ means that all modes are filtered except for the zeroth mode (since $k = 0$ yields $\sigma = 1$). We show in the next section that, if conserving quantities is important in your problem (and it should be) then the zeroth mode should not be modified.

Another popular choice is the erfc-log filter [387, 46] expressed as follows

$$\sigma_k(x) = \frac{1}{2} \text{erfc} \left(2\sqrt{F}\Omega \sqrt{-\frac{\log(1 - 4\Omega^2)}{4\Omega^2}} \right) \quad s \le k \le N$$

with

$$\Omega = |x| - \frac{1}{2}, \quad \text{erfc}(x) = 1 - \text{erf}(x).$$

Typical filter values are $s = \frac{2}{3}N$ and $F = 12$.

Figure 18.7 shows the filter function $\sigma_k(x)$ for the quadratic, exponential, and erfc-log filters. The dashed (light blue) line denotes the value of s at which point the filter activates. Note how the transition from 1 to 0 is smooth near s. Although there

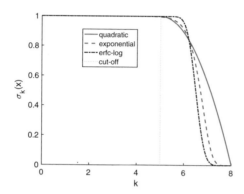

Fig. 18.7: The function $\sigma_k(x)$ as a function of wavenumber k for the quadratic (solid red), exponential (dashed blue), and erfc-log (dash-dotted black) filters.

exist many possibilities for generating $\sigma_k(x)$, Vandeven's theorem [387] yields some conditions for the filter to maintain spectral accuracy even when discontinuities exist in the domain. Of course, the spectral accuracy will diminish at the discontinuity itself, but Vandeven's theorem states that spectral accuracy can still be maintained away from the discontinuity. For example, Boyd [46] shows that the simpler erfc filter

$$\sigma_k(x) = \frac{1}{2}\text{erfc}\left(2\sqrt{F}\Omega\right) \quad s \le k \le N,$$

although looks quite similar to the erfc-log (see Fig. 18.8), does not satisfy Vandeven's theorem and therefore will not yield spectral accuracy away from discontinuities. However, both filters will work fine as long as there are no discontinuities present.

In Fig. 18.9 we show results for the exact same problem shown in Fig. 18.1 but with the inclusion of the erfc-log filter with weight $\mu = 0.05$. Note that the weight $\mu = 0.05$ means that the original $\sigma_k(x)$ filter function is only applied at 5% strength where 95% of the original unfiltered signal is maintained.

For comparison, we show the results with the quadratic filter in Fig. 18.10 with weight $\mu = 0.05$. Although this filter does not satisfy Vandeven's theorem, it is able to eliminate much of the dispersion errors from the non-dissipative CG method.

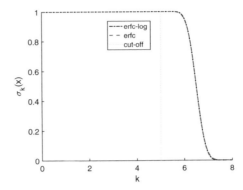

Fig. 18.8: The function $\sigma_k(x)$ as a function of wavenumber k for the erfc-log (dash-dotted black) and erfc (dashed blue) filters.

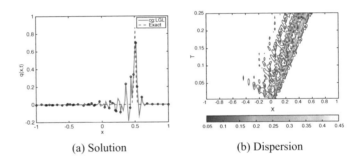

(a) Solution   (b) Dispersion

Fig. 18.9: The (a) solution and (b) dispersion plots for the 1D wave equation with the erfc-log filter with $\mu = 0.05$ for CG using 25 elements and 4th order polynomials.

18.4 Riemann Solvers and Upwinding

In Fig. 18.11 we show results for the exact same problem shown in Fig. 18.1 but for the DG method with a centered flux. Note that the behavior is similar for both the CG and DG methods with no dissipation.

We saw that the Gibbs phenomena in the CG method can be reduced by using some form of dissipation such as filters. We can use the same dissipative mechanisms for DG but we also have the option to use dissipation via the numerical flux. If we include a dissipative term in the centered flux formula then we recover the Rusanov flux. Figure 18.12 shows the results with the Rusanov flux. The dissipation from the Rusanov flux is sufficient to obviate the need for filtering in this case. To better understand the role of Riemann solvers, let us look at Fig. 18.13. In this figure, we note that there are 4 possible wave states (from left to right): 1) the left state q_L,

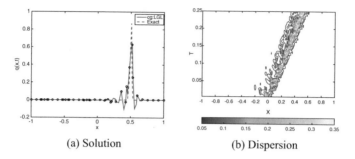

Fig. 18.10: The (a) solution and (b) dispersion plots for the 1D wave equation with the quadratic filter with $\mu = 0.05$ for CG using 25 elements and 4th order polynomials.

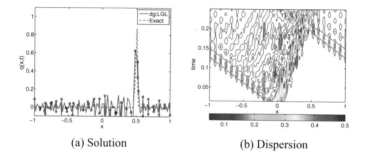

Fig. 18.11: The (a) solution and (b) dispersion plots for the 1D wave equation for the DG method with the centered flux using 25 elements and 4th order polynomials.

the left star region q_L^*, the right star region q_R^*, and the right state q_R. The left and right states (q_L and q_R) are the two original states in the initial Riemann problem. If we assume that $q_L > q_R$ then one likely scenario is that λ_L denotes the eigenvalue (wave speed) of rarefaction waves while λ_* represents the contact wave, with λ_R representing a compression wave (e.g., a shock wave). One possible approximate Riemann solver is the Harten-Lax-van Leer (HLL) solver which can be described as follows (see [379])

$$F^{HLL} = \begin{cases} F_L & \text{if} \quad 0 \leq \lambda_L \\ \frac{\lambda_R F_L - \lambda_L F_R + \lambda_L \lambda_R (q_R - q_L)}{\lambda_R - \lambda_L} & \text{if} \quad \lambda_L \leq 0 \leq \lambda_R. \\ F_R & \text{if} \quad 0 \geq \lambda_R \end{cases} \tag{18.8}$$

The HLL approximate Riemann solver only contains 3 wave states: the initial left state q_L, the combined star region (q_L^* and q_R^*), and the initial right state q_R. Therefore, although the HLL flux can capture expansion waves and shocks, it cannot represent the contact discontinuity (since the star regions are combined into one). Fortunately, an easy modification to this flux exists and is known as the HLLC flux. However,

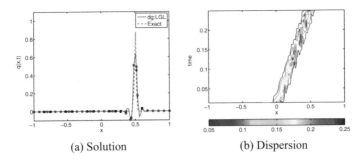

(a) Solution   (b) Dispersion

Fig. 18.12: The (a) solution and (b) dispersion plots for the 1D wave equation for the DG method with the Rusanov flux using 25 elements and 4th order polynomials.

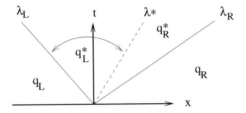

Fig. 18.13: Depiction of the Riemann problem with 4 wave states.

we will not describe it here since the form of the HLLC flux changes with the governing system of equations (e.g., Euler versus shallow water). In contrast, the HLL flux maintains the same simplified form regardless of the equations being solved. Another reason for describing the HLL flux is that the Rusanov flux is a special case of this flux. To see this, let us begin by first defining the eigenvalues for the HLL flux to be: $\lambda_L = \min(|u_L| - c_L, |u_R| - c_R)$ and $\lambda_R = \max(|u_L| + c_L, |u_R| + c_R)$, where u is the velocity and c are the characteristic speeds of the governing equations (e.g., acoustic wave speed for the Euler equations or the gravity wave speed for the shallow water equations). However, if we let $\lambda_{max} = \max(|u_L| + c_L, |u_R| + c_R)$ and then define $\lambda_L = -\lambda_{max}$ and $\lambda_R = +\lambda_{max}$ yields the following flux function

$$F^{Rusanov} = \frac{1}{2}(F_L + F_R) - \frac{\lambda_{max}}{2}(q_R - q_L) \qquad (18.9)$$

which, due to its simplicity, explains the reason for its popularity in many DG methods.

18.5 Limiters

18.5.1 Stabilization Limiters

For element-based Galerkin methods, another option in the arsenal of stabilization methods is limiters. The general idea behind limiters is to write the solution in modal form as in Eq. (18.5). Then, we must decide how to curtail each mode. One can view the limiting process, at least for element-based Galerkin methods, as an adaptive filter whereby the filter function is different for each element and the modes are modified based on some metric of the solution. However, before we review a few of the limiters typically used, let us first understand better the meaning of the amplitudes in the modal expansion given by Eq. (18.5).

18.5.1.1 Taylor Polynomials to Monomials

Let us begin by expanding the solution, within one element, using a Taylor series expansion about $x = 0$ within the element domain $x \in [-1, +1]$ as follows

$$q_N(x) = q(0) + q^{(1)}(0)x + q^{(2)}(0)\frac{x^2}{2} + \ldots + q^{(N)}(0)\frac{x^N}{N!} \qquad (18.10)$$

where $q^{(i)}$ denotes the ith derivative of the function. The Taylor expansion given by Eq. (18.10) can be written in terms of a monomial expansion as follows

$$q_N(x) = \tilde{q}_0 + \tilde{q}_1 x + \tilde{q}_2 x^2 + \ldots + \tilde{q}_N x^N \qquad (18.11)$$

where $\tilde{q}_0 = q(0)$, $\tilde{q}_1 = q_x(0)$, $\tilde{q}_2 = \frac{1}{2}q_{xx}(0)$, and $\tilde{q}_N = \frac{1}{N!}q^{(N)}(0)$. Equation (18.11) can be written compactly as follows

$$q_N(x) = \sum_{k=0}^{N} \tilde{q}_k x^k \qquad (18.12)$$

and is known as the Taylor polynomial expansion.

18.5.1.2 Monomials to Orthogonal Polynomials

The monomial basis $\{x^k \mid k = 0, \ldots, N\}$ is not orthogonal and, therefore, not optimal as a basis. We can orthogonalize this basis using the Gram-Schmidt orthogonalization (also known as QR factorization in linear algebra, (see, e.g., [380]). Let us briefly describe the Gram-Schmidt process using an example for $N = 2$.

Example 18.3 For $N = 2$ we start with the monomial basis: $A = (1, x, x^2)$, where $A \in \mathbb{R}^{1 \times 3}$. We seek $A = QR$ such that

$$(a_0|a_1|a_2) = (q_0|q_1|q_2) \begin{pmatrix} r_{00} & r_{01} & r_{02} \\ 0 & r_{11} & r_{12} \\ 0 & 0 & r_{22} \end{pmatrix}.$$

In a (reduced) QR decomposition, $Q \in \mathbb{R}^{1 \times 3}$ is a matrix with orthonormal columns and $R \in \mathbb{R}^{3 \times 3}$ is an upper triangular matrix (where, to simplify the exposition, we assume that all the elements of the decomposition live in real space). Following classical Gram-Schmidt we start with the first column: $a_0 = r_{00}q_0$ and letting $r_{00} = 1$ yields $q_0 = a_0 \equiv 1$.

Next, we write for the second column: $a_1 = r_{01}q_0 + r_{11}q_1$, which yields $r_{01} = (q_0, a_1) = 0$, where (a, b) is the inner product (integral) in L^2. The second column equality simplifies to $r_{11}q_1 = a_1$ with $r_{11} = 1$ and $q_1 = a_1 \equiv x$.

Finally, we write for the third column: $a_2 = r_{02}q_0 + r_{12}q_1 + r_{22}q_2$. From this relation we can derive the coefficients as follows: $r_{02} = (q_0, a_2) = \frac{2}{3}$, $r_{12} = (q_1, a_2) \equiv 0$, and finally $r_{22}q_2 = a_2 - r_{02}q_0 \equiv x^2 - \frac{2}{3}$. Rewriting the right-hand side gives $r_{22}q_2 = \frac{2}{3}\left(\frac{3}{2}x^2 - 1\right)$ and letting $r_{22} = \frac{2}{3}$ yields $q_2 = \frac{3}{2}x^2 - 1$.

Putting it all together gives us the QR factorization

$$(1, x, x^2) = \left(1, x, \frac{3}{2}x^2 - 1\right) \begin{pmatrix} 1 & 0 & \frac{2}{3} \\ 0 & 1 & 0 \\ 0 & 0 & \frac{2}{3} \end{pmatrix}. \tag{18.13}$$

Note that in the classical QR factorization, we usually enforce orthonormalization which we did not do here only to make the exposition simpler. Using Eq. (18.13), we can now construct the orthogonal polynomial expansion as follows

$$q_N(x) = \sum_{k=0}^{N} \tilde{q}_k x^k \equiv \sum_{k=0}^{N} \tilde{q}_k^{(P)} P_k(x) \tag{18.14}$$

where the new orthogonal polynomials P are, in fact, Legendre polynomials. Expressing the monomial basis as $M_k(x)$ we now rewrite the middle term in Eq. (18.14) as follows: $M_k(x)\tilde{q}_k$. Replacing $M_k(x)$ with QR from Gram-Schmidt yields the following

$$M_k(x)\tilde{q}_k = Q(R\tilde{q}_k) \tag{18.15}$$

which tells us that the new expansion coefficients, for the orthogonal polynomial expansion given in Eq. (18.14), are written as $\tilde{q}_k^{(P)} = R\tilde{q}_k$ (since $P_k(x) = Q$).

In Example 18.3, the Legendre expansion coefficients are found as follows

$$\begin{pmatrix} \tilde{q}_0^{(P)} \\ \tilde{q}_1^{(P)} \\ \tilde{q}_2^{(P)} \end{pmatrix} = \begin{pmatrix} 1 & 0 & \frac{2}{3} \\ 0 & 1 & 0 \\ 0 & 0 & \frac{2}{3} \end{pmatrix} \begin{pmatrix} \tilde{q}_0 \\ \tilde{q}_1 \\ \tilde{q}_2 \end{pmatrix}.$$

Simplifying, results in

$$\begin{pmatrix} \tilde{q}_0^{(P)} \\ \tilde{q}_1^{(P)} \\ \tilde{q}_2^{(P)} \end{pmatrix} = \begin{pmatrix} \tilde{q}_0 + \frac{2}{3}\tilde{q}_2 \\ \tilde{q}_1 \\ \frac{2}{3}\tilde{q}_2 \end{pmatrix}.$$

Writing the monomial coefficients $\tilde{q}_k$ in terms of Taylor coefficients yields

$$\begin{pmatrix} \tilde{q}_0^{(P)} \\ \tilde{q}_1^{(P)} \\ \tilde{q}_2^{(P)} \end{pmatrix} = \begin{pmatrix} q(0) + \frac{1}{3}q_{xx}(0) \\ q_x(0) \\ \frac{1}{3}q_{xx}(0) \end{pmatrix}.$$

18.5.1.3 Zeroth Mode of an Orthogonal Expansion

Revisiting the orthogonal polynomial expansion

$$q_N(x) = \sum_{k=0}^{N} \tilde{q}_k^{(P)} P_k(x) \tag{18.16}$$

we can show that the zeroth mode ($k = 0$) contains the total conservation of the quantity q. Taking the integral of Eq. (18.16) yields

$$\int_{-1}^{1} q_N(x)\, dx = \int_{-1}^{1} \sum_{k=0}^{N} \tilde{q}_k^{(P)} P_k(x)\, dx. \tag{18.17}$$

Exercise Evaluate the integral in Eq. (18.17). □

However, since $P_k(x)$ have been constructed to be orthogonal, then the integral above is the same as taking the inner product $(P_0(x), q_N(x))$ which yields

$$\int_{-1}^{1} q_N(x)\, dx = 2\tilde{q}_0^{(P)}.$$

Replacing the orthogonal polynomial coefficients with their related Taylor coefficients, for the specific example for $N = 2$ yields

$$\int_{-1}^{1} q_N(x)\, dx = 2\left(q(0) + \frac{1}{3}q_{xx}(0)\right). \tag{18.18}$$

This tells us that for Eq. (18.18) to truly represent the total conservation of q in the domain $x \in [-1, +1]$ we should get the same answer if we integrate the expression for the Taylor polynomial expansion given by Eq. (18.10). Let us write the Taylor polynomial expansion for our $N = 2$ example, yielding

$$q_N(x) = q(0) + q^{(1)}(0)x + q^{(2)}(0)\frac{x^2}{2} \tag{18.19}$$

Integrating Eq. (18.19) gives

$$\int_{-1}^{1} q_N(x)\,dx = 2q(0) + 0q_x(0) + \frac{2}{3}q_{xx}(0)$$

which is equal to the integral of the first term in the orthogonal polynomial expansion. Before we move on to the proof for general N modes, let us try to understand why the conservation of q is concentrated within the terms $q(0)$ and $q_{xx}(0)$ and not in the $q_x(0)$ term. Figure 18.14 shows the Taylor polynomials for $N = 4$. Note that in the domain $x \in [-1, +1]$ none of the odd powers will contribute to the total integral of q, whereas only the even powers will contribute. Therefore, for the case $N = 2$, we only get contributions from $x^{(0)}$ and $x^{(2)}$.

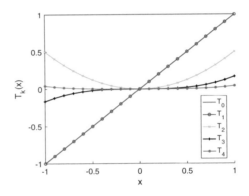

Fig. 18.14: The Taylor polynomials in $x \in [-1, +1]$ for $N = 4$.

Theorem 18.1 *The zeroth mode of an orthogonal expansion contains the total integral of the quantity q.*

Proof Let us write the Taylor basis in terms of an orthogonal basis using QR factorizaton as follows

$$(T_0, T_1, \ldots, T_N) = (\phi_0, \phi_1, \ldots, \phi_N)\begin{pmatrix} r_{00} & r_{01} & \cdots & r_{0N} \\ 0 & r_{11} & \cdots & r_{1N} \\ \vdots & & \ddots & \vdots \\ 0 & 0 & \cdots & r_{NN} \end{pmatrix}.$$

We can now write the first Taylor polynomial as follows: $T_0 = r_{00}\phi_0$. Similarly, the second Taylor polynomial is $T_1 = r_{01}\phi_0 + r_{11}\phi_1$ where $r_{01} = (\phi_0, T_1)$. Proceeding in this fashion, we note that $r_{0k} = (\phi_0, T_k)$ for all $k > 0$. Therefore, r_{0k} contains the integral contributions of all of the Taylor polynomials. From Eq. (18.15) we may write

$$
(T_0, T_1, \ldots, T_N)
\begin{pmatrix} \tilde{q}_0 \\ \tilde{q}_1 \\ \vdots \\ \tilde{q}_N \end{pmatrix}
= (\phi_0, \phi_1, \ldots, \phi_N)
\begin{pmatrix} r_{00} & r_{01} & \cdots & r_{0N} \\ 0 & r_{11} & \cdots & r_{1N} \\ \vdots & & \ddots & \vdots \\ 0 & 0 & \cdots & r_{NN} \end{pmatrix}
\begin{pmatrix} \tilde{q}_0 \\ \tilde{q}_1 \\ \vdots \\ \tilde{q}_N \end{pmatrix}.
$$

This can be simplified to

$$
(T_0, T_1, \ldots, T_N)
\begin{pmatrix} \tilde{q}_0 \\ \tilde{q}_1 \\ \vdots \\ \tilde{q}_N \end{pmatrix}
= (\phi_0, \phi_1, \ldots, \phi_N)
\begin{pmatrix} r_{00}\tilde{q}_0 + r_{01}\tilde{q}_1 + r_{02}\tilde{q}_2 + \ldots + r_{0N}\tilde{q}_N \\ r_{11}\tilde{q}_1 + r_{12}\tilde{q}_2 + \ldots + r_{1N}\tilde{q}_N \\ \vdots \\ r_{NN}\tilde{q}_{NN} \end{pmatrix}
$$

which shows that

$$
\int_{-1}^{+1} \sum_{k=0}^{N} \tilde{q}_k T_k(x)\,dx = r_{00}\tilde{q}_0 + r_{01}\tilde{q}_1 + r_{02}\tilde{q}_2 + \ldots + r_{0N}\tilde{q}_N.
$$

18.5.1.5 Minmod Limiter

Now that we understand the meaning behind the coefficients in a modal expansion, we can consider how to limit the solution. One of the most popular limiters is Harten's *minmod* limiter. The solution $\tilde{q}$ is modified as follows:

$$
\tilde{q}_e^{(1,lim)} = \text{minmod}\left(\tilde{q}_e^{(1)}, \Delta_+\tilde{q}^{(0)}, \Delta_-\tilde{q}^{(0)} \right) \tag{18.20}
$$

where $\Delta_+\tilde{q}^{(0)} = \tilde{q}_{e+1}^{(0)} - \tilde{q}_e^{(0)}$, $\Delta_-\tilde{q}^{(0)} = \tilde{q}_e^{(0)} - \tilde{q}_{e-1}^{(0)}$, and

$$
\text{minmod}(a, b, c) = \begin{cases} s\,\min(|a|, |b|, |c|) & \text{if } \text{sign}(a) = \text{sign}(b) = \text{sign}(c) = s \\ 0 & \text{otherwise.} \end{cases}
$$

Equation (18.20) states that the coefficient for the first mode ($\tilde{q}^{(1)}$) of element e is modified based on its solution and how it compares to the forward and backward differences of the neighboring elements ($e - 1$ and $e + 1$). Since the first mode represents a first derivative, then the minmod function compares the element e derivative value with the derivative $\Delta_+\tilde{q}^{(0)}$ that uses a stencil between elements $e + 1$ and e and the derivative $\Delta_-\tilde{q}^{(0)}$ that uses the stencil between elements e and $e - 1$. The zeroth mode $k = 0$ is not modified while the modes $k \geq 2$ are zeroed out so that this aggresive limiter only retains the zeroth and first modes. Since the first mode is the derivative, which is the slope of the solution, this class of limiters is known as a *slope limiter*. In this text, we only consider slope limiters. There is another class of limiters known as flux limiters (see, e.g., [251]) which we will not discuss here because we are only interested in limiters that can be applied equally to both CG and DG methods. Since the notion of fluxes[1] is not natural to CG methods, we can only limit the CG solution via slope limiters. It is also true that limiting the slope in DG methods is also more natural, because the original flux limiters were designed for finite volume methods where piecewise constant approximations of the solutions are used within each control volume. Since CG and DG methods use higher approximations, it is more natural to use slope limiters.

18.5.2 Positivity-Preserving Limiters

We already saw that limiters can be used to stabilize the numerical methods; however, another useful application of limiters is to construct positivity-preserving solutions. Although these ideas can be used for either CG or DG, it is easier to explain the approach by focusing on the DG method. In fact, much of the limiter technology built for finite volume methods in the early 1970s and 1980s can be used (e.g., [384, 385, 386, 383, 360, 320]). The challenge with those limiters (such as classical minmod limiters, etc.) is that they were designed for low-order methods (e.g., 2nd order finite volume methods). Since DG methods can be viewed as the high-order generalization of finite volume methods, then if we have chosen to use DG it is because we seek to use higher order than, say, second. Therefore, this situation demands the construction of a new set of limiters that: 1) maintain positivity and 2) do not destroy the high-order accuracy of the DG method.

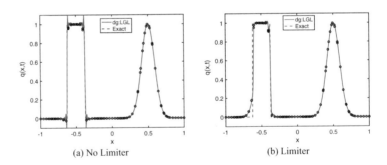

Fig. 18.15: Performance of DG positivity preserving limiter. The DG solution with 8th order polynomials and 50 elements after one-half revolution is shown for (a) no limiter and (b) the positivity preserving limiter.

---

[1] Fluxes exist in CG at the domain boundary but not between element faces.

Zhang and Shu [421] proposed the following linear scaling limiter for the compressible Euler equations

$$\tilde{q}_i^{(e)} = \theta \left(q_i^{(e)} - \bar{q}^{(e)} \right) + \bar{q}^{(e)}, \qquad \theta = min \left\{ \frac{\bar{q}^{(e)} - \epsilon}{\bar{q}^{(e)} - q_{min}}, 1 \right\} \qquad (18.21)$$

where $q_i^{(e)}$ is the DG solution inside the element e and on the quadrature point i, $\bar{q}^{(e)}$ is the mean value of q inside the element e, $q_{min} = min \left(q_i^{(e)} \right) \forall i = 0, \ldots, Q$, Q represents the order of the quadrature formula, and ϵ is a small number (e.g., 10^{-13}). This limiter is quite simple and rather effective, however, it will not stabilize the solution; its only role is to maintain positivity. From Eq. (18.21) one can readily see that if $0 < q_{min} < \bar{q}^{(e)}$, then $\theta = 1$ which yields $\tilde{q}_i^{(e)} = q_i^{(e)}$. On the other hand, if $q_{min} < 0 < \bar{q}^{(e)}$, then $\theta = \frac{\bar{q}^{(e)} - \epsilon}{\bar{q}^{(e)} - q_{min}}$ which then means that $\tilde{q}_i^{(e)} = \epsilon > 0$ when i corresponds to the minimum point inside of e. Clearly, this limiter maintains positivity as long as $\bar{q}^{(e)}$ is positive. This assumption is not too restrictive because we assume (e.g., for tracers) that the variable is initially positive everywhere and so the mean value inside each element should always remain positive. Once we integrate the equations forward in time as long as we have a positivity preserving limiter in space and use a strong stability preserving method in time (e.g., [341, 177, 178, 342, 351, 174, 176, 175]) then $\bar{q}^{(e)}$ should always remain positive. Moreover, it is easy to see from Eq. (18.21) that this limiter preserves conservation. This can be shown straightforwardly by recognizing that $q_i^{(e)}$ can be written in modal form as $q_i^{(e)} = \bar{q}^{(e)} + H.O.T$ where the high-order terms (H.O.T) are expanded via orthogonal polynomials. Since $P_0 = 1$ is the first orthogonal (Legendre) polynomial, then the integral of the remaining H.O.T terms, multiplied by $P_0 = 1$, will vanish.

What is not obvious to see (and of great importance) is that the limiter given in Eq. (18.21) also retains high-order convergence. In other words, if we choose to use high-order methods, the solution remains high-order. For a proof of the preservation of high-order convergence see [421]. Let us now demonstrate how this limiter works in practice.

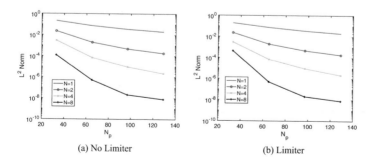

(a) No Limiter   (b) Limiter

Fig. 18.16: Convergence of DG positivity preserving limiter for the 1D wave equation with periodic boundary conditions. The convergence rates are shown for a smooth (Gaussian) problem with (a) no limiter and (b) with the positivity preserving limiter.

Let us solve the one-dimensional wave equation

$$\frac{\partial q}{\partial t} + \frac{\partial}{\partial x}(qu) = 0$$

where $u = 2$ with periodic boundary conditions and the initial condition

$$q(x, 0) = \exp\left(-64(x - x_g)^2\right) + \Delta q \qquad (18.22)$$

where

$$\Delta q = \begin{cases} 1 \text{ for } |x - x_s| \leq r_c \\ 0 \text{ for } |x - x_s| > r_c \end{cases}$$

with $x_g = -\frac{1}{2}$, $x_s = -\frac{1}{2}$, $r_c = \frac{1}{8}$, and $x \in [-1, +1]$. We use the SSP-RK3 time-integrator from Ch. 5 with a Courant number of $C = 0.25$.

Figure 18.15 shows the snapshot of the solution after a half revolution around the periodic domain using 8th degree polynomials. Note that in Fig. 18.15a we see that the high-order DG method has no problem resolving the smooth part of the wave but has major issues with maintaining positivity with the non-smooth square wave. When we apply the limiter, Fig. 18.15b, we see that the method now maintains positivity. Note however, that minor overshoots are still present in the solution. It is possible to eliminate these overshoots by a similar limiter based on a strict *maximum principle* but this assumes that we know what the maximum value can be (see [420]).

To see how the limiter affects the high-order accuracy of the DG method, we run a convergence study for various polynomial orders after one full revolution but only considering the smooth part of the curve (we do not expect high-order convergence for the non-smooth curve since it is not infinitely differentiable); in fact we use the initial condition defined in Eq. (5.32). Figure 18.16a shows the convergence rates with no limiter while Fig. 18.16b shows the convergence rates with the limiter; the convergence rates look very similar. This convergence rate study may not seem so interesting at first glance but in order to perform it we ran each polynomial order through a variety of different numbers of elements. For the under-resolved configurations the limiter was automatically activated in order to combat any negative numbers no matter how small. For these simulations we took $\epsilon = 0$ so we do indeed get perfect positivity for all the simulations shown in this figure. For CG methods, this limiter can be applied provided that the CG method is constructed in the unified CG/DG formulation described in Chs. 7 and 17 (using the local element-wise, or LEW, storage); the results are similar to those shown for the DG method (not shown).

Figure 18.17 shows the dispersion behavior for DG with the positivity preserving limiter of Shu and colleagues [408, 421] in addition to the Rusanov flux for the same initial condition used in Fig. 18.1. The addition of the positivity preserving limiter has allowed the solution to maintain positivity. This is important for some quantities that should remain positive (e.g., density or tracer variables). Looking at Fig. 18.18 we note that the elimination of dispersion errors can be achieved for DG with centered fluxes *provided* that an additional dissipation mechanism is used (in

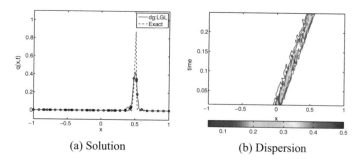

(a) Solution　　　　　　　　　　　(b) Dispersion

Fig. 18.17: The (a) solution and (b) dispersion plots for the 1D wave equation for the DG method with the Rusanov flux and a positivity preserving limiter using 25 elements and 4th order polynomials.

this case, the positivity preserving limiter). Using the positivity preserving limiter,

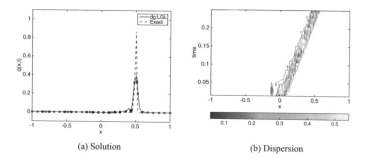

(a) Solution　　　　　　　　　　　(b) Dispersion

Fig. 18.18: The (a) solution and (b) dispersion plots for the 1D wave equation for the DG method with a centered flux and a positivity preserving limiter using 25 elements and 4th order polynomials.

we can see that it can also eliminate the dispersion errors in the CG solution as illustrated in Fig. 18.19.

18.6 Local Adaptive Viscosity Methods

The CG method is often criticized for being non-dissipative - it is understandable why this can be viewed as a deficiency. However, often times the strength and weakness of an idea are two sides of the same coin. The fact that the CG method is inherently

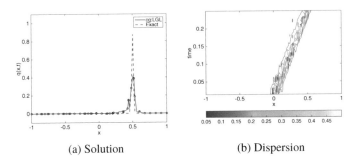

(a) Solution   (b) Dispersion

Fig. 18.19: The (a) solution and (b) dispersion plots for the 1D wave equation for the CG method with a positivity preserving limiter using 25 elements and 4th order polynomials.

non-dissipative should be embraced provided that there exist means to mitigate adverse behavior. In contrast, the DG method has natural dissipation built into it by virtue of the numerical flux (e.g., approximate Riemann solver) provided that some form of upwinding is used (for hyperbolic equations). If so then this concept generally provides sufficient dissipation to combat some steep gradients (see [162] for examples of CG versus DG for strong gradients). In a similar vein, the CG method needs the addition of a dissipative mechanism. Let us begin by introducing diffusion operators that are applicable to both the CG and DG methods.

18.6.1 Diffusion Operators

In order to motivate the validity of introducing diffusion operators to the original equations, let us revisit the difference equations we derived in Ch. 2 for the one-dimensional wave equation. We saw in that chapter that, using Taylor series, we can derive the following 2nd order method (in space)

$$\frac{\partial q_i}{\partial t} + \frac{f_{i+1}^n - f_{i-1}^n}{2\Delta x} = 0. \tag{18.23}$$

This difference equation can be derived either via a 2nd order Taylor series expansion (in a finite difference setting) or via finite elements using linear Lagrange polynomials. Now let us write the first order upwinding representation of this equation from Ch. 1 as follows

$$\frac{\partial q_i}{\partial t} + \frac{f_i^n - f_{i-1}^n}{\Delta x} = 0. \tag{18.24}$$

Next, let us rewrite Eq. (18.24) so that the spatial derivative looks like the one in Eq. (18.23). To do this, we add and subtract the term

$$\frac{f_{i+1}^n - f_{i-1}^n}{2\Delta x}$$

to Eq. (18.24). Simplifying, yields the following new equation

$$\frac{\partial q_i}{\partial t} + \frac{f_{i+1}^n - f_{i-1}^n}{2\Delta x} = \frac{f_{i+1}^n - 2f_i + f_{i-1}^n}{2\Delta x}, \tag{18.25}$$

where we can now see that the upwinding (first-order) representation of the derivative can be viewed as the addition of a diffusion operator[2] to the original centered 2nd order approximation. Although this is much more difficult to show for higher order derivative representations, the general principle still holds.

Exercise Derive Eq. (18.25). □

Exercise Derive a form similar to Eq. (18.25) but write the first derivatives using the third-order accurate representation

$$\frac{-f_{i+2} + 6f_{i+1} - 3f_i - 2f_{i-1}}{6\Delta x}.$$

Let us consider how to apply stabilization via diffusion operators. Generally, stabilization is applied to gridpoint models through the introduction of diffusion operators of the type

$$\nabla^{2K} q$$

where q is the function being filtered and K corresponds to the order of the Laplacian operator. For example, for $K = 1$ we recover a 2nd order operator (standard Laplacian) whereas $K = 2$ results in a 4th order operator, and so forth. Because continuous Galerkin methods are constructed to be C^0 it is not so straightforward to construct diffusion operators higher than $K > 1$ since for every value of K one must construct a weak form (integration by parts) CG Laplacian. The situation for DG is more difficult especially if we use the *flux formulation* for constructing elliptic operators presented in Chs. 9 and 13. We say that it is more difficult for DG with a flux formulation because, for each application of a Laplacian, we have to solve two PDEs. However, if we use the *primal formulation* such as, e.g., the Symmetric Interior Penalty Galerkin (SIPG) method as in Eq. (9.2) in Ch. 9, then we are able to construct the DG solution for elliptic operators using the same number of PDEs required in the CG method (one for each application of a Laplacian)[3].

To show how to construct general hyper-diffusion operators for CG methods, let us review the approach described in [152] and [224]. Assume we want to construct the Laplacian operator using integration by parts (i.e., Green's first identity) as follows

$$\int_{\Omega_e} \psi_i \nabla^2 q_N^{(e)} \, d\Omega_e = \int_{\Gamma_e} \psi_i \hat{\mathbf{n}} \cdot \nabla q_N^{(e)} \, d\Gamma_e - \int_{\Omega_e} \nabla \psi_i \cdot \nabla q_N^{(e)} \, d\Omega_e$$

---

[2] Note that the term on the right-hand side of Eq. (18.25) is, in fact, a 2nd order approximation to a diffusion operator.

[3] For DG, we still need to compute ∇q because this term is required in the flux integrals of the SIPG method.

where ψ are the basis functions, $i = 1, \ldots, M_N$ is the gridpoint we are constructing the Laplacian operator at, $e = 1, \ldots, N_e$ is the element, $\hat{\mathbf{n}}$ is the outward pointing unit normal vector at the boundary Γ_e of the domain Ω_e. Ignoring the boundary integral for simplicity[4] gives

$$\nabla^2 q_I = -M_{I,K}^{-1} L_{K,J} q_J \qquad (18.26)$$

where we have invoked the *DSS* operator as presented in Chs. 8 and 12. Using inexact integration yields

$$\nabla^2 q_I = -M_I^{-1} L_{I,J} q_J \qquad (18.27)$$

which is far more attractive than exact integration since it does not require solving a linear system of equations (because M_I is a vector, not a matrix). Using this idea of integration by parts results in the general $2K$ order diffusion operator:

$$\nabla^{2K} q_I = \left[\prod_{k=1}^{K} \left(-M_I^{-1} L_{I,J} q_J \right)_k \right] \qquad (18.28)$$

that requires K iterations to construct a $2K$ diffusion operator. Using $K = 1$ is an inexpensive option but can be too diffusive for most purposes; choosing $K = 2$ is perhaps the best compromise between cost and scale-selection and this is what is proposed in [224] and used effectively in [104].

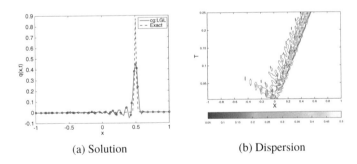

(a) Solution   (b) Dispersion

Fig. 18.20: The (a) solution and (b) dispersion plots for the 1D wave equation with 2nd order viscosity with $\nu_2 = 0.005$ for CG using 25 elements and 4th order polynomials.

---

[4] For CG, due to C^0 continuity, the boundary integral disappears at the interface between elements but not at the boundary of the domain.

In Fig. 18.20 we show results for the same problem shown in Fig. 18.1 but with the inclusion of a 2nd order Laplacian with viscosity coefficient $v_2 = 0.005$; i.e., we solve the following PDE

$$\frac{\partial q}{\partial t} + u\frac{\partial q}{\partial x} = v_2\frac{\partial^2 q}{\partial x^2}.$$

In Fig. 18.21 we show results with the inclusion of a 4th order Laplacian with viscosity coefficient $v_4 = 0.005$; here we solve the following PDE

$$\frac{\partial q}{\partial t} + u\frac{\partial q}{\partial x} = v_4\frac{\partial^4 q}{\partial x^4}.$$

The results in Figs. 18.20 and 18.21 show that the addition of viscosity indeed

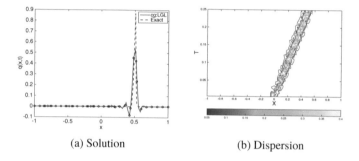

(a) Solution (b) Dispersion

Fig. 18.21: The (a) solution and (b) dispersion plots for the 1D wave equation with 4th order viscosity with $v_4 = 0.005$ for CG using 25 elements and 4th order polynomials.

reduces (2nd order) and eventually eliminates (4th order) the dispersion errors. We saw in Ch. 7, via a dissipation and dispersion analysis, that the lack of dissipation in classical CG methods needs to be rectified in order to avoid dispersion errors. Here, we see it more readily in the solutions for all space and time via an x-t diagram.

Although the same procedure shown here for CG can be carried out to include DG methods, the analysis is a bit more complicated because we cannot ignore the boundary integrals. However, the approach remains the same: we construct an approximation to the Laplacian operator as follows

$$\int_{\Omega_e} \psi_i \nabla^2 q_N^{(e)} d\Omega_e = -\int_{\Omega_e} \nabla\psi_i \cdot \nabla q_N^{(e)} d\Omega_e + \int_{\Gamma_e} \psi_i \left(\hat{\mathbf{n}} \cdot \nabla q_N^{(*)}\right) d\Gamma_e$$

$$+ \int_{\Gamma_e} q_N^{(*)} \left(\hat{\mathbf{n}} \cdot \nabla\psi_i\right) d\Gamma_e \qquad (18.29)$$

where, following the notation in Chs. 9, 13, and 14, the superscript (∗) denotes the numerical flux. From Eq. (18.29) we can define the matrix problem

$$\nabla^2 q_I = -M_{I,K}^{-1}\mathcal{L}_{K,J}q_J \qquad (18.30)$$

where the linear operator $\mathcal{L}$ denotes the entire operator on the right-hand side of Eq. (18.29). We note that for DG the matrix $M_{I,K}$ is in fact block-diagonal for exact integration and fully diagonal for inexact integration. Assuming inexact integration, we can now simplify the matrix problem to the following

$$\nabla^2 q_I = -M_I^{-1} \mathcal{L}_{I,J} q_J \tag{18.31}$$

which looks very similar to the one presented for CG given in Eq. (18.27). Once we construct this last equation, we may continue the process defined in Eq. (18.28), where we need to replace L by $\mathcal{L}$.

18.6.2 Streamline Upwind Petrov-Galerkin

One of the first successful attempts at addressing stabilization for continuous Galerkin methods was the streamline upwind Petrov-Galerkin (SUPG) method [48] where stabilization (diffusion) is added along the flow direction, in essence mimicking the idea of upwinding in finite volume and discontinuous Galerkin methods. Let us discuss this approach which will give us the foundation to understand the other stabilization methods described in this section.

For the one-dimensional advection-diffusion problem

$$\frac{\partial q}{\partial t} + u \frac{\partial q}{\partial x} - \nu \frac{\partial^2 q}{\partial x^2} = f(x,t) \tag{18.32}$$

with forcing $f(x,t)$, we can obtain the weak integral form by first multiplying by the test function ϕ as follows

$$\int_{\Omega_e} \phi_i \left(\frac{\partial q_N^{(e)}}{\partial t} + u_N^{(e)} \frac{\partial q_N^{(e)}}{\partial x} - \nu \frac{\partial^2 q_N^{(e)}}{\partial x^2} - f_N^{(e)}(x,t) \right) d\Omega_e = 0 \tag{18.33}$$

where the variables (q and u) are expanded as follows $q_N^{(e)} = \sum_{i=1}^{N+1} \psi(x_i) q_i^{(e)}(t)$ where ψ are the trial functions. For simplicity, let us assume that u is constant. A possibility for including upwinding into the finite element discretization is to define the test function as follows:

$$\phi(x) = \psi(x) + \tau u \frac{\partial \psi}{\partial x}(x)$$

and in multiple dimensions

$$\phi(\mathbf{x}) = \psi(\mathbf{x}) + \tau \mathbf{u}_N \cdot \nabla \psi(\mathbf{x})$$

where τ is a stabilization parameter to be defined later. Using this definition of the test function in Eq. (18.33) yields

$$\int_{\Omega_e} \psi_i \left(\frac{\partial q_N^{(e)}}{\partial t} + u_N^{(e)} \frac{\partial q_N^{(e)}}{\partial x} - \nu \frac{\partial^2 q_N^{(e)}}{\partial x^2} - f_N^{(e)}(x,t) \right) d\Omega_e$$

$$= \int_{\Omega_e} \tau u_N^{(e)} \frac{\partial \psi_i}{\partial x} R_N^{(e)}(x,t) d\Omega_e \tag{18.34}$$

where

$$R_N^{(e)}(x,t) = -\left(\frac{\partial q_N^{(e)}}{\partial t} + u_N^{(e)} \frac{\partial q_N^{(e)}}{\partial x} - \nu \frac{\partial^2 q_N^{(e)}}{\partial x^2} - f_N^{(e)}(x,t) \right) \tag{18.35}$$

is the residual of the original system. Equation (18.34) defines the SUPG form of Eq. (18.32). However, to better understand this method, let us simplify it to the *streamline upwind* (SU) method whereby only the advection term in Eq. (18.35) is considered which yields the following form

$$\int_{\Omega_e} \psi_i \left(\frac{\partial q_N^{(e)}}{\partial t} + u_N^{(e)} \frac{\partial q_N^{(e)}}{\partial x} - \nu \frac{\partial^2 q_N^{(e)}}{\partial x^2} - f_N^{(e)}(x,t) \right) d\Omega_e$$

$$= -\int_{\Omega_e} \tau \left(u_N^{(e)} \frac{\partial \psi_i}{\partial x} \right) \left(u_N^{(e)} \frac{\partial q_N^{(e)}}{\partial x} \right) d\Omega_e. \tag{18.36}$$

Including the test functions for q_N and rearranging the right-hand side of this equation yields

$$\int_{\Omega_e} \psi_i \psi_j d\Omega_e \frac{dq_j^{(e)}}{dt} + \int_{\Omega_e} \left(\psi_i u_N^{(e)} \frac{\partial \psi_j}{\partial x} + \nu \frac{\partial \psi_i}{\partial x} \frac{\partial \psi_j}{\partial x} \right) d\Omega_e \, q_j^{(e)}$$

$$+ \int_{\Omega_e} \tau \left(u_N^{(e)} \right)^2 \frac{\partial \psi_i}{\partial x} \frac{\partial \psi_j}{\partial x} d\Omega_e \, q_j^{(e)} = \int_{\Omega_e} \psi_i f_N^{(e)}(x,t) d\Omega_e \tag{18.37}$$

with $i, j = 0, \ldots, N$ and $e = 1, \ldots, N_e$, where, for simplicity, we assume homogeneous Dirichlet boundary conditions in the integration by parts for the diffusion term on the left-hand side. We can see what the stabilization is doing; it is an additional diffusion term that is scaled by the velocity and τ. In the case of the SU method, $\tau = 1$ whereas for SUPG τ is computed. For the one-dimensional advection-diffusion equation we consider in this section, there exists an analytical expression for τ [48] given as follows

$$\tau = \frac{1}{2} \frac{\Delta x}{\|u\|} \left(\coth(Pe) - \frac{1}{Pe} \right) \tag{18.38}$$

where $Pe = \frac{u\Delta x}{\nu}$ is the *Peclet number*. For general problems, when no analytic expression is available, a simple choice is given in [265] as follows

$$\tau = \frac{1}{2} \frac{\Delta x}{\|u\|} \frac{Pe}{Pe + 1}. \tag{18.39}$$

For our test problem (Eq. (18.32)) in $x \in [0,1]$ with $f(x,t) = 0$, the exact steady-state solution is

$$q_{\text{exact}}(x) = \frac{\exp \frac{u \cdot x}{\nu} - 1}{\exp \frac{u}{\nu} - 1}. \tag{18.40}$$

Exercise Show that Eq. (18.40) is the exact solution to the steady-state problem given in Eq. (18.32) for $f(x, t) = 0$.   □

If we use the exact solution for the boundary conditions, we arrive at the following solutions illustrated in Figs. 18.22 and 18.23 for various methods. For all simulations we use the CG method with $N = 1$ and $N_e = 20$ with $u = 1$ and variable v in order to change the Peclet number. If the Peclet number is small (the ratio of advection to diffusion is small) then the solution is well-behaved. For example, using the CG method with no stabilization for $Pe = 0.25$ yields the results shown in Fig. 18.22a which are quite good. In contrast, the numerical solution for $Pe = 2.5$ shown in Fig. 18.22b exhibits oscillations due to the lack of regularity in the solution. If we use

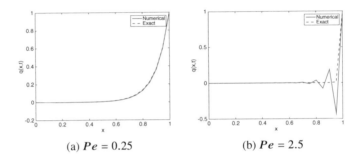

(a) $Pe = 0.25$ (b) $Pe = 2.5$

Fig. 18.22: The CG solution for (a) $Pe = 0.25$ and (b) $Pe = 2.5$ for the 1D advection-diffusion equation using 20 elements and 1st order polynomials.

the SU method with $\tau = 1$ we get the solution shown in Fig. 18.23a which, although smooth, is too diffusive. In Fig. 18.23b we show the solution using SUPG with the general τ parameter from Eq. (18.39) which is much better and at essentially the same complexity. Using the exact value of τ given in Eq. (18.38) yields the solution shown in Fig. 18.23c, which is exact up to machine precision.

Although the SU method can be successful in stabilizing the solution (although quite diffusive), its main criticism is that it is not consistent. Recall from Ch. 1 that consistency means that as the resolution is increased, i.e., $(\Delta x, \Delta t) \to 0$, that we are able to recover the original PDE. Clearly this is not the case for SU since the stabilization term never vanishes. In contrast, for SUPG since τ is a computed parameter then if we ensure that it vanishes for increasing resolution, then we can satisfy consistency. The methods that we describe below, although follow from the previous SUPG machinery, take a different approach (not necessarily relying on the Petrov-Galerkin machinery) to derive other types of consistent stabilization methods.

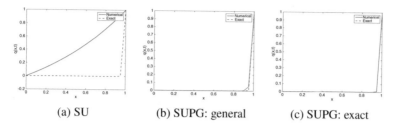

(a) SU   (b) SUPG: general   (c) SUPG: exact

Fig. 18.23: The CG solution for $Pe = 2.5$ with stabilization for (a) SU, (b) SUPG with the general τ parameter (SUPG: general), and (c) SUPG with the exact τ parameter (SUPG: exact) for the 1D advection-diffusion equation using 20 elements and 1st order polynomials.

18.6.3 Variational Multi-Scale Method

Another method for stabilizing element-based Galerkin methods is the *variational multi-scale* (VMS) method [209]. The idea behind VMS is very much akin to the SUPG method where the entire residual is part of the stabilization term. Let us describe the VMS method applied to high-order CG methods. This description follows closely the approach in [264].

Let us define the residual of the PDE we wish to stabilize as follows

$$R \equiv \frac{\partial q}{\partial t} + u \frac{\partial q}{\partial x} - \nu \frac{\partial^2 q}{\partial x^2} - f(x, t) = 0. \tag{18.41}$$

Recall that the classical CG discretization of this PDE is the following

$$\int_{\Omega_e} \psi_i \left(\frac{\partial q_N^{(e)}}{\partial t} + u_N^{(e)} \frac{\partial q_N^{(e)}}{\partial x} - \nu \frac{\partial^2 q_N^{(e)}}{\partial x^2} - f_N^{(e)}(x, t) \right) d\Omega_e = 0 \tag{18.42}$$

where ψ are the test functions and q is approximated by an Nth degree trial function[5]. We have already seen that Eq. (18.42) is susceptible to oscillatory behavior for large Peclet numbers. The idea behind VMS is to handle the unresolved scales appropriately that pollute the finite element solution. To ameliorate this situation, the VMS method introduces two function spaces: $\mathcal{V} = \mathcal{V}_N \oplus \tilde{\mathcal{V}}$, where $\mathcal{V}_N$ is the Nth degree polynomial space and $\tilde{\mathcal{V}}$ is the space that fully completes it. I.e, $\mathcal{V}_N$ approximates the grid resolvable scales while $\tilde{\mathcal{V}}$ represents the unresolved (sub-grid) scales. The task is now to approximate the unresolved scales and account for them in some sense. By partitioning the test functions and solution vectors using these two spaces allows us to write: $\psi = \psi^h + \tilde{\psi}$ and $q = q^h + \tilde{q}$ where we have changed notation slightly in order to adhere to standard VMS notation[6]. Substituting these

---

[5] This slight abuse of notation is necessary in order to simplify the exposition.
[6] The variable $q^h = \sum_{i=0}^{N} \psi_i^h(x) q_i^{(e)}(t)$ and we assume that $u = $ constant.

expressions into the classical CG discretization (Eq. (18.42)) yields

$$\int_{\Omega_e} \psi_i^h \left(\frac{\partial q^h}{\partial t} + u \frac{\partial q^h}{\partial x} - \nu \frac{\partial^2 q^h}{\partial x^2} - f(x,t) \right) d\Omega_e$$
$$+ \int_{\Omega_e} \psi_i^h \left(\frac{\partial \tilde{q}}{\partial t} + u \frac{\partial \tilde{q}}{\partial x} - \nu \frac{\partial^2 \tilde{q}}{\partial x^2} \right) d\Omega_e = 0 \tag{18.43}$$

and

$$\int_{\Omega_e} \tilde{\psi}_i \left(\frac{\partial q^h}{\partial t} + u \frac{\partial q^h}{\partial x} - \nu \frac{\partial^2 q^h}{\partial x^2} - f(x,t) \right) d\Omega_e$$
$$+ \int_{\Omega_e} \tilde{\psi}_i \left(\frac{\partial \tilde{q}}{\partial t} + u \frac{\partial \tilde{q}}{\partial x} - \nu \frac{\partial^2 \tilde{q}}{\partial x^2} \right) d\Omega_e = 0. \tag{18.44}$$

Substituting Eq. (18.41) into Eq. (18.44) gives us an approximation for the unresolved scales as follows

$$\int_{\Omega_e} \tilde{\psi} \mathcal{L}(\tilde{q}) d\Omega_e = - \int_{\Omega_e} \tilde{\psi} R(q^h) d\Omega_e \tag{18.45}$$

where

$$\mathcal{L}(\tilde{q}) = \frac{\partial \tilde{q}}{\partial t} + u \frac{\partial \tilde{q}}{\partial x} - \nu \frac{\partial^2 \tilde{q}}{\partial x^2}.$$

Writing Eq. (18.45) in the strong differential form

$$\mathcal{L}(\tilde{q}) = -R(q^h) \tag{18.46}$$

and approximating $\tilde{q}$ in terms of bubble functions $b(x)$ and the residual $R(q^h)$ yields, after some algebra (see, e.g., p. 90 in [263] for a detailed derivation), the following approximation of the unresolved scales

$$\tilde{q} = -\tau R(q^h) \tag{18.47}$$

which we substitute into Eq. (18.43). However, substituting directly would complicate matters because we would need to take mixed space-time derivatives of $R(q^h)$. The solution to this dilemma is to first rewrite Eq. (18.43) in the following way

$$\int_{\Omega_e} \psi^h \left(\frac{\partial q^h}{\partial t} + u \frac{\partial q^h}{\partial x} - \nu \frac{\partial^2 q^h}{\partial x^2} - f(x,t) \right) d\Omega_e + \int_{\Omega_e} \psi^h \mathcal{L}(\tilde{q}) d\Omega_e = 0 \tag{18.48}$$

and then use integration by parts in order to derive the adjoint operator of $\mathcal{L}$. For the moment let us only focus on the second integral term in Eq. (18.48). Applying integration by parts to the second integral in Eq. (18.48) yields the following expression

$$\int_{\Omega_e} \psi^h \mathcal{L}(\tilde{q}) d\Omega_e = - \int_{\Omega_e} \left(u \frac{\partial \psi^h}{\partial x} + \nu \frac{\partial^2 \psi^h}{\partial x^2} \right) \tilde{q} \, d\Omega_e \tag{18.49}$$

where

$$\mathcal{L}^* = -u\frac{\partial}{\partial x} - v\frac{\partial^2}{\partial x^2} \tag{18.50}$$

is the adjoint operator of

$$\mathcal{L} = u\frac{\partial}{\partial x} - v\frac{\partial^2}{\partial x^2} \tag{18.51}$$

where we assumed all boundary integrals are zero and made a further assumption that $\frac{\partial \tilde{q}}{\partial t} = 0$.

Exercise Show that Eq. (18.50) is in fact the adjoint of $\mathcal{L}$ given in Eq. (18.51). Assume that all boundary terms are zero and that $\frac{\partial \tilde{q}}{\partial t} = 0$.

Exercise Work out the previous exercise but for the general multi-dimensional case. ☐

Using the adjoint operator we derived in Eq. (18.49) and the approximation for $\tilde{q}$ in Eq. (18.47) we arrive at the VMS representation of Eq. (18.43)

$$\int_{\Omega_e} \psi^h \left(\frac{\partial q^h}{\partial t} + u\frac{\partial q^h}{\partial x} - v\frac{\partial^2 q^h}{\partial x^2} - f(x,t) \right) d\Omega_e - \int_{\Omega_e} \tau \mathcal{L}^*(\psi^h)R(q^h)d\Omega_e = 0 \tag{18.52}$$

where, as in the SUPG method, the crux is finding a good stabilization parameter τ. In [264] a simple parameter for use with high-order CG methods for the advection-diffusion equation is derived and given as follows

$$\tau = \frac{h}{2u}\left(\coth\left(Pe^{(e)} \right) - \frac{1}{Pe^{(e)}} \right) \tag{18.53}$$

where h is the element length, u is the velocity, and $Pe^{(e)}$ is the element Peclet number defined as $Pe^{(e)} = \frac{uh}{2v}|_{\Omega_e}$ where v is the kinematic viscosity of the problem. The attraction of the VMS method, as well as the SUPG method, is that the dissipation (last term in Eq. (18.52)) acts only when the residual is non-zero. When the method is able to satisfy the governing PDE this term vanishes thereby satisfying consistency.

In order to better understand the VMS method, it is useful to derive explicitly the resulting equations for a simpler problem. Let us make the following assumptions: $\frac{\partial q}{\partial t} = 0$ (steady-state problem), $f(x,t) = 0$ (homogeneous PDE), and $v = 0$ (the inviscid problem). Under these assumptions, Eq. (18.52) becomes

$$\int_{\Omega_e} \psi^h u\frac{\partial q^h}{\partial x}d\Omega_e + \int_{\Omega_e} \tau\left(u\frac{\partial \psi^h}{\partial x} \right)\left(u\frac{\partial q^h}{\partial x} \right)d\Omega_e = 0 \tag{18.54}$$

which is nothing more than the SU or SUPG formulation of the steady-state advection equation. For the viscous case ($v > 0$) the VMS formulation is similar to SUPG but not identical.

18.6.4 Dynamic Sub-Grid Scales

The class of method that we refer to as *dynamic sub-grid scales* (DSGS) is essentially an artificial diffusion formulation based on physical arguments and is related to numerous other forms of large-eddy simulation (LES) closures including those in [252, 347, 100, 272, 391]. DSGS is also related to many of the entropy viscosity methods first proposed in [364] and others [302, 21, 226, 182, 282, 281, 1]. Although many types of DSGS methods exist, we only discuss one method because it can be equally applied to both continuous and discontinuous Galerkin methods while many other methods are specifically focused on either the CG or DG method. The method that we describe below is due to Nazarov and Hoffman [282] which was initially developed for (low-order) finite element methods. This method was extended to high-order element-based Galerkin methods in [268] and is the approach we now describe.

To remain consistent with the discussion of the previous methods, let us continue with the one-dimensional advection-diffusion equation defined in Eq. (18.32). However, we begin by augmenting the original PDE with a dissipation term as follows:

$$
\int_{\Omega_e} \psi_i \left(\frac{\partial q_N^{(e)}}{\partial t} + u_N^{(e)} \frac{\partial q_N^{(e)}}{\partial x} - \nu \frac{\partial^2 q_N^{(e)}}{\partial x^2} - f_N^{(e)}(x,t) \right) d\Omega_e
$$

$$
= \int_{\Omega_e} \psi_i \mu(x,t) \frac{\partial^2 q_N^{(e)}}{\partial x^2} d\Omega_e \qquad (18.55)
$$

where $\mu(x,t)$ must be computed for each element Ω_e and changes in time. A physical argument for computing μ is to introduce artificial diffusion only when the residual within each element is large. One such possibility is to define

$$
\mu = \min \left(\mu_{max}^{(e)}, \mu_{res}^{(e)} \right) \qquad (18.56)
$$

where $\mu_{max}^{(e)} = 0.5 \bar{\Delta} \| \lambda^{(e)} \|_{L^\infty}$ with $\lambda^{(e)}$ being the maximum wave-speed in the element Ω_e (i.e., maximum eigenvalue of the Jacobian) and $\bar{\Delta}$ is a characteristic length (either the length of the element or the distance between two grid points), and

$$
\mu_{res}^{(e)} = \bar{\Delta}^2 \frac{\| R \left(q_N^{(e)} \right) \|_{L^\infty}}{\| q_N^{(e)} - \bar{q}_N^{(g)} \|_{L^\infty}} \qquad (18.57)
$$

where $\bar{q}_N^{(g)}$ is the global mean value of the solution vector, $q_N^{(e)}$ is the element-wise solution vector, and $R \left(q_N^{(e)} \right)$ is the element-wise residual with $\| \cdot \|_{L^\infty}$ representing the element-wise L^∞ norm. From Eq. (18.57), we can see that the DSGS method is indeed consistent because, as $R \left(q_N^{(e)} \right) \to 0$, then so does μ. In other words, as the PDE becomes well resolved, then there should be no need for stabilization and the original PDE is recovered. To highlight this point, we see in Fig. 18.24b that when the flow is under-resolved, the DSGS method automatically computes the stabilization parameter μ to be large, especially where it is needed. However, once the flow is

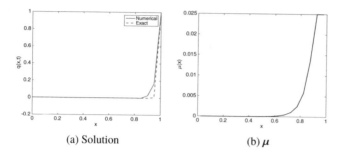

(a) Solution   (b) μ

Fig. 18.24: The CG solution with DSGS stabilization for under-resolved flow with $Pe = 2.5$ and 20 elements for (a) the solution vector and (b) the stabilization parameter μ for the 1D advection-diffusion equation using 1st order polynomials.

better resolved, the stabilization parameter dramatically decreases in magnitude as shown in Fig. 18.25b. The results shown in Fig. 18.24 use 20 elements whereas the results shown in Fig. 18.25 use 200 elements which account for the difference in the Peclet number (both use $u = 1$ and $\nu = 0.01$).

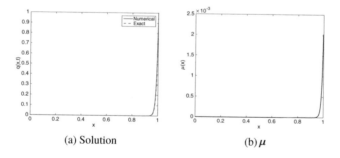

(a) Solution   (b) μ

Fig. 18.25: The CG solution with DSGS stabilization for well-resolved flow with $Pe = 0.25$ and 200 elements for (a) the solution vector and (b) the stabilization parameter μ for the 1D advection-diffusion equation using 1st order polynomials.

18.7 Provably Stable Methods

The methods described previously for stabilizing element-based Galerkin methods can be characterized as *a posteriori* methods, meaning that an additional mechanism is introduced into the original discretization that stabilizes the method. The advantage of these *a posteriori* methods is that they are equally applicable to both the continuous and discontinuous Galerkin methods described throughout this text.

The provably stable methods described in this section are known in the literature as *energy stable* and *entropy stable* methods and have their roots in early work by Tadmor with application to finite volume methods [362, 363, 364, 365]. The general idea behind *provably stable* methods is to derive a new set of variables and fluxes that can be proven to remain stable under some norm. These ideas have proliferated into other discretization methods including finite difference methods [130] and discontinuous Galerkin methods [140, 1, 240, 142, 242, 294, 76, 406, 73, 74, 245, 407] whereby either the energy (called energy stable methods) or entropy (called entropy stable methods) is conserved for smooth solutions and dissipated for non-smooth solutions, which maintains stability regardless of the order of accuracy. Although the current work on provably stable methods does not yet cover continuous Galerkin methods, it should not be impossible to extend these ideas to continuous methods. The reason why discontinuous Galerkin methods have been the method of choice is due to the simple analysis required (only one element needs to be analyzed to prove stability whereas for CG we need to extend to more elements in order to incorporate external boundary conditions). The other reason why DG methods have been mostly considered is due to the fact that the so called *summation by parts* (SBP) property of certain finite difference methods have been shown to apply to specific construction of DG methods (inexact integration on Lobatto points using tensor-products, i.e., the focus of this textbook). The SBP property means that the continuous integration by parts property also holds at the discrete level. This SBP property is required to prove conservation and is necessary to construct provably stable methods (both CG and DG satisfy the SBP property) and piggy-backing on the work of this rather extensive community has allowed the construction of provably stable DG methods (some works on SBP finite difference methods can be found in these citations [355, 125, 359, 97]).

18.7.1 Classical DG Solution of the 1D Burgers Equation

To briefly describe the provably stable approach, let us consider the one-dimensional Burgers equation written in conservation form as

$$\frac{\partial u}{\partial t} + \frac{\partial f}{\partial x} = 0 \qquad (18.58)$$

where $f = \frac{1}{2}u^2$. Discretizing Eq. (18.58) element-wise by the DG method yields

$$\int_{\Omega_e} \psi_i \frac{\partial u_N^{(e)}}{\partial t} d\Omega_e + \int_{\Omega_e} \psi_i \frac{\partial f_N^{(e)}}{\partial x} d\Omega_e = 0 \qquad (18.59)$$

where $q_N^{(e)} = \sum_{j=0}^{N} \psi_j(x) q_j(t)$ is the approximation of q in the element Ω_e and $i = 0, \ldots, N$. Integrating by parts and introducing the numerical flux yields

$$\int_{\Omega_e} \psi_i \frac{\partial u_N^{(e)}}{\partial t} d\Omega_e + \left[\psi_i f_N^{(*)} \right]_{\Gamma_e} - \int_{\Omega_e} \frac{\partial \psi_i}{\partial x} f_N^{(e)} d\Omega_e = 0 \qquad (18.60)$$

where for convenience we use the Rusanov flux

$$f_N^{(*)} = \frac{1}{2} \left[f_N^{(e)} + f_N^{(k)} - \hat{\mathbf{n}} \lambda^{(e,k)} \left(u_N^{(k)} - u_N^{(e)} \right) \right]$$

with $\lambda^{(e,k)} = \max \left(|u_N^{(k)}|, |u_N^{(e)}| \right)_{\Gamma_e}$ where k is the face neighbor of e. Following the DG discretization described in Ch. 6, the global matrix-vector form of Eq. (18.60) becomes

$$M_{IJ} \frac{dq_J}{dt} + F_{IJ} f_J - \tilde{D}_{IJ} f_J = 0 \qquad (18.61)$$

where M and $\tilde{D}$ are block diagonal and F is the flux matrix which couples neighboring elements and $I, J = 1, \ldots, N_p$ where $N_p = N_e(N+1)$ are the total number of gridpoints in the grid.

Following [140] we solve the following problem: let $x \in [0, 2]$ and $t \in [0, \infty)$ with periodic boundary conditions and the initial condition defined as

$$u(x, 0) = \sin \pi x + 0.01.$$

We use $N = 6$ polynomials with $N_e = 21$ elements using the 2nd order Runge-Kutta method from Ch. 20 with $\Delta t = 0.001$. Note that we use an odd number of elements to make the test more stringent since the discontinuity (located at $x = 1$) is positioned in the middle of an element. If we choose an even number of elements, then the test case is less stringent since the discontinuity will be positioned exactly at the interface of two elements, which can be handled much easier by the numerical flux.

Figure 18.26 shows the solution at $t = 0.45$ time units, which clearly shows that the method goes unstable.

18.7.2 Energy Stable Solution of the 1D Burgers Equation

Since the classical DG discretization of the conservative form of the Burgers equation is unstable, we must now consider how to achieve stability. For the 1D Burgers equation, it is sufficient to use the spectral filters presented in Sec. 18.3 or the limiters presented in Sec. 18.5; however, these approaches may not necessarily be sufficient for nonlinear systems of equations such as the Euler equations. In fact, often times local adaptive viscosity methods such as those presented in Sec. 18.6

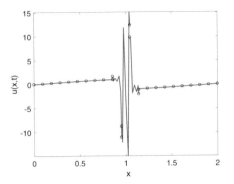

Fig. 18.26: The solution for the 1D Burgers equation in conservation form for the weak form DG method with Rusanov flux using 21 elements and 6th order polynomials. This simulation goes unstable.

are used. In this section, we describe a different approach based on provably stable methods.

The *skew-symmetric form*[7] of the 1D Burgers equation is written as follows

$$\frac{\partial u}{\partial t} + \alpha \frac{\partial f}{\partial x} + (1-\alpha)\left[\frac{\partial}{\partial x}\left(\frac{1}{2}u\right)u + \left(\frac{1}{2}u\right)\frac{\partial u}{\partial x}\right] = 0 \qquad (18.62)$$

and the weak form for each element becomes

$$\int_{\Omega_e} \psi_i \frac{\partial u_N^{(e)}}{\partial t} d\Omega_e + \alpha \left\{\left[\psi_i f_N^{(*)}\right]_{\Gamma_e} - \int_{\Omega_e} \frac{\partial \psi_i}{\partial x} f_N^{(e)} d\Omega_e\right\} \qquad (18.63)$$

$$+ (1-\alpha)\left\{\left[\psi_i f_N^{(*)}\right]_{\Gamma_e} - \int_{\Omega_e} \frac{\partial}{\partial x}\left(\psi_i u_N^{(e)}\right)\left(\frac{1}{2}u_N^{(e)}\right) d\Omega_e\right\}$$

$$+ (1-\alpha)\left\{\int_{\Omega_e} \left(\frac{1}{2}u_N^{(e)}\right)\frac{\partial}{\partial x}\left(u_N^{(e)}\right) d\Omega_e\right\} = 0$$

where the terms on the second line result from integration by parts of the first term in square brackets of Eq. (18.62). Simplifying Eq. (18.63) results in

$$\int_{\Omega_e} \psi_i \frac{\partial u_N^{(e)}}{\partial t} d\Omega_e + \left[\psi_i f_N^{(*)}\right]_{\Gamma_e} - \alpha \int_{\Omega_e} \frac{\partial \psi_i}{\partial x} f_N^{(e)} d\Omega_e \qquad (18.64)$$

$$+ (1-\alpha)\left[-\int_{\Omega_e} \frac{\partial}{\partial x}\left(\psi_i u_N^{(e)}\right)\left(\frac{1}{2}u_N^{(e)}\right) d\Omega_e + \int_{\Omega_e} \left(\frac{1}{2}u_N^{(e)}\right)\frac{\partial}{\partial x}\left(u_N^{(e)}\right)\right] d\Omega_e = 0$$

---

[7] The *skew-symmetric form* typically refers to writing the discrete operators using an average of the weak and strong forms.

which can be discretized as is or, alternatively, using the product rule in the first term in the square brackets on the second line, yields

$$\int_{\Omega_e} \psi_i \frac{\partial u_N^{(e)}}{\partial t} d\Omega_e + \left[\psi_i f_N^{(*)} \right]_{\Gamma_e} - \int_{\Omega_e} \frac{\partial \psi_i}{\partial x} f_N^{(e)} d\Omega_e$$

$$+ (1 - \alpha) \int_{\Omega_e} u_N^{(e)} \frac{\partial}{\partial x} \left(u_N^{(e)} \right) d\Omega_e = 0. \qquad (18.65)$$

In [140] it is shown that by replacing ψ with u in Eq. (18.64), the value $\alpha = \frac{2}{3}$ allows for the cancellation of all volume integrals. This then means that for Eq. (18.64) to be energy stable (conserves energy for smooth solutions and dissipates energy for non-smooth solutions) we only need to satisfy certain constraints on the numerical flux. The Rusanov flux satisfies these constraints and thereby using inexact integration with the Rusanov flux along with $\alpha = \frac{2}{3}$ results in an energy stable method.

Running the energy stable method on the same test as in Sec. 18.7.1 reveals that the solution now remains stable for all time. Figure 18.27 shows snapshots of the solution at various times. Note that the solution in all the panels look exactly the same, with the exception that the extrema are decreasing, i.e., because a discontinuity (non-smooth solution) is present, the energy stable method dissipates energy. What is remarkable about this approach is that the difference between the original conservative formulation and the energy stable form is quite minimal. Comparing Eqs. (18.60) and (18.65) we note that the only difference is in an additional volume integral. Unlike in the local adaptive viscosity, limiter, or filter approach there are no tunable parameters except for α which is derived once for each equation and does not change regardless of the number of elements, polynomial order, or time-step. However, as can be seen in Fig. 18.27, although provably stable methods guarantee that the solution remains stable for all time, they do not guarantee that the solution will be non-oscillatory; we still require using filters or limiters to this end.

18.7.3 Entropy Stable Methods

For general equations, the energy stable approach introduced in Sec. 18.7.2 will not work (see [73]). Instead, we must turn to entropy stable methods whereby we reformulate the equations using an entropy variable with corresponding entropy flux functions. Following [73], let us introduce this idea for a general one-dimensional nonlinear conservation law written as follows

$$\frac{\partial \mathbf{u}}{\partial t} + \frac{\partial f(\mathbf{u})}{\partial x} = 0 \qquad (18.66)$$

where $\mathbf{u}(x, t) = (u_1(x, t), u_2(x, t), \ldots, u_n(x, t))$ are the n variables of the equations and $f(\mathbf{u})$ are the associated fluxes for each of the n equations associated with the

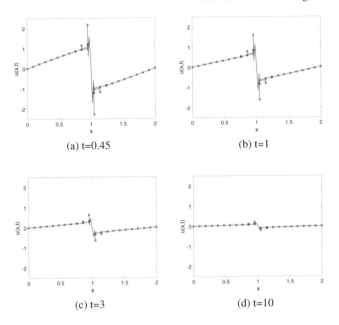

(a) t=0.45

(b) t=1

(c) t=3

(d) t=10

Fig. 18.27: The energy stable DG solution for the 1D Burgers equation using 21 elements and 6th order polynomials at times (a) t=0.45, (b) t=1, (c) t=3, and (d) t=10.

variable $\mathbf{u} = u_i$ for $i = 1, \ldots, n$. Next, we need to construct a convex entropy function $U(u)$ which satisfies

$$\frac{\partial^2 U}{\partial \mathbf{u}^2} A(\mathbf{u}) = \left(\frac{\partial^2 U}{\partial \mathbf{u}^2} A(\mathbf{u})\right)^T \tag{18.67}$$

where $A(\mathbf{u}) = \frac{\partial f(\mathbf{u})}{\partial u}$ is the Jacobian of the system of equations. Once U has been selected, then we can define the entropy variables as follows

$$\mathbf{v} = \frac{\partial U(\mathbf{u})}{\partial \mathbf{u}}. \tag{18.68}$$

Equation (18.67) ensures that the map $\mathbf{u} \leftrightarrow \mathbf{v}$ exists. From the convexity condition (18.67) we obtain the following equation

$$\mathbf{v}^T \frac{\partial f(\mathbf{u})}{\partial \mathbf{u}} = \frac{\partial F(\mathbf{u})}{\partial \mathbf{u}}^T \tag{18.69}$$

for the entropy flux function $F(\mathbf{u})$, along with the following definition for an entropy potential

$$\psi(\mathbf{v}) = \mathbf{v}^T f(\mathbf{u}(\mathbf{v})) - F(\mathbf{u}(\mathbf{v})) \tag{18.70}$$

where

$$\frac{\partial \psi(\mathbf{v})}{\partial v} = f(\mathbf{u}(\mathbf{v})).$$

Taking the inner product of Eq. (18.66) with $\mathbf{v}$ yields

$$\mathbf{v}^T \frac{\partial \mathbf{u}}{\partial t} + \mathbf{v}^T \frac{\partial f(\mathbf{u})}{\partial x} = 0$$

which, after substituting Eqs. (18.68) and (18.69) (along with the chain rule) yields

$$\frac{\partial U}{\partial t} + \frac{\partial F(U)}{\partial x} \leq 0 \tag{18.71}$$

which defines either the entropy conservation (= 0) for smooth solutions or the entropy inequality (< 0) for non-smooth solutions. Note that Eq. (18.71) is a single scalar equation that we aim to satisfy. Once this expression has been derived, along with the proper conserving numerical flux for $F(U)$, this then allows us to construct the numerical scheme for the original PDE (18.66).

Chapter 19
Adaptive Mesh Refinement

19.1 Introduction

One of the advantages of element-based Galerkin methods is that they are geometrically flexible. We define *geometric flexibility* as the capacity of a method to handle unstructured meshes. This is important if, say, we wish to develop a general partial differential equation (PDE) solver for use in arbitrary domains (e.g., on complex geometries such as shells [39], flow over an aircraft [256], or flow over complex terrain [162]). Unstructured meshes arise from having to solve real-world problems and often times the grids are generated by scanning a physical object (e.g., the surface of the human body, surface of the planet, etc.). However, geometric flexibility can also mean that a numerical method for solving PDEs is amenable to adaptive mesh refinement (AMR).

In Ch. 1 we saw an example of an adaptive simulation for a rising thermal bubble and that allowing the grid to refine where the solution needs finer resolution allows a model to be both more accurate and efficient. The accuracy stems from the high-resolution in areas that need this fine granularity whereas the efficiency stems from the fact that in regions where the resolution is not needed AMR constructs a coarse grid. By using fine grids where they are needed and coarse grids where they are not allows a model to use fewer degrees of freedom compared to standard uniform resolution simulations. In [230] an efficiency gain of a factor of five is reported with adaptive methods versus uniform resolution. While it may not always be possible to achieve such gains, one thing is clear: adaptive methods can offer significant savings. Moreover, it is conjectured that AMR will allow the possibility to study complex processes that have been beyond our reach. There exists a large volume of work in the general area of mesh generation [172, 415, 119, 422, 135, 330, 336, 10, 261, 304], adaptive mesh refinement [257, 32, 186, 290, 255, 30, 308, 269, 346, 309, 36, 39, 256, 345, 27, 214, 149, 131, 9, 192, 189, 207, 305, 312, 366, 114, 26, 287, 211, 321, 352, 10,

© The Editor(s) (if applicable) and The Author(s), under exclusive license
to Springer Nature Switzerland AG 2020
F. X. Giraldo, *An Introduction to Element-Based Galerkin Methods on
Tensor-Product Bases*, Texts in Computational Science and Engineering 24,
https://doi.org/10.1007/978-3-030-55069-1_19

138, 19, 279, 31, 56, 117, 122, 145, 37, 278, 230, 414, 231, 267, 193, 40] and moving
mesh methods [113,49,67,50,242] but we will only focus on non-conforming AMR.

This chapter is organized as follows: Sec. 19.2 describes the differences between
conforming and non-conforming grids. Then in Secs. 19.3, 19.4, 19.5 we describe
the three types of AMR which are h-, p-, and r-refinement methods, respectively. In
each of these sections, we show a simple example for these three types of AMR in 1D.
We end this chapter in Sec. 19.6 with an example of non-conforming h-refinement
in 2D.

19.2 Conforming vs non-conforming mesh

The first question that needs to be answered when constructing the AMR algorithm
is whether we seek to construct a conforming or non-conforming strategy.

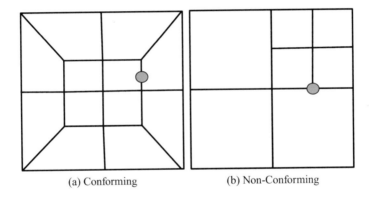

(a) Conforming   (b) Non-Conforming

Fig. 19.1: (a) Conforming and (b) non-conforming adaptive mesh refinement grids.

Figure 19.1 shows examples of conforming (panel a) and non-conforming (panel
b) grids (or meshes)[1]. A *conforming mesh* is defined as a mesh where the faces of the
grid are only shared by two elements. We can see this to be the case for the mesh in
Fig. 19.1a where, in particular, we point the reader's attention to the face containing
the blue dot. In contrast, Fig. 19.1b shows an example of a non-conforming mesh.
Once again, pointing out the face with the blue dot we see that this face is shared by
three elements. The reason why it is referred to as *non-conforming* is that the face with
the blue dot is defined by one element on one side (bottom) while it is defined by two
elements on the other side (top) (see, e.g., [17]). In the conforming case, the entire
burden of handling the changing mesh falls on the mesh adaptation algorithm, which
has to make sure that the grid remains conforming after adaptation. The upside of this
approach is an easy communication between neighboring elements. Since the faces
are conforming, the AMR algorithm does not introduce any additional complication

---

[1] We use grid and mesh interchangably in this section.

to the PDE solver. In other words, if a PDE solver can handle a static conforming grid then it can also handle a dynamically adaptive conforming grid with absolutely no changes. However, constructing a dynamically adaptive conforming mesh refinement strategy is non-trivial and usually requires a significant effort to build a robust algorithm especially if one considers doing this for tensor-product elements such as quadrilaterals and hexahedra. Nonetheless, much work has been done on this topic and can be found in the literature (see, e.g., [309, 36, 422, 330, 138, 10, 117, 261]).

In contrast, in the non-conforming case the mesh adaptation algorithm is kept simple: each element marked for adaptation is then divided into a pre-defined number of children elements. For example, a very simple strategy is to divide, e.g., quadrilateral elements into four children elements[2] (see, e.g, [257,25,239,192,336,243,305,321,230,231,267]). This leads to a situation where, if only one of two neighbor elements is refined, the non-refined neighbor shares a face with two children elements. This requires the PDE solver to handle non-conforming faces. This approach shifts the burden from the mesh adaptation algorithm to the PDE solver.

We describe the general approach developed by Kopriva [234] for constructing h-refinement AMR for discontinuous methods. In this text, we extend this approach to both continuous and discontinuous methods. First, we begin with a summary of mesh refinement methods.

19.3 H-Refinement Method

H-refinement is a form of mesh enrichment, that relies on decreasing the local error of the discretization by the introduction of additional smaller elements. For example, beginning with a 1D domain consisting of one element, we can then increase the accuracy of the simulation by subdividing this element into two smaller ones. We can continue this recursive process by subdividing the two elements into four elements and so on. In one-dimension, this approach forms a binary tree that is easy to construct and traverse. Figure 19.2 shows a binary tree data structure that can be used to refine a grid. In each of the three panels, the red font displays the new element (e) and gridpoints at each of the levels of refinement. Specifically, Fig. 19.2a shows the initial level 0 grid, while Fig. 19.2b shows the level 1 grid resulting from one level of refinement, and Fig. 19.2c shows the level 2 grid resulting from two levels of refinement. The *intma* data structure associated with this grid is shown in Table 19.1. The necessary data structures required for adding h-refinement to an existing EBG code are the following: *intma* constructed in a tree data-structure, *parent, children, refinement_level, active*, and *sfc*. We have already described the *intma* data structure (Fig. 19.2 and Table 19.1). For each element e we need to know if it has a *parent*; this will be the case if the element e has $refinement\_level > 0$. Note

---

[2] The strategy in two-dimensions is known as quadtrees while an analogous strategy in three-dimensions would use octrees. In the octree approach, each hexahedron is divided into 8 smaller hexahedra.

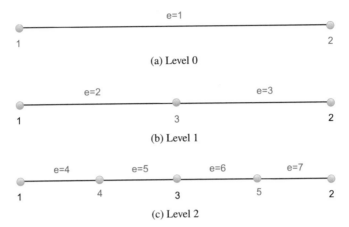

Fig. 19.2: Binary tree-based mesh refinement.

Table 19.1: intma(i,e) array for the grid shown in Fig. 19.2.

refinement_level	e	i=1	i=N
0	1	1	2
1	2	1	3
1	3	3	2
2	4	1	4
2	5	4	3
2	6	3	5
2	7	5	2

that $dim(parent) \in N_e$ where N_e are the total number of elements in the grid and is the dimension of the data structure *intma*. The data structure *child* contains the list of children that an element has; this is only true if the element e is a parent. Note that in 1D $dim(child) = (2, N_e)$ where the 2 stores the two children of the element. In addition, $dim(refinement\_level) = N_e$ and stores the refinement level of each element and $dim(active) = N_e$ and is 0 for inactive or 1 for active elements. Finally, the data structure sfc, standing for *space filling curve*, is the data structure that stores the list of active elements in a consecutive fashion in order to make the code more efficient and to simplify the discretization process. However, sfc is the data structure that requires the most explanation.

19.3.1 Data Structures

Table 19.2 illustrates the *parent* and *child* arrays corresponding to Fig. 19.2. Element $e = 1$ has a value $parent(1) = 0$ because it is a root element (level 0) and therefore has no parent. However, its two children are $child(1 : 2, e) = (2, 3)$. Note that element $e = 2$ then has a value $parent(2) = 1$ because element 1 is the parent of element 2;

Table 19.2: *parent* and *child* arrays for the grid shown in Fig. 19.2.

e	parent(e)	child(1,e)	child(2,e)
1	0	2	3
2	1	4	5
3	1	6	7
4	2	0	0
5	2	0	0
6	3	0	0
7	3	0	0

similarly for element e=3. Elements 2 and 3 are at *refinement_level* = 1. Element e=4 has a value *parent*(4) = 2 because element 2 is its parent. Note that element e=4 has no children because it is at the highest refinement level (level 2).

The remaining two data structures that require explanation are: sfc and *active* with the following dimensions $dim(active) = N_e$ and $dim(sfc) = N_e$. We also need to introduce the variable N_e^a which denotes the number of elements that are currently active. In our example here $N_e = 7$ although N_e^a could be equal to 1, 2, 3, or 4, depending on which elements are *active*. For example, if we are only using the level 0 grid then $active(1) = 1$ and $active(2:7) = 0$. In this case, $sfc(1) = 1$ and $N_e^a = 1$.

However, if we are at the level 1 grid, then $active(2) = active(3) = 1$ and $active = 0$ for all other elements. In this case, $N_e^a = 2$ with $sfc(1:2) = 2, 3$. At the level 2 grid $active(4) = active(5) = active(6) = active(7) = 1$ and $active = 0$ for all other elements. In this case, $N_e^a = 4$ with $sfc(1:4) = 4, 5, 6, 7$. Of course, we may also have a mixture of level 1 and level 2 grids and both possibilities have $N_e^a = 3$. One possibility is the following: $active(2) = active(6) = active(7) = 1$ with $active = 0$ for all other elements. In this case, $sfc(1:3) = 2, 6, 7$; this configuration is shown in Fig. 19.3a. The second possibility is the following: $active(3) = active(4) = active(5) = 1$ with $active = 0$ for all other elements. In this case $sfc(1:3) = 4, 5, 3$ and is shown in Fig. 19.3b.

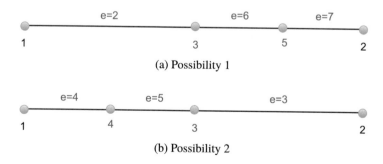

(a) Possibility 1

(b) Possibility 2

Fig. 19.3: Combined Level 1 and 2 mesh refinement.

19.3.2 H-Refinement Algorithm

The h-refinement algorithm is described in three separate snippets of pseudocode. The first one shown in Alg. 19.1 goes through all the active elements and decides which element should be marked for either coarsening or refinement. V_N in Alg. 19.1

Algorithm 19.1 H-refinement: Marking Elements

---

function H_REFINEMENT_MARK
    for $ee = 1 : N_e^a$ **do**
      $e = sfc(ee)$
      $\tilde{q}_N^{(e)} = V_N^{-1} \cdot q_N^{(e)}$ ▷ only needed for nodal solutions
      $test = \dfrac{\|\tilde{q}(N)^{(e)}\|_2}{\|\tilde{q}(0:N)^{(e)}\|_2}$
      if $(test < \epsilon_c$ & $refinement\_level(e) > 0)$ **then**
         $iparent = parent(e)$
         $coarsen(iparent) \mathrel{+}= 1$
      else $(test > \epsilon_r$ & $refinement\_level(e) < refinement\_level_{max})$
         $refine(e) = 1$
      end if
    end for
end function

---

is the Vandermonde matrix. The predefined values ϵ_c and ϵ_r are the tolerances used in this example to determine whether an element needs coarsening or refinement, respectively. Note that we also check whether the refinement level of the element is either at its minimum or maximum before the element is flagged. This allows us to predict *a priori* the maximum cost of the AMR algorithm.

Algorithm 19.2 H-refinement: Coarsening Elements

---

function H_REFINEMENT_COARSEN
    for $ee = 1 : N_e^a$ **do**
      $e = sfc(ee)$
      if $(coarsen(e) == 2)$ **then**
         $iparent = e$
         $ichild1 = child(1, e)$
         $ichild2 = child(2, e)$
         $q(:, iparent) = P^{G1} \cdot q(:, ichild1) + P^{G2} \cdot q(:, ichild2)$
         $active(iparent) = 1$
         $active(ichild1) = 0$
         $active(ichild2) = 0$
      end if
    end for
end function

Algorithm 19.2 loops through all the elements and only coarsens an element if both children need coarsening ($coarsen(e) == 2$). Then the solutions of the children are gathered by the projection gather matrices (P^{G1}, P^{G2}) and the respective elements are activated and deactivated. We define the gather matrices in Sec. 19.3.3.

Algorithm 19.3   H-refinement: Refining Elements

---

function H_REFINEMENT_REFINE
    for $ee = 1 : N_e^a$ **do**
        $e = sfc(ee)$
        if $(refine(e) == 1)$ **then**
            $iparent = e$
            $ichild1 = child(1, e)$
            $ichild2 = child(2, e)$
            $q(:, ichild1) = P^{S1} \cdot q(:, iparent)$
            $q(:, ichild2) = P^{S2} \cdot q(:, iparent)$
            $active(iparent) = 0$
            $active(ichild1) = 1$
            $active(ichild2) = 1$
        end if
    end for
end function

Algorithm 19.3 loops through active elements only, looking for an element that has been marked for refinement. Once identified, we apply the projection scatter operation (P^{S1}, P^{S2}) and then activate and deactivate as necessary. Let us now define the gather and scatter projection matrices.

19.3.3 Gather and Scatter Matrices

To address the challenge of projecting the data required for either gathering (for coarsening) or scattering (for refining) we make use of projection operators that are nothing more than interpolation operators. The variable q^P from the parent element is projected onto its two children q^{C1} and q^{C2}. In order to perform this scatter operation we design two projection matrices $\mathbf{P}^{S1}$ and $\mathbf{P}^{S2}$ such that

$$q^{C1} = \mathbf{P}^{S1} q^P$$
$$q^{C2} = \mathbf{P}^{S2} q^P. \tag{19.1}$$

Similarly for the gather operation we need the matrices $\mathbf{P}^{G1}$ and $\mathbf{P}^{G2}$ which satisfy

$$q^P = \mathbf{P}^{G1} q^{C1} + \mathbf{P}^{G2} q^{C2}.$$

The projection matrices are constructed using the integral projection technique [234], derived for different size ratios and different polynomial orders in neighboring elements. Here we describe the method applied to a specific h-non-conforming face in a 2:1 balance (each element is subdivided into 2 elements when refined or two smaller elements are combined into one larger element when coarsened).

Let $\xi \in [-1, 1]$ denote the coordinate in the standard element space corresponding to the parent element. Define $z^{(1)}, z^{(2)} \in [-1, 1]$ as the coordinates of the standard elements corresponding to the two children elements. Let

$$z^{(1)} = \frac{\xi - o^{(1)}}{s}, \qquad z^{(2)} = \frac{\xi - o^{(2)}}{s}$$

be a map $\xi \to z^{(k)}$, $k = 1, 2$ from the parent space to children spaces, where $o^{(k)}$ is the offset parameter for the child k and s is the scale parameter. In our case $o^{(1)} = -0.5$, $o^{(2)} = 0.5$, $s = 0.5$, and the inverse map $o^{(k)} \to \xi$ is defined as follows

$$\xi = s \cdot z^{(k)} + o^{(k)}, \quad k = 1, 2. \tag{19.2}$$

We can now expand the variables using a polynomial basis as follows

$$q^P(\xi) = \sum_{j=0}^{N} q_j^P \psi_j(\xi), \tag{19.3}$$

$$q^{Ck}(z^{(k)}) = \sum_{j=0}^{N} q_j^{Ck} \psi_j(z^{(k)}), \quad k = 1, 2. \tag{19.4}$$

By substituting (19.2) into (19.3) we get

$$q^P(z^{(k)}) = \sum_{j=0}^{N} q_j^P \psi_j(s \cdot z^{(k)} + o^{(k)}), \quad k = 1, 2. \tag{19.5}$$

In order to perform the (L^2) *projection* from the parent to the two children elements we require the following to be satisfied

$$\int_{-1}^{1} \left(q^{Ck}(z^{(k)}) - q^P(z^{(k)}) \right) \psi_i(z^{(k)}) dz^{(k)} = 0, \quad k = 1, 2. \tag{19.6}$$

Substitution of (19.4) and (19.5) into (19.6) and rearranging yields

$$\sum_{j=0}^{N} \left(\int_{-1}^{1} \psi_j(z^{(k)}) \psi_i(z^{(k)}) dz^{(k)} \right) q_j^{Ck} = \sum_{j=0}^{N} \left(\int_{-1}^{1} \psi_j(s \cdot z^{(k)} + o^{(k)}) \psi_i(z^{(k)}) dz^{(k)} \right) q_j^P.$$

$$\tag{19.7}$$

Since $z^{(k)} \in [-1, 1]$ regardless of k, we can write $z = z^{(k)}$ in order to simplify the notation. The terms in brackets can be represented in matrix form as follows

$$\mathbf{M}_{ij} = \int_{-1}^{1} \psi_i(z)\psi_j(z)dz, \tag{19.8}$$

$$\mathbf{S}_{ij}^{(k)} = \int_{-1}^{1} \psi_i(z)\psi_j(s \cdot z + o^{(k)})dz, \quad k = 1, 2, \tag{19.9}$$

which simplifies Eq. (19.7) to

$$\mathbf{M}_{ij}q_j^{Ck} = \mathbf{S}_{ij}^{(k)}q_j^{P}.$$

Note that $\mathbf{M}_{ij}$ is the standard 1D mass matrix, which is easily invertible. If we let $\mathbf{P}_{ij}^{Sk} = \mathbf{M}_{ij}^{-1}\mathbf{S}_{ij}^{(k)}$, then

$$q_i^{Ck} = \mathbf{P}_{ij}^{Sk}q_j^{P}$$

and we call $\mathbf{P}^{Sk}$ the projection scatter matrix. The integrals defined in Eqs. (19.8) and (19.9) are evaluated using the quadrature formulas of the same order used in the Galerkin formulation.

Similarly the gather operation from two children to the parent is performed via a projection gather matrix which we now describe. We require that on the parent element the following relation be satisfied

$$\int_{-1}^{1} \left(q^P(\xi) - \tilde{q}^P(\xi) \right) \psi_i(\xi)d\xi = 0, \tag{19.10}$$

where q^P is the continuous projection of the variables q^{C1} and q^{C2} from the children elements to the parent, and $\tilde{q}^P$ is defined as follows

$$\tilde{q}^P(\xi) = \begin{cases} q^{C1}(z^{(1)}) = q^{C1}\left(\frac{\xi - o^{(1)}}{s} \right) & for \quad -1 \le \xi \le 0^-, \\ q^{C2}(z^{(2)}) = q^{C2}\left(\frac{\xi - o^{(2)}}{s} \right) & for \quad 0^+ \le \xi \le 1. \end{cases} \tag{19.11}$$

Note that $\tilde{q}^P(\xi)$ allows for a discontinuity at $\xi = 0$, which is not of interest here but will be important when we use this approach to handle non-conforming faces in multiple dimensions. Substituting Eq. (19.11) into Eq. (19.10) yields

$$\int_{-1}^{0} \left(q^P(\xi) - q^{C1}\left(\frac{\xi - o^{(1)}}{s} \right) \right) \psi_i(\xi)d\xi + \int_{0}^{1} \left(q^P(\xi) - q^{C2}\left(\frac{\xi - o^{(2)}}{s} \right) \right) \psi_i(\xi)d\xi = 0.$$

Using an expansion analogous to Eqs. (19.4) and (19.5), and rearranging yields

$$\left(\int_{-1}^{1} \psi_i(\xi)\psi_j(\xi)d\xi \right)q_j^P - \left(\int_{-1}^{0} \psi_i(\xi)\psi_j\left(\frac{\xi - o^{(1)}}{s}\right)d\xi \right)q_j^{C1}$$

$$- \left(\int_{0}^{1} \psi_i(\xi)\psi_j\left(\frac{\xi - o^{(2)}}{s}\right)d\xi \right)q_j^{C2} = 0. \quad (19.12)$$

Introducing the variable change $\xi = s \cdot z + o^{(k)}$ and $d\xi = s \cdot dz$ to the second and third integrals, allows us to write

$$\left(\int_{-1}^{1} \psi_i(\xi)\psi_j(\xi)d\xi \right)q_j^P - s\sum_{k=1}^{2}\left(\int_{-1}^{1} \psi_i(s \cdot z + o^{(k)})\psi_j(z)dz \right)q_j^{Ck} = 0. \quad (19.13)$$

The term in brackets to the left of q_j^{Ck} is the transpose of $S_{ij}^{(k)}$ defined in Eq. (19.9). We can now write the integrals in matrix notation as follows

$$\mathbf{M}_{ij}q_j^P - s\sum_{k=1}^{2}\mathbf{S}_{ij}^{(k)^T}q_j^{Ck} = 0,$$

where $\mathbf{M}_{ij}$ is the mass matrix as defined in Eq. (19.8). Finally, if we define

$$\mathbf{P}_{ij}^{Gk} = s \cdot \mathbf{M}_{ij}^{-1}\mathbf{S}_{lj}^{(k)^T}$$

to be the projection gather matrix, then the gathered solution on the parent element is defined as follows

$$q_i^P = \sum_{k=1}^{2}\mathbf{P}_{ij}^{Gk}q_j^{Ck}.$$

If we assume that the same coarsening and refinement strategy will be used throughout the domain, then we only need to construct one copy of each of the following matrices: P^{Gk} and P^{Sk}. Throughout this section, we assume that $k = 1, 2$ and therefore only four projection matrices need to be created, regardless of the number of elements in the domain.

19.3.4 H-Refinement Example

Let us now solve the 1D wave equation with nodal DG (Lobatto points) using the same test case used in Chs. 5 and 6; the only difference is in the initial condition defined as

$$q(x, 0) = \exp\left(-64x^2\right). \quad (19.14)$$

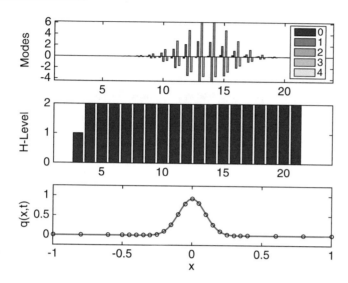

Fig. 19.4: H-refinement with nodal DG using 4th order polynomials.

We use this steeper Gaussian in order to ensure that the AMR algorithm clusters the grid refinement in a smaller region. The solution is shown after one complete revolution with an initial setup of 10 elements of polynomial order 4 using the SSP-RK3 time-integrator with Courant number $C = 0.25$. This simulation uses 2 maximum refinement levels which yields a maximum of 40 elements although the simulation averages around 20 elements. The refinement flags for this simulation are defined as follows: $\epsilon_c = 10^{-4}$ and $\epsilon_r = 10^{-2}$.

In Fig. 19.4, the top panel shows the modes (modes 0 through 4), the middle panel shows the h-level of refinement as described in Fig. 19.2, and the bottom panel shows the solution (where the circles denote the element boundaries). The AMR algorithm activates and refines in regions where the high mode numbers are active. In elements where this is not the case, the AMR algorithm coarsens to the level 0 (original) grid (away from the Gaussian bump). The horizontal axis for the top and middle panels shows the element identification number in the sfc data structure, while the horizontal axis for the bottom panel shows the physical coordinate location. Although the coordinate location and the element number do not correspond exactly, they are sufficiently similar in order to be able to see that the position of the Gaussian bump is correlated to where the AMR algorithm activates by virtue of the values of the modes 0 through 4.

19.4 P-Refinement Method

The idea behind the p-refinement method is to alter the order of the interpolation functions for each element based on the solution [16, 184, 185, 186, 14, 15, 246, 122, 37, 382, 381]. Therefore, p-refinement can be viewed as a focused filtering strategy, or a limiting strategy, since the modal representation of the solution is modified based on some pre-defined conditions. The difference between p-refinement and these two other strategies just described is that based on how the interpolating (trial) functions are modified, we then either add or remove the degrees of freedom. This way, when we coarsen the polynomial order, we are able to increase the efficiency of the computations. Therefore, as in the h-refinement case, the challenge is to construct flexible data structures that allow coarsening or refining of the polynomial order inside each element. In what follows, we begin with the easiest possible p-refinement version using modal DG; this is described in Sec. 19.4.1. Then in Sec. 19.4.2 we describe the additional steps required by both the nodal CG and DG methods for p-refinement.

19.4.1 P-Refinement for Modal DG

Modal DG is tailor-made for p-refinement because the solution variable, q, is approximated by hierarchical basis functions as such

$$q_N(x, t) = \sum_{i=0}^{N} \tilde{q}_i(t) P_i(x)$$

where $P_i(x)$ are, e.g., Legendre polynomials and $\tilde{q}_i(t)$ are the amplitudes associated with the frequencies $P_i(x)$. With this polynomial representation, it is rather straightforward to apply p-refinement. Algorithm 19.4 details this approach where the crux of the problem is to define some conditions *a priori* such as ϵ_c, ϵ_r, p_{min}, and p_{max}. The general idea of Alg. 19.4 is to determine whether the highest mode (Nth) contains a significant amount of information. If it is deemed small (ϵ_c) then it is unnecessary to carry this mode and so we reduce the number of modes by one. In contrast, if the information is large (ϵ_r) then it means that this mode may be saturated and that higher order modes should be added - this is handled by the *Else* statement. The constants p_{min} and p_{max} are the minimum and maximum polynomial orders defined for the simulation, and the conditionals involving these parameters are to ensure that we do not exceed these bounds.

Figure 19.5 shows an example of Alg. 19.4 added to a modal DG formulation for the test problem defined in Eq. (19.14). The top panel shows the values of the modes inside each element, the middle panel shows the polynomial order of each element, and the bottom panel shows the solution after one complete revolution. At initial time, the simulation is defined with $N_e = 10$ elements all of order $N = 8$ with $\epsilon_c = \epsilon_r = 1 \times 10^{-4}$, and $(p_{min}, p_{max}) = (1, 8)$. As the simulation evolves

---

Algorithm 19.4 P-refinement for Modal DG

---

function P_REFINEMENT_MODAL_DG
    for $e = 1 : N_e$ **do**
        N=ElementOrder(e);
        $test = \dfrac{\|\tilde{q}(N)^{(e)}\|_2}{\|\tilde{q}(0:N)^{(e)}\|_2}$
        if $(test < \epsilon_c \;\&\; N > p_{min})$ **then**
            $N \mathrel{-}= 1$
        else $(test > \epsilon_r \;\&\; N < p_{max})$
            $N \mathrel{+}= 1$
        end if
        ElementOrder(e)=N;
    end for
end function

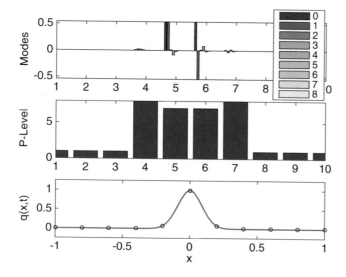

Fig. 19.5: P-refinement with modal DG.

in time, the p-refinement algorithm continues to coarsen the polynomial space. In fact, the middle panel in Fig. 19.5 reveals that six out of the ten elements use linear polynomials whereas only two use the maximum polynomial order. Clearly this situation changes at each time-step but this configuration is representative of what happens throughout the simulation. P-refinement is particularly attractive for simulations when it is expected that a region in the domain may not require high-order for some physical reason. Such a situation arises for the shallow water equations for either tsunami or storm-surge modeling, e.g., where either the dry regions or regions where the wave has not yet passed would not benefit from high-order and so for efficiency reasons it is beneficial to curtail the polynomial order to the minimum value.

19.4.2 P-Refinement for Nodal CG and DG

The p-refinement algorithm for both nodal CG and DG is shown in Alg. 19.5. The results for nodal CG and DG are identical to the results shown in Fig. 19.5, in fact, the DG results should always be identical (modal and nodal). Comparing this algorithm to that for modal DG (Alg. 19.4) we note that the only differences are those labeled with comments. Although the differences do not appear significant,

---

Algorithm 19.5 P-refinement for Nodal CG and DG

function P_REFINEMENT_NODAL_CGDG
   for $e = 1 : N_e$ **do**
      N=ElementOrder(e);
      $\tilde{q}_N^{(e)} = V_N^{-1} \cdot q_N^{(e)}$ ▷ required by nodal CG/DG
      $test = \frac{\|\tilde{q}(N)^{(e)}\|_2}{\|\tilde{q}(0:N)^{(e)}\|_2}$
      if $(test < \epsilon_c$ & $N > p_{min})$ **then**
         $N \mathrel{-}= 1$
      else $(test > \epsilon_r$ & $N < p_{max})$
         $N \mathrel{+}= 1$
      end if
      ElementOrder(e)=N;
      $q_N^{(e)} = V_N \cdot \tilde{q}_N^{(e)}$ ▷ required by nodal CG/DG
   end for
   if CG **then**
      ApplyDSS($q_N^{(e)}$)   ▷ required by CG
   end if
end function

---

the mapping from modal to nodal space via the Vandermonde matrix (V) and from nodal to modal space via its inverse (V^{-1}) requires a bit of computation which we now describe. The last part of Alg. 19.5 applies the DSS operator only for the CG method, in a similar manner described in Alg. 5.7.

19.4.3 Additional Matrices for Nodal P-Refinement

The reason why we need to construct V_N and V_N^{-1} for various polynomial orders N can be explained by revisiting Figs. 2.4 and 2.5 where 8th degree modal and nodal bases are illustrated. In the modal case, eliminating the highest mode is as simple as ignoring the last mode in Fig. 2.4. However, in the nodal case we need to ignore this last mode and then remap all of the nodal values numbered from 0 to 8 to the smaller space 0 through 7 where the positions of the nodes in the element now change (e.g., they can be Chebyshev, Legendre, or Lobatto points which are not hierarchical).

Therefore in order to apply p-refinement to an existing nodal CG or DG code, we need to construct a collection of matrices defined as follows

$$dim(V_N) = 0 : i \times 0 : i \times 0 : i, \ i = p_{min}, \ldots, p_{max}$$

where p_{min} and p_{max} are the minimum and maximum polynomial orders allowed in the simulation. Each Vandermonde matrix (for a specific value of N) is constructed using the procedure described in Sec. 3.3.2. In addition, to the Vandermonde matrix, we also need to compute Lagrange basis functions and their derivatives for each order as described in Algs. 3.1 and 3.2. For each order N we need to pick an integration order Q and store the quadrature weights in w_N which we define as

$$dim(w_N) = 0 : Q \times 0 : i, \ i = p_{min}, \ldots, p_{max},$$

where Q is a function of i (e.g., $Q = i + 1$ yields exact integration for the mass matrix for most of the integration points described in Ch. 4). Element mass matrices $M_N^{(e)}$ for each order N also need to be stored. This concludes the list of additional required matrices for p-refinement with nodal CG or DG. Clearly, there is a bit more complication in implementing p-refinement with nodal methods but these complications are not prohibitive - at least not in 1D. In two-dimensions (see Sec. 19.6), the situation is complicated by the introduction of non-conforming faces. Although not a major obstacle, it does require some careful attention as well as some additional source code.

19.5 R-Refinement Method

Unlike in both h- and p-refinement, in *r-refinement* the total number of degrees of freedom (DOF) always remains constant but their position change with time. The r is meant to represent the *relocation* of the DOF (see, e.g., [412, 244, 271, 270, 102, 113, 401, 290, 20, 214, 49, 67, 372, 50]). For a review of various methods used in r-refinement the reader is referred to, in particular, [412, 118, 377, 119, 376, 366].

There are numerous possibilities for constructing the r-refinement algorithm (also known in the literature as r-adaptivity), such as using the Monge-Ampere equations (see, e.g., [50]) or by using other partial differential equations, PDEs, (see, e.g., [377]). An approach using elliptic PDEs is to first write the following problem:

$$\nabla^2 \xi_i = P_i(\mathbf{x}) \tag{19.15}$$

where, in multi-dimensions $i = 1, \ldots, 3$ and $\mathbf{x} = (x, y, z)$, or variants thereof. The vector $\mathbf{x}$ denotes the physical coordinates of the grid whereas ξ_i denotes the coordinates in the structured and uniform computational domain with $P_i(\mathbf{x})$ being a forcing function that allows us to move the points in the interior of the domain in a specific way. Since the homogeneous version of this PDE (the Laplace equation) satisfies the *maximum principle*, then it means that we are guaranteed to obtain smooth meshes whereby the extrema will always lie on the boundary. Once we add the right-hand side forcing function, this is no longer guaranteed but with judicious choices of $P_i(\mathbf{x})$

we can obtain well-behaved meshes. Equation (19.15) cannot be solved as written because we are, in fact, looking for the physical coordinates **x**. Writing the chain rule in 2D (for simplicity)

$$\frac{\partial}{\partial x_i} = \frac{\partial \xi}{\partial x_i} \frac{\partial}{\partial \xi} + \frac{\partial \eta}{\partial x_i} \frac{\partial}{\partial \eta}$$

where $x_i = (x, y)$, we can write the second derivatives as follows

$$\frac{\partial^2}{\partial x_i \partial x_j} = \frac{\partial \xi}{\partial x_i} \frac{\partial \xi}{\partial x_j} \frac{\partial^2}{\partial \xi^2} + \left(\frac{\partial \xi}{\partial x_i} \frac{\partial \eta}{\partial x_j} + \frac{\partial \xi}{\partial x_i} \frac{\partial \eta}{\partial x_j} \right) \frac{\partial^2}{\partial \xi \partial \eta} + \frac{\partial \eta}{\partial x_i} \frac{\partial \eta}{\partial x_j} \frac{\partial^2}{\partial \eta^2}.$$

The metric terms $\frac{\partial \xi_i}{\partial x_j}$ are then computed using the approach defined in Sec. 12.4. In two-dimensions, the elliptic equations to be solved are written as follows

$$\left(\alpha \frac{\partial^2}{\partial \xi^2} + 2\beta \frac{\partial^2}{\partial \xi \partial \eta} + \gamma \frac{\partial^2}{\partial \eta^2} \right) x_i = P_i(\xi, \eta) \qquad (19.16a)$$

$$\alpha = \left(\frac{\partial \xi}{\partial x} \right)^2 + \left(\frac{\partial \xi}{\partial y} \right)^2, \quad \beta = \left(\frac{\partial \xi}{\partial x} \frac{\partial \eta}{\partial y} + \frac{\partial \xi}{\partial y} \frac{\partial \eta}{\partial x} \right), \quad \gamma = \left(\frac{\partial \eta}{\partial x} \right)^2 + \left(\frac{\partial \eta}{\partial y} \right)^2 \qquad (19.16b)$$

where we have replaced the computational variables $\xi_i = (\xi, \eta)$ in the original problem given in Eq. (19.15) by the physical variables $x_i = (x, y)$ (see [377, 132]). The power of this approach is that we can now solve for the physical grid variables x_i as a function of the known computational variables ξ_i. Moreover, Eq. (19.16) defines a non-linear elliptic equation which, through a judicious choice of the forcing function P_i, can be used to generate relatively smooth grids that are still able to cluster points near specific regions.

Although grid generation (and r-refinement) is much more powerful when constructed with partial differential equations, let us show an example using an algebraic grid generation approach which is far simpler to describe. In Fig. 19.6 we show the results using this idea whereby we cluster the points by a simple grid stretching approach based on the solution. We move the points based on the analytic function

$$x = L\xi + A(x_c - L\xi)(1 - \xi)\xi$$

(see, [377]) where $\xi \in [0, 1]$, $A = 2.5$, $L = 2$ and x_c is the physical point where we wish to cluster points around. Here, we choose this point to be the maximum value of the solution at the given time. As shown in Fig. 19.6b, the minimum grid spacing occurs where the solution is a maximum (see Fig. 19.6a); here we only show the initial condition from Eq. (19.14) with 50 elements using linear polynomials.

19.6 H-Refinement Example in 2D

So far, we have only discussed adaptive mesh refinement (AMR) algorithms in one-dimension which, while overly simple, is sufficient to convey the general ideas

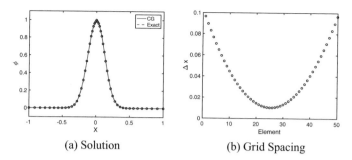

(a) Solution   (b) Grid Spacing

Fig. 19.6: The (a) initial condition and (b) grid spacing for r-refinement of the 1D wave equation.

behind each of the methods. However, the disadvantage of only discussing AMR in one-dimension is that many of the challenges in multi-dimensional mesh refinement are ignored, particularly for non-conforming AMR. In order to address some of these challenges, let us discuss non-conforming h-refinement in 2D[3]. There are two popular approaches to non-conforming h-refinement: one is known as block-structured AMR whereby block regions of the domain are refined (see, e.g., [32, 30, 27, 207, 287, 211, 352]) and the other relies on localized refinement through the use of tree-based data structures (binary trees in 1D, quadtrees in 2D, and octrees in 3D). In what follows, we shall only discuss the tree-based approach and begin by describing the quadtree data structure.

19.6.1 Quadtree Data Structure

To simplify the exposition of non-conforming h-refinement, let us begin by describing the quadtree data structure often used to keep track of the active elements in a mesh refinement simulation (see, e.g., [257, 415, 262, 269, 27, 192, 305, 339, 54, 37, 230, 231, 304]).

First we generate an initial coarse mesh, which has to represent the geometrical features of the domain. In Fig. 19.7a we present a simple four element initial mesh. Let us call it the level 0 mesh, where each element is a root for a tree of children elements. If we decide to refine element 4, we replace this element with four children, which belong to the level 1 mesh. We represent it graphically on the right panel of Fig. 19.7b. Active elements (i.e., the set of elements which pave the domain entirely) are marked in black and blue; the black numbered elements are the original elements and

---

[3] Note that for p-refinement we can use the same ideas presented here for non-conforming h-refinement.

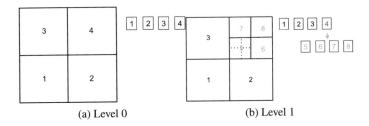

(a) Level 0   (b) Level 1

Fig. 19.7: Non-conforming h-refinement in 2D with the (a) level 0 and (b) level 1 grids, along with their corresponding tree-based data structure (right panels).

the blue are the new elements. Element 4 is now inactive (red numbered), replaced by the four newly created elements: 5, 6, 7 and 8.

If we further choose to refine element 5, and thus introduce level 2 elements, we render this element inactive and replace it with four children.

19.6.2 Space filling curve and data storage

In order to traverse all active elements in the mesh in a memory contiguous manner, we rely on space filling curves (SFC) (see, e.g., [354, 26, 103, 55, 402, 18]) just as we did for the 1D example. To each active element we assign an index, which defines the position of the element in the space filling array. With this in mind, we can now store the solution data in a local element-wise fashion. Let us call the solution variable q which we represent by the array $q(1 : N + 1, 1 : N + 1, 1 : N_e^a)$, where N is the polynomial order and N_e^a is the number of active elements in the space filling curve. After the mesh changes, the data in q is rearranged to reflect the new element ordering. This way we can traverse the space filling curve simply by looping over the last index of the array q. To further simplify the exposition, let us only consider the refinement of an element into four smaller children with the additional restriction on the interface between levels of refinement that we discuss below.

19.6.3 2:1 balance

A mesh, where each face is shared by at most three elements, is called a 2:1 *balanced mesh*, and a face which is shared by two elements on one side, and one element on the other, is called 2:1 balanced face. Element refinement may lead to a situation, where an element, that has a 2:1 balanced face and lies on the refined side of the face, is marked for refinement, while its neighbor across this face is not. This would violate the 2:1 balance causing the face to be shared by a total of four elements (one on one side, and three on the other). An example of such a situation is shown in Fig. 19.7b if we decided to refine element 5. Refining element 5 would not violate the 2:1 balance rule for elements 6 and 7; however, it would violate the 2:1 balance rule for elements 2 and 3. In order to avoid this situation, a special balancing procedure needs to be introduced.

In the situation presented in Fig. 19.7b, the solution of the 2:1 balance problem is to refine elements 2 and 3 first, before refining element 5, even though elements 2 and 3 might not be originally marked for refinement. This simple strategy will always respect the 2:1 balance rule. One might imagine, however, that for a more complex mesh, refining elements 2 and 3 might also cause a conflict with other faces owned by them. Such a phenomenon is called the *ripple effect*, where refinement of one element can cause an entire area not directly neighboring the element in question to be refined [358] and is discussed in detail in, e.g., [230].

It is easy to show that in 2D, the ripple propagation is limited by the lowest level element in the mesh. In a 2:1 balanced mesh the level difference between neighboring elements can be at most 1. The conflict can occur only when refining an n level element which has a neighbor of level $(n-1)$. Therefore we need to bring the $(n-1)$ level element to level n before refining the original element to level $(n+1)$. If in turn the $(n-1)$ level element causes a conflict with an $(n-2)$ level element, we need to follow the balancing procedure recursively. In the worst case scenario we will propagate the ripple down to the level 0 element, which by definition is a root of the element tree. Therefore by refining one level n element we may be forced, in the worst case, to refine n other elements. This will cause $4n$ new elements to be created in the areas possibly not indicated by the refinement criterion. $4n$ is typically a relatively small number since $n < 10$.

In the case of element coarsening, we adopt a different strategy. If coarsening of an element would cause a conflict, we do not perform this operation. In order to keep the 2:1 balance we avoid propagating a coarsening ripple to higher levels. The rationale behind this strategy is that it is better to have more refined elements than we need, rather than lack the resolution in the areas where it is necessary.

19.6.4 Handling of non-conforming faces for DG

The discussion of the previous sections now allows us to better understand the special situation required in multi-dimensions due to non-conforming faces. Let us now describe the construction of fluxes in the DG method across a 2:1 balanced mesh.

19.6.4.1 Projection onto 2:1 faces

In the DG method we need to evaluate the numerical flux through all the element faces. For non-conforming elements, we need to address the problem of projecting the data between two sides of a non-conforming face.

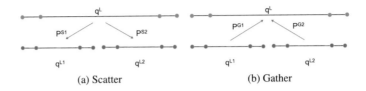

(a) Scatter   (b) Gather

Fig. 19.8: The (a) scatter and (b) gather operations along a non-conforming face.

Consider the situation shown in Fig. 19.8a. The variable q^L from a parent face is projected onto two children faces and becomes q^{L1} and q^{L2}. In order to perform this scatter operation we design two projection matrices $\mathbf{P}^{S1}$ and $\mathbf{P}^{S2}$ such that

$$q^{L1} = \mathbf{P}^{S1} q^L$$
$$q^{L2} = \mathbf{P}^{S2} q^L. \tag{19.17}$$

Similarly for the gather operation we need the matrices $\mathbf{P}^{G1}$ and $\mathbf{P}^{G2}$ which satisfy

$$q^L = \mathbf{P}^{G1} q^{L1} + \mathbf{P}^{G2} q^{L2}.$$

Here, we use the projection matrices previously defined in Sec. 19.3.3. By limiting the non-conforming faces to a 2:1 ratio, the scatter and gather projection matrices are the same for all faces and need to be computed only once, which makes the algorithm quite simple and efficient.

19.6.4.2 Projection between parent and children elements

Once an element is refined, its data must be projected onto its children elements. Similarly, when a group of elements are coarsened they need to be used in order to project the data onto the parent elements. In order to perform these two operations we use the 2D version of the integral projection technique discussed in Sec. 19.6.4.1. Figure 19.9 shows schematically the projections from a parent element with coordinates $(\xi, \eta) \in [-1, 1]^2$ to four children, each with separate coordinates $(z_1^{(k)}, z_2^{(k)}) \in [-1, 1]^2$, where $k = 1, \dots, 4$ enumerates the children elements. For this projection we construct the scatter matrix P_{2D}^{Sk}. The inverse operation is performed using the gather matrix P_{2D}^{Gk}.

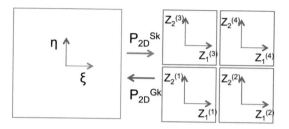

Fig. 19.9: The gather and scatter operations in 2D.

Let us define the map

$$z_1^{(k)} = \frac{\xi - o_1^{(k)}}{s}, \quad z_2^{(k)} = \frac{\eta - o_2^{(k)}}{s}$$

where $o_1^{(k)}$ and $o_2^{(k)}$ are offset parameters corresponding to each child element k and coordinate z_1 and z_2. The inverse mapping is now

$$\xi = s \cdot z_1^{(k)} + o_1^{(k)}, \quad \eta = s \cdot z_2^{(k)} + o_2^{(k)}$$

where $k = 1, \ldots, 4$, the scale parameter $s = 0.5$ and offsets $o_i^{(k)} = \pm 0.5$ depend on the direction i and element number k.

Each element has a polynomial basis defined in which we expand the projected variable,

$$q^P(\xi, \eta) = \sum_{j=1}^{M_N} q_j^P \psi_j(\xi, \eta), \qquad (19.18)$$

$$q^{Ck}(z_1^{(k)}, z_2^{(k)}) = \sum_{j=1}^{M_N} q_j^{Ck} \psi_j(z_1^{(k)}, z_2^{(k)}), \qquad (19.19)$$

where q^P is the parent element variable and q^{Ck} is the k-th child element variable projected from q^P, and M_N is the number of nodal points in the element[4]. We can substitute the inverse map into (19.18) and represent the parent variable in terms of the children coordinate system as follows

$$q^P(\xi, \eta) = q^P(z_1^{(k)}, z_2^{(k)}) = \sum_{j=1}^{M_N} q_j^P \psi_j(s \cdot z_1^{(k)} + o_1^{(k)}, s \cdot z_2^{(k)} + o_2^{(k)}), \qquad k = 1, \ldots, 4.$$

$$(19.20)$$

---

[4] For simplicity assume that $M_N = (N + 1)^2$ where N denotes the order of the interpolation points in each direction.

For each $k = 1, \ldots, 4$ we require that

$$\int\limits_{-1}^{1} \int\limits_{-1}^{1} \left(q^{Ck}(z_1^{(k)}, z_2^{(k)}) - q^P(z_1^{(k)}, z_2^{(k)}) \right) \psi_j(z_1^{(k)}, z_2^{(k)}) dz_1^{(k)} dz_2^{(k)} = 0 \qquad (19.21)$$

which is the L^2 *projection* of the solution spaces from the children elements (Ck) to the parent (P). Substituting expansions (19.18), (19.19), rearranging and employing matrix notation yields

$$\mathbf{M}_{ij} q_j^{Ck} - \mathbf{S}_{ij}^{(k)} q_j^P = 0, \qquad (19.22)$$

where

$$\mathbf{M}_{ij} = \int\limits_{-1}^{1} \int\limits_{-1}^{1} \psi_j(z_1, z_2) \psi_i(z_1, z_2) dz_1 dz_2, \qquad (19.23)$$

$$\mathbf{S}_{ij}^{(k)} = \int\limits_{-1}^{1} \int\limits_{-1}^{1} \psi_j(s \cdot z_1 + o_1^{(k)}, s \cdot z_2 + o_2^{(k)}) \psi_i(z_1, z_2) dz_1 dz_2. \qquad (19.24)$$

The projection matrix is once again constructed by inverting the mass matrix and left-multiplying the inverse to Eq. (19.22). This yields

$$q_i^{Ck} = (\mathbf{P}_{2D}^{Sk})_{ij} q_j^P, \quad k = 1, \ldots, 4,$$

where

$$\mathbf{P}_{2D}^{Sk} = \mathbf{M}^{-1} \mathbf{S}^{(k)}. \qquad (19.25)$$

Similarly as in the case of the gather projection in the 1D non-conforming face case, the 2D gather projection matrix is constructed by multiplying the inverse of the mass matrix and transpose of the 2D $\mathbf{S}^{(k)}$ matrix

$$\mathbf{P}_{2D}^{Gk} = s \cdot \mathbf{M}^{-1} \mathbf{S}^{(k)^T}, \qquad (19.26)$$

which yields

$$q_i^P = \sum_{k=1}^{4} (\mathbf{P}_{2D}^{Gk})_{ij} q_j^{Ck}. \qquad (19.27)$$

This approach is easily extendable to 3D projections of hexahedral elements since the same tensor-product operations are being applied in 1D, 2D, or 3D. The only difference is that in 3D the integrals in Eqs. (19.23) and (19.24) would comprise all three spatial directions of the reference element.

19.6.5 Handling of non-conforming faces for CG

So far we have discussed the handling of non-conforming faces for the DG method. For the CG method, all of the DG data structures can be reused if we use a DG-like data storage scheme as shown in Fig. 19.10 (i.e., the LEW storage from Ch. 7). Figure 19.10a shows a schematic of the numbering strategy for two adjacent elements; in Fig. 19.10a the indices (l, e) refer to the local element-wise number $(l = 1, \ldots, 9)$ and the element $(e = 1, 2)$. Each element uses its own (local) numbering scheme even though gridpoints may coincide in physical space. Figure 19.10b shows how these local numbers are defined in terms of the global index $(g = 1, \ldots, 15)$, which allows us to apply the direct stiffness summation (DSS) operation to enforce continuity. Therefore, if we use the DG storage but augment it with a map from the local to the global space, then we can apply all of the DG operators so far described. The only item that needs to be addressed is how to handle the map from the local to the global space, which is what we now describe. Fischer et. al. [129] explain the procedure for

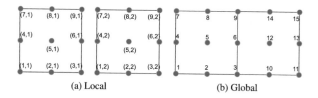

(a) Local   (b) Global

Fig. 19.10: The (a) local (DG storage) and (b) global (CG storage) numbering of points.

moving the data between local (DG) and global (CG) storage. In their formulation, a Boolean connectivity matrix $\mathbf{Q}$ maps global values of $\mathbf{u}^{CG}$ to locally stored values $\mathbf{u}^{DG}$. $\mathbf{Q}^T$ sums the local values of $\mathbf{u}^{DG}$ from corresponding nodes and stores the result in a global vector $\mathbf{v}^{CG}$. Note that $\mathbf{v}^{CG} \neq \mathbf{u}^{CG}$ but the following relations are true

$$\mathbf{u}^{DG} = \mathbf{Q}\,\mathbf{u}^{CG}, \qquad \mathbf{v}^{CG} = \mathbf{Q}^T \mathbf{u}^{DG}. \tag{19.28}$$

We can perform the DSS operation in the three steps outlined in Alg. 19.6. For simplicity, we have assumed in the second step that inexact integration is used which results in a diagonal mass matrix although this is not strictly necessary. In practice, the matrix $\mathbf{Q}$ should not be constructed. Instead, we evaluate its action, as well as the action of its transpose. The details of the DSS operation for the non-conforming interface is discussed next. Since the matrix $\mathbf{Q}$ is a Boolean map between the local and global node numbering, we implement this map as follows $g = intma(l, e)$, where g is a global node number, e is the element number and l is a local node number. The actions of $\mathbf{Q}$ and $\mathbf{Q}^T$ for conforming elements are detailed in the first loops of Algs. 19.7 and 19.8.

Algorithm 19.6 Direct stiffness summation

---

function APPLY_DSS($\mathbf{u}^{DG}$)
    $\mathbf{v}^{CG} \leftarrow \mathbf{Q}^T \mathbf{u}^{DG}$ ▷ gather data to CG storage
    for $j = 1 : N_g$ **do**
        $u_j^{CG} = v_j^{CG}/M_j^{CG}$ ▷ multiply by inverse mass: $M^{CG} = \mathbf{Q}^T M^{DG}$
    end for
    $\mathbf{u}^{DG} \leftarrow \mathbf{Q}\,\mathbf{u}^{CG}$ ▷ scatter back to DG storage
    return q
end function

19.6.5.1 Direct stiffness summation for non-conforming faces

Non-conforming faces occur when a face is shared by more than two elements. Fig. 19.11 shows (a) local and (b) global numbering of nodes in such a situation for first order elements. In panel (b) the black circle (gridpoint 8) marks a *hanging node*. The value of the solution there is constrained by the C^0 continuity requirement of the CG method. The DSS operation satisfies this constraint in the non-conforming element situation by an appropriate construction of the $\mathbf{Q}$ matrix, which is described in detail in [129]. Let us now outline the implementation of its action in the non-conforming case. Algorithm 19.7 provides pseudocode for evaluating the action of $\mathbf{Q}$ (scatter

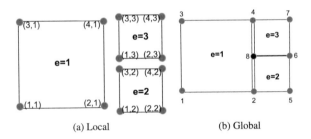

          (a) Local   (b) Global

Fig. 19.11: Non-conforming elements: (a) local numbering of points (DG storage), (b) global numbering of points (global storage). Global numbering point no. 8 is a so called *hanging node*, as its value is constrained by the continuity condition between the three elements.

operation) while Alg. 19.8 outlines the action of $\mathbf{Q}^T$ (gather operation). The first set of loops in both algorithms perform a typical conforming scatter/gather operation using the *intma* array. In the non-conforming case we need to introduce adjustments to accommodate the hanging nodes. The index f in both algorithms loops over all non-conforming faces whereby we need to identify the parent and children elements. The parent element is the large element sharing the face, while the children are the small elements on the other side of the non-conforming face. The naming convention comes from the fact that the parent element will ensure continuity, which will be enforced on the children faces. In Fig. 19.11a the parent is the element $p = e = 1$, while the children are the smaller elements $c1 = e = 2$ and $c2 = e = 3$. The array *imap* finds the element local number of the ith point that belongs to a given element $(p, c1, c2)$ and lies on the face f. In the example shown in Fig. 19.11b $\mathbf{v} = [u_2^{CG}, u_4^{CG}]$ and for the gather operation $\mathbf{u}_1 = [u_{(1,2)}^{DG}, u_{(3,2)}^{DG}]$ and $\mathbf{u}_2 = [u_{(1,3)}^{DG}, u_{(3,3)}^{DG}]$. The matrices

---

Algorithm 19.7 Scatter operation $\mathbf{Q}$ in a non-conforming case

for $e = 1 : N_e^a$ do   ▷ conforming faces
    for $l = 1 : N_p$ do
      $g = \text{intma}(l,e)$
      $u^{DG}(l, e) = u^{CG}(g)$
    end for
end for

for $\forall f \in \text{NCF}$ do   ▷ non-conforming faces
    $p \leftarrow \text{parent}(f)$
    $c1 \leftarrow \text{child}(1,f)$
    $c2 \leftarrow \text{child}(2,f)$
    for i=1:N+1 do   ▷ find nodes on f
      l=imap(i,f,p)
      g=intma(l,p)
      $v(i) \leftarrow u^{CG}(g)$
    end for
    $\mathbf{u}_1 = \mathbf{L}_1 \mathbf{v}$ ▷ interpolate parent values onto children
    $\mathbf{u}_2 = \mathbf{L}_2 \mathbf{v}$
    for i=1:N+1 do
      ic1=imap(i,f,c1)
      ic2=imap(i,f,c2)
      $u^{DG}(ic1, c1) = u_1(i)$ ▷ store children in DG-storage
      $u^{DG}(ic2, c2) = u_2(i)$
    end for
end for

$(L_1)_{ij} = \psi_i(\xi_j^{c1})$ and $(L_2)_{ij} = \psi_i(\xi_j^{c2})$ are the interpolation matrices from the parent side of the face to two children sides. Their entries are the values of the 1D parent face basis functions ψ_i at the children side nodal points $\xi_j^{c1,c2}$. In the Fig. 19.11 example, $\xi^{c1} = [\xi_{(1,2)}, \xi_{(3,2)}]$ and $\xi^{c2} = [\xi_{(1,3)}, \xi_{(3,3)}]$ or using the global numbering $\xi^{c1} = [\xi_2, \xi_8], \xi^{c2} = [\xi_8, \xi_4]$. The ξ coordinate is measured in the parent element coordinate system. All the interpolations are evaluated on standard elements in the reference element space $\xi \in [-1, 1]$ and, therefore, the matrices $\mathbf{L}_{1,2}$ only need to be computed once and can then be used for all elements. In the scatter operation we copy the interpolated vectors $\mathbf{u}_1$ and $\mathbf{u}_2$ to the local storage (bottom of Alg. 19.7).

For the gather operation outlined in Alg. 19.8, we use the transpose of the interpolation matrix to compute the contributions from the children elements to the parent side of the face. The contributions are summed, but before they can be applied to the solution gathered in the CG storage we need to account for the fact that some contribution from the corner nodes were already added in the conforming step. In the Fig. 19.11 example, data from points $(1, 2)$ and $(3, 3)$ are gathered to global points 2 and 4, respectively. The summed contributions $\mathbf{v}_{1+2}$ account for contributions of local points $(1, 2), (3, 2), (1, 3), (3, 3)$ to points 2 and 4 in global storage. We avoid double counting by subtracting $u^{DG}_{(1,2)}$ and $u^{DG}_{(3,3)}$ from u^{CG}_2 and u^{CG}_4 respectively, which is shown by the comment *correct corner* in Alg. 19.8. For the gather operation, the

---

Algorithm 19.8 Gather operation $\mathbf{Q}^T$ in a non-conforming case

---

$\mathbf{v}^{CG} = 0$
for $e = 1 : N_e$ **do**   ▷ conforming faces
    for $l = 1 : N_p$ **do**
        $g = intma(l, e)$
        $v^{CG}(g) \mathrel{+}= u^{DG}(l, e)$
    end for
end for

for $\forall f \in$ NCF **do**   ▷ non-conforming faces
    $p \leftarrow$ parent(f)
    $c1 \leftarrow$ child(1,f)
    $c2 \leftarrow$ child(2,f)
    for i=1:N+1 **do**   ▷ find nodes on f
        ic1=imap(i,f,c1)
        ic2=imap(i,f,c2)
        $u_1(i) \leftarrow u^{DG}(ic1, c1)$ ▷ get values from face nodes
        $u_2(i) \leftarrow u^{DG}(ic2, c2)$
    end for
    $\mathbf{v}_{1+2} = \mathbf{L}_1^T \mathbf{u}_1 + \mathbf{L}_2^T \mathbf{u}_2$ ▷ interpolate & sum
    ic=imap(1,f,c1)
    g=intma(ic,c1)
    $v^{CG}(g) = v^{CG}(g) - u^{DG}(ic, c1)$ ▷ Correct corner #1
    ic=imap(N+1,f,c2)
    g=intma(ic,c2)
    $v^{CG}(g) = v^{CG}(g) - u^{DG}(ic, c2)$ ▷ Correct corner #2
    for i=1:N+1 **do**
        l=imap(i,f,p)
        g=intma(l,p)
        $v^{CG}(g) = v^{CG}(g) + v_{1+2}(i)$
    end for
end for

---

summed solution $\mathbf{v}_{1+2}$ is added to appropriate locations in the global CG storage.

In the non-conforming case the DSS algorithm outlined in Alg. 19.6 remains unchanged. The only difference lies in the construction of the $\mathbf{Q}$ matrix. In the example shown in Fig. 19.11 we first perform the gather operation, to account for the contributions of nodes $(1, 2), (3, 2), (1, 3), (3, 3)$ to global nodes 2, 4. Once complete, we divide the gathered solution by the gathered mass. Since both the solution and the mass are gathered in the same way, this division is equivalent to a weighted average of multiple contribution from local nodes to a single global node. The weight of each local node is its corresponding mass matrix entry. The final DSS step is to scatter the data back to local storage using the **Q** matrix. Since the scatter operation is essentially an interpolation, we can rest assured that the solution is continuous across the non-conforming interfaces.

A good check for the correct implementation of the **Q** matrix is to construct it explicitly for a small problem and inspect its entries[5]. The **Q** matrix for the conforming example presented in Fig. 19.10 is as follows

$$
u^{DG} = \mathbf{Q}u^{CG} \quad \rightarrow \quad
\begin{bmatrix} u_{(1,1)} \\ u_{(2,1)} \\ u_{(3,1)} \\ u_{(4,1)} \\ u_{(5,1)} \\ u_{(6,1)} \\ u_{(7,1)} \\ u_{(8,1)} \\ u_{(9,1)} \\ u_{(1,2)} \\ u_{(2,2)} \\ u_{(3,2)} \\ u_{(4,2)} \\ u_{(5,2)} \\ u_{(6,2)} \\ u_{(7,2)} \\ u_{(8,2)} \\ u_{(9,2)} \end{bmatrix}^{DG}
=
\begin{bmatrix}
1 & & & & & & & & & & & & & & \\
 & 1 & & & & & & & & & & & & & \\
 & & 1 & & & & & & & & & & & & \\
 & & & 1 & & & & & & & & & & & \\
 & & & & 1 & & & & & & & & & & \\
 & & & & & 1 & & & & & & & & & \\
 & & & & & & 1 & & & & & & & & \\
 & & & & & & & 1 & & & & & & & \\
 & & & & & & & & 1 & & & & & & \\
1 & & & & & & & & & & & & & & \\
 & & & & & & & & & 1 & & & & & \\
 & & & & & & & & & & 1 & & & & \\
 & & & & & 1 & & & & & & & & & \\
 & & & & & & & & & & & 1 & & & \\
 & & & & & & & & & & & & 1 & & \\
 & & & & & & & 1 & & & & & & & \\
 & & & & & & & & & & & & & 1 & \\
 & & & & & & & & & & & & & & 1
\end{bmatrix}
\begin{bmatrix} u_1 \\ u_2 \\ u_3 \\ u_4 \\ u_5 \\ u_6 \\ u_7 \\ u_8 \\ u_9 \\ u_{10} \\ u_{11} \\ u_{12} \\ u_{13} \\ u_{14} \\ u_{15} \end{bmatrix}^{CG} .
$$

For comparison, the **Q** matrix for the non-conforming case shown in Fig. 19.11 is the following

$$
u^{DG} = \mathbf{Q}u^{CG} \quad \rightarrow \quad
\begin{bmatrix} u_{(1,1)} \\ u_{(2,1)} \\ u_{(3,1)} \\ u_{(4,1)} \\ u_{(1,2)} \\ u_{(2,2)} \\ u_{(3,2)} \\ u_{(4,2)} \\ u_{(1,3)} \\ u_{(2,3)} \\ u_{(3,3)} \\ u_{(4,3)} \end{bmatrix}^{DG}
=
\begin{bmatrix}
1 & & & & & & \\
 & 1 & & & & & \\
 & & 1 & & & & \\
 & & & 1 & & & \\
1 & & & & & & \\
 & & & & 1 & & \\
\tfrac{1}{2} & \tfrac{1}{2} & & & & & \\
 & & & & & 1 & \\
\tfrac{1}{2} & \tfrac{1}{2} & & & & & \\
 & & & & & 1 & \\
 & & & & 1 & & \\
 & & & & & & 1
\end{bmatrix}
\begin{bmatrix} u_1 \\ u_2 \\ u_3 \\ u_4 \\ u_5 \\ u_6 \\ u_7 \end{bmatrix}^{CG} .
$$

Note that the non-Boolean entries come from the interpolation between parent and children elements, occurring at the hanging node (global node 8) which is formed by

---

[5] Thanks to Lucas Wilcox for this simple yet very useful tip.

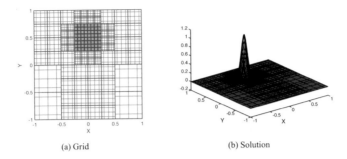

(a) Grid   (b) Solution

Fig. 19.12: 2D test case for non-conforming adaptive h-refinement showing the (a) grid and (b) solution.

local node (3,2) and (1,3). Further note that global node 8 is not stored but, rather, its value is constructed from the global node 2 and 4 solutions. This is necessary in order to maintain C^0 continuity at element interfaces. It should also be mentioned that the CG method conserves the variable $\mathbf{u}$ globally regardless of whether conforming (see, e.g., [370]) or non-conforming (see, e.g., [231]) elements are used.

19.6.6 2D Test Problem

To illustrate the value of non-conforming adaptive mesh refinement with h-refinement, let us solve the 2D advection equation with periodic boundary conditions and the initial condition

$$q(x, y, 0) = \exp\left(-\sigma \left[(x - x_c)^2 + (y - y_c)^2\right]\right) \tag{19.29}$$

where $\sigma = 128$ which defines a much steeper Gaussian wave and vigorously tests the adaptive mesh refinement algorithm. Figure 19.12 shows the mesh and the solution after a quarter revolution using 4 levels of refinement[6].

---

[6] The code that produced this result can be found at
https://github.com/fxgiraldo/Element-based-Galerkin-Methods/tree/master/Chapters/Chapter_19/2D_Wave_DG_Nodal_Lobatto_Hrefinement

Chapter 20
Time Integration

20.1 Introduction

This text is concerned with the construction of discretization methods arising from the continuous spatial operators such as the divergence, gradient, and Laplacian operators. However, if we wish to use element-based Galerkin methods to solve time-dependent partial differential equations (PDEs), then we must know something about time-integrators. The most popular way of solving time-dependent PDEs is through the use of the *Method of Lines* (MOL). To understand how the MOL works, let us assume that we wish to solve the conservation law

$$\frac{\partial q}{\partial t} + \boldsymbol{\nabla} \cdot (q\mathbf{u}) = 0. \tag{20.1}$$

Upon constructing a discrete representation of the continuous operator $\boldsymbol{\nabla}$ using either the CG or DG methods described in Chs. 5, 6, 7, 15, 16, and 17 we can now write Eq. (20.1) as follows

$$\frac{dq}{dt} = S(q). \tag{20.2}$$

S denotes the discrete spatial representation of the continuous operators and the time-derivative on the left-hand side has been replaced by a total derivative to denote that Eq. (20.2) now represents a system of nonlinear ordinary differential equations (ODEs). The role of time-integrators is to solve the ODEs given by Eq. (20.2).

In what follows, we describe a few types of time-integrators, including: explicit, fully-implicit, implicit-explicit, semi-Lagrangian, and multirate methods. Although this list is not exhaustive, it does cover most of the time-integrators of interest.

© The Editor(s) (if applicable) and The Author(s), under exclusive license
to Springer Nature Switzerland AG 2020
F. X. Giraldo, *An Introduction to Element-Based Galerkin Methods on
Tensor-Product Bases*, Texts in Computational Science and Engineering 24,
https://doi.org/10.1007/978-3-030-55069-1_20

20.2 Explicit Methods

The easiest methods to implement for solving the system of nonlinear ODEs in Eq. (20.2) are explicit methods. The simplest such method is the forward Euler method which is written as follows

$$\frac{q^{n+1} - q^n}{\Delta t} = S(q^n) \tag{20.3}$$

where q^n and q^{n+1} denote the solution at the current time level (n) and the next time level (n+1). Rewriting Eq. (20.3)

$$q^{n+1} = q^n + \Delta t S(q^n) \tag{20.4}$$

allows us to see why this class of methods is called *explicit* - because the solution at the next time level that we are looking for (q^{n+1}) is given explicitly in terms of known values (all the terms on the right-hand side). However, these methods must adhere to a rather restrictive Courant-Lewy-Friedrichs (CFL) stability condition as discussed in Ch. 1. We can extend the explicit time-integration methodology to higher order approximations using either the general Runge-Kutta or multi-step approaches which we outline below.

20.2.1 Single-step Multi-stage Methods

Explicit Runge-Kutta (RK) methods belong to the general class of single-step multi-stage methods which are generally written in terms of a *Butcher tableau* [61,63] as given in Eq. (20.5)

$$\begin{array}{c|c} c & A \\ \hline & b^T \end{array} \tag{20.5}$$

where $A = a_{ij}$, $i, j = 1, \dots s$, and $c_i = \sum_j a_{ij}$ represent the time when $S(q)$ is evaluated; i.e., at each stage the function S is evaluated at $t + c_i \Delta t$. The RK method is applied in the following two-step approach:

$$Q^{(i)} = q^n + \Delta t \sum_{j=1}^{i-1} a_{ij} S(Q^{(j)}), \quad i = 1, \dots, s, \tag{20.6a}$$

$$q^{n+1} = q^n + \Delta t \sum_{i=1}^{s} b_i S(Q^{(i)}), \tag{20.6b}$$

where $Q^{(i)}$ denote the stage value solutions (at the time interval c_i) and q^{n+1} is the final solution, where s are the number of stage values and $\sum_i b_i = 1$. RK methods are known as single-step multi-stage methods because we only need to store a *single* step of the solution (q^n) in order to determine the solution at the next time level

(q^{n+1}) using a series of multi-steps defined by the number of stage values s [7]. RK methods have many attractions but one obvious one is that each application of a single time-step is self-contained in that we only need to know the starting value (q^n) and the time-step (Δt) and so we can modify the time-step size without having to change anything in the algorithm. For this reason, RK methods lend themselves quite naturally to adaptive time-stepping whereby Δt can change, on-the-fly, in order to yield a certain level of accuracy [223, 350]. Moreover, for a given RK method of order s, it is easy to build another RK method of order $s - 1$ which can then be used for error control. Finally, RK methods are endowed with a property known as *dense output* which means that we can know the solution (up to the order of the method) at any point in the time-interval $t \in (t^n, t^{n+1})$. A good way of understanding this point is that RK methods are endowed with an interpolation function in time (in an analogous manner to element-based Galerkin methods).

Table 20.1: Butcher tableau for the classical fourth-order Runge-Kutta method.

$$
\begin{array}{c|cccc}
0 & 0 & & & \\
\frac{1}{2} & \frac{1}{2} & & & \\
\frac{1}{2} & 0 & \frac{1}{2} & & \\
1 & 0 & 0 & 1 & \\
\hline
& \frac{1}{6} & \frac{1}{3} & \frac{1}{3} & \frac{1}{6}
\end{array}
$$

As a specific example of RK methods, let us describe the classical explicit fourth-order Runge-Kutta method. The Butcher tableau for this method is given in Table 20.1 which results in the following stage solutions

$$Q^{(1)} = q^n,$$
$$Q^{(2)} = q^n + \Delta t \frac{1}{2} S(Q^{(1)}),$$
$$Q^{(3)} = q^n + \Delta t \frac{1}{2} S(Q^{(2)}),$$
$$Q^{(4)} = q^n + \Delta t S(Q^{(3)}),$$

with the corresponding final solution

$$q^{n+1} = q^n + \Delta t \left[\frac{1}{6} S(Q^{(1)}) + \frac{1}{3} S(Q^{(2)}) + \frac{1}{3} S(Q^{(3)}) + \frac{1}{6} S(Q^{(4)}) \right]. \qquad (20.7)$$

Examples of other explicit Runge-Kutta methods can be found in [61, 341, 343, 187, 91, 403], including strong-stability preserving (SSP) methods [341, 177, 178, 325, 174, 176, 175, 90].

20.2.2 Multi-step Methods

Explicit *multi-step* methods can be written in the following general form

$$q^{n+1} = \sum_{k=0}^{K-1} \alpha_k q^{n-k} + \gamma \Delta t \sum_{k=0}^{K-1} \beta_k S(q^{n-k}), \qquad (20.8)$$

where K is the order of the method, and q^{n-k} for $k = 0, \ldots, K - 1$ are the solution states at the previous time-levels. Some sample multi-step methods are shown in Table 20.2 which include the *forward Euler*, *Leapfrog*, and second order *Adams-Bashforth* methods. As an example, the 2nd order Adams-Bashforth method is de-

Table 20.2: Sample Multi-step Methods where Euler is the 1st order forward Euler, LF is the 2nd order leapfrog, AB2 is the 2nd order Adams-Bashforth, and BDF2 is the 2nd order backwards difference formula.

Method	K	α_k	γ	β_k
Euler	1	1	1	1
LF	2	0,1	2	1, 0
AB2	2	1,0	1	$\frac{3}{2}, -\frac{1}{2}$
BDF2	2	$\frac{4}{3}, -\frac{1}{3}$	$\frac{2}{3}$	2,-1

fined as follows

$$q^{n+1} = q^n + \Delta t \left[\frac{3}{2} S(q^n) - \frac{1}{2} S(q^{n-1}) \right]. \qquad (20.9)$$

The coefficients in Eq. (20.8) (with some examples shown in Table 20.2) must satisfy the following conditions: $\sum_k \alpha_k = \sum_k \beta_k = 1$. Some more explicit multi-step methods can be found in [179, 180, 143, 60, 324, 91, 350, 329, 89].

20.3 Fully-Implicit Methods

An example of the simplest fully-implicit method is the first-order backward Euler method given as follows

$$q^{n+1} = q^n + \Delta t S(q^{n+1}) \qquad (20.10)$$

where the source function S is now evaluated at the unknown value q^{n+1}. The method given in Eq. (20.10) is called *fully-implicit* because the solution q^{n+1} is implicitly defined (not explicitly) since q^{n+1} is embedded within the source function S. Let us now discuss how we approach the solution of the fully-implicit problem. If S is a linear function, then the right-hand side vector $S(q^{n+1})$ can be written as the matrix-vector multiplication $S(q^{n+1}) = Lq^{n+1}$ where L is a matrix. Using this representation, we can then write Eq. (20.10) as follows

$$(I - \Delta t L) q^{n+1} = q^n \tag{20.11}$$

which now reveals the matrix-vector form of the problem, with $A = I - \Delta t L$ being the matrix problem that we need to solve.

However, if S is a non-linear function, then we need to linearize the problem before attempting to solve it. In the non-linear case, we start by writing Eq. (20.10) as the functional

$$F(q^{n+1}) = q^{n+1} - q^n - \Delta t S(q^{n+1}) \equiv 0. \tag{20.12}$$

Next, we apply a second-order Taylor series as follows

$$F(q^{(k+1)}) = F(q^{(k)}) + \frac{\partial F(q^{(k)})}{\partial q} \Delta q + O(\Delta q^2) \equiv 0 \tag{20.13}$$

where $\Delta q = (q^{(k+1)} - q^{(k)})$ and we have replaced n with k to denote an iteration value with the initial guess $q^{(0)} = q^n$. The argument for doing this is as follows: we do not know the final state q^{n+1} but we know that when we arrive at this state the functional in Eq. (20.12) should be satisfied. Then if we iterative on $q^{(k+1)}$ from a known starting value, then using Eq. (20.13) will allow us to converge to the solution quadratically with respect to Δq. Rearranging Eq. (20.13) gives the (linear) final form

$$\frac{\partial F(q^{(k)})}{\partial q} (q^{(k+1)} - q^{(k)}) = -F(q^{(k)}) \tag{20.14}$$

for the solution variable $q^{(k+1)}$, with the stopping criterion $\| q^{(k+1)} - q^{(k)} \|_p < \epsilon_{stop}$ where p is some norm and ϵ_{stop} is the stopping tolerance. In classical *Newton*-type methods, the largest cost is due to the formation of the Jacobian matrix $J^{(k)} = \frac{\partial F(q^{(k)})}{\partial q}$. However, this cost can be substantially mitigated by the introduction of *Jacobian-free Newton-Krylov* methods (see, e.g., [227]) whereby we recognize that the Jacobian can be approximated as follows

$$J^{(k)} = \frac{F(q^{(k)} + \epsilon \Delta q) - F(q^{(k)})}{\epsilon \Delta q}$$

(where ϵ is, e.g., machine zero) and direct substitution into Eq. (20.14) leads to

$$\frac{F(q^{(k)} + \epsilon \Delta q) - F(q^{(k)})}{\epsilon} = -F(q^{(k)}), \tag{20.15}$$

which we can write in the following residual form

$$\text{Res} = \frac{F(q^{(k)} + \epsilon \Delta q) - F(q^{(k)})}{\epsilon} + F(q^{(k)}). \tag{20.16}$$

Note that Eq. (20.16) only requires the construction of the vector F, with no need for constructing a matrix. This problem statement is in a perfect form for solving it approximately using any Krylov subspace method (e.g., GMRES, BiCGStab, etc. [326, 380]). One of the advantages of the Jacobian-free Newton-Krylov (JFNK) method is that we can exploit the iterative nature of both the Krylov and Newton methods, meaning that, often times, the Krylov stopping criterion (condition used for checking convergence of Eq. (20.16)) need not be so stringent as long as the final solution in the Newton solver satisfies the functional defined by Eq. (20.13).

In contrast to explicit methods, which are conditionally stable, fully-implicit methods are unconditionally stable meaning that they can use as large a time-step as desired. But one must bear in mind that since the order of accuracy of the time-integrator is $O(\Delta t^K)$ we should increase K if we wish to increase Δt. There is also a more pragmatic reason for not using very large Δt and this has to do with the condition number of the linear Krylov solution being proportional to the time-step size so that taking a large time-step size translates into an increase in the number of Krylov iterations which then may not necessarily yield a more efficient time-to-solution. For this reason it is especially important to choose the proper Krylov method (e.g., the cost of GMRES increases quadratically with the number of iterations/Krylov vectors). Preconditioners become all the more important for this class of time-integration methods if one wishes to build competitive (using the time-to-solution metric) strategies. Preconditioning is a very important topic that, unfortunately, has a literature that is too vast and ever-changing to do it justice here. Although far from an exhaustive list, the interested reader is referred to the following sample works [41,127,380,297,194,311,28,195,388,70,71,299,300]. Just as for the case of explicit methods, with fully-implicit methods we can derive both single-step multi-stage or multi-step methods which we now describe.

20.3.1 Single-step Multi-stage Methods

We can write the single-step multi-stage fully-implicit method in the following general form:

$$Q^{(i)} = q^n + \Delta t \sum_{j=1}^{i} \tilde{a}_{ij} S(Q^{(j)}), \quad i = 1, \dots, s, \qquad (20.17\text{a})$$

$$q^{n+1} = q^n + \Delta t \sum_{i=1}^{s} \tilde{b}_i S(Q^{(i)}), \qquad (20.17\text{b})$$

where we note that there is a term $Q^{(i)}$ also on the right-hand side of Eq. (20.17a) which makes the problem implicit and possibly non-linear. The coefficients are defined by the Butcher tableau

$$\begin{array}{c|c} \tilde{c} & \tilde{A} \\ \hline & \tilde{b}^{\mathcal{T}} \end{array} \qquad (20.18)$$

where $\tilde{A} = \tilde{a}_{ij}$, $i, j = 1, \ldots s$, $\sum_i \tilde{b}_i = 1$, and $\tilde{c}_i = \sum_j \tilde{a}_{ij}$; i.e., the form for the fully-implicit Runge-Kutta method is, in fact, quite similar to that for the explicit RK method. The class of implicit RK (IRK) methods is rather large and, for brevity, have only included a specific class known as *Diagonally-Implicit Runge-Kutta* (DIRK) methods. In DIRK methods, the coefficient matrix $\tilde{A}$ is not full but only lower triangular which makes it cheaper to solve since, at each stage i, we only need to solve for $Q^{(i)}$. For SDIRK (Singly-Diagonally-Implicit Runge-Kutta) methods, the values of $diag(A)$ are the same which vastly simplifies the construction of preconditioners for matrix-free methods (see, e.g., [160]). For examples of other types of IRK methods the reader is referred to, e.g., [298].

Table 20.3: Butcher tableau for a second-order implicit Runge-Kutta method.

$$
\begin{array}{c|ccc}
0 & 0 & & \\
2-\sqrt{2} & 1-\frac{1}{\sqrt{2}} & 1-\frac{1}{\sqrt{2}} & \\
1 & \frac{1}{2\sqrt{2}} & \frac{1}{2\sqrt{2}} & 1-\frac{1}{\sqrt{2}} \\
\hline
& \frac{1}{2\sqrt{2}} & \frac{1}{2\sqrt{2}} & 1-\frac{1}{\sqrt{2}}
\end{array}
$$

In Table 20.3 we give a sample second-order SDIRK method [64]. This second-order method has the following stage values

$$Q^{(1)} = q^n, \tag{20.19a}$$

$$Q^{(2)} = q^n + \Delta t \left[\left(1 - \frac{1}{\sqrt{2}}\right) S(Q^{(1)}) + \left(1 - \frac{1}{\sqrt{2}}\right) S(\mathbf{Q^{(2)}}) \right], \tag{20.19b}$$

$$Q^{(3)} = q^n + \Delta t \left[\frac{1}{2\sqrt{2}} S(Q^{(1)}) + \frac{1}{2\sqrt{2}} S(Q^{(2)}) + \left(1 - \frac{1}{\sqrt{2}}\right) S(\mathbf{Q^{(3)}}) \right], \tag{20.19c}$$

with the corresponding final solution

$$q^{n+1} = q^n + \Delta t \left[\frac{1}{2\sqrt{2}} S(Q^{(1)}) + \frac{1}{2\sqrt{2}} S(Q^{(2)}) + \left(1 - \frac{1}{\sqrt{2}}\right) S(Q^{(3)}) \right]. \tag{20.20}$$

The implicitness, and possible non-linearity, of the method is evident by the presence of the stage solutions $Q^{(2)}$ and $Q^{(3)}$ on the right-hand sides of Eqs. (20.19b) and (20.19c), respectively (in bold font). For a listing of other implicit single-step multi-stage methods the reader is referred to the following references [58, 59, 60, 62].

20.3.2 Multi-step Methods

Implicit multi-steps methods can be written in the following general form

$$q^{n+1} = \sum_{k=0}^{K-1} \alpha_k q^{n-k} + \gamma \Delta t \sum_{k=-1}^{K-1} \tilde{\beta}_k S(q^{n-k}), \qquad (20.21)$$

where the index in the second sum starting from $k = -1$ is what produces the implicitness. A popular fully-implicit method is the second-order backward-difference formula (BDF2) which we write as follows

$$q^{n+1} = \frac{4}{3} q^n - \frac{1}{3} q^{n-1} + \frac{2\Delta t}{3} S(q^{n+1}). \qquad (20.22)$$

The popularity of the BDF2 method given in Eq. (20.22) stems from the fact that it is *A-stable* (absolute stability means that the method is stable for any time-step, for all eigenvalues on the left-hand plane of the imaginary axis), is second-order, and only requires the solution of the source function S at one time-level, while only having to carry one additional time-level at $n-1$. The BDF2 method is extremely popular in the solution of the incompressible Navier-Stokes equations (see, e.g., [218, 105, 219]).

Figure 20.1 shows the absolute stability region of BDF1, BDF2, and BDF3; the stability regions are the areas outside of each curve. Note that BDF1 and BDF2 are absolutely stable (A-stable) on the left-hand plane whereas the stability curve of BDF3 crosses this region (vertical dotted line). The stability regions were computed using the following ODE

$$\frac{dq}{dt} = \lambda q$$

with $z = \lambda \Delta t$, where we plot the hull of the stability function.

20.4 Implicit-Explicit (IMEX) Methods

Implicit-Explicit (IMEX) methods are formed from a combination of both types of methods discussed in Secs. 20.2 (explicit) and 20.3 (implicit). The difference between IMEX methods and fully-implicit methods is that the linearization of the problem is performed on the continuous equations in order to construct a mixed implicit-explicit time-discretization that is already linearized. To define the IMEX time-discretization, we begin by decomposing $S(q) = N(q) + L(q)$ where the decomposition of S is performed in order to maintain the stiff terms in the linear operator L while the non-linear non-stiff terms are placed in N. By performing this decomposition at the continuous (with respect to time) level, we can then apply explicit methods to $N(q) = S(q) - L(q)$ and implicit methods to $L(q)$ resulting in an implicit but linear problem for q. Let us now describe both the single-step multi-stage and multi-step methods for the IMEX approach.

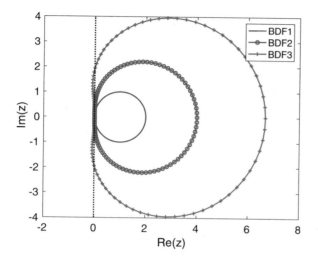

Fig. 20.1: Stability region of backward difference formulas (BDF) of orders 1, 2, and 3; the stability region is defined outside the curves. BDFk methods of order $k > 2$ are not A-stable as evident by BDF3 since it crosses the imaginary axis (dotted line).

20.4.1 Single-step Multi-stage Methods

The general form of the IMEX discretization using single-step multi-stage methods is as follows

$$Q^{(i)} = q^n + \Delta t \sum_{j=1}^{i-1} a_{ij} \mathcal{N}(Q^{(j)}) + \Delta t \sum_{j=1}^{i} \tilde{a}_{ij} \mathcal{L}(Q^{(j)}) \quad i = 1, \ldots, s, \tag{20.23a}$$

$$q^{n+1} = q^n + \Delta t \sum_{i=1}^{s} \left[b_i \mathcal{N}(Q^{(i)}) + \tilde{b}_i \mathcal{L}(Q^{(i)}) \right], \tag{20.23b}$$

with the corresponding double Butcher tableau

$$\begin{array}{c|c} c & A \\ \hline & b^{\mathcal{T}} \end{array} \qquad \begin{array}{c|c} \tilde{c} & \tilde{A} \\ \hline & \tilde{b}^{\mathcal{T}} \end{array} \tag{20.24}$$

where $b = \tilde{b}$ is necessary in order to conserve all linear invariants [160].

One possible second-order IMEX method [160] is defined by the following double Butcher tableau in Table 20.4 with $a_{32} = \frac{1}{6}\left(3 + 2\sqrt{2}\right)$ or $a_{32} = \frac{1}{2}$ which comes with both a lower order method (for error control) as well as a dense output formula (for interpolating the solution within the time-step interval). The notation ARK(i,e,o) represents an additive Runge-Kutta method with i number of implicit stages, e number of explicit stages, and an overall o order of accuracy (see [13, 293]). We refer to the method with $a_{32} = \frac{1}{6}\left(3 + 2\sqrt{2}\right)$ as ARKA(2,3,2) and the method with $a_{32} = \frac{1}{2}$ as ARKB(2,3,2).

Table 20.4: Butcher tableau for a second-order IMEX Runge-Kutta ARK(2,3,2) method.

$$
\begin{array}{c|ccc}
0 & 0 & & \\
2 - \sqrt{2} & 2 - \sqrt{2} & 0 & \\
1 & 1 - a_{32} & a_{32} & 0 \\
\hline
 & \frac{1}{2\sqrt{2}} & \frac{1}{2\sqrt{2}} & 1 - \frac{1}{\sqrt{2}}
\end{array}
\qquad
\begin{array}{c|ccc}
0 & 0 & & \\
2 - \sqrt{2} & 1 - \frac{1}{\sqrt{2}} & 1 - \frac{1}{\sqrt{2}} & \\
1 & \frac{1}{2\sqrt{2}} & \frac{1}{2\sqrt{2}} & 1 - \frac{1}{\sqrt{2}} \\
\hline
 & \frac{1}{2\sqrt{2}} & \frac{1}{2\sqrt{2}} & 1 - \frac{1}{\sqrt{2}}
\end{array}
$$

The ARK(2,3,2) family of methods have stage solutions given as follows

$$
Q^{(1)} = q^n, \tag{20.25a}
$$

$$
\begin{aligned}
Q^{(2)} = q^n &+ \Delta t \left[\left(2 - \sqrt{2}\right) \mathcal{N}(Q^{(1)})\right] \\
&+ \Delta t \left[\left(1 - \frac{1}{\sqrt{2}}\right) \mathcal{L}(Q^{(1)}) + \left(1 - \frac{1}{\sqrt{2}}\right) \mathcal{L}(Q^{(2)})\right]
\end{aligned} \tag{20.25b}
$$

$$
\begin{aligned}
Q^{(3)} = q^n &+ \Delta t \left[(1 - a_{32}) \mathcal{N}(Q^{(1)}) + a_{32} \mathcal{N}(Q^{(2)})\right] \\
&+ \Delta t \left[\frac{1}{2\sqrt{2}} \mathcal{L}(Q^{(1)}) + \frac{1}{2\sqrt{2}} \mathcal{L}(Q^{(2)}) + \left(1 - \frac{1}{\sqrt{2}}\right) \mathcal{L}(Q^{(3)})\right]
\end{aligned} \tag{20.25c}
$$

with the corresponding final solution

$$
q^{n+1} = q^n + \Delta t \left[\frac{1}{2\sqrt{2}} \mathcal{S}(Q^{(1)}) + \frac{1}{2\sqrt{2}} \mathcal{S}(Q^{(2)}) + \left(1 - \frac{1}{\sqrt{2}}\right) \mathcal{S}(Q^{(3)})\right]. \tag{20.26}
$$

In Eq. (20.26) we have used the fact that $b = \tilde{b}$ and $\mathcal{S}(q) = \mathcal{N}(q) + \mathcal{L}(q)$. The second-order SDIRK IMEX method outlined in Eqs. (20.25) and (20.26) is only one of a vast collection of possible single-step multi-stage methods. For other choices the interested reader is referred to [322, 13, 228, 292, 222, 293, 216, 42, 43, 88].

20.4.2 Multi-step Methods

The general form for IMEX multi-step methods can be written as follows

$$q^{n+1} = \sum_{k=0}^{K-1} \alpha_k q^{n-k} + \gamma \Delta t \sum_{k=0}^{K-1} \beta_k \mathcal{N}(q^{n-k}) + \gamma \Delta t \sum_{k=-1}^{K-1} \tilde{\beta}_k \mathcal{L}(q^{n-k}). \qquad (20.27)$$

Some sample IMEX multi-step methods are given in Table 20.5. For a list of other possible IMEX multi-step methods, the reader is referred to [143, 60, 134, 324, 115].

Table 20.5: Sample Multi-step Methods where Euler is the 1st order forward-backward Euler, CN-LF is the 2nd order Crank-Nicholson (implicit) with leapfrog (explicit), and BDF2 is the 2nd order backwards-difference-formula with 2nd order extrapolation for the explicit part.

Method	K	α_k	γ	β_k	$\tilde{\beta}_k$
Euler	1	1	1	1	1,0
CN-LF	2	0,1	2	1, 0	$\frac{1}{2}$, 0, $\frac{1}{2}$
BDF2	2	$\frac{4}{3}, -\frac{1}{3}$	$\frac{2}{3}$	2,-1	1, 0, 0

20.4.3 Stability of IMEX Methods

To analyze the stability of IMEX methods, we use the linear partitioned ODE written as

$$\frac{dq}{dt} = ik_s q(t) + ik_f q(t)$$

where k_s and k_f denote the wave speeds of the slow and fast components whereby we treat the slow components explicitly and the fast components implicitly. The stability region is obtained by finding for which values of k_s and k_f the amplification factor is less than or equal to one (see Eqs. (3.1) and (3.2) in [13] for details)[1]. Figure 20.2 shows the amplification factor for the BDF2, ARKA(2,3,2), ARKB(2,3,2), and ARK(1,2,1) methods for wave numbers $k_s \in [-2, 2]$ for the slow waves and $k_f \in [0, 20]$ for the fast waves. The ARK(1,2,1) method is similar to the standard forward-backward Euler method except that it has been modified to satisfy $b = \tilde{b}$ with the Butcher tableau given in Table 20.6.

---

[1] The source code to produce these plots can be found at
https://github.com/fxgiraldo/Element-based-Galerkin-Methods/tree/master/Chapters/
Chapter_20/IMEX_stability_analysis.

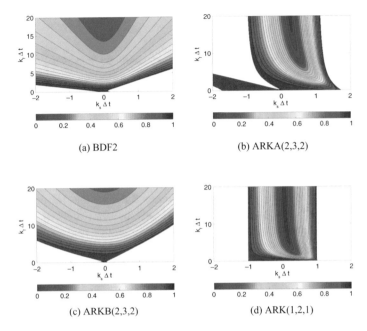

Fig. 20.2: Stability regions for (a) BDF2, (b) ARKA(2,3,2), (c) ARKB(2,3,2), and (d) ARK(1,2,1) for $k_s \in [-2, 2]$ and $k_f \in [0, 20]$.

Table 20.6: Butcher tableau for a first-order IMEX Runge-Kutta ARK(1,2,1) method.

$$
\begin{array}{c|cc}
0 & 0 & \\
1 & 1 & 0 \\
\hline
  & 0 & 1
\end{array}
\qquad
\begin{array}{c|cc}
0 & 0 & \\
1 & 0 & 1 \\
\hline
  & 0 & 1
\end{array}
$$

Note that BDF2 and ARKB(2,3,2) have the typical wedge-shape stability region where the explicit part of the stability region (horizontal axis) allows larger wave speeds for the slow component than expected as we handle larger wave speeds for the fast component (vertical axis). However, this occurs because the waves are damped (e.g., blue region). In contrast, the ARKA(2,3,2) and ARK(1,2,1) methods are attractive because at the stability limit of the method (red region) we are able to get a solution that is not damped but at the cost of adhering to smaller wave speeds for the explicit (slow) component ($|k_s| < 2$); the advantage of ARKA(2,3,2) over ARK(1,2,1) is that the method is second order for nonlinear terms and third order for linear terms. These are the sorts of trade-offs that we must contend with when we choose numerical methods for solving PDEs.

20.5 Semi-Lagrangian Methods

Recall that in the MOL, the spatial and temporal discretizations are handled separately. This is in contrast to *semi-Lagrangian* methods where the spatial and temporal discretizations are constructed in unison. In this section, we discuss two types of semi-Lagrangian methods: non-conservative and conservative formulations. Non-conservative semi-Lagrangian methods are the classical methods that are most widely used in the weather modeling community [405, 353, 348, 319, 303, 317, 152, 38, 409, 413, 98, 382, 381] and, for this reason, we discuss them in detail. Although there are many classes of conservative semi-Lagrangian methods, we only discuss those methods that are used within a Galerkin formulation; these methods are known as Lagrange-Galerkin methods [274, 57, 307, 253, 150, 151, 153, 206, 190, 161, 155, 313, 314, 33].

20.5.1 Non-Conservative Semi-Lagrangian Method

Let us describe the time-integration of the following PDE

$$\frac{\partial q}{\partial t} + \mathbf{u} \cdot \nabla q = 0 \tag{20.28}$$

which describes the transport of the quantity q. This hyperbolic partial differential equation can be replaced by the following two ordinary differential equations

$$\frac{dq}{dt} = 0 \tag{20.29a}$$

$$\frac{d\mathbf{x}}{dt} = \mathbf{u} \tag{20.29b}$$

where $\frac{d}{dt} = \frac{\partial}{\partial t} + \mathbf{u} \cdot \nabla$ is the *total* (or Lagrangian) *derivative*, and the second ODE is the trajectory equation which allows us to solve for the feet of the characteristics. Integrating Eq. (20.29) from time n to $n + 1$ yields

$$\int_{t^n}^{t^{n+1}} \frac{dq}{dt} dt = 0 \rightarrow q(\mathbf{x}(t^{n+1}), t^{n+1}) - q(\mathbf{x}(t^n), t^n) = 0 \tag{20.30a}$$

$$\int_{t^n}^{t^{n+1}} \frac{d\mathbf{x}}{dt} dt = \int_{t^n}^{t^{n+1}} \mathbf{u} \, dt \rightarrow \mathbf{x}(t^{n+1}) - \mathbf{x}(t^n) = \int_{t^n}^{t^{n+1}} \mathbf{u} \, dt \tag{20.30b}$$

where the right-hand side of the trajectory equation must be approximated with a sufficiently accurate quadrature rule. To simplify the exposition, let us assume a second order RK method for the trajectory equation. With this in mind, we can now write the discrete form of Eq. (20.30) as follows

$$q^{n+1} = \tilde{q}^n \tag{20.31a}$$

$$\tilde{\mathbf{x}}^{n+\frac{1}{2}} = \mathbf{x}^{n+1} - \Delta t \mathbf{u}(x(t^{n+1}), t^{n+1}) \qquad (20.31\text{b})$$

$$\tilde{\mathbf{x}}^{n} = \tilde{\mathbf{x}}^{n+\frac{1}{2}} - \Delta t \mathbf{u}(\tilde{x}^{n+\frac{1}{2}}, t^{n+\frac{1}{2}}) \qquad (20.31\text{c})$$

where values such as $\tilde{q}$ denote that they are taken along characteristic paths, which are denoted by the dashed arrows in Fig. 20.3. Let us now review some semi-Lagrangian notation.

Figure 20.3 illustrates the semi-Lagrangian trajectories for two elements with linear polynomial order. The values of x at time $n + 1$ are known as the arrival points (blue points) and define the gridpoints used in the simulation (in this case, they are the global gridpoints of, say, the CG method). Equation (20.31b) allows us to solve for the midpoint trajectory point $\tilde{\mathbf{x}}^{n+\frac{1}{2}}$ if we know the velocity $\mathbf{u}(x(t^{n+1}), t^{n+1})$, which is defined at the arrival time which we do not yet know. Here, we use a second order extrapolation as follows

$$\mathbf{u}(x(t^{n+1}), t^{n+1}) = 2\mathbf{u}(x(t^{n}), t^{n}) - \mathbf{u}(x(t^{n-1}), t^{n-1})$$

defined at the arrival points but at previous time levels. Once we know the midpoint trajectory point $\tilde{\mathbf{x}}^{n+\frac{1}{2}}$ we can use Eq. (20.31c) to find the departure point $\tilde{\mathbf{x}}^{n}$. Note that to do so requires two steps: first we need to extrapolate the velocity field to the time level $n + \frac{1}{2}$ as follows

$$\mathbf{u}(x(t^{n+\frac{1}{2}}), t^{n+\frac{1}{2}}) = \frac{3}{2}\mathbf{u}(x(t^{n}), t^{n}) - \frac{1}{2}\mathbf{u}(x(t^{n-1}), t^{n-1})$$

at all the arrival points (which are the known gridpoints). Then, we need to interpolate $\mathbf{u}(x(t^{n+\frac{1}{2}}), t^{n+\frac{1}{2}})$ at the midpoint trajectory point $\tilde{\mathbf{x}}^{n+\frac{1}{2}}$ which we do using the basis functions of our element. To find the midpoint and departure points (red points in

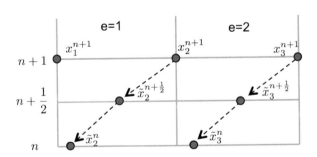

Fig. 20.3: Semi-Lagrangian trajectories for linear polynomials with two elements.

Fig. 20.3), we need to identify in which element these characteristic points lie. Once we have identified the midpoint/departure point element then we need to find the reference element coordinate $\tilde{\xi}$ which corresponds to the physical grid point $\tilde{\mathbf{x}}$. Upon doing so, we can then find the interpolated departure point value using the standard basis function expansion

$$q_d \equiv \tilde{q}_N^n = \sum_{i=0}^{N} \psi_i \left(\tilde{\xi} \right) q_i^n$$

where q_i^n are the solutions at the gridpoints i from the previous time level n.

By scrutinizing the steps required in constructing the semi-Lagrangian method it becomes obvious why these methods are not classified as an MOL approach. This is evident when we consider the original PDE given by Eq. (20.28) which contains both spatial and temporal derivatives. Then we replace this PDE by the two ODEs given in Eq. (20.29). We then recognize that in order to do this requires that the time-integration method know something about the order of accuracy of the underlying spatial discretization method; in this case it occurs through the basis functions of the polynomial expansion.

20.5.1.1 Algorithm for the Non-Conservative Semi-Lagrangian Method

The algorithm for implementing the non-conservative semi-Lagrangian method can be decomposed into its two main parts: the departure point calculation and the interpolation of the solution variable at the departure points. Algorithm 20.1 outlines the steps for constructing the departure point values. In this example, the velocity is assumed constant but in the general case we need to use RK time-stepping with the extrapolation mentioned previously (see, e.g., [161]).

Algorithm 20.1 Departure point calculation.

---

function DEPARTURE_POINTS($x, \Delta t, u$)
    for $i = 1 : N_p$ **do**   ▷ loop through gridpoints
        $x_d(i) = x(i) - \Delta t u$
        if $(x_d(i) < x_{min})$ **then**
            $x_d(i) = x_{max} + x_d(i) - x_{min}$ ▷ impose periodicity
        else if $(x_d(i) > x_{max})$ **then**
            $x_d(i) = x_{min} + x_d(i) - x_{max}$ ▷ impose periodicity
        end if
    end for
end function

Algorithm 20.2 outlines the steps for constructing the interpolation of the solution vector q at the departure points x_d. This algorithm is quite general and is independent of the dimension of the problem or whether or not the velocity is constant or time-dependent. Once the departure point values are determined from Alg. 20.1, all we need to do is interpolate the solution at the departure points. For multi-dimensional problems, it may be necessary to use a nonlinear iteration to determine the departure point values in terms of the reference element coordinates. Such a procedure is outlined in [151]. Algorithm 20.1 uses periodic boundary conditions where x_{min} and x_{max} are the extrema of the physical coordinates of the one-dimensional grid. For multi-dimensional problems, the application of this boundary condition is similar to the one-dimensional example. Algorithm 20.2 yields the interpolated solution q at

---

Algorithm 20.2 Interpolation calculation.

function INTERPOLATE_VALUES(q)
    for $i = 1 : N_p$ **do**   ▷ loop through gridpoints
        for $e = 1 : N_e$ **do**   ▷ loop through elements
            if $(x_d(i) \in [x_0^{(e)}, x_N^{(e)}]$ **then**
                $\xi_d = 2\frac{x_d(i) - x_0^{(e)}}{x_N^{(e)} - x_0^{(e)}} - 1$ ▷ departure point in reference coordinate
                Construct $\psi^d = \psi(\xi_d)$ ▷ basis function at departure point (Alg. 3.1)
                $q_d(i) = \sum_{k=0}^{N} \psi_k^d q_k^{(e)}$
            end if
        end for
    end for
end function

---

the departure points which we denote as $q_d = \tilde{q}^n$. In an explicit semi-Lagrangian approach, the solution at the departure point is transported to the arrival point at the new time-step as indicated by Eq. (20.31a). This method is described in detail in [151, 161, 155].

20.5.2 Conservative Semi-Lagrangian Method

A conservative semi-Lagrangian method can be constructed for the following PDE

$$\frac{\partial q}{\partial t} + \nabla \cdot (q\mathbf{u}) = 0 \tag{20.32}$$

which describes a conservation law for the quantity q. We now multiply Eq. (20.32) by a test function and integrate to construct the following Galerkin problem: find $q_N \in S$ such that

$$\int_{\Omega_e} \psi_i \left(\frac{\partial q_N^{(e)}}{\partial t} + \nabla \cdot \left(q_N^{(e)} \mathbf{u}_N^{(e)} \right) \right) d\Omega_e = 0, \ \forall \psi \in S \tag{20.33}$$

where S is an appropriate discrete function space for either the CG or DG method. Substituting the following two identities

$$\psi \frac{\partial q}{\partial t} = \frac{\partial \psi q}{\partial t} - q \frac{\partial \psi}{\partial t}$$

$$\psi \nabla \cdot (q\mathbf{u}) = \nabla \cdot (\psi q \mathbf{u}) - q \nabla \cdot (\psi \mathbf{u})$$

into Eq. (20.33) yields

$$\int_{\Omega_e} \left(\frac{\partial \psi_i q_N^{(e)}}{\partial t} + \nabla \cdot \left(\psi_i q_N^{(e)} \mathbf{u}_N^{(e)} \right) - q_N^{(e)} \frac{d\psi_i}{dt} \right) d\Omega_e = 0. \qquad (20.34)$$

Exercise Starting with Eq. (20.33), derive Eq. (20.34).   □

If we constrain the basis functions to be constant along characteristics then we can write Eq. (20.34) as follows

$$\int_{\Omega_e} \left(\frac{\partial \psi_i q_N^{(e)}}{\partial t} + \nabla \cdot \left(\psi_i q_N^{(e)} \mathbf{u}_N^{(e)} \right) \right) d\Omega_e = 0. \qquad (20.35)$$

We can now invoke the *Reynolds' transport theorem* to write Eq. (20.35) as follows

$$\frac{d}{dt} \int_{\Omega_e} \psi_i q_N^{(e)} \, d\Omega_e = 0 \rightarrow \int_{\Omega_e(t^{n+1})} \psi_i q_N^{(e)} \, d\Omega_e = \int_{\Omega_e(t^n)} \psi_i q_N^{(e)} \, d\Omega_e \qquad (20.36)$$

where the bounds of integration $\Omega_e(t^{n+1})$ and $\Omega_e(t^n)$ denote the volumes of the element at the arrival and departure points, respectively. Note that the integration takes into account the fact that the domain does not necessarily remain constant but may change with the characteristic path; in fact, it is this property that allows this approach to be conservative. In practice, this approach is difficult to implement because a mesh generation step is involved to construct the volumes at the departure point values (see, e.g., [150]). For this reason, many have proposed to use quadrature rather than exact integration which, although affects conservation, does not adversely affect the stability of the method [274, 307, 150]. The departure point trajectory is computed from the following equation

$$\frac{d\mathbf{x}}{dt} = \mathbf{u}$$

which is identical to the one we already defined for the non-conservative semi-Lagrangian method.

20.5.2.2 Algorithm for the Conservative Semi-Lagrangian Method

Since we use the same trajectory equation, we can reuse Alg. 20.1. The difference between the non-conservative and conservative semi-Lagrangian (*Lagrange-Galerkin*) method is that now we need to construct quadrature rules for the following integrals

$$\int_{\Omega_e(t^{n+1})} \psi_i q_N^{(e)} \, d\Omega_e = \int_{\Omega_e(t^n)} \psi_i q_N^{(e)} \, d\Omega_e \tag{20.37}$$

which can be replaced by the following matrix-vector problem

$$M_{i,j}^{n+1} q_j^{n+1} = \tilde{M}_{ij}^n \tilde{q}_j^n \tag{20.38}$$

with $i, j = 0, \ldots, N$, where we have dropped the superscript (e) without loss of meaning. In Eq. (20.38), the mass matrices in one-dimension are defined as follows

$$M_{i,j}^{n+1} = \sum_{k=0}^{Q} w_k \Delta x^{n+1} \psi_{i,k} \psi_{j,k}.$$

$$\tilde{M}_{i,j}^n = \sum_{k=0}^{Q} w_k \Delta \tilde{x}^n \psi_{i,k} \psi_{j,k},$$

where $\Delta x^{n+1} = x_N - x_0$ is the element length at the arrival points (blue points in Fig. 20.3) and $\Delta \tilde{x}^n = x_N^d - x_0^d$ is the element length at the departure points where x^d denotes the departure points (red points in Fig. 20.3).

Algorithm 20.3 outlines the steps for constructing the right-hand side of Eq. (20.38) where we assume that we have used Alg. 20.1 for constructing the departure points x_d and Alg. 20.2 for constructing the solution at the departure point values q_d. After invoking Alg. 20.3, we need to call the proper DSS operation for either CG

Algorithm 20.3 Constructing the RHS of Lagrange-Galerkin method.

---
function LAGRANGE_GALERKIN(x_d, q)
    for $e = 1 : N_e$ **do**   ▷ loop through elements
        $I_0 = intma(0, e);\ I_N = intma(N, e)$
        $\Delta \tilde{x}^n = x_d(I_N) - x_d(I_0)$
        for $i = 0 : N$ **do**   ▷ loop through element DOF
            $\tilde{R}(i, e) = \sum_{k=0}^{Q} w_k \Delta \tilde{x}^n \psi(i, k) q_d(k)$
        end for
    end for
end function

---

or DG. After this point, we need to invert by the global mass matrix at the arrival points as such

$$q^{n+1} = \left(M_G^{n+1} \right)^{-1} \tilde{R}_G \tag{20.39}$$

where the subscript G denotes a global matrix or vector which has undergone the DSS operation.

20.5.3 Semi-Lagrangian Example

Let us use the same passive transport problem that we used in Chs. 5 and 6 which we can write as follows

$$\frac{\partial q}{\partial t} + \frac{\partial}{\partial x}(qu) = 0$$

which is defined in detail in Sec. 5.10. Recall that in this example, $u = $ constant so that this PDE is equivalent to writing

$$\frac{\partial q}{\partial t} + u\frac{\partial q}{\partial x} = 0.$$

The point of mentioning this is that both the non-conservative and conservative semi-Lagrangian methods can be applied to the same PDE for this particular example (or any whereby $\nabla \cdot \mathbf{u} = 0$).

Exercise Show that the non-conservative and conservative semi-Lagrangian methods are applicable to the same PDE provided that $\nabla \cdot \mathbf{u} = 0$. □

Figure 20.4 shows the convergence rates for the semi-Lagrangian method outlined in Alg. 20.2 using various polynomial orders after one complete revolution; we use the initial condition given in Eq. (5.32) with periodic boundary conditions. The convergence rates are exactly the same for Courant numbers of 0.5 or 2.5. This is only the case, for this particular example, because the trajectory calculations are exact. However, in general, as shown by Falcone and Ferretti [123] the order of accuracy of the semi-Lagrangian method is given by

$$O\left(\Delta t^K, \frac{\Delta x^P}{\Delta t}\right) \tag{20.40}$$

where K is the order of the time-integrator (for the trajectory calculation) and P is the order of interpolation of the solution at the departure points. The value of Eq. (20.40) is that it tells us what is the optimal time-step for a specific choice of K and P. Moreover, this equation also tells us that as $\Delta t \to 0$, the semi-Lagrangian method loses accuracy and, therefore, is not advisable to use small Δt.

20.5.4 Semi-Lagrangian Method for Balance Laws

We have only discussed the application of semi-Lagrangian methods to homogeneous PDEs of the type

$$\frac{dq}{dt} = 0$$

or

$$\frac{d}{dt}\int_{\Omega_e} \psi_i q_N^{(e)} \, d\Omega_e = 0.$$

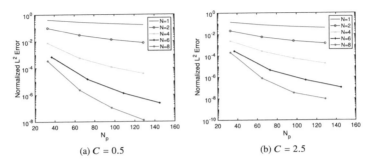

Fig. 20.4: Convergence rate for the non-conservative CG semi-Lagrangian method with polynomial order N for Courant numbers (a) $C = 0.5$ and (b) $C = 2.5$.

We can extend the methods described in this section to general balance laws of the type

$$\frac{dq}{dt} = S(q)$$

or

$$\frac{d}{dt} \int_{\Omega_e} \psi_i q_N^{(e)} \, d\Omega_e = S(q)$$

whereby we can treat the left-hand side as described in this section and treat the right-hand side by the implicit methods discussed in Sec. 20.3. In the case that $S(q)$ is a linear operator, then the solution procedure is straightforward and follows the approach discussed in Sec. 20.4 for implicit-explicit methods. If $S(q)$ is nonlinear then we need to combine the solution procedure with Newton's method as in the JFNK approach discussed in Sec. 20.3.

20.6 Multirate Methods

To describe *multirate methods*, let us begin with Eq. (20.2) and write it in the partitioned form

$$\frac{dq}{dt} \equiv S(q) = N(q) + \mathcal{L}(q) \qquad\qquad (20.41)$$

where, as in the IMEX problem presented in Sec. 20.4, we assume that the terms in $N(q)$ contain the slow modes and those in $\mathcal{L}(q)$ the fast modes. In Sec. 20.4 we treated the slow modes explicitly and the fast modes implicitly. Multirate methods are typically used in one of two ways: using either a geometric partition [91, 329, 89, 171, 275, 92, 337, 338, 133, 181, 68] whereby different time-steps are used in different regions of the domain or a process partition [144, 120, 228, 187, 332, 333, 403, 45, 334, 335, 331, 24] where processes (or terms) are handled with different time-steps.

For example, the geometric partition multirate may arise from using adaptive mesh refinement (as in Ch. 19) where the small elements use a time-step that is smaller than the time-step size used for the larger elements. An example of the process partition multirate arises in the same situation that gives rise to the use of IMEX solvers, i.e., some processes have slow modes while others have faster modes. In what follows, we focus only on the process partitioned multirate method although these methods can be applied equally to the geometric partition multirate. We focus on the *multirate infinitesimal step* methods.

20.6.1 Multirate Infinitesimal Step

To describe the implemenation of the multirate infinitesimal step method [228, 333, 403, 335], let us consider Eq. (20.41). Algorithm 20.4 shows that the slow modes in $N(q)$ are evaluated in the outer loop (i-loop) with time-step Δt (via the term r_i) while the fast modes in $\mathcal{L}(q)$ are evaluated in the inner loop (j-loop) with time-step $\tilde{c}_i \Delta t$.

Algorithm 20.4 2-Partition General Explicit Multirate RK.

---

function MULTIRATE-RK(q^n)
  $Q^{(1)} = q^n$
  for $i = 2 : I + 1$ **do**   ▷ Loop through slow process
    $r_i = \sum_{j=1}^{i-1} \tilde{a}_{i,j} N(Q^{(1)})$
    $V_{i,1} = Q^{(i-1)}$
    for $j = 2 : J + 1$ **do**   ▷ Loop through fast process
      $V_{i,j} = V_{i,j-1} + \Delta t \sum_{k=1}^{j-1} \tilde{a}_{j,k} \left[r_i + \tilde{c}_i \mathcal{L}(V_{i,k}) \right]$
    end for
    $Q^{(i)} = V_{i,J+1}$
  end for
  $q^{n+1} = Q^{(I+1)}$
end function

---

where

$$\tilde{a}_{i,j} = \begin{cases} a_{i,j} - a_{i-1,j} & i < s + 1 \\ b_j - a_{s,j} & i = s + 1 \end{cases}$$

and

$$\tilde{c}_i = \begin{cases} c_i - c_{i-1} & i < s + 1 \\ 1 - c_s & i = s + 1 \end{cases}$$

with $A = a_{i,j}$, b, and c being the Butcher tableau coefficients and s denotes the number of stages. The only constraint on this algorithm is that the coefficients c increase monotonically. If they do not, we can still use this approach but Alg. 20.4 would have to be modified. To obtain the theoretical rates of convergence, special order conditions need to be satisfied as outlined in, e.g., [333]. For second order base methods, second order convergence is guaranteed, for third order we need to satisfy an additional order condition, and for fourth order we need to satisfy more order conditions (e.g., see [24]).

In the case that the inner loop of Alg. 20.4 does not offer the desired ratio of fast to slow time-step ratio, we can introduce substepping to the inner loop (fast method) as shown in Alg. 20.5. If $\mathcal{L}(q)$ can be further decomposed into fast and faster parts, then the inner loop can be solved using an IMEX method which obviates the need for substepping.

Algorithm 20.5 2-Partition General Explicit Multirate RK with Substepping.

---

function MULTIRATE-RK-SUBSTEP(q^n)
   $Q^{(1)} = q^n$
   for $i = 2 : I + 1$ **do**   ▹ Loop through slow process
     $r_i = \sum_{j=1}^{i-1} \widetilde{a}_{i,j} \mathcal{N}(Q^{(1)})$
     $q_m = Q^{(i-1)}$
     for $m = 1 : M$ **do**   ▹ Substepping loop
       $V_{i,1} = q_m$
       for $j = 2 : J + 1$ **do**   ▹ Loop through fast process
         $V_{i,j} = V_{i,j-1} + \frac{\Delta t}{M} \sum_{k=1}^{j-1} \widetilde{a}_{j,k} \left[r_i + \widetilde{c}_i \mathcal{L}(V_{i,k}) \right]$
       end for
       $q_m = V_{i,J+1}$
     end for
     $Q^{(i)} = q_m$
   end for
   $q^{n+1} = Q^{(I+1)}$
end function

Algorithm 20.5 allows the multirate method to achieve the desired ratio of slow to fast time-step by choosing $M = R_s$ where $R_s = \frac{\Delta t_{slow}}{\Delta t_{fast}}$. However, this value of M might be overly conservative since the substeps of an RK method are not uniform. Instead, we can compute an M value for each outer loop as follows: given a $\Delta t = \Delta t_{slow}$ we can define the ideal fast time-step to be

$$\Delta t_{fast}^{ideal} = \frac{\Delta t}{R_s}$$

and the actual (or real) fast time-step is

$$\Delta t_{fast}^{real} = \widetilde{c}_i \Delta t.$$

This means that the optimum number of substeps is

$$M_i = \frac{\Delta t_{fast}^{real}}{\Delta t_{fast}^{ideal}} \equiv \widetilde{c}_i R_s$$

per outer (large time-step, i.e., the i-loop) stage. Algorithm 20.5 can now be used with this computed value of M at each i-stage[2].

20.6.2 Convergence Rate

To test this approach, let us use the 2-partition non-autonomous[3] ordinary differential equation (ODE)

$$\frac{dq}{dt} = \sin t + cq(t)$$

where $t \in [0, \pi]$, $c = -100$, and the slow mode is $\mathcal{N}(q) = \sin t$ and the fast mode is $\mathcal{L}(q) = cq(t)$. The result of the 2-partition multirate method for this ODE using the Runge-Kutta method given in Table 20.3 is shown in Fig. 20.5 (ARKA(2,3,2)) where the L^2 error is computed as follows

$$||q||_{L^2} = \frac{||q - q^{exact}||_{L^2}}{||q^{exact}||_{L^2}}$$

where the exact solution is

$$q^{exact} = \frac{\exp(ct) + (\exp(ct) - c\sin t - \cos t)}{1 + c^2}.$$

The rate of convergence is of order 3 when no substepping is used because this particular RK method is 3rd order accurate for linear equations and 2nd order for nonlinear equations (see [160]). With $M = 10$ substeps we achieve order 2 convergence but note that the error is actually smaller than the no substepping case and, moreover, a much larger time-step ratio between the slow and fast waves can be used.

20.6.3 Speedup

One question that immediately comes to mind is: how much speedup should we expect from using this multirate method? To answer this question, we must first find the optimal explicit single-rate time-integration method and then compare the speedup of its multirate counterpart. To find the best explicit single-rate time-integration method, we must first perform a stability analysis.

---

[2] This idea is due to Jeremy Kozdon.
[3] An autonomous ODE is only a function of the dependent variable q whereas a non-autonomous ODE is a function of both the dependent variable q and the independent variable t. Thanks to Emil Constantinescu for his help computing the convergence rates.

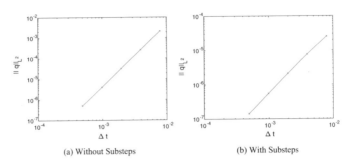

Fig. 20.5: Multirate Method using 3-stage 2nd-order RK base method (a) without substeps and (b) with $M = 10$ substeps.

20.6.3.1 Stability Analysis

The polynomial of z which satisfies stability is given as (see, e.g., [61] p. 230)

$$P = \frac{det(\mathbf{I}_s - zA + z\mathbf{1}_s b^T)}{det(\mathbf{I}_s - zA)} \quad (20.42)$$

where $z = x + iy$ with $i = \sqrt{-1}$ and $\mathbf{I}_s$ is the rank-s identity matrix, $\mathbf{1}_s$ is the vector of ones of length s, and $A = a_{i,j}$ and b are the Butcher coefficients with $c_i = \sum_{j=1}^{s} a_{ij}$. Once this polynomial is obtained (e.g., via a symbolic toolbox) we then plot contour values where the hull of the plot reveals the stability region. For implicit Runge-Kutta methods, we can use Eq. (20.42) but P then becomes a rational function.

Doing this for various time-integrators, results in Fig. 20.6 which shows that ARKA(2,3,2) (dashed blue line) from [160] has one of the largest stability regions of all the methods along the imaginary axis. However, classical RK4 (dashed black line) and the low-storage 5-stage 4th-order (LSRK(5,4), dashed cyan line) methods [223] are better. The strong-stability preserving (SSP) methods SSP(5,3)R [323] and SSP(5,3)H [202] do not do as well, nor does the low-storage 14-stage 4th-order method [286]. In what follows, we choose the LSRK(5,4) method since it only requires two storage registers and has an excellent stability region relative to the number of stages.

20.6.3.2 Complexity Analysis

Using LSRK(5,4) as our best single-rate method, we can now compute the expected speedup that its multirate version will yield. Revisiting Eq. (20.41), the total cost of a single-rate method is

$$C_{total} N_{time}$$

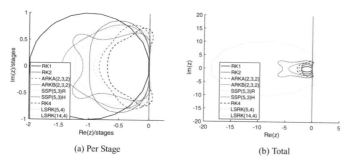

(a) Per Stage   (b) Total

Fig. 20.6: Stability region of various explicit RK methods (a) normalized by the number of stages of each method (per stage) and (b) showing the total stability region (total).

where C_{total} is the cost of building the total right-hand side $\mathcal{N}(q) + \mathcal{L}(q)$, N_{time} is the number of time-steps (assuming the small time-step Δt_{fast}). The cost of its multirate counterpart is

$$C_{slow} \frac{N_{time}}{R_s} + C_{fast} N_{time}$$

where $R_s = \frac{\Delta t_{slow}}{\Delta t_{fast}}$ is the ratio of slow to fast process speeds in the problem. The speedup is then defined as

$$S = \frac{C_{total} N_{time}}{C_{slow} \frac{N_{time}}{R_s} + C_{fast} N_{time}} \equiv \frac{C_{total}}{C_{slow} \frac{1}{R_s} + C_{fast}}.$$

The best case scenario is obtained when R_s gets large which we write as

$$S = \lim_{R_s \to \infty} \frac{C_{total}}{C_{slow} \frac{1}{R_s} + C_{fast}} = \frac{C_{total}}{C_{fast}} \tag{20.43}$$

which tells us that the maximum speedup will be a direct ratio of the cost of constructing the total right-hand side to the cost of constructing the fast components. Therefore, the more processes we are able to add to the total right-hand side the better the speedup. For some applications, this could yield a significant speedup. Take for example the case of low Mach number compressible Navier-Stokes where, the fastest waves in the system are the acoustic waves (traveling at 300 meters per second) while the velocity can be traveling anywhere between 3 to 30 meters per second (yielding an $R_s = 100$ or 10) where then the fast right-hand side operator ($\mathcal{L}(q)$) only contains the acoustic waves while the total right-hand operator contains everything else including, say, diffusion or hyper-diffusion (which is usually quite expensive in high-order element-based Galerkin methods).

We end this section by making a simplification to Eq. (20.43) by assuming that the cost of building the total right-hand side is $C_{total} = C_{fast} + C_{slow}$ which allows us to write

$$S = \frac{C_{fast} + C_{slow}}{C_{fast}} \equiv 1 + R_c \qquad (20.44)$$

where $R_c = \frac{C_{slow}}{C_{fast}}$ is the ratio of the cost of building the slow component to the fast component.

Chapter 21
1D Hybridizable Discontinuous Galerkin Method

21.1 Introduction

One of the main criticisms of the DG method, compared to the CG method, is that it requires more degrees of freedom to solve a problem for the same number of elements and polynomial degree per element. The total degrees of freedom for the DG method is

$$N_p^{DG} = [N_e(N+1)]^d$$

for the local element-wise (LEW) storage scheme described in Chs. 6 and 7, where N_e are the number of elements, N is the polynomial degree within each element, and d is the number of spatial dimensions. In contrast, the total degrees of freedom for the CG method is

$$N_p^{CG} = (N_e \cdot N + 1)^d$$

for the global gridpoint (GGP) storage scheme described in Ch. 7. For simplicity, we assume that the number of points along each spatial dimension is constant and that tensor-product bases are used. The comparison of the number of points between the CG and DG methods is illustrated in Table 21.1 for one, two, and three dimensions. The ratio R_p^d gives the ratio between the number of points (N_p) between the DG and CG methods for spatial dimension d. The ratio of the degrees of freedom (DOF) between CG and DG is greatest at low order and decreases with increasing N.

In Fig. 21.1 we show the location of the points for both CG (in blue) and DG (in red) for $N_e = 3$ for two configurations: $N = 1$ and $N = 2$. Both Table 21.1 and Fig. 21.1 confirm that the difference in the DOF between CG and DG is greatest at low order. Once we begin to increase the order of the polynomials within each element, the difference between GGP and LEW storage decreases[1].

---

[1] See Ch. 7 for a discussion on global gridpoint (GGP) and local element-wise (LEW) storage.

© The Editor(s) (if applicable) and The Author(s), under exclusive license
to Springer Nature Switzerland AG 2020
F. X. Giraldo, *An Introduction to Element-Based Galerkin Methods on
Tensor-Product Bases*, Texts in Computational Science and Engineering 24,
https://doi.org/10.1007/978-3-030-55069-1_21

Table 21.1: Comparison of degrees of freedom between the CG and DG methods along one, two, and three dimensions.

N_e	N	R_p^{1d}	R_p^{2d}	R_p^{3d}
3	1	1.50	2.25	3.38
3	2	1.29	1.65	2.13
3	4	1.15	1.33	1.54
3	8	1.08	1.17	1.26

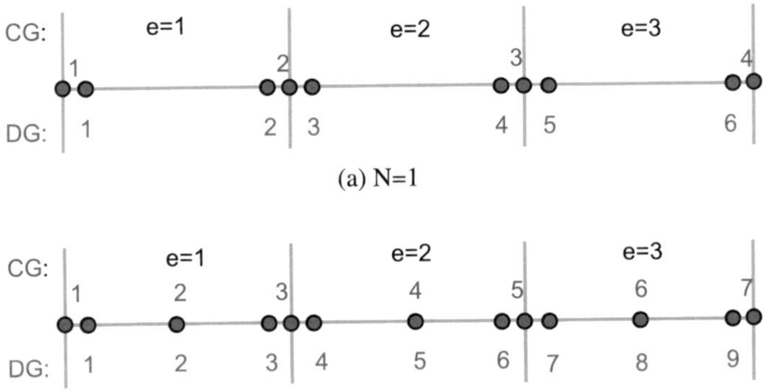

(a) N=1

(b) N=2

Fig. 21.1: Layout for the degrees of freedom for CG with GGP storage (in blue) and DG with LEW storage (in red) for $N_e = 3$ and (a) $N = 1$ and (b) $N = 2$.

Nonetheless, the DG method always requires more DOF than CG (with GGP storage) and to make the DG method competitive with the CG method requires a different strategy for decreasing the DOF and this is exactly what the hybridizable discontinuous Galerkin (HDG) method aims to do (see, e.g., [327, 79, 82, 80, 81, 283, 85, 83, 301, 273, 284, 75, 52, 51, 411, 217]). However, the HDG method can only serve to reduce the degrees of freedom for implicit time-integration which we now discuss. Unfortunately, for explicit time-integration, the HDG method is of no use in reducing the degrees of freedom as the reader will shortly understand the reason for this. In what follows, we lump the classical HDG method with its variant, the embedded discontinuous Galerkin (EDG) [183, 84, 285, 95, 215, 136], and simply refer to both as HDG. For the purpose of our discussion, the complexity analysis that we perform is representative of the EDG method, with the classical HDG method having a slightly larger complexity due to the solution space of the trace variables. In the classical HDG method in two dimensions, vertex points are duplicated whereas in EDG they are not. In 3D, face points are duplicated for HDG but not for EDG. However, to simplify the discussion we will focus on the EDG method (even though we refer to it as HDG).

21.2 Implicit Time-Integration for CG/DG

To understand the value of the HDG method, we need to discuss the discretization of CG/DG methods in an implicit in time approach. To simplify the exposition, let us focus our discussion on the solution of the one-dimensional wave equation

$$\frac{\partial q}{\partial t} + \frac{\partial f}{\partial x} = 0 \tag{21.1}$$

where $f = qu$ with the velocity component u, initial condition $q(x,0) = q_0(x)$ and periodic boundary conditions. In addition, let us begin by first describing the spatial discretization of this equation using the CG/DG method. Once we have discretized Eq. (21.1) in space then we can focus on the time-discretization using: i) the first-order backward Euler method since it is the easiest implicit method to implement, followed by ii) a second-order implicit Runge-Kutta method. The goal, as we will show, is to write the semi-discrete problem as follows

$$\frac{dq}{dt} = S(q)$$

where S represents the spatially discretized operators.

21.2.1 Continuous and Discontinuous Galerkin Discretizations

Using a combined CG/DG discretization, we can obtain the semi-discrete form of Eq. (21.1) as follows

$$M_{ij}^{(e)} \frac{dq_j^{(e)}}{dt} + F_{ij}^{(e,k)} f_j^{(e)} - \tilde{D}_{ij}^{(e)} f_j^{(e)} = 0 \tag{21.2}$$

for $e = 1, \ldots N_e$ and $i, j = 0, \ldots N$ where the superscript (e) denotes a local element-wise array and (e,k) denotes an array that accounts for the element e as well as its neighbor k where the interface (e,k) is defined by the number of face neighbors N_{FN} that an element has. For a tensor-product grid, $N_{FN} = 2d$ where d is the spatial dimension (e.g., in 1D there are two neighbors, in 2D there are four, and in 3D there are six). The construction of the matrices M, F, and $\tilde{D}$ follows the approach described in Ch. 7 in Eq. (7.3). Next, if we use the fact that $f = qu$ then we can rewrite Eq. (21.2) as follows

$$M_{ij}^{(e)} \frac{dq_j^{(e)}}{dt} + u \left(F_{ij}^{(e,k)} - \tilde{D}_{ij}^{(e)} \right) q_j^{(e)} = 0. \tag{21.3}$$

Let us now define the matrices in Eq. (21.3). The mass matrix is simply the inner product of the basis functions as follows

$$M_{ij}^{(e)} = \int_{\Omega_e} \psi_i \psi_j d\Omega_e$$

and the weak form differentiation matrix is

$$\tilde{D}_{ij}^{(e)} = \int_{\Omega_e} \frac{\partial \psi_i}{\partial x} \psi_j d\Omega_e$$

which arises from integration by parts. It is easiest to describe the flux term vector if we write it in the following way

$$F_{ij}^{(e,k)} f_j^{(e)} = \sum_{k=1}^{N_{FN}} \mathbf{n}^{(e,k)} \psi_i f^{(*,e,k)}$$

where, previously, we defined the normal vector as $\hat{\mathbf{n}}$ but here we use $\mathbf{n}$ in order to avoid confusion with the hybridized variables discussed below. Using the Rusanov numerical flux presented in Chs. 6 and 18

$$f^{(*,e,k)} = \frac{1}{2}\left[f^{(e)} + f^{(k)} - \mathbf{n}^{(e,k)}|u| \left(q^{(k)} - q^{(e)} \right) \right] \tag{21.4}$$

and assuming $u = $ constant > 0 allows us to rearrange the flux function as follows

$$f^{(*,e,k)} = \frac{u}{2}\left[\left(1 + \mathbf{n}^{(e,k)} \right) q^{(e)} + \left(1 - \mathbf{n}^{(e,k)} \right) q^{(k)} \right]. \tag{21.5}$$

Multiplying Eq. (21.5) by the normal vector and rearranging yields

$$\mathbf{n}^{(e,k)} f^{(*,e,k)} = \frac{u}{2}\left[\left(\mathbf{n}^{(e,k)} + 1 \right) q^{(e)} + \left(\mathbf{n}^{(e,k)} - 1 \right) q^{(k)} \right].$$

Rearranging one more time yields the final form

$$\mathbf{n}^{(e,k)} f^{(*,e,k)} = \frac{u}{2} \mathbf{n}^{(e,k)} \left(q^{(e)} + q^{(k)} \right) + \frac{u}{2} \left(q^{(e)} - q^{(k)} \right) \tag{21.6}$$

which now allows us to write the flux vector as follows

$$F_{ij}^{(e,k)} f_j^{(e)} = C^{(e,k)} q_j + \mathcal{J}^{(e,k)} q_j \tag{21.7}$$

where the first term on the right-hand side of Eq. (21.6) corresponds to the *centered* matrix

$$C^{(e,k)} = \frac{u}{2} \mathbf{n}^{(e,k)} \left(\delta_{i_e,i_e} + \delta_{i_e,i_k} \right) \tag{21.8}$$

and the second term corresponds to the *jump* matrix

$$\mathcal{J}^{(e,k)} = \frac{u}{2} \left(\delta_{i_e,i_e} - \delta_{i_e,i_k} \right) \tag{21.9}$$

where δ is the Kronecker delta function and i_e and i_k are the matrix indices of the ith gridpoint in the elements e and k. Note that we have also assumed that the interpolation points are Lobatto points which simplifies the discussion since $\psi = 1$ at the faces of the elements. The matrices $C^{(e,k)}$ and $\mathcal{J}^{(e,k)}$ can be combined into a single matrix but here we write both separately in order to better understand the role of each term in the Rusanov flux.

For the case shown in Fig. 21.1a (with $N_e = 3$ and $N = 1$) the collection of all element matrix values are as follows: the mass matrix is

$$M^{(e)} = \frac{\Delta x^{(e)}}{6} \begin{pmatrix} 2 & 1 & 0 & 0 & 0 & 0 \\ 1 & 2 & 0 & 0 & 0 & 0 \\ 0 & 0 & 2 & 1 & 0 & 0 \\ 0 & 0 & 1 & 2 & 0 & 0 \\ 0 & 0 & 0 & 0 & 2 & 1 \\ 0 & 0 & 0 & 0 & 1 & 2 \end{pmatrix},$$

the differentiation matrix is

$$\tilde{D}^{(e)} = \frac{u}{2} \begin{pmatrix} -1 & -1 & 0 & 0 & 0 & 0 \\ 1 & 1 & 0 & 0 & 0 & 0 \\ 0 & 0 & -1 & -1 & 0 & 0 \\ 0 & 0 & 1 & 1 & 0 & 0 \\ 0 & 0 & 0 & 0 & -1 & -1 \\ 0 & 0 & 0 & 0 & 1 & 1 \end{pmatrix},$$

the centered matrix from the flux term is

$$C^{(e,k)} = \frac{u}{2} \begin{pmatrix} -1 & 0 & 0 & 0 & 0 & -1 \\ 0 & 1 & 1 & 0 & 0 & 0 \\ 0 & -1 & -1 & 0 & 0 & 0 \\ 0 & 0 & 0 & 1 & 1 & 0 \\ 0 & 0 & 0 & -1 & -1 & 0 \\ 1 & 0 & 0 & 0 & 0 & 1 \end{pmatrix},$$

and the jump matrix from the flux term is

$$\mathcal{J}^{(e,k)} = \frac{u}{2} \begin{pmatrix} 1 & 0 & 0 & 0 & 0 & -1 \\ 0 & 1 & -1 & 0 & 0 & 0 \\ 0 & -1 & 1 & 0 & 0 & 0 \\ 0 & 0 & 0 & 1 & -1 & 0 \\ 0 & 0 & 0 & -1 & 1 & 0 \\ -1 & 0 & 0 & 0 & 0 & 1 \end{pmatrix}.$$

Note that we have slightly modified the weak form differentiation matrix by including the velocity u so that $\tilde{D}^{(e)} \rightarrow u\tilde{D}^{(e)}$. With all of these matrices defined, we can rewrite the final form of Eq. (21.3) as follows

$$M_{ij}^{(e)} \frac{dq_j^{(e)}}{dt} + \left(C_{ij}^{(e,k)} + \mathcal{J}_{ij}^{(e,k)} - \tilde{D}_{ij}^{(e)} \right) q_j^{(e)} = 0 \tag{21.10}$$

with $i, j = 0, \ldots, N$ and $e = 1, \ldots, N_e$, which is valid for both the CG (with LEW storage) and DG methods. Note that while $M^{(e)}$ and $\tilde{D}^{(e)}$ are block-diagonal (confined to the element), the flux term matrices $C^{(e,k)}$ and $\mathcal{J}^{(e,k)}$ span across adjacent elements. It is most clear for rows 1 and 6 where the first and last columns are non-zero due to the periodic boundary conditions.

We can rewrite Eq. (21.10) in the form

$$\frac{dq}{dt} = S(q) \tag{21.11}$$

by defining S as follows

$$S(q) = - \left(M^{(e)} \right)^{-1} \left(C^{(e,k)} + \mathcal{J}^{(e,k)} - \tilde{D}^{(e)} \right) q^{(e)}. \tag{21.12}$$

In the case of CG, we need to apply the DSS operation as discussed in Sec. 7.4 to the entire right-hand side of Eq. (21.12), including the mass matrix inverse. Now we are ready to discretize the CG/DG semi-discrete problem using implicit time-integration methods. In what follows, we drop the superscript (e) from the vector q without loss of meaning.

21.2.2 Backward Euler

Discretizing Eq. (21.11) by the backward Euler method yields

$$q^{n+1} = q^n + \Delta t S \left(q^{n+1} \right). \tag{21.13}$$

Rearranging Eq. (21.13) allows us to write

$$A_{ij} q_j^{n+1} = b_i \tag{21.14}$$

where

$$A_{ij} = M_{ij}^{(e)} + \Delta t \left(C_{ij}^{(e,k)} + \mathcal{J}_{ij}^{(e,k)} - \tilde{D}_{ij}^{(e)} \right)$$

and

$$b_i = M_{ij}^{(e)} q_j^n.$$

In the backward Euler method, Eq. (21.14) is solved at each time-step where b_i changes with time due to the solution vector q^n while, for this linear equation, the matrix A is constant. Note that the matrix A is not block-diagonal due to the contributions from the flux matrices $C^{(e,k)}$ and $\mathcal{J}^{(e,k)}$. Therefore, the solution procedure for Eq. (21.14) requires either an iterative solver (e.g., the conjugate gradient or GMRES method) or a direct solver (e.g., LU decomposition). In the case that the matrix A is not linear, then we would require using the JFNK approach presented in Sec. 20.3.

21.2.3 Runge-Kutta Method

Let us rewrite the implicit Runge-Kutta method presented in Sec. 20.3.1 as follows

$$Q^{(i)} = q^n + \Delta t \sum_{j=1}^{i} \tilde{a}_{ij} S(Q^{(j)}), \quad i = 1, \ldots, s, \tag{21.15a}$$

$$q^{n+1} = q^n + \Delta t \sum_{i=1}^{s} \tilde{b}_i S(Q^{(i)}), \tag{21.15b}$$

where S denotes the spatial discretization defined in Eq. (21.12). Using the second-order SDIRK method presented in Sec. 20.3.1, we rewrite Eq. (20.19) as follows

$$Q^{(1)} = q^n, \tag{21.16a}$$

$$Q^{(2)} = q^n + \Delta t \left[\tilde{a}_{21} S(Q^{(1)}) + \tilde{a}_{22} S(Q^{(2)}) \right], \tag{21.16b}$$

$$Q^{(3)} = q^n + \Delta t \left[\tilde{a}_{31} S(Q^{(1)}) + \tilde{a}_{32} S(Q^{(2)}) + \tilde{a}_{33} S(Q^{(3)}) \right], \tag{21.16c}$$

with the corresponding solution for the next time-step

$$q^{n+1} = q^n + \Delta t \left[\tilde{b}_1 S(Q^{(1)}) + \tilde{b}_2 S(Q^{(2)}) + \tilde{b}_3 S(Q^{(3)}) \right] \tag{21.17}$$

where the coefficients $\tilde{a}_{ij}$ and $\tilde{b}_i$ are given in Table 20.3. Let us now group terms together onto the left-hand side in order to illustrate the solution procedure required to solve the CG/DG methods implicitly. We rewrite Eq. (21.16) as follows:

$$Q^{(1)} = q^n, \tag{21.18a}$$

$$A^{(2)} Q^{(2)} = q^n + \Delta t \, \tilde{a}_{21} \, R(Q^{(1)}), \tag{21.18b}$$

$$A^{(3)} Q^{(3)} = q^n + \Delta t \left[\tilde{a}_{31} \, R(Q^{(1)}) + \tilde{a}_{32} \, R(Q^{(2)}) \right], \tag{21.18c}$$

where the matrix A is defined as follows

$$A^{(i)} = M^{(e)} + \Delta t \, \tilde{a}_{ii} \mathcal{D},$$

$$\mathcal{D} = F^{(e,k)} - \tilde{\mathcal{D}}^{(e)},$$

where

$$F^{(e,k)} = C^{(e,k)} + \mathcal{J}^{(e,k)},$$

and R is

$$R = -\mathcal{D}q.$$

From these equations, the value of using SDIRK methods is clear since if all the $\tilde{a}_{ii}$ are the same then, for a linear problem, the matrix A is also the same for all stages which means that we can either i) invert it once if using direct methods or ii) use the same preconditioner for all stages if using an iterative solver. In either approach, the update solution is carried out by Eq. (21.17).

21.2.3.1 Algorithm for the CG/DG Solution

In Alg. 21.1 we highlight the salient steps for one time-step of the implicit CG/DG solution procedure. In the case of CG, the flux matrix $F^{(e,k)}$ vanishes for interior faces but both $M^{(e)}$ and $\tilde{D}^{(e)}$ become global matrices after DSS. In the case of DG, the mass and differentiation matrices are local but the flux matrix does not vanish and results in a global matrix. For these reasons, regardless of whether CG or DG is used, the matrix A is global and its inversion is the most costly portion of the algorithm.

Algorithm 21.1 Implicit CG/DG Solution

---

function CGDG_SDIRK_SOLUTION
   $\mathcal{D} = F^{(e,k)} - \tilde{D}^{(e)}$
   $Q^{(1)} = q^n$
   $R^{(1)} = -\mathcal{D} Q^{(1)}$
   $A^{(i)} = M^{(e)} + \Delta t \, \tilde{a}_{ii} \mathcal{D}$ ▹ for SDIRK all $\tilde{a}_{ii}$ are the same

   for $i = 2 : stages$ **do**   ▹ get stage solutions
      $R_{sum} = 0$
      for $j = 1 : i - 1$ **do**
         $R_{sum} \mathrel{+}= \tilde{a}_{ij} R^{(j)}$
      end for
      $\mathcal{R} = M^{(e)} Q^{(1)} + \Delta t R_{sum}$
      $A^{(i)} Q^{(i)} = \mathcal{R}$ ▹ global solve for $Q^{(i)}$
      $R^{(i)} = -\mathcal{D} Q^{(i)}$
   end for

   $R_{sum} = 0$
   for $i = 1 : stages$ **do**   ▹ solution update
      $R_{sum} \mathrel{+}= \tilde{b}_i R^{(i)}$
   end for
   $q^{n+1} = q^n + \Delta t \left(M^{(e)} \right)^{-1} R_{sum}$
end function

21.3 Implicit Time-Integration for HDG

21.3.1 Spatial Discretization with HDG

To describe the HDG method, let us revisit Eq. (21.2) which we rewrite below

$$M_{ij}^{(e)} \frac{dq_j^{(e)}}{dt} + F_{ij}^{(e,k)} f_j^{(e)} - \tilde{D}_{ij}^{(e)} f_j^{(e)} = 0 \qquad (21.19)$$

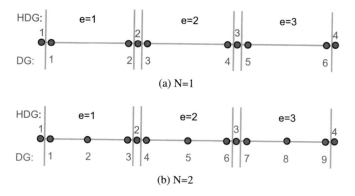

Fig. 21.2: Layout for the degrees of freedom for HDG for $N_e = 3$ and (a) $N = 1$ and (b) $N = 2$.

where we saw that the flux vector is given by Eq. (21.7) with $C^{(e,k)}$ and $\mathcal{J}^{(e,k)}$ given by Eqs. (21.8) and (21.9). The reason for discussing these matrices is that we saw in Sec. 21.2.1 that for DG while the mass and differentiation matrices are block diagonal, the matrix resulting from the flux terms ($C^{(e,k)}$ and $\mathcal{J}^{(e,k)}$) is not and, in fact, is what couples the problem globally. This can be seen from Fig. 21.1b where the points at the faces of each element couple the problem globally for both CG and DG. Therefore a way to circumvent this issue is to seek additional degrees of freedom that only live at the element faces, which we call the HDG solution. However, we still require the standard DG solution to live within each element as shown in Fig. 21.2 where the red points are the DG solutions and the blue points are the HDG solutions.

Recall that the matrices $C^{(e,k)}$ and $\mathcal{J}^{(e,k)}$ are derived from the flux term as shown for the specific example of Rusanov given by Eq. (21.4). In this equation, it is the terms $f^{(k)}$ and $q^{(k)}$ that couple the elements Ω_e and Ω_k. Therefore, if we *hybridize* the flux in order to remove or, at least lessen, the level of coupling then we are able to reduce the global degrees of freedom. To see how this works, let us now replace the Rusanov flux given by Eq. (21.4) with the *upwind* flux written as follows

$$\hat{f}^{(*,e,k)} = \left[f^{(e)} - \mathbf{n}^{(e,k)} \tau \left(\hat{q}^{(e,k)} - q^{(e)} \right) \right] \tag{21.20}$$

where $q^{(e)}$ is the solution that lives within the element Ω_e, $\hat{q}^{(e,k)}$ is the solution that only lives on the boundary between elements Ω_e and Ω_k (i.e., *trace*, which we called the HDG solution in Fig. 21.2), and τ is a stabilization parameter which is usually taken to be the wave speed as in the Rusanov flux, i.e., let us assume $\tau = |u|$. The symbol $\hat{\ }$ over f and q denotes that these values have been *hybridized*. At first glance we note that we are actually increasing the degrees of freedom (DOF) by introducing the additional trace variable $\hat{q}$ but this will then give us the opportunity to reduce the DOF of the implicit problem. Once again, assuming $u = \text{constant} > 0$ allows us to rearrange the upwind flux as follows

$$\hat{f}^{(*,e,k)} = u \left[\left(1 + \mathbf{n}^{(e,k)} \right) q^{(e)} - \mathbf{n}^{(e,k)} \hat{q}^{(e,k)} \right]. \tag{21.21}$$

Multiplying Eq. (21.21) by the normal vector and rearranging yields

$$\mathbf{n}^{(e,k)} \hat{f}^{(*,e,k)} = u \left(\mathbf{n}^{(e,k)} + 1 \right) q^{(e)} - u \hat{q}^{(e,k)} \tag{21.22}$$

which allows us to partition the interior element-wise solution from the trace solution as follows

$$F_{ij}^{(e)} f_j^{(e)} = u \left(\mathbf{n}^{(e,k)} + 1 \right) q_j^{(e)} \quad \text{and} \quad \hat{F}_{ij}^{(e,k)} \hat{f}_j = -u \hat{q}_j^{(e,k)} \tag{21.23}$$

with the flux vector now defined as follows

$$F_{ij}^{(e,k)} f_j^{(e)} = F_{ij}^{(e)} f_j^{(e)} + \hat{F}_{ij}^{(e,k)} \hat{f}_j. \tag{21.24}$$

21.3.1.1 Conservation Condition for HDG

At this point, we have all of the necessary information in order to discretize Eq. (21.19). However, since we have introduced the new trace variable $\hat{q}$ we then need to impose an additional condition in order to close the system. Since the role of $\hat{q}$ is to decouple the discontinuous DG solution that lives within each element Ω_e from the mesh skeleton (i.e., faces of the elements) this means that $\hat{q}^{(e,k)}$ must be continuous across elements Ω_e and Ω_k which then implies that the following integral must be satisifed

$$\int_{\Gamma_e} \phi_i \left(\mathbf{n}^{(L)} \cdot \hat{f}_N^{(L)} + \mathbf{n}^{(R)} \cdot \hat{f}_N^{(R)} \right) d\Gamma_e = 0 \tag{21.25}$$

where Γ_e denotes the element faces (or trace), ϕ is a basis function that only lives on the trace, and the superscripts L and R denote the left and right states, respectively, of the face Γ_e. Eq. (21.25) is the *conservation condition* of the HDG method and in one dimension simplifies to the following jump term

$$\left(\mathbf{n}^{(L)} \hat{f}_N^{(L)} + \mathbf{n}^{(R)} \hat{f}_N^{(R)} \right)_{\Gamma_e} = 0 \qquad \left\{ \forall \Gamma_e : \Gamma = \bigcup_{e=1}^{N_e} \Gamma_e \right\} \tag{21.26}$$

where we assume Lobatto points which yields $\phi = 1$ at the element faces.

Before we continue, let us rewrite Eq. (21.22) for both the left and right states of the element Ω_e

$$\mathbf{n}^{(L)} \hat{f}^{(L)} = u \left(\mathbf{n}^{(L)} + 1 \right) q^{(L)} - u \hat{q} \tag{21.27a}$$

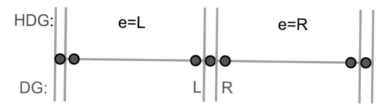

Fig. 21.3: Schematic of the conservation condition for the HDG method.

$$\mathbf{n}^{(R)} \hat{f}^{(R)} = u \left(\mathbf{n}^{(R)} + 1\right) q^{(R)} - u\hat{q} \qquad (21.27b)$$

which we show pictorially in Fig. 21.3 (the middle of the figure defines the intersection of the left and right elements). Substituting Eq. (21.27) into (21.26) yields

$$u \left(\mathbf{n}^{(L)} + 1\right) q_N^{(L)} - u\hat{q} + u \left(\mathbf{n}^{(R)} + 1\right) q_N^{(R)} - u\hat{q} = 0 \qquad (21.28)$$

for each element face where, e.g., $q_N^{(L)}$ and $q_N^{(R)}$ are the face values of the left and right elements and $\hat{q}$ is the continuous trace solution between them. Eq. (21.28) can now be partitioned into the DG and HDG portions as follows

$$C_{ij} q_j + \hat{C}_{ij} \hat{q}_j = 0 \qquad (21.29)$$

where

$$C_{ij} = u \left(\mathbf{n}^{(L)} + 1\right) \delta_{i,L} + u \left(\mathbf{n}^{(R)} + 1\right) \delta_{i,R} \qquad (21.30)$$

and

$$\hat{C}_{ij} = -2u\,\delta_{i,j} \qquad (21.31)$$

where the Kronecker delta $\delta_{i,k} = 1$ when $i = k$ and L and R are the corresponding global gridpoint indices in the global matrix which we describe below.

21.3.1.2 HDG Matrices

For the grid configuration shown in Fig. 21.2a the HDG matrices that are required (in addition to the standard DG matrices already defined) are the flux matrices

$$F^{(e)} = 2u \begin{pmatrix} 0\,0\,0\,0\,0\,0 \\ 0\,1\,0\,0\,0\,0 \\ 0\,0\,0\,0\,0\,0 \\ 0\,0\,0\,1\,0\,0 \\ 0\,0\,0\,0\,0\,0 \\ 0\,0\,0\,0\,0\,1 \end{pmatrix}, \qquad (21.32)$$

$$\hat{F}^{(e,k)} = -u \begin{pmatrix} 1 & 0 & 0 & 0 \\ 0 & 1 & 0 & 0 \\ 0 & 1 & 0 & 0 \\ 0 & 0 & 1 & 0 \\ 0 & 0 & 1 & 0 \\ 1 & 0 & 0 & 0 \end{pmatrix}, \tag{21.33}$$

as well as the matrices for the conservation condition

$$C = 2u \begin{pmatrix} 0 & 0 & 0 & 0 & 0 & 1 \\ 0 & 1 & 0 & 0 & 0 & 0 \\ 0 & 0 & 0 & 0 & 0 & 0 \\ 0 & 0 & 0 & 1 & 0 & 0 \end{pmatrix}, \tag{21.34}$$

$$\hat{C} = -2u \begin{pmatrix} 1 & 0 & 0 & 0 \\ 0 & 1 & 0 & 0 \\ 0 & 0 & 1 & 0 \\ 0 & 0 & 0 & -\frac{1}{2u} \end{pmatrix}. \tag{21.35}$$

Note that the matrices live in the following spaces

$$F^{(e)} \in \mathbb{R}^{N_P^{DG} \times N_P^{DG}}, \quad \hat{F}^{(e,k)} \in \mathbb{R}^{N_P^{DG} \times N_P^{HDG}},$$

$$C \in \mathbb{R}^{N_P^{HDG} \times N_P^{DG}}, \quad \hat{C} \in \mathbb{R}^{N_P^{HDG} \times N_P^{HDG}},$$

where, in this particular example, $N_P^{DG} = 6$ and $N_P^{HDG} = 4$. The effect of periodic boundary conditions can only be observed in the last row (first column) of $\hat{F}^{(e,k)}$ for the HDG matrix problem (where $\hat{q}_4 = \hat{q}_1$) - this is the case because it is only through this matrix that the elements are coupled. However, the HDG conservation condition also *feels* the boundary conditions through both the C matrix (first row, last column) and the $\hat{C}$ matrix (last row). The boundary condition in the C matrix is a DG-like boundary condition whereas for the $\hat{C}$ matrix it is more CG-like since the continuity condition for the trace variables states that the solution for the first gridpoint (blue point 1) in Fig. 21.2a is equal to the solution of the last gridpoint (blue point 4). Therefore, the value of the diagonal term on the last row does not matter, as long as this matrix is non-singular. Upon computing the solution, we set $\hat{q}_4 = \hat{q}_1$.

21.3.1.3 Algorithm for Constructing HDG Matrices

Let us now describe the algorithm for constructing the HDG matrices: the flux matrices $F^{(e)}$ and $\hat{F}^{(e,k)}$, and the conservation condition matrices C and $\hat{C}$.

Algorithm 21.2 delineates the necessary steps required to construct the HDG matrices. The element connectivity arrays need further explanation. The *intma* array is the usual LEW connectivity array (that permits duplicate degrees of freedom at element interfaces) and the *intma_cg* array is the standard GGP connectivity

---

Algorithm 21.2 Construction of the HDG Matrices

---

function HDG_MATRICES
  $C = 0$; $\hat{C} = 0$; $\hat{F}^{(e,k)} = 0$ ▷ initialize
  for $e = 1 : N_e$ **do**   ▷ loop over elements

      $L = e - 1$ ▷ left element
      if $e = 1$ **then**
         $L = N_e$ ▷ apply periodicity
      end if
      $\mathbf{n}^{(e,L)} = -1$ ▷ left normal vector
      $I = intma(0, e)$ ▷ DG gridpoint
      $J = cg\_to\_hdg(intma\_cg(N, L))$ ▷ HDG gridpoint
      $F_{I,I}^{(e)} = un^{(e,L)}(1 + \mathbf{n}^{(e,L)})$ ▷ local flux matrix
      $\hat{F}_{I,J}^{(e,k)} \mathrel{-}= un^{(e,L)}(1 + \mathbf{n}^{(e,L)})$ ▷ global flux matrix

      $I = cg\_to\_hdg(intma\_cg(0, e))$ ▷ HDG gridpoint
      $J = intma(0, e)$ ▷ DG gridpoint
      $C_{I,J} \mathrel{+}= un^{(e,L)}(1 + \mathbf{n}^{(e,L)})$ ▷ local conservation matrix
      $\hat{C}_{I,I} \mathrel{-}= un^{(e,L)}(1 + \mathbf{n}^{(e,L)})$ ▷ global conservation matrix

      $R = e + 1$ ▷ right element
      if $e = N_e$ **then**
         $R = 1$ ▷ apply periodicity
      end if
      $\mathbf{n}^{(e,R)} = +1$ ▷ right normal vector
      $I = intma(N, e)$ ▷ DG gridpoint
      $J = cg\_to\_hdg(intma\_cg(0, R))$ ▷ HDG gridpoint
      $F_{I,I}^{(e)} = un^{(e,R)}(1 + \mathbf{n}^{(e,R)})$ ▷ local flux matrix
      $\hat{F}_{I,J}^{(e,k)} \mathrel{-}= un^{(e,R)}(1 + \mathbf{n}^{(e,R)})$ ▷ global flux matrix

      $I = cg\_to\_hdg(intma\_cg(N, e))$ ▷ HDG gridpoint
      $J = intma(N, e)$ ▷ DG gridpoint
      $C_{I,J} \mathrel{+}= un^{(e,R)}(1 + \mathbf{n}^{(e,R)})$ ▷ local conservation matrix
      $\hat{C}_{I,I} \mathrel{-}= un^{(e,R)}(1 + \mathbf{n}^{(e,R)})$ ▷ global conservation matrix
  end for
end function

---

array (no duplication of degrees of freedom)[2]. The array *cg_to_hdg* is strictly not necessary but simplifies and reduces the global solution of the HDG problem.

21.3.2 HDG Solution Procedure

Now that we have defined all of the matrices, we can construct the HDG solution procedure for implicit time-integration beginning with Eq. (21.19) where we have

---

[2] LEW and GGP storage are described in Ch. 7.

Algorithm 21.3 Construction of the CG $\rightarrow$ HDG Array

---

function CG_TO_HDG
    $i = 0$
    for $e = 1 : N_e$ **do**   ▷ loop over elements
        $I = intma\_cg(0, e)$
        $i += 1$
        $cg\_to\_hdg(I) = i$
    end for
    $I = intma\_cg(N, N_e)$
    $i += 1$
    $cg\_to\_hdg(I) = 1$ ▷ periodic BC
end function

---

substituted for the flux matrix given by Eq. (21.24) which results in

$$M_{ij}^{(e)} \frac{dq_j^{(e)}}{dt} + \left(F_{ij}^{(e)} - \tilde{D}_{ij}^{(e)} \right) q_j^{(e)} + \hat{F}_{ij}^{(e,k)} \hat{q}_j = 0 \tag{21.36}$$

where we have used the fact that $f = qu$ and $u = $ constant, and then embedded the velocity u into the matrices F, $\tilde{D}$, and $\hat{F}$ in order to simplify the exposition. Defining $\mathcal{D} = F - \tilde{D}$ allows us to simplify Eq. (21.36) as follows

$$M_{ij}^{(e)} \frac{dq_j^{(e)}}{dt} + \mathcal{D}_{ij}^{(e)} q_j^{(e)} + \hat{F}_{ij}^{(e,k)} \hat{q}_j = 0 \tag{21.37}$$

which we will now use to derive the time-integration strategies for both backward Euler and a general implicit Runge-Kutta method.

21.3.2.1 Backward Euler

Recall that the general form for the backward Euler method is given as follows

$$q^{n+1} = q^n + \Delta t \mathcal{S} \left(q^{n+1} \right). \tag{21.38}$$

Writing the HDG matrix form given by Eq. (21.37) into the backward Euler method given by Eq. (21.38) yields

$$M_{ij}^{(e)} q_j^{n+1} = M_{ij}^{(e)} q_j^n - \Delta t \left(\mathcal{D}_{ij}^{(e)} q_j^{n+1} + \hat{F}_{ij}^{(e,k)} \hat{q}_j \right), \tag{21.39}$$

where we drop the superscript (e) from q to simplify the exposition. Let us simplify Eq. (21.39) as follows

$$A_{ij} q_j^{n+1} = M_{ij}^{(e)} q_j^n - \Delta t \hat{F}_{ij}^{(e,k)} \hat{q}_j \tag{21.40}$$

where

$$A_{ij} = M_{ij}^{(e)} + \Delta t \mathcal{D}_{ij}^{(e)}. \tag{21.41}$$

The matrix A in Eq. (21.41) is block-diagonal, meaning that it is trivial to invert since it is only defined in a per element manner. Therefore, we can write Eq. (21.40) as follows

$$q^{n+1} = A^{-1}\left(M^{(e)}q^n - \Delta t \hat{F}^{(e,k)}\hat{q}\right) \tag{21.42}$$

where we have dropped the subscripts for convenience. Substituting into the conservation condition given by Eq. (21.29) gives

$$CA^{-1}\left(M^{(e)}q^n - \Delta t \hat{F}^{(e,k)}\hat{q}\right) + \hat{C}\hat{q} = 0. \tag{21.43}$$

Rearranging Eq. (21.43) gives

$$\left(\hat{C} - \Delta t C A^{-1}\hat{F}^{(e,k)}\right)\hat{q} = -CA^{-1}M^{(e)}q^n. \tag{21.44}$$

Let us now review the dimensions of the matrices in Eq. (21.44).

Moving from left to right in Eq. (21.44), we recall that $\hat{C} \in \mathbb{R}^{N_P^{HDG} \times N_P^{HDG}}$ and $C \in \mathbb{R}^{N_P^{HDG} \times N_P^{DG}}$. Furthermore, recall that $A \in \mathbb{R}^{N_P^{DG} \times N_P^{DG}}$, while $\hat{F} \in \mathbb{R}^{N_P^{DG} \times N_P^{HDG}}$. This means that the matrix product $CA^{-1}\hat{F} \in \mathbb{R}^{N_P^{HDG} \times N_P^{HDG}}$ and so we can solve for $\hat{q}$ by solving the matrix-vector problem

$$\hat{A}\,\hat{q} = -CA^{-1}M^{(e)}q^n \tag{21.45}$$

where

$$\hat{A} = \hat{C} - \Delta t C A^{-1}\hat{F}^{(e,k)} \tag{21.46}$$

which, although global, is a much smaller matrix problem since N_P^{HDG} only contains the DOF related to the number of trace variables (the DOF associated with the mesh skeleton). Once we solve for $\hat{q}$ from Eq. (21.45), we substitute $\hat{q}$ into Eq. (21.42) and solve for q^{n+1} in an element-by-element manner which is a purely local problem.

21.3.2.2 Runge-Kutta Method

To demonstrate the HDG solution for an implicit Runge-Kutta method, let us write the HDG implicit problem in the following general form

$$M^{(e)}\frac{dq}{dt} = R(q, \hat{q}) \tag{21.47}$$

where

$$R(q, \hat{q}) = -\left(\mathcal{D}^{(e)}q + \hat{F}^{(e,k)}\hat{q}\right). \tag{21.48}$$

Now we can write the implicit Runge-Kutta method for Eq. (21.47) as follows

$$M^{(e)}Q^{(i)} = M^{(e)}q^n + \Delta t \sum_{j=1}^{i} \tilde{a}_{ij} R\left(Q^{(j)}, \hat{Q}^{(j)}\right), \quad i = 1, \ldots, s, \tag{21.49a}$$

$$q^{n+1} = q^n + \Delta t \left(M^{(e)}\right)^{-1} \sum_{i=1}^{s} \tilde{b}_i R\left(Q^{(i)}, \hat{Q}^{(i)}\right) \tag{21.49b}$$

where Q and $\hat{Q}$ are the stage solutions for the DG and trace variables, respectively. It will simplify the proceeding exposition if we rewrite Eq. (21.49a) as follows

$$M^{(e)}Q^{(i)} = M^{(e)}q^n + \Delta t \sum_{j=1}^{i-1} \tilde{a}_{ij} R\left(Q^{(j)}, \hat{Q}^{(j)}\right) + \Delta t \tilde{a}_{ii} R\left(Q^{(i)}, \hat{Q}^{(i)}\right), \quad i = 1, \dots, s, \tag{21.50}$$

which we can further simplify to

$$M^{(e)}Q^{(i)} = \mathcal{R}^{(i-1)} + \Delta t \tilde{a}_{ii} R\left(Q^{(i)}, \hat{Q}^{(i)}\right), \quad i = 1, \dots, s, \tag{21.51}$$

where

$$\mathcal{R}^{(i-1)} = M^{(e)}q^n + \Delta t \sum_{j=1}^{i-1} \tilde{a}_{ij} R\left(Q^{(j)}, \hat{Q}^{(j)}\right) \tag{21.52}$$

represents the known part of the right-hand side. Note that we do not actually have to store this matrix for all values of i but rather only for the current value. Using the second-order SDIRK method presented in Sec. 20.3.1, we can now define the implicit HDG solution stage by stage which we now describe.

Stage 1

For the first stage we have the following equations

$$M^{(e)}Q^{(1)} = M^{(e)}q^n, \tag{21.53a}$$

$$CQ^{(1)} + \hat{C}\hat{Q}^{(1)} = 0 \tag{21.53b}$$

where the last equation is the conservation condition required in order to extract a solution from the trace variables. Using (21.52) we can rewrite Eq. (21.53a) as follows

$$M^{(e)}Q^{(1)} = \mathcal{R}^{(0)}.$$

Doing so now allows us to rewrite Eq. (21.53b) as follows

$$\hat{C}\hat{Q}^{(1)} = -C\left(M^e\right)^{-1}\mathcal{R}^{(0)}.$$

This equation allows us to solve for $\hat{Q}^{(1)}$; note that $Q^{(1)}$ is already known.

Stage 2

For the second stage we have the following equations

$$M^{(e)}Q^{(2)} = M^{(e)}q^n + \Delta t \left[\tilde{a}_{21} R\left(Q^{(1)}, \hat{Q}^{(1)}\right) + \tilde{a}_{22} R\left(Q^{(2)}, \hat{Q}^{(2)}\right)\right], \tag{21.54a}$$

$$CQ^{(2)} + \hat{C}\hat{Q}^{(2)} = 0. \tag{21.54b}$$

We can now rewrite Eq. (21.54a) as follows

$$M^{(e)}Q^{(2)} = \mathcal{R}^{(1)} - \Delta t \tilde{a}_{22}\left(\mathcal{D}^{(e)}Q^{(2)} + \hat{F}^{(e,k)}\hat{Q}^{(2)}\right), \tag{21.55}$$

where we have simplified the right-hand side using Eqs. (21.52) and (21.48). Collecting terms yields

$$\left(M^{(e)} + \Delta t \tilde{a}_{22}\mathcal{D}^{(e)}\right)Q^{(2)} = \mathcal{R}^{(1)} - \Delta t \tilde{a}_{22}\hat{F}^{(e,k)}\hat{Q}^{(2)}, \tag{21.56}$$

and then substituting Eq. (21.56) into Eq. (21.54b) yields

$$\hat{C}\hat{Q}^{(2)} = -C\left(A^{(2)}\right)^{-1}\left[\mathcal{R}^{(1)} - \Delta t \tilde{a}_{22}\hat{F}^{(e,k)}\hat{Q}^{(2)}\right] \tag{21.57}$$

where

$$A^{(2)} = M^{(e)} + \Delta t \tilde{a}_{22}\mathcal{D}^{(e)}$$

which is completely local (block diagonal) and thereby trivial to invert. Equation (21.57) can now be rewritten in the following form

$$\left(\hat{C} - \Delta t \tilde{a}_{22}C\left(A^{(2)}\right)^{-1}\hat{F}^{(e,k)}\right)\hat{Q}^{(2)} = -C\left(A^{(2)}\right)^{-1}\mathcal{R}^{(1)}. \tag{21.58}$$

Equation (21.58) represents the global system that we have to solve. Once $\hat{Q}^{(2)}$ is obtained, we can solve for $Q^{(2)}$ locally (element-by-element) using Eq. (21.56).

Stage 3

For the third stage we have the following equations

$$M^{(e)}Q^{(3)} = M^{(e)}q^n \tag{21.59a}$$
$$+ \Delta t \left[\tilde{a}_{31}R\left(Q^{(1)}, \hat{Q}^{(1)}\right) + \tilde{a}_{32}R\left(Q^{(2)}, \hat{Q}^{(2)}\right) + \tilde{a}_{33}R\left(Q^{(3)}, \hat{Q}^{(3)}\right)\right],$$

$$CQ^{(3)} + \hat{C}\hat{Q}^{(3)} = 0. \tag{21.59b}$$

We can simplify Eq. (21.59b) as follows

$$M^{(e)}Q^{(3)} = \mathcal{R}^{(2)} - \Delta t \tilde{a}_{33}\left(\mathcal{D}^{(e)}Q^{(3)} + \hat{F}^{(e,k)}\hat{Q}^{(3)}\right), \tag{21.60}$$

where we have once again used Eqs. (21.52) and (21.48). Collecting terms yields

$$\left(M^{(e)} + \Delta t \tilde{a}_{33}\mathcal{D}^{(e)}\right)Q^{(3)} = \mathcal{R}^{(2)} - \Delta t \tilde{a}_{33}\hat{F}^{(e,k)}\hat{Q}^{(3)}, \tag{21.61}$$

and then substituting Eq. (21.61) into Eq. (21.59b) yields

$$\hat{C}\hat{Q}^{(3)} = -C\left(A^{(3)}\right)^{-1}\left[\mathcal{R}^{(2)} - \Delta t\tilde{a}_{33}\hat{F}^{(e,k)}\hat{Q}^{(3)}\right] \tag{21.62}$$

where

$$A^{(3)} = M^{(e)} + \Delta t\tilde{a}_{33}\mathcal{D}^{(e)}.$$

Equation (21.62) can now be rewritten in the following form

$$\left(\hat{C} - \Delta t\tilde{a}_{33}C\left(A^{(3)}\right)^{-1}\hat{F}^{(e,k)}\right)\hat{Q}^{(3)} = -C\left(A^{(3)}\right)^{-1}\mathcal{R}^{(2)}. \tag{21.63}$$

Equation (21.63) represents the global system that we have to solve. Once $\hat{Q}^{(3)}$ is obtained, we can solve for $Q^{(3)}$ locally (element-by-element) using Eq. (21.61).

General i-Stage

From the solution procedure for this particular 3-stage SDIRK method, a pattern arises that allows us to generalize the HDG solution to an implicit RK method of any order. We notice the following for the stage i

$$M^{(e)}Q^{(i)} = \mathcal{R}^{(i-1)} - \Delta t\tilde{a}_{ii}\left(\mathcal{D}^{(e)}Q^{(i)} + \hat{F}^{(e,k)}\hat{Q}^{(i)}\right), \tag{21.64a}$$

$$CQ^{(i)} + \hat{C}\hat{Q}^{(i)} = 0. \tag{21.64b}$$

Collecting $Q^{(i)}$ terms yields

$$A^{(i)}Q^{(i)} = \mathcal{R}^{(i-1)} - \Delta t\tilde{a}_{ii}\hat{F}^{(e,k)}\hat{Q}^{(i)}, \tag{21.65}$$

where

$$A^{(i)} = M^{(e)} + \Delta t\tilde{a}_{ii}\mathcal{D}^{(e)},$$

and $\mathcal{R}^{(i-1)}$ is defined by Eq. (21.52). Substituting Eq. (21.65) into Eq. (21.64b) yields

$$\hat{C}\hat{Q}^{(i)} = -C\left(A^{(i)}\right)^{-1}\left[\mathcal{R}^{(i-1)} - \Delta t\tilde{a}_{ii}\hat{F}^{(e,k)}\hat{Q}^{(i)}\right]. \tag{21.66}$$

Equation (21.66) can now be rewritten in the following form

$$\left(\hat{C} - \Delta t\tilde{a}_{ii}C\left(A^{(i)}\right)^{-1}\hat{F}^{(e,k)}\right)\hat{Q}^{(i)} = -C\left(A^{(i)}\right)^{-1}\mathcal{R}^{(i-1)} \tag{21.67}$$

where

$$\hat{A} = \hat{C} - \Delta t\tilde{a}_{ii}C\left(A^{(i)}\right)^{-1}\hat{F}^{(e,k)}$$

represents the HDG global matrix. Equation (21.67) represents the global system that we have to solve. Once $\hat{Q}^{(i)}$ is obtained, we can solve for $Q^{(i)}$ locally (element-by-element) using Eq. (21.65).

Solution Update

The solution at the end of the time-step is obtained by the following update step

$$q^{n+1} = q^n + \Delta t \left(M^{(e)} \right)^{-1} \left[\tilde{b}_1 R \left(Q^{(1)}, \hat{Q}^{(1)} \right) + \tilde{b}_2 R \left(Q^{(2)}, \hat{Q}^{(2)} \right) + \tilde{b}_3 R \left(Q^{(3)}, \hat{Q}^{(3)} \right) \right]$$

(21.68)

where the coefficients $\tilde{a}_{ij}$ and $\tilde{b}_i$ are given in Table 20.3.

21.3.2.3 Algorithm for the HDG Solution

In Alg. 21.4 we highlight the steps for one time-step of the HDG solution procedure. The local DG matrix $A^{(i)}$ only needs to be constructed for one value of i if we are using an SDIRK method since all the diagonals $\tilde{a}_{ii}$ are equal (if SDIRK is not used then we have to move any term containing $\tilde{a}_{ii}$ inside of the i-loop). The matrix $\hat{A}$ is the global HDG matrix and it need not necessarily be computed. In fact, if we are using an iterative solver then we just construct the action of $\hat{A}$ on $\hat{Q}^{(i)}$; this is the most costly part of the algorithm. Once we compute $\hat{Q}^{(i)}$, we then solve for $Q^{(i)}$ locally by inverting $A^{(i)}$, which is trivial to do so since it is block-diagonal.

Looking at Algorithms 21.1 and 21.4 one can see that these two algorithms can be combined into one code. In fact, the results shown in Sec. 21.4 are given by a single unified code. Let us now suggest the following exercise.

Exercise Combine Algs. 21.1 and 21.4 into a single algorithm with as few lines as possible. □

21.4 HDG Example

Now that we have a better understanding of the HDG method, let us show results for the CG (using GGP storage), DG, and HDG solutions for the one-dimensional wave equation given by Eq. (21.1) with periodic boundary conditions. We use the initial condition defined in Eq. (5.32).

Figure 21.4 shows the spatial convergence rates for the CG, DG, and HDG methods for various polynomial orders (N) using Lobatto points with a varying number of points (N_p) all using a 5th order SDIRK time-integrator. The results show that all three methods yield similar convergence rates which is expected because all the methods are of the same order of accuracy. The importance of this result is that it confirms that the HDG method yields the proper convergence rates.

Algorithm 21.4 Implicit HDG Solution

---

function HDG_SDIRK_SOLUTION

$\quad \mathcal{D} = F^{(e)} - \tilde{D}^{(e)}$

$\quad A^{(i)} = M^{(e)} + \Delta t \tilde{a}_{ii} \mathcal{D}$ $\triangleright$ construct local DG matrix

$\quad \hat{A} = \hat{C} - \Delta t \tilde{a}_{ii} C \left(A^{(i)} \right)^{-1} \hat{F}^{(e,k)}$ $\triangleright$ construct global HDG matrix

$\quad Q^{(1)} = q^n$

$\quad \hat{C} \hat{Q}^{(1)} = -C Q^{(1)}$ $\triangleright$ solve for $\hat{Q}^{(1)}$

$\quad \hat{Q}^{(1)}_{N_p^{HDG}} = \hat{Q}^{(1)}_1$ $\triangleright$ apply periodicity

$\quad R^{(1)} = -\left(\mathcal{D} Q^{(1)} + \hat{F}^{(e,k)} \hat{Q}^{(1)} \right)$

$\quad$ **for** $i = 2 : stages$ **do** $\triangleright$ get stage solutions

$\quad\quad R_{sum} = 0$

$\quad\quad$ **for** $j = 1 : i - 1$ **do**

$\quad\quad\quad R_{sum} += \tilde{a}_{ij} R^{(j)}$

$\quad\quad$ **end for**

$\quad\quad \mathcal{R} = M^{(e)} Q^{(1)} + \Delta t R_{sum}$

$\quad\quad \hat{A} \hat{Q}^{(i)} = -C \left(A^{(i)} \right)^{-1} \mathcal{R}$ $\triangleright$ global solve for $\hat{Q}^{(i)}$

$\quad\quad \hat{Q}^{(i)}_{N_p^{HDG}} = \hat{Q}^{(i)}_1$ $\triangleright$ apply periodicity

$\quad\quad A^{(i)} Q^{(i)} = \mathcal{R} - \Delta t \tilde{a}_{ii} \hat{F}^{(e,k)} \hat{Q}^{(i)}$ $\triangleright$ local solve for $Q^{(i)}$

$\quad\quad R^{(i)} = -\left(\mathcal{D} Q^{(i)} + \hat{F}^{(e,k)} \hat{Q}^{(i)} \right)$

$\quad$ **end for**

$\quad R_{sum} = 0$

$\quad$ **for** $i = 1 : stages$ **do** $\triangleright$ solution update

$\quad\quad R_{sum} += \tilde{b}_i R^{(i)}$

$\quad$ **end for**

$\quad q^{n+1} = q^n + \Delta t \left(M^{(e)} \right)^{-1} R_{sum}$

end function

Figure 21.5 shows the temporal convergence rates for the CG, DG, and HDG methods using polynomial order $N = 8$ with $N_e = 16$ elements for various orders of the SDIRK time-integrators and different time-steps (Δt). The results show that all three methods yield the expected temporal convergence rates for each of the SDIRK methods as compared to the theoretical results from Fig. 21.5d.

The results shown in Figs. 21.4 and 21.5 are used to confirm that it is indeed fair to compare all three methods since they have been implemented correctly. In the next section, we discuss the complexity (number of floating point operations) for the CG, DG, and HDG methods.

21.5 Complexity Analysis of HDG

Let us now review the number of operations required by each of the methods: CGc (CG with GGP storage), DG, and HDG. The most expensive part of Alg. 21.1 for

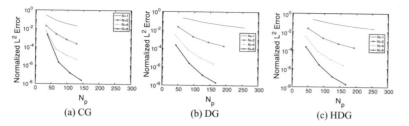

Fig. 21.4: Spatial convergence rates for the 1D wave equation for polynomial orders $N = 1$, $N = 2$, $N = 4$, and $N = 6$, using a total number of gridpoints N_p for (a) CG, (b) DG, and (c) HDG using a 5th order SDIRK implicit time-integrator with $\Delta t = 10^{-3}$.

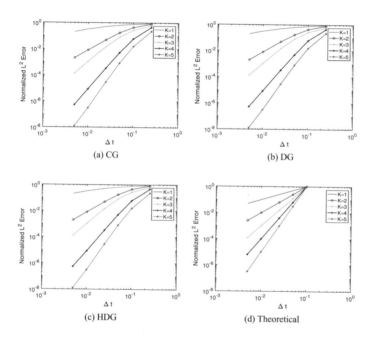

Fig. 21.5: Temporal convergence rates for the 1D wave equation for implicit time-integrators of orders $K = 1$, $K = 2$, $K = 3$, $N = 4$, and $K = 5$ using polynomial order $N = 8$ and $N_e = 16$ for (a) CG, (b) DG, (c) HDG, and (d) theoretical rate of convergence for various time-steps.

the CG/DG methods is the global solution for $Q^{(i)}$ obtained from

$$A^{(i)} Q^{(i)} = \mathcal{R}$$

where the dimension of the problem is determined by $\dim\left(A^{(i)}\right) = N_p \times N_p$, $\dim\left(Q^{(i)}\right) = N_p$, and $\dim(\mathcal{R}) = N_p$. N_p can be defined by either $N_p^{CGc} = (N_e \cdot N + 1)^d$ or $N_p^{DG} = [N_e(N+1)]^d$ where d denotes the spatial dimension, and for simplicity we assume tensor-product grids with equal polynomial orders and number of elements along all spatial dimensions. Therefore, the main cost requires the inversion of a matrix of size $N_p^{CGc} \times N_p^{CGc} = (N_e \cdot N + 1)^{2d}$ or $N_p^{DG} \times N_p^{DG} = [N_e(N+1)]^{2d}$.

In contrast, the most expensive parts of Alg. 21.4 for the HDG method are: 1) the global solution for $\hat{Q}^{(i)}$

$$\hat{A}\,\hat{Q}^{(i)} = -C\left(A^{(i)}\right)^{-1}\mathcal{R}$$

and 2) the local solution for $Q^{(i)}$

$$A^{(i)}\,Q^{(i)} = \mathcal{R} - \Delta t \tilde{a}_{ii} \hat{F}^{(e,k)}\hat{Q}^{(i)}.$$

The dimension of the global solution for $\hat{Q}^{(i)}$ is determined by $\dim\left(\hat{A}\right) = N_p^{HDG} \times N_p^{HDG}$, $\dim\left(\hat{Q}^{(i)}\right) = N_p^{HDG}$, $\dim(C) = N_p^{HDG} \times N_p^{DG}$, $\dim\left(A^{(i)}\right) = N_p^{DG} \times N_p^{DG}$, and $\dim(\mathcal{R}) = N_p^{DG}$; where we have yet to define the dimension of N_p^{HDG}.

The dimension of the local solution for $Q^{(i)}$ is determined by $\dim\left(A^{(i)}\right) = N_p^{DG} \times N_p^{DG}$, $\dim\left(Q^{(i)}\right) = N_p^{DG}$, $\dim(\mathcal{R}) = N_p^{DG}$, and $\dim\left(\hat{F}^{(e,k)}\hat{Q}^{(i)}\right) = N_p^{DG}$; however, the cost of inverting $A^{(i)}$ is much lower since $A^{(i)}$ is block diagonal and only involves the solution of a system of dimension $(N+1)^d \times (N+1)^d \equiv (N+1)^{2d}$ for each of the $(N_e)^d$ elements in the mesh, where d is the spatial dimension of the problem. In the worst case scenario (purely serial implementation) this results in the inversion of a matrix of size $N_p^{HDG-L} = N_e^d(N+1)^{2d}$ for the local solution. We reserve further discussion of this part of the algorithm for Sec. 21.5.4. For now, let us only consider the cost of the global problems.

Let us now define the dimension of the global problem N_p^{HDG}. Since N_p^{HDG} represents the number of C^0 variables that live on the mesh skeleton (which we refer to as the *trace*) then we first need to determine the number of faces defining the mesh skeleton. If we assume a tensor-product grid whereby the number of elements, N_e, and polynomial order, N, are the same along all spatial dimensions then we arrive at the following number of faces $N_F = dN_e^{(d-1)}(N_e + 1)$ where d is the spatial dimension. The number of trace variables then is $N_p^{HDG} = d(N_e \cdot N + 1)^{(d-1)}(N_e + 1)$.

21.5.1 HDG versus CGc

Let us compare the DOF for the HDG and CG methods. To do so, let us solve for the polynomial order N for which $N_p^{HDG} < N_p^{CGc}$. The inequality is

$$d(N_e \cdot N + 1)^{(d-1)}(N_e + 1) < (N_e \cdot N + 1)^d$$

which can be rearranged as follows

$$
d(N_e + 1) < (N_e \cdot N + 1) \rightarrow d\left(1 + \frac{1}{N_e}\right) < \left(N + \frac{1}{N_e}\right) \rightarrow d + \frac{d-1}{N_e} < N.
$$

Therefore, we observe that for $N_e \gg 1$ we see that for $N > d$ the HDG method will have fewer degrees of freedom than the CGc method. This means that for $N > 1$, HDG will have fewer degrees of freedom than CGc in one dimension, $N > 2$ for two dimensions, and $N > 3$ in three dimensions.

21.5.2 HDG versus DG

Comparing the DOF between the HDG and DG methods, we find that HDG will have fewer DOF for N and N_e satisfying the following inequality

$$
d(N_e \cdot N + 1)^{(d-1)}(N_e + 1) < [N_e (N + 1)]^d .
$$

To simplify this inequality, we factor N_e, and simplify

$$
dN_e^{(d-1)}\left(N + \frac{1}{N_e}\right)^{(d-1)} N_e \left(1 + \frac{1}{N_e}\right) < N_e^d (N + 1)^d
$$

which can then be written as

$$
d\left(N + \frac{1}{N_e}\right)^{(d-1)} \left(1 + \frac{1}{N_e}\right) < (N + 1)^d .
$$

The last inequality can be approximated and simplified as follows

$$
d (N + 1)^{(d-1)} (1 + 0) < (N + 1)^d \rightarrow d < (N + 1) \rightarrow d - 1 < N \qquad (21.69)
$$

where we assume that $N_e \gg 1$. Therefore, we observe that for $N > 0, 1, 2$ the HDG method will have fewer degrees of freedom than the DG method for one, two, and three dimensions, respectively.

Exercise Derive Eq. (21.69), explaining the justification for the approximations. □

21.5.3 Degrees of Freedom as a Function of N and N_e

The DOF we just computed for HDG, CGc, and DG, although correct, do not tell the full story. What we mean here is that there are other costs in the methods (in particular, for the HDG method) that we have not taken into account so far. To get a more accurate sense of the costs, we now turn to computing the actual DOF for all

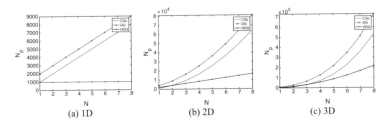

Fig. 21.6: Comparison of the number of degrees of freedom (N_p) for various polynomial orders (N) for CGc, DG, and HDG for (a) one, (b) two, and (c) three dimensions.

three methods taking into account the main differences in the algorithms described previously in this chapter. But first let us compare the DOF given by the following equations:

$$N_p^{CGc} = (N_e \cdot N + 1)^d, \quad N_p^{DG} = [N_e(N+1)]^d, \quad N_p^{HDG} = d(N_e \cdot N + 1)^{d-1}(N_e + 1)$$

which we show in Fig. 21.6 (for $N_e = 1000$). The results in Fig. 21.6 show that our previous estimates are correct, that is, that HDG has fewer DOF than CGc for $N > d$ and for DG for $N > d - 1$. However, as we noted previously, this only considers the size (DOF) of the global implicit problem. There are other costs incurred by the HDG method which we now describe.

21.5.4 Operation Count as a Function of N and N_e

In Sec. 21.5.3 we compared the degrees of freedom between CGc, DG, and HDG and noted that the implicit global problem for HDG has a lower dimension than both CGc and DG. However, as stated earlier in Sec. 21.5, the actual cost of the HDG method is higher than the dimension of the global problem indicates. This is due to the fact that there are two types of solutions involved for HDG: a global solution as well as a local one. Let us now compare the operation count required for all three methods. We already have the proper dimensions for both the CGc and DG methods and only need to include the additional costs incurred by the HDG method.

We have already computed the dimension of the global problem for HDG and denoted it by N_p^{HDG}. We also noted that the dimension of the local problem is not N_p^{DG} but rather N_p^{HDG-L}. In order to complete the complexity analysis we will have to make an assumption on the type of method we use to invert the matrix problems. From linear algebra, recall that the cost of LU factorization is $O(\frac{2}{3}m^3)$ and the cost of GMRES is $O(2mn^2)$ where m is the degrees of freedom of the matrix and n is the number of iterations. In the worst case scenario, GMRES would require $n = m$ iterations resulting in $O(m^3)$ operations, similar to LU factorization.

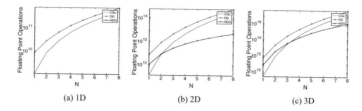

Fig. 21.7: Comparison of floating point operations (FLOPS) for various polynomial orders (N) for solving the CGc, DG, and HDG matrix systems for (a) one, (b) two, and (c) three dimensions.

So, let us assume that we use LU factorization for all of the matrix solutions[3]. Therefore the solution of the global problem for all of the methods is $O(m^3)$ (where we have omitted the fraction for simplicity since it will not change the analysis) where $m = (N_p^{CGc}, N_p^{DG}, N_p^{HDG}, N_p^{HDG-L})$. Therefore, the operation count for CGc is

$$O\left(\left(N_p^{CGc}\right)^3\right) \approx (N_e \cdot N + 1)^{3d},$$

for DG it is

$$O\left(\left(N_p^{DG}\right)^3\right) \approx [N_e(N + 1)]^{3d},$$

while that for HDG is

$$O\left(\left(N_p^{HDG}\right)^3\right) + O\left(\left(N_p^{HDG-L}\right)^3\right) \approx \left[d(N_e \cdot N + 1)^{(d-1)}(N_e + 1)\right]^3 + N_e^d(N + 1)^{6d}.$$

$$(21.70)$$

Exercise Explain why the last term in Eq. (21.70) is written as N_e^d and not as N_e^{3d}.□

Figure 21.7 shows the total floating point operations (FLOPS) required for solving the implicit problem for the CGc, DG, and HDG methods in one, two, and three dimensions as a function of polynomial order. These results bring us to the same conclusions in the comparison of the DOF for all three methods, i.e., that for $N > d$ the HDG method will require fewer FLOPS than both CGc and DG.

---

[3] For large matrix dimensions, GMRES will eventually beat LU factorization especially with the use of a good preconditioner. Unfortunately, both iterative methods and preconditioners are beyond the scope of this text. However, the reader is referred to articles such as [70, 71] which describe preconditioners specifically designed for hyperbolic and elliptic equations resulting from CG/DG spatial discretizations.

Appendix A
Classification of Partial Differential Equations

A.1 Scalar Second Order Equation

Let us write the following scalar second order equation

$$au_{xx} + bu_{xy} + cu_{yy} + du_x + eu_y + fu = g \qquad (A.1)$$

to introduce partial differential equations (PDEs). Because the topic of this text is on the numerical solution of PDEs it is important to review some fundamentals of PDE theory. Let us begin with the simplest case possible where (a, b, c, d, e, f, g) are constants although this is strictly not necessary. Equation (A.1) defines the following map: $\mathbb{R}^2 \to \mathbb{R}$ meaning that we take (x, y) as input and get a scalar as output. Note that we can introduce two different variables, say $\xi = \xi(x, y)$ and $\eta = \eta(x, y)$, that can then be used to rewrite Eq. (A.1).

From the chain rule we write

$$dx = \frac{\partial x}{\partial \xi} d\xi + \frac{\partial x}{\partial \eta} d\eta$$

$$dy = \frac{\partial y}{\partial \xi} d\xi + \frac{\partial y}{\partial \eta} d\eta \qquad (A.2)$$

with the inverse mapping

$$d\xi = \frac{\partial \xi}{\partial x} dx + \frac{\partial \xi}{\partial y} dy$$

$$d\eta = \frac{\partial \eta}{\partial x} dx + \frac{\partial \eta}{\partial y} dy. \qquad (A.3)$$

F. X. Giraldo, *An Introduction to Element-Based Galerkin Methods on Tensor-Product Bases*, Texts in Computational Science and Engineering 24, https://doi.org/10.1007/978-3-030-55069-1

Using this mapping, we can now write the first derivatives in Eq. (A.1) in terms of the new variables (ξ, η) as follows

$$\frac{\partial u}{\partial x} = \frac{\partial u}{\partial \xi}\frac{\partial \xi}{\partial x} + \frac{\partial u}{\partial \eta}\frac{\partial \eta}{\partial x}$$

$$\frac{\partial u}{\partial y} = \frac{\partial u}{\partial \xi}\frac{\partial \xi}{\partial y} + \frac{\partial u}{\partial \eta}\frac{\partial \eta}{\partial y}$$

where the derivative operators are defined as follows

$$\frac{\partial}{\partial x} = \xi_x \frac{\partial}{\partial \xi} + \eta_x \frac{\partial}{\partial \eta} \tag{A.4}$$

$$\frac{\partial}{\partial y} = \xi_y \frac{\partial}{\partial \xi} + \eta_y \frac{\partial}{\partial \eta}. \tag{A.5}$$

In order to compute second derivatives, let us do so in the following way: let

$$u_{xx} = \frac{\partial}{\partial x}(u_x) = \frac{\partial}{\partial x}\left(u_\xi \xi_x + u_\eta \eta_x\right)$$

and using the chain rule, gives

$$u_{xx} = \frac{\partial}{\partial x}(u_\xi)\xi_x + u_\xi \xi_{xx} + \frac{\partial}{\partial x}(u_\eta)\eta_x + u_\eta \eta_{xx}.$$

Substituting Eqs. (A.4) and (A.5) into this expression yields

$$u_{xx} = \xi_x \left(\xi_x u_{\xi\xi} + \eta_x u_{\xi\eta}\right) + \eta_x \left(\xi_x u_{\xi\eta} + \eta_x u_{\eta\eta}\right) + u_\xi \xi_{xx} + u_\eta \eta_{xx}.$$

Grouping like terms of derivatives, yields

$$u_{xx} = u_{\xi\xi}\xi_x^2 + 2u_{\xi\eta}\xi_x \eta_x + u_{\eta\eta}\eta_x^2 + u_\xi \xi_{xx} + u_\eta \eta_{xx}.$$

Using a similar approach we can write all the second derivatives as follows

$$u_{xx} = u_{\xi\xi}\xi_x^2 + 2u_{\xi\eta}\xi_x \eta_x + u_{\eta\eta}\eta_x^2 + u_\xi \xi_{xx} + u_\eta \eta_{xx}$$
$$u_{xy} = u_{\xi\xi}\xi_x \xi_y + u_{\xi\eta}\left(\xi_x \eta_y + \eta_x \xi_y\right) + u_{\eta\eta}\eta_x \eta_y + u_\xi \xi_{xy} + u_\eta \eta_{xy}$$
$$u_{yy} = u_{\xi\xi}\xi_y^2 + 2u_{\xi\eta}\xi_y \eta_y + u_{\eta\eta}\eta_y^2 + u_\xi \xi_{yy} + u_\eta \eta_{yy}.$$

Substituting these expressions into Eq. (A.1) yields

$$Au_{\xi\xi} + Bu_{\xi\eta} + Cu_{\eta\eta} + \dots + Fu = G \tag{A.6}$$

where

$$A = a\xi_x^2 + b\xi_x \xi_y + c\xi_y^2 \tag{A.7}$$

$$B = 2a\xi_x \eta_x + b\left(\xi_x \eta_y + \eta_x \xi_y\right) + 2c\xi_y \eta_y \tag{A.8}$$

$$C = a\eta_x^2 + b\eta_x\eta_y + c\eta_y^2.$$ (A.9)

We can now write Eqs. (A.7) and (A.9) compactly by using the variable ζ as a proxy for ξ and η, as follows

$$a\zeta_x^2 + b\zeta_x\zeta_y + c\zeta_y^2.$$

These two relations are important because computing the roots of A and C allows us to solve the resulting PDE quite simply. Let us now see how we find the roots of these expressions and what sort of solutions they yield. Dividing by ζ_y^2 and equating to zero yields

$$a\left(\frac{\zeta_x}{\zeta_y}\right)^2 + b\left(\frac{\zeta_x}{\zeta_y}\right) + c = 0.$$ (A.10)

Along constant values of $\zeta(x, y)$ we have $d\zeta = \zeta_x dx + \zeta_y dy = 0$ which can be written as

$$\frac{\zeta_x}{\zeta_y} = -\frac{dy}{dx}$$

which, upon substituting into Eq. (A.10) yields

$$a\left(\frac{dy}{dx}\right)^2 - b\left(\frac{dy}{dx}\right) + c = 0$$ (A.11)

which has roots

$$\frac{dy}{dx} = \frac{b \pm \sqrt{b^2 - 4ac}}{2a}.$$ (A.12)

The following three possibilities may arise

$$b^2 - 4ac \begin{cases} > 0 \text{ two real roots} & \rightarrow \text{ hyperbolic} \rightarrow \text{ initial value problem} \\ = 0 \text{ one real root} & \rightarrow \text{ parabolic} \quad \rightarrow \text{ initial value problem} \\ < 0 \text{ two imaginary roots} \rightarrow \text{ elliptic} \quad \rightarrow \text{ boundary value problem.} \end{cases}$$
(A.13)

Two real roots means that there are two *characteristics* which results in a hyperbolic problem which is known as an initial value or *Cauchy* problem and is the primary PDE that we are interested in solving in this text. The reason why we pay so much attention to hyperbolic problems in this text is because this class of PDE governs wave motion which is the primary interest in a large class of scientific and engineering problems ranging from geophysical fluid dynamics to aerodynamics. An example of a hyperbolic PDE is the one-dimensional wave equation

$$u_t + u_x = 0.$$

A second order version of the wave equation can be written as

$$u_{xx} - u_{yy} = 0.$$

If there is only one real root (one characteristic) then the equation is parabolic and is also an initial value problem. This class of PDE governs, e.g., the transient solution of heat propagation. An example of a parabolic PDE is the heat equation

$$u_t = u_{xx}.$$

PDES of this type may also be solved quite straightforwardly with element-based Galerkin methods. The time-dependent terms are solved by the same class of methods used for hyperbolic equations discussed in Ch. 20 whereas the space-dependent terms are solved by methods similar to those used for elliptic PDEs.

   If there are only imaginary roots (imaginary characteristics) then the equation is elliptic and the PDE represents a boundary value problem. An example of an elliptic PDE is the well-known Poisson equation

$$u_{xx} + u_{yy} = f(x, y)$$

where the homogeneous version of this equation ($f(x, y) = 0$) defines *Laplace's* equation. Elliptic PDEs may be solved by element-based Galerkin methods and we address this class of PDEs in chapters 8, 9, 12, 13, and 14.

Appendix B
Jacobi Polynomials

In this chapter, we review the Jacobi polynomials that are of interest to element-based Galerkin discretizations. We saw in Ch. 3 that the solution to the Sturm-Liouville operator

$$\frac{d}{dx}\left[\left(1 - x^2\right) w(x) \frac{d}{dx} P_N^{(\alpha,\beta)}(x)\right] + \lambda_N w(x) P_N^{(\alpha,\beta)}(x) = 0 \qquad (B.1)$$

in the space $x \in [-1, +1]$ is the special case of *Jacobi polynomials* which are defined by α and β with the associated weighting function $w(x) = (1 - x)^\alpha (1 + x)^\beta$ and eigenvalues $\lambda_N = N(N + \alpha + \beta + 1)$. The importance of this equation is that its solution is the set of orthogonal polynomials with real eigenvalues; these orthogonal polynomials, $P_N^{(\alpha,\beta)}(x)$, are the Jacobi polynomials and are orthonormal with respect to the weighting function $w(x)$. They satisfy the orthogonality condition

$$\int_{-1}^{+1} w(x) P_i^{(\alpha,\beta)}(x) P_j^{(\alpha,\beta)}(x) dx = \delta_{ij} \ \ \forall \ (i, j) = 0, ..., N$$

where δ_{ij} is the Kronecker delta function which is only non-zero for $i = j$. These orthonormal functions form a complete basis in the function space defined by the boundary conditions, i.e., the domain in which the Sturm-Liouville operator is solved. For the purposes of constructing polynomials for use in element-based Galerkin methods, we concentrate on Chebyshev, Legendre, and Lobatto polynomials.

B.1 Chebyshev Polynomials

Chebyshev polynomials are a special case of Jacobi polynomials and are defined as $T_N(x) = P_N^{(\alpha,\beta)}(x)$, with $\alpha = \beta = -\frac{1}{2}$. The Chebyshev polynomials are the solution

F. X. Giraldo, *An Introduction to Element-Based Galerkin Methods on Tensor-Product Bases*, Texts in Computational Science and Engineering 24, https://doi.org/10.1007/978-3-030-55069-1

to the Sturm-Liouville operator

$$\frac{d}{dx}\left[\left(1-x^2\right)\frac{d}{dx}\phi(x)\right] + \frac{\lambda_N}{\sqrt{1-x^2}}\phi(x) = 0.$$

The Chebyshev polynomials are written as follows

$$\phi_0(x) = 1$$
$$\phi_1(x) = x,$$
$$\phi_N(x) = 2x\phi_{N-1} - \phi_{N-2}(x) \quad \forall N \geq 2$$

where, writing $T_N(x) = \phi_N(x)$, allows us to write the orthogonality condition as

$$\int_{-1}^{+1} w(x)T_i(x)T_j(x)dx = \delta_{ij} \quad \forall\ (i, j) = 0, ..., N$$

where the weighting function is $w(x) = \frac{1}{\sqrt{1-x^2}}$. Algorithm B.1 outlines the algorithm for building Chebyshev polynomials.

Algorithm B.1   Construction of Chebyshev Polynomials

---
function CHEBYSHEV_POLYNOMIALS
  $T_1 = x;\ T_0 = 1;\ T_{-1} = 0$
  for $i = 2 : N$ **do**
    $T_i = 2xT_{i-1} - T_{i-2}$
  end for
end function

B.2 Legendre Polynomials

Legendre polynomials are a special case of Jacobi polynomials and are defined as $L_N(x) = P_N^{(\alpha,\beta)}(x)$ where $\alpha = \beta = 0$. The Legendre polynomials are the solution to the Sturm-Liouville operator

$$\frac{d}{dx}\left[\left(1-x^2\right)\frac{d}{dx}\phi(x)\right] + \lambda_N\phi(x) = 0.$$

The Legendre polynomials can be obtained by the following recurrence relations

$$\phi_0(x) = 1$$
$$\phi_1(x) = x,$$
$$\phi_N(x) = \frac{2N-1}{N}x\phi_{N-1} - \frac{N-1}{N}\phi_{N-2}(x) \quad \forall N \geq 2.$$

If we let $P_N(x) = \phi_N(x)$ then we can state the Legendre polynomials orthogonality condition as

$$\int_{-1}^{+1} w(x)P_i(x)P_j(x)dx = \delta_{ij} \quad \forall \, (i, j) = 0, ..., N$$

where the weighting function in the integral is $w(x) = 1$. Algorithm B.2 outlines the algorithm for constructing the Legendre polynomials.

Algorithm B.2   Construction of Legendre Polynomials

---

function LEGENDRE_POLYNOMIALS
    $L_0 = 1; \; L_{-1} = 0$
    $L_0' = 0; \; L_{-1}' = 0$
    for $i = 1 : N$ **do**
        $a = (2i - 1)/i$
        $b = (i - 1)/i$
        $L_i = axL_{i-1} - bL_{i-2}$
        $L_i' = a(L_{i-1} + xL_{i-1}') - bL_{i-2}'$
    end for
end function

In Algorithm B.2 the variable L_i are the Legendre polynomials and L_i' are the derivatives.

B.3 Lobatto Polynomials

Lobatto polynomials are obtained by constructing a modified set of orthogonal polynomials comprised of the derivatives of the Legendre polynomials augmented by the end points of the domain $x \in [-1, +1]$. We can write the Lobatto polynomials as follows

$$\phi_N^{Lob}(x) = (1 + x)(1 - x)\frac{d}{dx}\phi_{N-1}^{Leg}(x) \equiv (1 - x^2)\frac{d}{dx}\phi_{N-1}^{Leg}(x) \quad \forall N \geq 2.$$

Since the recurrence relation for the Legendre polynomials is defined as follows

$$\phi_N(x) = \frac{2N - 1}{N}x\phi_{N-1} - \frac{N - 1}{N}\phi_{N-2}(x),$$

we can use this relation to obtain the derivative as

$$\frac{d}{dx}\phi_N^{Leg}(x) = \frac{2N - 1}{N}\phi_{N-1} - \frac{2N - 1}{N}x\frac{d}{dx}\phi_{N-1} - \frac{N - 1}{N}\frac{d}{dx}\phi_{N-2}(x).$$

The derivatives of the Legendre polynomials $\frac{d}{dx}\phi_N(x)$ are denoted by L_i' in Algorithm B.2.

Appendix C
Boundary Conditions

The aim of this chapter is to discuss the use of different boundary conditions in a consistent manner. In previous chapters, we described periodic, Dirichlet, and impermeable wall boundary conditions separately but for most problems of interest we need to use a combination of these and the question which arises is how to do this in a consistent manner.

To simplify the exposition, let us consider the multidimensional conservation law

$$\frac{\partial q}{\partial t} + \nabla \cdot (q\mathbf{u}) = \nabla \cdot (\nu \nabla q) \tag{C.1}$$

where we assume ν is constant.

Applying the CG method to Eq. (C.1) results in

$$\int_{\Omega_e} \psi_i \frac{\partial q_N^{(e)}}{\partial t} d\Omega_e + \int_{\Gamma_e} \psi_i \hat{\mathbf{n}} \cdot (q\mathbf{u})_N^{(*)} d\Gamma_e - \int_{\Omega_e} \nabla \psi_i \cdot (q\mathbf{u})_N^{(e)} d\Omega_e$$
$$= \int_{\Gamma_e} \psi_i \hat{\mathbf{n}} \cdot \left(\nu \nabla q_N^{(*)} \right) d\Gamma_e - \int_{\Omega_e} \nabla \psi_i \cdot \left(\nu \nabla q_N^{(e)} \right) \tag{C.2}$$

where we can impose both impermeable wall and Neumann boundary conditions consistently. If Dirichlet boundary conditions are required then we can impose them in a strong way using the lift operator described in Chs. 8 and 12.

Applying the DG method with LDG, results in the following continuous problem

$$\mathbf{Q} = \nabla q \tag{C.3a}$$

$$\frac{\partial q}{\partial t} + \nabla \cdot (q\mathbf{u}) = \nabla \cdot (\nu \mathbf{Q}) \tag{C.3b}$$

F. X. Giraldo, *An Introduction to Element-Based Galerkin Methods on Tensor-Product Bases*, Texts in Computational Science and Engineering 24, https://doi.org/10.1007/978-3-030-55069-1

where $\mathbf{Q}$ is an auxiliary variable that stores the gradient of q. Applying the LDG method as described in Ch. 13 yields

$$\int_{\Omega_e} \mathbf{Q}_N^{(e)} \cdot \tau_i d\Omega_e = \int_{\Gamma_e} (\hat{\mathbf{n}} \cdot \tau_i) q_N^{(*)} d\Gamma_e - \int_{\Omega_e} (\nabla \cdot \tau_i) q_N^{(e)} d\Omega_e \qquad \text{(C.4)}$$

$$\int_{\Omega_e} \psi_i \frac{\partial q_N^{(e)}}{\partial t} d\Omega_e + \int_{\Gamma_e} \psi_i \hat{\mathbf{n}} \cdot (q\mathbf{u})_N^{(*)} d\Gamma_e - \int_{\Omega_e} \nabla \psi_i \cdot (q\mathbf{u})_N^{(e)} d\Omega_e$$
$$= \int_{\Gamma_e} \psi_i \hat{\mathbf{n}} \cdot \left(\nu \mathbf{Q}_N^{(*)}\right) d\Gamma_e - \int_{\Omega_e} \nabla \psi_i \cdot \left(\nu \mathbf{Q}_N^{(e)}\right) d\Omega_e \quad \text{(C.5)}$$

where we need boundary conditions on $q^{(*)}$, $\hat{\mathbf{n}} \cdot (q\mathbf{u})^{(*)}$, and $\mathbf{Q}^{(*)}$. To be specific, let us assume the Rusanov flux for the advective term and the centered flux for the diffusive term as follows

$$(q\mathbf{u})^{(*)} = \frac{1}{2} \left[(q\mathbf{u})^{(R)} + (q\mathbf{u})^{(L)} - \hat{\mathbf{n}}|\mathbf{u}| \left(q^{(R)} - q^{(L)} \right) \right] \qquad \text{(C.6a)}$$

$$q^{(*)} = \frac{1}{2} \left(q^{(R)} + q^{(L)} \right), \qquad \text{(C.6b)}$$

$$\mathbf{Q}^{(*)} = \frac{1}{2} \left(\mathbf{Q}^{(R)} + \mathbf{Q}^{(L)} \right) \qquad \text{(C.6c)}$$

where, at the boundary, we assume that the superscript (L) denotes the interior element and (R) the value at the boundary (exterior to the mesh).

We can now describe the following four types of boundary conditions: transmissive, inflow/outflow, impermeable wall, and gradients.

C.1 Transmissive Boundary Condition

This boundary condition is defined as $q^{(R)} = q^{(L)}$ which, lets the values of the interior element *transmit* through the boundary. This is the boundary condition we apply when we have no boundary condition to impose.

C.2 Inflow/Outflow Boundary Condition

For this boundary condition, we impose a Dirichlet condition and let $q^{(R)} = g(\mathbf{x})$ where $g(\mathbf{x})$ is the required Dirichlet boundary condition. We also need to impose $q^{(*)} = g(\mathbf{x})$ in Eq. (C.4).

C.3 Impermeable Wall Boundary Condition

For this boundary condition, we need to satisfy

$$\hat{\mathbf{n}} \cdot (q\mathbf{u})^{(*)} = 0 \qquad\qquad (\text{C.7})$$

at the boundary. This can be done simply by eliminating the term

$$\int_{\Gamma_e} \psi_i \hat{\mathbf{n}} \cdot (q\mathbf{u})_N^{(*)} \, d\Gamma_e = 0$$

on the first row of Eq. (C.5). However, sometimes we may need the value of $\mathbf{u}$ at the boundary to build the flux function (say for systems of equations). To compute $\mathbf{u}$ on the boundary in order to satisfy $\hat{\mathbf{n}} \cdot \mathbf{u} = 0$ we can construct the velocity on the right to be

$$\mathbf{u}^{(R)} = \left(\mathbf{I} - 2\hat{\mathbf{n}}\hat{\mathbf{n}}^T \right) \mathbf{u}^{(L)} \qquad\qquad (\text{C.8})$$

where the term in brackets is the Householder reflector. It is now easy to see that if we use transmissive boundary conditions for q such that $q^{(R)} = q^{(L)}$ and Eq. (C.8) for $\mathbf{u}^{(R)}$ then we satisfy Eq. (C.7).

C.4 Gradient Boundary Condition

For this boundary condition, we require

$$\hat{\mathbf{n}} \cdot \boldsymbol{\nabla} q^{(*)} = h(\mathbf{x}) \qquad\qquad (\text{C.9})$$

where $h(\mathbf{x})$ is the Neumann boundary condition that we need to satisfy. Therefore, on the physical boundary, the first term on the second row of Eq. (C.5) becomes

$$\int_{\Gamma_e} v\psi_i \hat{\mathbf{n}} \cdot \mathbf{Q}_N^{(*)} d\Gamma_e = \int_{\Gamma_e} v\psi_i h(\mathbf{x}) d\Gamma_e$$

since $\mathbf{Q} = \boldsymbol{\nabla} q$. Looking at Eq. (C.4) we note that we also need a boundary condition on $q^{(*)}$. If a Dirichlet condition is required then we set $q^{(*)} = g(\mathbf{x})$, and if not then we impose the transmissive condition $q^{(R)} = q^{(L)}$ which we then use in Eq. (C.6b).

C.5 Summary

Following Bassi and Rebay [22] we recommend imposing the boundary conditions $q^{(*)} = g(\mathbf{x})$, $\hat{\mathbf{n}} \cdot \boldsymbol{\nabla} q^{(*)} = h(\mathbf{x})$ whenever possible. When this option is not available then the boundary condition should be imposed on the solution variable as such $q^{(R)} = g(\mathbf{x})$. Situations may also arise where all three of these boundary conditions need to be invoked. In fact, these boundary conditions, in addition to the impermeable boundary condition, are quite common. Examples when these arise can be found in [22] for the compressible Navier-Stokes equations.

Appendix D
Contravariant Form

The aim of this chapter is to describe the advantages of using the contravariant form of the equations. If we have selected to use element-based Galerkin (EBG) methods then it most likely means that we are interested in using unstructured grids, which can be done quite easily in covariant form whereby we focus on solving the equations in an element-by-element fashion regardless of the coordinate system. Take, for example, Fig. 1.1 which shows an unstructured mesh of the Indian Ocean. Using EBG methods, we can solve (say) the shallow water equations in covariant form in a straightforward fashion simply by ensuring that the shallow water equations are discretized in each element (whereby we then invoke the DSS operator to construct the global solution). However, there are situations when we know something specific about one of the directions that we then can exploit. An example of such a situation arises in solving the 3D Euler equations on a spherical manifold as is done in nonhydrostatic atmospheric modeling.

Let us write the 3D Euler equations in the following compact vector form

$$\frac{\partial \mathbf{q}}{\partial t} + \nabla \cdot \mathbf{F}(\mathbf{q}) = S(\mathbf{q}) \tag{D.1}$$

where

$$\mathbf{q} = \begin{pmatrix} \rho \\ \mathbf{U} \\ E \end{pmatrix}, \quad \mathbf{F}(\mathbf{q}) = \begin{pmatrix} \mathbf{U} \\ \frac{\mathbf{U} \otimes \mathbf{U}}{\rho} + P\mathbf{I}_d \\ \frac{(E+P)\mathbf{U}}{\rho} \end{pmatrix}, \quad S(\mathbf{q}) = \begin{pmatrix} 0 \\ -\rho g\hat{\mathbf{r}} \\ 0 \end{pmatrix},$$

ρ is the density, $\mathbf{U} = \rho\mathbf{u}$ is the momentum (with $\mathbf{u}$ being the velocity), $E = \rho e$ where $e = c_v T + \frac{1}{2}\mathbf{u} \cdot \mathbf{u} + gz$ is the total energy, P is the pressure and T is the temperature, c_v is the specific heat for constant volume, g is the gravitational constant, $\mathbf{I}_d$ is the rank-d identity matrix, and $\hat{\mathbf{r}}$ is the unit vector along which gravity acts (but pointing

F. X. Giraldo, *An Introduction to Element-Based Galerkin Methods on Tensor-Product Bases*, Texts in Computational Science and Engineering 24, https://doi.org/10.1007/978-3-030-55069-1

upward). We have already seen the gradient operator which is defined as

$$\nabla = \frac{\partial}{\partial x}\hat{\mathbf{i}} + \frac{\partial}{\partial y}\hat{\mathbf{j}} + \frac{\partial}{\partial z}\hat{\mathbf{k}}.$$

Following Kopriva [236], going from the covariant (in terms of x, y, and z coordinates) to contravariant form (in terms of the reference element coordinates ξ, η, and ζ) allows us to write Eq. (D.1) as follows

$$\frac{\partial \mathbf{q}}{\partial t} + \frac{1}{J^{(e)}}\nabla_\xi \cdot \left(J^{(e)}\mathbf{F}^\xi\right) = S(\mathbf{q}) \tag{D.2}$$

where $J^{(e)}$ is the element Jacobian, and the contravariant differential operator is defined as

$$\nabla_\xi = \frac{\partial}{\partial \xi}\mathbf{v}^1 + \frac{\partial}{\partial y}\mathbf{v}^2 + \frac{\partial}{\partial z}\mathbf{v}^3 \tag{D.3}$$

where $\mathbf{v}^i$ are the contravariant unit vectors defined in Eq. (12.22). Furthermore, the contravariant fluxes $\mathbf{F}^\xi$ are constructed as follows

$$\mathbf{F}^\xi = \nabla\xi \cdot \mathbf{F}$$

where $\mathbf{F}$ are the covariant fluxes and

$$\mathbf{F}^\xi = \nabla\xi \cdot \mathbf{F}, \quad \mathbf{F}^\eta = \nabla\eta \cdot \mathbf{F}, \quad \mathbf{F}^\zeta = \nabla\zeta \cdot \mathbf{F} \tag{D.4}$$

where $\nabla\xi = \mathbf{v}^i$.

We can now expand Eq. (D.2) as follows

$$\frac{\partial \mathbf{q}}{\partial t} + \frac{1}{J^{(e)}}\frac{\partial}{\partial \xi}\left(J^{(e)}\mathbf{F}^\xi\right) + \frac{1}{J^{(e)}}\frac{\partial}{\partial \eta}\left(J^{(e)}\mathbf{F}^\eta\right) + \frac{1}{J^{(e)}}\frac{\partial}{\partial \zeta}\left(J^{(e)}\mathbf{F}^\zeta\right) = S(\mathbf{q}). \tag{D.5}$$

If we wish to solve these equations on the 3-sphere (three-dimensional sphere) whereby ξ and η are along the spherical manifold and ζ is along the radial direction, then we can separate the space defined by the manifold from the radial direction. This is important because only certain processes act along the radial direction (e.g., gravity) and because typically we would like to treat the radial direction differently from the manifold because typically the grid resolution is much finer along this direction. Therefore, we are able to mimic spherical coordinates even with, say, Cartesian coordinates. This is possible thanks due to the use of the contravariant form of the equations. As an example of what we mean, let us assume that we want to discretize the terms along ζ implicitly while those along ξ and η explicitly. In this situation, we would write the IMEX Runge-Kutta method (see Sec. 20.4.1)

$$\mathbf{Q}^{(i)} = \mathbf{q}^n + \Delta t \sum_{j=0}^{i-1} a_{ij} S^{(\xi,\eta)}\left(\mathbf{Q}^{(j)}\right) + \Delta t \sum_{j=0}^{i} \tilde{a}_{ij} S^{(\zeta)}\left(\mathbf{Q}^{(j)}\right)$$

where

$$S^{(\xi,\eta)}(\mathbf{Q}) = -\frac{1}{J^{(e)}}\frac{\partial}{\partial \xi}\left(J^{(e)}\mathbf{F}^\xi(\mathbf{Q})\right) - \frac{1}{J^{(e)}}\frac{\partial}{\partial \eta}\left(J^{(e)}\mathbf{F}^\eta(\mathbf{Q})\right) + S(\mathbf{Q})$$

and

$$S^{(\zeta)}(\mathbf{Q}) = -\frac{1}{J^{(e)}}\frac{\partial}{\partial \zeta}\left(J^{(e)}\mathbf{F}^\zeta(\mathbf{Q})\right),$$

with the final update given by

$$\mathbf{q}^{n+1} = \mathbf{q}^n + \Delta t \sum_{i=1}^{s}\left[b_i S^{(\xi,\eta)}\left(\mathbf{Q}^{(i)}\right) + \tilde{b}_i S^{(\zeta)}\left(\mathbf{Q}^{(i)}\right)\right].$$

Note that the contravariant form can also be used for the non-conservative form of the equations. As an example, let us apply this approach to the 3D shallow water equations written as follows

$$\frac{\partial h}{\partial t} + \nabla \cdot (h\mathbf{u}) = 0, \tag{D.6a}$$

$$\frac{\partial \mathbf{u}}{\partial t} + \mathbf{u} \cdot \nabla \mathbf{u} + g\nabla h = \mathbf{0} \tag{D.6b}$$

where $\nabla = \left(\frac{\partial}{\partial x}, \frac{\partial}{\partial y}, \frac{\partial}{\partial z}\right)$ and $\mathbf{u} = (u, v, w)$. Let us expand the first equation using the product rule to arrive at

$$\frac{\partial h}{\partial t} + \mathbf{u} \cdot \nabla h + h\nabla \cdot \mathbf{u} = 0. \tag{D.7}$$

Equations (D.6) take the following form in terms of contravariant coordinates

$$\frac{\partial h}{\partial t} + \mathbf{u}^\xi \cdot \nabla_\xi h + \frac{h}{J^{(e)}}\nabla_\xi \cdot \left(J^{(e)}\mathbf{u}^\xi\right) = 0 \tag{D.8a}$$

$$\frac{\partial u}{\partial t} + \mathbf{u}^\xi \cdot \nabla_\xi u + g\nabla_\xi h \cdot \frac{\partial \xi}{\partial x} = 0 \tag{D.8b}$$

$$\frac{\partial v}{\partial t} + \mathbf{u}^\xi \cdot \nabla_\xi v + g\nabla_\xi h \cdot \frac{\partial \xi}{\partial y} = 0 \tag{D.8c}$$

$$\frac{\partial w}{\partial t} + \mathbf{u}^\xi \cdot \nabla_\xi w + g\nabla_\xi h \cdot \frac{\partial \xi}{\partial z} = 0 \tag{D.8d}$$

where

$$\mathbf{u}^\xi = u^\xi \mathbf{v}^1 + u^\eta \mathbf{v}^2 + u^\zeta \mathbf{v}^3$$

is the contravariant velocity obtained using Eq. (D.4) but with $\mathbf{u}$ instead of $\mathbf{F}$, and the contravariant metric terms are defined as follows

$$\frac{\partial \xi}{\partial x} = \frac{\partial \xi}{\partial x}\mathbf{v}^1 + \frac{\partial \eta}{\partial x}\mathbf{v}^2 + \frac{\partial \zeta}{\partial x}\mathbf{v}^3,$$

$$\frac{\partial \xi}{\partial y} = \frac{\partial \xi}{\partial y}\mathbf{v}^1 + \frac{\partial \eta}{\partial y}\mathbf{v}^2 + \frac{\partial \zeta}{\partial y}\mathbf{v}^3,$$

$$\frac{\partial \xi}{\partial z} = \frac{\partial \xi}{\partial z}\mathbf{v}^1 + \frac{\partial \eta}{\partial z}\mathbf{v}^2 + \frac{\partial \zeta}{\partial z}\mathbf{v}^3,$$

and ∇_ξ is defined in Eq. (D.3). The form given in Eq. (D.8) can be derived by starting with Eq. (D.6) and then using the chain rule from the covariant to contravariant space. The exception is the divergence term in Eq. (D.7) (last term) which has to be treated as described in [236].

Exercise Convert the shallow water equations given by Eq. (D.6) to the contravariant form given by Eq. (D.8). For simplicity, assume two dimensions only.   □

Expanding the derivatives in Eqs. (D.8) yields

$$\frac{\partial h}{\partial t} + u^\xi \frac{\partial h}{\partial \xi} + u^\eta \frac{\partial h}{\partial \eta} + u^\zeta \frac{\partial h}{\partial \zeta} + \frac{h}{J^{(e)}}\left[\frac{\partial}{\partial \xi}\left(J^{(e)}u^\xi\right) + \frac{\partial}{\partial \eta}\left(J^{(e)}u^\eta\right) + \frac{\partial}{\partial \zeta}\left(J^{(e)}u^\zeta\right)\right] = 0$$
(D.9a)

$$\frac{\partial u}{\partial t} + u^\xi \frac{\partial u}{\partial \xi} + u^\eta \frac{\partial u}{\partial \eta} + u^\zeta \frac{\partial u}{\partial \zeta} + g\left[\frac{\partial h}{\partial \xi}\frac{\partial \xi}{\partial x} + \frac{\partial h}{\partial \eta}\frac{\partial \eta}{\partial x} + \frac{\partial h}{\partial \zeta}\frac{\partial \zeta}{\partial x}\right] = 0 \qquad \text{(D.9b)}$$

$$\frac{\partial v}{\partial t} + u^\xi \frac{\partial v}{\partial \xi} + u^\eta \frac{\partial v}{\partial \eta} + u^\zeta \frac{\partial v}{\partial \zeta} + g\left[\frac{\partial h}{\partial \xi}\frac{\partial \xi}{\partial y} + \frac{\partial h}{\partial \eta}\frac{\partial \eta}{\partial y} + \frac{\partial h}{\partial \zeta}\frac{\partial \zeta}{\partial y}\right] = 0 \qquad \text{(D.9c)}$$

$$\frac{\partial w}{\partial t} + u^\xi \frac{\partial w}{\partial \xi} + u^\eta \frac{\partial w}{\partial \eta} + u^\zeta \frac{\partial w}{\partial \zeta} + g\left[\frac{\partial h}{\partial \xi}\frac{\partial \xi}{\partial z} + \frac{\partial h}{\partial \eta}\frac{\partial \eta}{\partial z} + \frac{\partial h}{\partial \zeta}\frac{\partial \zeta}{\partial z}\right] = 0. \qquad \text{(D.9d)}$$

Next, if we solve the 3D shallow water equations on the 2-sphere and align the surface of the spherical manifold with the (ξ, η) contravariant directions then, we can solve Eqs. (D.9) with the boundary condition $u^\zeta = 0$, which results in

$$\frac{\partial h}{\partial t} + u^\xi \frac{\partial h}{\partial \xi} + u^\eta \frac{\partial h}{\partial \eta} + \frac{h}{J^{(e)}}\left[\frac{\partial}{\partial \xi}\left(J^{(e)}u^\xi\right) + \frac{\partial}{\partial \eta}\left(J^{(e)}u^\eta\right)\right] = 0 \qquad \text{(D.10a)}$$

$$\frac{\partial u}{\partial t} + u^\xi \frac{\partial u}{\partial \xi} + u^\eta \frac{\partial u}{\partial \eta} + g\left[\frac{\partial h}{\partial \xi}\frac{\partial \xi}{\partial x} + \frac{\partial h}{\partial \eta}\frac{\partial \eta}{\partial x} + \frac{\partial h}{\partial \zeta}\frac{\partial \zeta}{\partial x}\right] = 0 \qquad \text{(D.10b)}$$

$$\frac{\partial v}{\partial t} + u^\xi \frac{\partial v}{\partial \xi} + u^\eta \frac{\partial v}{\partial \eta} + g\left[\frac{\partial h}{\partial \xi}\frac{\partial \xi}{\partial y} + \frac{\partial h}{\partial \eta}\frac{\partial \eta}{\partial y} + \frac{\partial h}{\partial \zeta}\frac{\partial \zeta}{\partial y}\right] = 0 \qquad \text{(D.10c)}$$

$$\frac{\partial w}{\partial t} + u^\xi \frac{\partial w}{\partial \xi} + u^\eta \frac{\partial w}{\partial \eta} + g\left[\frac{\partial h}{\partial \xi}\frac{\partial \xi}{\partial z} + \frac{\partial h}{\partial \eta}\frac{\partial \eta}{\partial z} + \frac{\partial h}{\partial \zeta}\frac{\partial \zeta}{\partial z}\right] = 0. \qquad \text{(D.10d)}$$

This then allows us to solve the shallow water equations on the surface of the sphere without the need to use spherical coordinates. This approach presented here results in the Cartesian method presented in [154, 159, 161, 168, 157, 267, 40] without the need to use orthogonal projectors.

D.1 Summary

Many applications of the element-based Galerkin method use the covariant form of the equations, i.e., the equations are written in the physical (covariant) coordinate system and then the derivatives in the covariant space are constructed using the element basis functions and metric terms. However, this section shows that using the alternative contravariant form yields a certain elegance to the equations since the equations are expressed fully in their local element coordinates. The contravariant form can be exploited to design special discretizations when the grid is constructed to be aligned with directions of particular significance. The example described here comes from geophysical fluid dynamics whereby flow on spherical manifolds typically are treated differently along the manifold and along the radial direction. By using the contravariant form of the equations, these directions can be decoupled explicitly regardless of the underlying physical (covariant) coordinate used.

References

1. Abbassi, H., Mashayek, F., Jacobs, G.B.: Shock capturing with entropy-based artificial viscosity for staggered grid discontinuous spectral element method. Computers & Fluids **98**(SI), 152–163 (2014). DOI {10.1016/j.compfluid.2014.01.022}
2. Abdi, D., Giraldo, F.X.: Efficient construction of unified continuous and discontinuous Galerkin formulations for the 3D Euler Equations. Journal of Computational Physics **320**(1), 46–68 (2016)
3. Abdi, D.S., Wilcox, L.C., Warburton, T.C., Giraldo, F.X.: A GPU-accelerated continuous and discontinuous Galerkin non-hydrostatic atmospheric model. The International Journal of High Performance Computing Applications **33**(1), 81–109 (2017). DOI 10.1177/1094342017694427. URL https://doi.org/10.1177/1094342017694427
4. Abdi, D.S., Giraldo, F.X., Constantinescu, E.M., Carr, L.E., Wilcox, L.C., Warburton, T.C.: Acceleration of the Implicit-Explicit nonhydrostatic unified model of the atmosphere on manycore processors. The International Journal of High Performance Computing Applications, 1094342017732395 (2017). DOI 10.1177/1094342017732395. URL https://doi.org/10.1177/1094342017732395
5. Ahmad, N., Lindeman, J.: Euler solutions using flux–based wave decomposition. International Journal for Numerical Methods in Fluids **54**(1), 47–72 (2007). URL http://dx.doi.org/10.1002/fld.1392
6. Alevras, D.: Simulating tsunamis in the Indian Ocean with real bathymetry by using a high-order triangular discontinuous Galerkin oceanic shallow water model. Master's thesis, Naval Postgraduate School (2009)
7. Alexander, R.: Diagonally implicit Runge-Kutta methods for stiff ODEs. SIAM Journal on Numerical Analysis **14**(6), 1006–1021 (1977)
8. Allaneau, Y., Jameson, A.: Connections between the filtered discontinuous Galerkin method and the flux reconstruction approach to high order discretizations. Computer Methods in Applied Mechanics and Engineering **200**(49-52), 3628–3636 (2011). DOI {10.1016/j.cma.2011.08.019}
9. Almgren, A., Bell, J., Colella, P., Howell, L., Welcome, M.: A conservative adaptive projection method for the variable density incompressible Navier-Stokes equations. Journal of Computational Physics **142**(1), 1–46 (1998). DOI {10.1006/jcph.1998.5890}
10. Anderson, B.D., Benzley, S.E., Owen, S.J.: Automatic all quadrilateral mesh adaption through refinement and coarsening. In: Clark, BW (ed.) Proceedings of the 18th International Meshing Roundtable, pp. 557+. Springer-Verlag Berlin, Heidelberger Platz 3, D-14197 Berlin, Germany (2009). DOI {10.1007/978-3-642-04319-2_32}. 18th International Meshing Roundtable (IMR), Salt Lake City, UT, OCT 25-28, 2009
11. Anderson, D., Tannehill, J., Pletcher, R.: Computational Fluid Mechanics and Heat Transfer. Hemisphere Publishing Corporation (1984)
12. Arnold, D., Brezzi, F., Cockburn, B., Marini, L.: Unified analysis of discontinuous Galerkin methods for elliptic problems. SIAM Journal on Numerical Analysis **39**(5), 1749–1779 (2002). DOI 10.1137/S0036142901384162. URL http://link.aip.org/link/?SNA/39/1749/1
13. Ascher, U., Ruuth, S., Spiteri, R.: Implicit-explicit Runge-Kutta methods for time-dependent partial differential equations. Applied Numerical Mathematics **25**, 151–167 (1997)
14. Babuska, I., Suri, M.: The optimal convergence rate of the p-version of the finite-element method. SIAM Journal on Numerical Analysis **24**(4), 750–776 (1987)
15. Babuska, I., Suri, M.: The p-version and h-p-version of the finite-element method, an overview. Computer Methods in Applied Mechanics and Engineering **80**(1-3), 5–26 (1990). DOI {10.1016/0045-7825(90)90011-A}. 1st International Conference on Spectral and High Order Methods for Partial Differential Equations (ICOSAHOM 89), Como, Italy, JUN 26-29, 1989
16. Babuska, I., Szabo, B.A., Katz, I.N.: The p-version of the finite-element method. SIAM Journal on Numerical Analysis **18**(3), 515–545 (1981). DOI {10.1137/0718033}

17. Babuska, I., Zlamal, M.: Nonconforming elements in finite-element method with penalty. SIAM Journal on Numerical Analysis **10**(5), 863–875 (1973)
18. Bader, M.: Space-Filling Curves: An Introduction with Applications in Scientific Computing, vol. 9. Springer (2012)
19. Bader, M., Böck, C., Schwaiger, J., Vigh, C.A.: Dynamically adaptive simulations with minimal memory requirement - solving the shallow water equations using Sierpinski curves. SIAM Journal on Scientific Computing **32**(1), 212–228 (2010)
20. Baines, M., Wathen, A.: Moving finite-element methods for evolutionary problems. 1. theory. Journal of Computational Physics **79**(2), 245–269 (1988). DOI {10.1016/0021-9991(88)90016-2}
21. Barter, G.E., Darmofal, D.L.: Shock capturing with PDE-based artificial viscosity for DGFEM: Part I. Formulation. Journal of Computational Physics **229**(5), 1810–1827 (2010). DOI {10.1016/j.jcp.2009.11.010}
22. Bassi, F., Rebay, S.: A high-order accurate discontinuous finite element method for the numerical solution of the compressible Navier–Stokes equations. Journal of Computational Physics **131**(2), 267–279 (1997). DOI http://dx.doi.org/10.1006/jcph.1996.5572. URL http://www.sciencedirect.com/science/article/B6WHY-45KV05X-F/2/46999bcaea51c7ed1db8b2291f532f4a
23. Bassi, F., Rebay, S.: High-order accurate discontinuous finite element solution of the 2d Euler equations. Journal of Computational Physics **138**(2), 251–285 (1997). URL http://www.sciencedirect.com/science/article/B6WHY-45S9288-K/2/8ff35ee84d7e01497e6d15a4c91aa328
24. Bauer, T.P., Knoth, O.: Extended multirate infinitesimal step methods: Derivation of order conditions. Journal of Computational and Applied Mathematics p. 112541 (2019). DOI https://doi.org/10.1016/j.cam.2019.112541. URL http://www.sciencedirect.com/science/article/pii/S0377042719305461
25. Begue, C., Bernardi, C., Debit, N., Maday, Y., Kariadakis, G.E., Mavriplis, C., Patera, A.T.: Non-conforming spectral element-finite element approximations for partial-differential equations. Computer Methods in Applied Mechanics and Engineering **75**(1-3), 109–125 (1989). DOI {10.1016/0045-7825(89)90018-2}
26. Behrens, J., Rakowsky, N., Hiller, W., Handorf, D., Lauter, M., Papke, J., Dethloff, K.: amatos: parallel adaptive mesh generator for atmospheric and oceanic simulation. Ocean Modelling **10**(1-2), 171–183 (2005). DOI {10.1016/j.ocemod.2004.06.003}
27. Bell, J., Berger, M., Saltzman, J., Welcome, M.: 3-Dimensional adaptive mesh refinement for hyperbolic conservation-laws. SIAM Journal on Scientific Computing **15**(1), 127–138 (1994). DOI {10.1137/0915008}
28. Benzi, M., Golub, G.: A preconditioner for generalized saddle point problems. SIAM Journal on Matrix Analysis and Applications **26**(1), 20–41 (2004). DOI {10.1137/S0895479802417106}
29. Berger, M., Leveque, R.: Adaptive mesh refinement using wave-propagation algorithms for hyperbolic systems. SIAM Journal on Numerical Analysis **35**(6), 2298–2316 (1998). DOI {10.1137/S0036142997315974}
30. Berger, M.J., Colella, P.: Local adaptive mesh refinement for shock hydrodynamics. Journal of Computational Physics **82**(1), 64–84 (1989)
31. Berger, M.J., George, D.L., LeVeque, R.J., Mandli, K.T.: The GeoClaw software for depth-averaged flows with adaptive refinement. Advances in Water Resources **34**(9, SI), 1195–1206 (2011). DOI {10.1016/j.advwatres.2011.02.016}
32. Berger, M.J., Oliger, J.: Adaptive mesh refinement for hyperbolic partial differential equations. Journal of Computational Physics **53**(3), 484–512 (1984)
33. Bermejo, R., Saavedra, L.: Modified Lagrange-Galerkin methods of first and second order in time for convection-diffusion problems. Numerische Mathematik **120**(4), 601–638 (2012). DOI {10.1007/s00211-011-0418-8}
34. Bernard, P.E., Remacle, J.F., Comblen, R., Legat, V., Hillewaert, K.: High-order discontinuous Galerkin schemes on general 2d manifolds applied to the shallow water equations. Journal of Computational Physics **228**(17), 6514–6535 (2009). DOI {10.1016/j.jcp.2009.05.046}

35. Black, K.: A conservative spectral element method for the approximation of compressible fluid flow. Kybernetika **35**(1), [133]–146 (1999). URL http://eudml.org/doc/33415

36. Blacker, T.D., Stephenson, M.B.: PAVING - a new approach to automated quadrilateral mesh generation. International Journal for Numerical Methods in Engineering **32**(4), 811–847 (1991). DOI {10.1002/nme.1620320410}

37. Blaise, S., St-Cyr, A.: A dynamic hp-adaptive discontinuous Galerkin method for shallow-water flows on the sphere with application to a global tsunami simulation. Monthly Weather Review **140**(3), 978–996 (2012). DOI {10.1175/MWR-D-11-00038.1}

38. Bonaventura, L.: A semi-implicit semi-Lagrangian scheme using the height coordinate for a nonhydrostatic and fully elastic model of atmospheric flows. Journal of Computational Physics **158**(2), 186–213 (2000). URL http://www.sciencedirect.com/science/article/B6WHY-45F4WH7-J/2/d7acbc129a6562646ccf37bb88727a45

39. Bonet, J., Pica, A., Peiro, J., Wood, R.D.: Adaptive mesh refinement for faceted shells. Communications in Applied Numerical Methods **8**(5), 319–329 (1992). DOI {10.1002/cnm.1630080506}

40. Bonev, B., Hesthaven, J.S., Giraldo, F.X., Kopera, M.A.: Discontinuous Galerkin scheme for the spherical shallow water equations with applications to tsunami modeling and prediction. Journal of Computational Physics **362**, 425–448 (2018). DOI {10.1016/j.jcp.2018.02.008}

41. Bornemann, F., Deuflhard, P.: The cascadic multigrid method for elliptic problems. Numerische Mathematik **75**(2), 135–152 (1996). DOI {10.1007/s002110050234}

42. Boscarino, S.: Error analysis of IMEX Runge-Kutta methods derived from differential-algebraic systems. SIAM Journal on Numerical Analysis **45**(4), 1600–1621 (2007). DOI 10.1137/060656929

43. Boscarino, S.: On an accurate third order implicit-explicit Runge-Kutta method for stiff problems. Applied Numerical Mathematics **59**(7), 1515 – 1528 (2009). DOI DOI:10.1016/j.apnum.2008.10.003. URL http://www.sciencedirect.com/science/article/B6TYD-4TPF48B-2/2/5d73bef9d9a9f5d57a739b8a51043882

44. Bottasso, C., Micheletti, S., Sacco, R.: A multiscale formulation of the discontinuous Petrov-Galerkin method for advective-diffusive problems. Computer Methods in Applied Mechanics and Engineering **194**(25-26), 2819–2838 (2005). DOI {10.1016/j.cma.2004.07.024}

45. Bouzarth, E.L., Minion, M.L.: A multirate time integrator for regularized Stokeslets. Journal of Computational Physics **229**(11), 4208–4224 (2010). DOI {10.1016/j.jcp.2010.02.006}

46. Boyd, J.: Two comments on filtering (artificial viscosity) for Chebyshev and Legendre spectral and spectral element methods: preserving boundary conditions and interpretation of the filter as a diffusion. Journal of Computational Physics **143**(1), 283–288 (1998). DOI {10.1006/jcph.1998.5961}

47. Boyd, J.: Chebyshev and Fourier spectral methods. Dover Publications Inc., Mineola, New York (2001.)

48. Brooks, A.N., Hughes, T.J.R.: Streamline upwind/Petrov-Galerkin formulations for convective dominated flows with particular emphasis on the incompressible Navier-Stokes equations. Computer Methods in Applied Mechanics and Engineering **32**, 199–259 (1982)

49. Budd, C.J., Huang, W.H., Russell, R.D.: Moving mesh methods for problems with blow-up. SIAM Journal on Scientific Computing **17**(2), 305–327 (1996). DOI {10.1137/S1064827594272025}

50. Budd, C.J., Williams, J.F.: Moving mesh generation using the parabolic Monge-Ampére equation. SIAM Journal on Scientific Computing **31**, 3438–3465 (2009)

51. Bui-Thanh, T.: From Godunov to a unified hybridized discontinuous Galerkin framework for partial differential equations. Journal of Computational Physics **295**, 114–146 (2015)

52. Bui-Thanh, T.: From Rankine-Hugoniot condition to a constructive derivation of HDG methods, pp. 483–491. Lecture Notes in Computational Sciences and Engineering. Springer (2015)

53. Burden, R., Faires, J.: Numerical Analysis. Brooks/Cole, Boston, MA. (2011)

54. Burstedde, C., Ghattas, O., Gurnis, M., Isaac, T., Stadler, G., Warburton, T., Wilcox, L.C.: Extreme-scale AMR. In: Proc. 2010 ACM/IEEE Int. Conference for High Performance Computing, Networking, Storage and Analysis, 1, pp. 1–12 (2010)

55. Burstedde, C., Ghattas, O., Gurnis, M., Stadler, G., Tan, E., Tu, T., Wilcox, L.C., Zhong, S.: Scalable adaptive mantle convection simulation on petascale supercomputers. In: Proc. 2008 ACM/IEEE Supercomputing, p. 62. IEEE Press (2008)

56. Burstedde, C., Wilcox, L.C., Ghattas, O.: p4est: scalable algorithms for parallel adaptive mesh refinement on forests of octrees. SIAM Journal on Scientific Computing **33**(3), 1103–1133 (2011)

57. Buscaglia, G., Dari, E.: Implementation of the Lagrange-Galerkin method for the incompressible Navier-Stokes equations. International Journal for Numerical Methods in Fluids **15**(1), 23–36 (1992). DOI {10.1002/fld.1650150103}

58. Butcher, J.: A stability property of implicit Runge-Kutta methods. BIT Numerical Mathematics **15**(4), 358–361 (1975)

59. Butcher, J.: A generalization of singly-implicit methods. BIT Numerical Mathematics **21**(2), 175–189 (1981). DOI 10.1007/BF01933162

60. Butcher, J.: Stability properties for a general class of methods for ordinary differential equations. SIAM Journal on Numerical Analysis **18**(1), 37–44 (1981)

61. Butcher, J.: The numerical analysis of ordinary differential equations: Runge-Kutta and general linear methods. Wiley-Interscience (1987)

62. Butcher, J.: Diagonally-implicit multi-stage integration methods. Applied Numerical Mathematics **11**(5), 347–363 (1993). DOI http://dx.doi.org/10.1016/0168-9274(93)90059-Z

63. Butcher, J.: General Linear Methods. Acta Numerica **15**, 157–256 (2006)

64. Butcher, J., Chen, D.: A new type of singly-implicit Runge-Kutta method. Applied Numerical Mathematics **34**(2-3), 179–188 (2000)

65. Cangiani, A., Manzini, G.: Flux reconstruction and solution post-processing in mimetic finite difference methods. Computer Methods in Applied Mechanics and Engineering **197**(9-12), 933–945 (2008). DOI {10.1016/j.cma.2007.09.019}

66. Canuto, C., Hussaini, M., Quarteroni, A., Zang, T.: Spectral Methods. Springer (2006)

67. Cao, W.M., Huang, W.Z., Russell, R.D.: An r-adaptive finite element method based upon moving mesh PDEs. Journal of Computational Physics **149**(2), 221–244 (1999). DOI {10.1006/jcph.1998.6151}

68. Carciopolo, L.D., Bonaventura, L., Scotti, A., Formaggia, L.: A conservative implicit multirate method for hyperbolic problems. Computational Geosciences **23**(4), 647–664 (2019). DOI {10.1007/s10596-018-9764-2}

69. Carpenter, M.H., Gottlieb, D., Abarbanel, S.: The stability of numerical boundary treatments for compact high-order finite-difference schemes. Journal of Computational Physics **108**(2), 272 – 295 (1993). DOI http://dx.doi.org/10.1006/jcph.1993.1182. URL http://www.sciencedirect.com/science/article/pii/S0021999183711824

70. Carr III, L.E., Borges, C.F., Giraldo, F.X.: An element-based spectrally optimized approximate inverse preconditioner for the Euler equations. SIAM Journal on Scientific Computing **34**(4), B392–B420 (2012). DOI {10.1137/11083229X}

71. Carr III, L.E., Borges, C.F., Giraldo, F.X.: Matrix-free polynomial-based nonlinear least squares optimized preconditioning and its application to discontinuous Galerkin discretizations of the Euler equations. Journal of Scientific Computing **66**(3), 917–940 (2016). DOI {10.1007/s10915-015-0049-9}

72. Castillo, P., Cockburn, B., Perugia, I., Schötzau, D.: An a priori error analysis of the local discontinuous Galerkin method for elliptic problems. SIAM Journal on Numerical Analysis **38**(5), 1676–1706 (2000)

73. Chan, J.: On discretely entropy conservative and entropy stable discontinuous Galerkin methods. Journal of Computational Physics **362**, 346–374 (2018). DOI {10.1016/j.jcp.2018.02.033}

74. Chan, J., Wilcox, L.C.: On discretely entropy stable weight-adjusted discontinuous Galerkin methods: curvilinear meshes. Journal of Computational Physics **378**, 366–393 (2019)

75. Chen, H., Lu, P., Xu, X.: A robust multilevel method for hybridizable discontinuous Galerkin method for the Helmholtz equation. Journal of Computational Physics **264**, 133–151 (2014)

76. Chen, T., Shu, C.W.: Entropy stable high order discontinuous Galerkin methods with suitable quadrature rules for hyperbolic conservation laws. Journal of Computational Physics **345**, 427–461 (2017). DOI {10.1016/j.jcp.2017.05.025}

77. Cheong, H.B.: Double Fourier series on a sphere: applications to elliptic and vorticity equations. Journal of Computational Physics **157**(1), 327 – 349 (2000). DOI http://dx.doi.org/10.1006/jcph.1999.6385. URL http://www.sciencedirect.com/science/article/pii/S0021999199963854

78. Choi, S.J., Giraldo, F.X., Kim, J., Shin, S.: Verification of a non-hydrostatic dynamical core using the horizontal spectral element method and vertical finite difference method: 2d aspects. Geoscientific Model Development **7**(6), 2717–2731 (2014). DOI {10.5194/gmd-7-2717-2014}

79. Cockburn, B., Dong, B., Guzmán, J.: A superconvergent LDG-hybridizable Galerkin method for second-order elliptic problems. Mathematics of Computation **77**, 1887–1916 (2008). URL http://www.ams.org/mcom/0000-000-00/S0025-5718-08-02123-6/S0025-5718-08-02123-6.pdf

80. Cockburn, B., Dong, B., Guzman, J., Restelli, M., Sacco, R.: A hybridizable discontinuous Galerkin method for steady state convection-diffusion-reaction problems. SIAM Journal on Scientific Computing **31**, 3827–3846 (2009)

81. Cockburn, B., Gopalakrishnan, J.: The derivation of hybridizable discontinuous Galerkin methods for Stokes flow. SIAM Journal on Numerical Analysis **47**(2), 1092–1125 (2009)

82. Cockburn, B., Gopalakrishnan, J., Lazarov, R.: Unified hybridization of discontinuous Galerkin, mixed, and continuous Galerkin methods for second order elliptic problems. SIAM Journal on Numerical Analysis **47**(2), 1319–1365 (2009)

83. Cockburn, B., Gopalakrishnan, J., Sayas, F.J.: A projection-based error analysis of HDG methods. Mathematics of Computation **79**(271), 1351–1367 (2010)

84. Cockburn, B., Guzman, J., Soon, S.C., Stolarski, H.K.: An analysis of the embedded discontinuous Galerkin method for second-order elliptic problems. SIAM Journal on Numerical Analysis **47**(4), 2686–2707 (2009). DOI {10.1137/080726914}

85. Cockburn, B., Nguyen, N.C., Peraire, J.: A comparison of HDG methods for Stokes flow. Journal of Scientific Computing **45**(1), 215–237 (2010)

86. Cockburn, B., Shu, C.: The local discontinuous Galerkin method for time-dependent convection-diffusion systems. SIAM Journal on Numerical Analysis **35**(6), 2440–2463 (1998)

87. Cohen, G., Joly, P., Roberts, J., Tordjman, N.: Higher order triangular finite elements with mass lumping for the wave equation. SIAM Journal on Numerical Analysis **38**(6), 2047–2078 (2001)

88. Conde, S., Gottlieb, S., Grant, Z.J., Shadid, J.N.: Implicit and implicit-explicit strong stability preserving Runge-Kutta methods with high linear order. Journal of Scientific Computing **73**(2-3, SI), 667–690 (2017). DOI {10.1007/s10915-017-0560-2}

89. Constantinescu, E., Sandu, A.: On extrapolated multirate methods. In: H.G. Bock, F. Hoog, A. Friedman, A. Gupta, H. Neunzert, W.R. Pulleyblank, T. Rusten, F. Santosa, A.K. Tornberg, V. Capasso, R. Mattheij, H. Neunzert, O. Scherzer, A.D. Fitt, J. Norbury, H. Ockendon, E. Wilson (eds.) Progress in Industrial Mathematics at ECMI 2008, *Mathematics in Industry*, vol. 15, pp. 341–347. Springer Berlin Heidelberg (2010). URL http://dx.doi.org/10.1007/978-3-642-12110-4_52. 10.1007/978-3-642-12110-4_52

90. Constantinescu, E., Sandu, A.: Optimal explicit strong-stability-preserving general linear methods. SIAM Journal on Scientific Computing **32**(5), 3130–3150 (2010). DOI 10.1137/090766206

91. Constantinescu, E.M., Sandu, A.: Multirate timestepping methods for hyperbolic conservation laws. Journal of Scientific Computing **33**(3), 239–278 (2007). DOI {10.1007/s10915-007-9151-y}

92. Constantinescu, E.M., Sandu, A.: Extrapolated multirate methods for differential equations with multiple time scales. Journal of Scientific Computing **56**(1), 28–44 (2013). DOI {10.1007/s10915-012-9662-z}

93. Cools, R.: Monomial cubature rules since Stroud: a compilation - part 2. Journal of Computational and Applied Mathematics **112**, 21–27 (1999)

94. Cools, R., Rabinowitz, P.: Monomial cubature rules since Stroud: a compilation. Journal of Computational and Applied Mathematics **48**, 309–326 (1993)
95. Cotter, C.J., Kuzmin, D.: Embedded discontinuous Galerkin transport schemes with localised limiters. Journal of Computational Physics **311**, 363–373 (2016). DOI {10.1016/j.jcp.2016.02.021}
96. Courant, R., Friedrichs, K., Lewy, H.: On the partial difference equations of mathematical physics. IBM Journal: translation from the original paper in Mathematische Annale, 100,32-24 **100**, 215–234 (1928)
97. Crean, J., Hicken, J.E., Fernandez, D.C.D.R., Zingg, D.W., Carpenter, M.H.: Entropy-stable summation-by-parts discretization of the Euler equations on general curved elements. Journal of Computational Physics **356**, 410–438 (2018). DOI {10.1016/j.jcp.2017.12.015}
98. Davies, T., Cullen, M., Malcolm, A., Mawson, M., Staniforth, A., White, A., Wood, N.: A new dynamical core for the Met Office's global and regional modelling of the atmosphere. Quarterly Journal of the Royal Meteorological Society **131**(608, Part B), 1759–1782 (2005). DOI {10.1256/qj.04.101}
99. De Grazia, D., Mengaldo, G., Moxey, D., Vincent, P.E., Sherwin, S.J.: Connections between the discontinuous Galerkin method and high-order flux reconstruction schemes. International Journal for Numerical Methods in Fluids **75**(12), 860–877 (2014). DOI {10.1002/fld.3915}
100. Deardorff, J.W.: A numerical study of three-dimensional turbulent channel flow at large Reynolds numbers. Journal of Fluid Mechanics **41**, 452–480 (1970)
101. Debnath, L., Mikusinski, P.: Introduction to Hilbert Spaces with Applications. Academic Press, San Diego (1990)
102. Demkowicz, L.: Some remarks on moving finite-element methods. Computer Methods in Applied Mechanics and Engineering **46**(3), 339–349 (1984). DOI {10.1016/0045-7825(84)90109-9}
103. Dennis, J., Fournier, A., Spotz, W.F., St-Cyr, A., Taylor, M.A., Thomas, S.J., Tufo, H.: High-resolution mesh convergence properties and parallel efficiency of a spectral element atmospheric dynamical core. International Journal of High Performance Computing Applications **19**(3), 225–235 (2005). DOI {10.1177/1094342005056108}
104. Dennis, J.M., Edwards, J., Evans, K.J., Guba, O., Lauritzen, P.H., Mirin, A.A., St-Cyr, A., Taylor, M.A., Worley, P.H.: Cam-se: A scalable spectral element dynamical core for the community atmosphere model. International Journal of High Performance Computing Applications **26**(1), 74–89 (2012). DOI 10.1177/1094342011428142. URL http://hpc.sagepub.com/content/26/1/74.abstract
105. Deville, M., Fischer, P., Mund, E.: High-Order Methods for Incompressible Flow. Cambridge University Press, New York (2002)
106. Diamessis, P., Domaradzki, J., Hesthaven, J.: A spectral multidomain penalty method model for the simulation of high Reynolds number localized incompressible stratified turbulence. Journal of Computational Physics **202**(1), 298–322 (2005). DOI {10.1016/j.jcp.2004.07.007}
107. Dick, E.: Accurate Petrov-Galerkin methods for transient convective diffusion problems. International Journal for Numerical Methods in Engineering **19**(10), 1425–1433 (1983). DOI {10.1002/nme.1620191002}
108. DiPietro, D.A., Ern, A.: Mathematical Aspects of Discontinuous Galerkin Methods. Springer, Heidelberg, Germany (2012)
109. Dolejší, V., Feistauer, M.: A semi-implicit discontinuous Galerkin finite element method for the numerical solution of inviscid compressible flow. Journal of Computational Physics **198**(2), 727–746 (2004). URL http://www.sciencedirect.com/science/article/B6WHY-4BT1RG4-2/2/8c7614808baa4a20cb4fcf6bfba2b705
110. Douglas, J., Dupont, T.: Interior penalty procedures for elliptic and parabolic Galerkin methods. Lecture Notes in Physics **58**, 207–216 (1976)
111. Dubiner, M.: Spectral methods on triangles and other domains. Journal of Scientific Computing **6**(4), 345–90 (1991)
112. Dudhia, J.: A nonhydrostatic version of the Penn State NCAR mesoscale model - validation tests and simulation of an Atlantic cyclone and cold-front. Monthly Weather Review **121**(5), 1493–1513 (1993)

113. Dukowicz, J.K.: A simplified adaptive mesh technique derived from the moving finite-element method. Journal of Computational Physics **56**(2), 324–342 (1984). DOI {10.1016/0021-9991(84)90098-6}
114. Dupont, F., Lin, C.: The adaptive spectral element method and comparisons with more traditional formulations for ocean modeling. Journal of Atmospheric and Oceanic Technology **21**(1), 135–147 (2004). DOI {10.1175/1520-0426(2004)021<0135:TASEMA>2.0.CO;2}
115. Durran, D., Blossey, P.: Implicit–explicit multistep methods for fast-wave–slow-wave problems. Monthly Weather Review **140**(4), 1307–1325 (2012)
116. Durran, D., Klemp, J.: A compressible model for the simulation of moist mountain waves. Monthly Weather Review **111**(12), 2341–2361 (1983)
117. Ebeida, M.S., Patney, A., Owens, J.D., Mestreau, E.: Isotropic conforming refinement of quadrilateral and hexahedral meshes using two-refinement templates. International Journal for Numerical Methods in Engineering **88**(10), 974–985 (2011). DOI {10.1002/nme.3207}
118. Eisman, P.R.: Grid generation for fluid mechanics computation. Annual Review of Fluid Mechanics **17**, 487–522 (1985)
119. Eisman, P.R.: Adaptive grid generation. Computer Methods in Applied Mechanics and Engineering **64**, 321–376 (1987)
120. Engstler, C., Lubich, C.: Multirate extrapolation methods for differential equations with different time scales. Computing **58**(2), 173–185 (1997). DOI {10.1007/BF02684438}
121. Escobar-Vargas, J.A., Diamessis, P.J., Giraldo, F.X.: High-order discontinuous element-based schemes for the inviscid shallow water equations: spectral multidomain penalty and discontinuous Galerkin methods. Applied Mathematics and Computation **218**(9), 4825–4848 (2012). DOI {10.1016/j.amc.2011.10.046}
122. Eskilsson, C.: An hp-adaptive discontinuous Galerkin method for shallow water flows. International Journal for Numerical Methods in Fluids **67**(11), 1605–1623 (2011)
123. Falcone, M., Ferretti, R.: Convergence analysis for a class of high-order semi-Lagrangian advection schemes. SIAM Journal on Numerical Analysis **35**, 909–940 (1998)
124. Feistauer, M., Kučera, V.: On a robust discontinuous Galerkin technique for the solution of compressible flow. Journal of Computational Physics **224**(1), 208–221 (2007). URL http://www.sciencedirect.com/science/article/B6WHY-4N146DY-1/1/7935061df2de55fe5bef7f3df343a49f
125. Fernandez, D.C.D.R., Hicken, J.E., Zingg, D.W.: Review of summation-by-parts operators with simultaneous approximation terms for the numerical solution of partial differential equations. Computers & Fluids **95**, 171–196 (2014). DOI {10.1016/j.compfluid.2014.02.016}
126. Fey, M.: Multidimensional upwinding. Part I. The method of transport for solving the Euler equations. Journal of Computational Physics **143**(1), 159–180 (1998). DOI {10.1006/jcph.1998.5958}
127. Fischer, P.: An overlapping Schwarz method for spectral element solution of the incompressible Navier-Stokes equations. Journal of Computational Physics **133**(1), 84–101 (1997)
128. Fischer, P.: Projection techniques for iterative solution of Ax=b with successive right-hand sides. Computer Methods in Applied Mechanics and Engineering **163**(1-4), 193–204 (1998)
129. Fischer, P.F., Kruse, G.W., Loth, F.: Spectral element methods for transitional flows in complex geometries. Journal of Scientific Computing **17**(1-4), 81–98 (2002)
130. Fisher, T.C., Carpenter, M.H.: High-order entropy stable finite difference schemes for nonlinear conservation laws: Finite domains. Journal of Computational Physics **252**, 518–557 (2016)
131. Flaherty, J., Loy, R., Shephard, M., Szymanski, B., Teresco, J., Ziantz, L.: Adaptive local refinement with octree load balancing for the parallel solution of three-dimensional conservation laws. Journal of Parallel and Distributed Computing **47**(2), 139–152 (1997). DOI {10.1006/jpdc.1997.1412}
132. Fletcher, C.: Computational Techniques for Fluid Dynamics - Vol I: Fundamentals and General Techniques, 1st edn. Springer-Verlag (1987)
133. Fok, P.W.: A linearly fourth order multirate Runge-Kutta method with error control. Journal of Scientific Computing **66**(1), 177–195 (2016). DOI {10.1007/s10915-015-0017-4}

134. Frank, J., Hundsdorfer, W., Verwer, J.: On the stability of implicit-explicit linear multistep methods. Applied Numerical Mathematics **25**(2-3), 193–205 (1997)

135. Frey, P., George, P.: Mesh Generation: Application to Finite Elements, 1st edn. ISTE Publishing Company (2000)

136. Fu, G., Shu, C.W.: Analysis of an embedded discontinuous Galerkin method with implicit-explicit time-marching for convection-diffusion problems. International Journal of Numerical Analysis and Modeling **14**(4-5), 477–499 (2017)

137. Gabersek, S., Giraldo, F.X., Doyle, J.D.: Dry and moist idealized experiments with a two-dimensional spectral element model. Monthly Weather Review **140**(10), 3163–3182 (2012). DOI {10.1175/MWR-D-11-00144.1}

138. Garimella, R.: Conformal refinement of unstructured quadrilateral meshes. In: Clark, BW (ed.) Proceedings of the 18th International Meshing Roundtable, pp. 31–44. Springer-Verlag Berlin, Heidelberger Platz 3, D-14197 Berlin, Germany (2009). DOI {10.1007/978-3-642-04319-2_3}. 18th International Meshing Roundtable (IMR), Salt Lake City, UT, OCT 25-28, 2009

139. Gassmann, A., Herzog, H.: A consistent time–split numerical scheme applied to the nonhydrostatic compressible equations. Monthly Weather Review **135**(1), 20–36 (2007)

140. Gassner, G.J.: A skew-symmetric discontinuous Galerkin spectral element discretization and its relation to SBP-SAT finite difference methods. SIAM Journal on Scientific Computing **35**(3), A1233–A1253 (2013). DOI {10.1137/120890144}

141. Gassner, G.J., Loercher, F., Munz, C.D., Hesthaven, J.S.: Polymorphic nodal elements and their application in discontinuous Galerkin methods. Journal of Computational Physics **228**(5), 1573–1590 (2009). DOI {10.1016/j.jcp.2008.11.012}

142. Gassner, G.J., Winters, A.R., Kopriva, D.A.: Split form nodal discontinuous Galerkin schemes with summation-by-parts property for the compressible Euler equations. Journal of Computational Physics **327**, 39–66 (2016). DOI {10.1016/j.jcp.2016.09.013}

143. Gear, C.: Hybrid methods for initial value problems in ordinary differential equations. SIAM Journal on Numerical Analysis **2**(1), 69–86 (1965)

144. Gear, C., Wells, D.: Multirate linear multistep methods. BIT Numerical Mathematics **24**(4), 484–502 (1984). DOI {10.1007/BF01934907}

145. George, D.L.: Adaptive finite volume methods with well-balanced Riemann solvers for modeling floods in rugged terrain: Application to the Malpasset dam-break flood (France, 1959). International Journal for Numerical Methods in Fluids **66**(8), 1000–1018 (2011). DOI {10.1002/fld.2298}

146. Gerritsma, M., Phillips, T.: Discontinuous spectral element approximations for the velocity-pressure-stress formulation of the Stokes problem. International Journal for Numerical Methods in Engineering **43**(8), 1401–1419 (1998). DOI {10.1002/(SICI)1097-0207(19981230)43:8<1401::AID-NME475>3.3.CO;2-R}

147. Gibbs, J.W.: Letter to the editor, fourier series. Nature **59**, 200 (1898)

148. Gibbs, J.W.: Letter to the editor, fourier series. Nature **59**, 606 (1899)

149. Giraldo, F.X.: A space marching adaptive remeshing technique applied to the 3d Euler equations for supersonic flow. Ph.D. thesis, University of Virginia (1995)

150. Giraldo, F.X.: Lagrange-Galerkin methods on spherical geodesic grids. Journal of Computational Physics **136**(1), 197–213 (1997)

151. Giraldo, F.X.: The Lagrange-Galerkin spectral element method on unstructured quadrilateral grids. Journal of Computational Physics **147**(1), 114–146 (1998)

152. Giraldo, F.X.: Trajectory calculations for spherical geodesic grids in Cartesian space. Monthly Weather Review **127**(7), 1651–1662 (1999). DOI {10.1175/1520-0493(1999)127<1651:TCFSGG>2.0.CO;2}

153. Giraldo, F.X.: Lagrange-Galerkin methods on spherical geodesic grids: the shallow water equations. Journal of Computational Physics **160**(1), 336–368 (2000)

154. Giraldo, F.X.: A spectral element shallow water model on spherical geodesic grids. International Journal for Numerical Methods in Fluids **35**(8), 869–901 (2001). DOI {10.1002/1097-0363(20010430)35:8<869::AID-FLD116>3.0.CO;2-S}

155. Giraldo, F.X.: Strong and weak Lagrange-Galerkin spectral element methods for the shallow water equations. Computers & Mathematics with Applications **45**(1-3), 97–121 (2003). DOI {10.1016/S0898-1221(03)80010-X}
156. Giraldo, F.X.: Semi-implicit time-integrators for a scalable spectral element atmospheric model. Quarterly Journal of the Royal Meteorological Society **131**(610), 2431–2454 (2005). URL http://dx.doi.org/10.1256/qj.03.218
157. Giraldo, F.X.: High-order triangle-based discontinuous Galerkin methods for hyperbolic equations on a rotating sphere. Journal of Computational Physics **214**(2), 447–465 (2006). URL http://www.sciencedirect.com/science/article/B6WHY-4HJS5JW-1/2/f93b33d52f2c51233e9a166f45b5d223
158. Giraldo, F.X.: Hybrid Eulerian-Lagrangian semi-implicit time-integrators. Computers & Mathematics with applications **52**(8-9), 1325–1342 (2006). DOI {10.1016/j.camwa.2006.11.009}
159. Giraldo, F.X., Hesthaven, J.S., Warburton, T.: Nodal high-order discontinuous Galerkin methods for the spherical shallow water equations. Journal of Computational Physics **181**(2), 499–525 (2002). URL http://www.sciencedirect.com/science/article/B6WHY-46SY2J5-7/2/809295d1a21106dd028d25d6b37c33b4
160. Giraldo, F.X., Kelly, J.F., Constantinescu, E.M.: Implicit-explicit formulations of a three-dimensional nonhydrostatic unified model of the atmosphere (NUMA). SIAM Journal on Scientific Computing **35**(5), B1162–B1194 (2013). DOI {10.1137/120876034}
161. Giraldo, F.X., Perot, J.B., Fischer, P.F.: A spectral element semi-Lagrangian (SESL) method for the spherical shallow water equations. Journal of Computational Physics **190**(2), 623–650 (2003). DOI {10.1016/S0021-9991(03)00300-0}
162. Giraldo, F.X., Restelli, M.: A study of spectral element and discontinuous Galerkin methods for the Navier–Stokes equations in nonhydrostatic mesoscale atmospheric modeling: equation sets and test cases. Journal of Computational Physics **227**(8), 3849–3877 (2008). URL http://www.sciencedirect.com/science/article/B6WHY-4RDS48D-1/1/2c21804d8d1855a714d1c33b77db2db5
163. Giraldo, F.X., Restelli, M.: High-order semi-implicit time-integrators for a triangular discontinuous Galerkin oceanic shallow water model. International Journal for Numerical Methods in Fluids **63**, 1077–1102 (2009)
164. Giraldo, F.X., Restelli, M.: High-order semi-implicit time-integrators for a triangular discontinuous Galerkin oceanic shallow water model. International Journal for Numerical Methods in Fluids **63**(9), 1077–1102 (2010). DOI {10.1002/fld.2118}
165. Giraldo, F.X., Restelli, M., Läuter, M.: Semi-implicit formulations of the Navier-Stokes equations: application to nonhydrostatic atmospheric modeling. SIAM Journal on Scientific Computing **32**(6), 3394–3425 (2010). DOI {10.1137/090775889}
166. Giraldo, F.X., Rosmond, T.E.: A scalable spectral element eulerian atmospheric model (SEE-AM) for NWP: dynamical core tests. Monthly Weather Review **132**(1), 133–153 (2004)
167. Giraldo, F.X., Taylor, M.A.: A diagonal-mass-matrix triangular-spectral-element method based on cubature points. Journal of Engineering Mathematics **56**(3), 307–322 (2006). DOI {10.1007/s10665-006-9085-7}
168. Giraldo, F.X., Warburton, T.: A nodal triangle-based spectral element method for the shallow water equations on the sphere. Journal of Computational Physics **207**, 129–150 (2005)
169. Giraldo, F.X., Warburton, T.: A high-order triangular discontinuous Galerkin oceanic shallow water model. International Journal for Numerical Methods in Fluids **56**(7), 899–925 (2008). URL http://dx.doi.org/10.1002/fld.1562
170. Godunov, S.: A difference method for numerical calculation of discontinuous solutions of the equations of hydrodynamics. Mathematicheskii Sbornik **47**(3), 271–306 (1959)
171. Goedel, N., Schomann, S., Warburton, T., Clemens, M.: GPU accelerated Adams-Bashforth multirate discontinuous Galerkin FEM simulation of high-frequency electromagnetic fields. IEEE Transactions on Magnetics **46**(8), 2735–2738 (2010). DOI {10.1109/TMAG.2010.2043655}. 17th International Conference on the Computation of Electromagnetic Fields (COMPUMAG 09), Florianopolis, Brazil, Nov 22-26, 2009

172. Gordon, W.N., Hall, C.A.: Construction of curvilinear coordinate systems and application to mesh generation. International Journal for Numerical Methods in Engineering **7**, 461–477 (1973)
173. Gottlieb, D., Hesthaven, J.: Spectral methods for hyperbolic problems. Journal of Computational and Applied Mathematics **128**(1-2, SI), 83–131 (2001). DOI {10.1016/S0377-0427(00)00510-0}
174. Gottlieb, S.: On high order strong stability preserving Runge–Kutta and multi step time discretizations. Journal of Scientific Computing **25**(1), 105–128 (2005)
175. Gottlieb, S., Ketcheson, D., Shu, C.: High order strong stability preserving time discretizations. Journal of Scientific Computing **38**(3), 251–289 (2009)
176. Gottlieb, S., Ruuth, S.: Optimal strong-stability-preserving time-stepping schemes with fast downwind spatial discretizations. Journal of Scientific Computing **27**(1-3), 289–303 (2006)
177. Gottlieb, S., Shu, C.W.: Total variation diminishing Runge-Kutta schemes. Mathematics of Computation **67**(221), 73–85 (1998)
178. Gottlieb, S., Shu, C.W., Tadmor, E.: Strong stability-preserving high-order time discretization methods. SIAM Review **43**(1), 89–112 (2001)
179. Gragg, W.: Repeated extrapolation to the limit in the numerical solution of ordinary differential equations. Ph.D. thesis, University of California Los Angeles (1964)
180. Gragg, W., Stetter, H.: Generalized multistep predictor-corrector methods. Journal of the ACM **11**(2), 188–209 (1964). DOI 10.1145/321217.321223
181. Guenther, M., Hachtel, C., Sandu, A.: Multirate GARK schemes for multiphysics problems. In: Bartel, A and Clemens, M and Gunther, M and TerMaten, EJW (ed.) Scientific Computing in Electrical Engineering (SCEE 2014), *Mathematics in Industry*, vol. 23, pp. 115–121. Springer-Verlag Berlin, Heidelberger Platz 3, D-14197 Berlin, Germany (2016). DOI {10.1007/978-3-319-30399-4_12}. 10th International Conference on Scientific Computing in Electrical Engineering (SCEE), Wuppertal, Germany, JUL 22-25, 2014
182. Guermond, J.L., Pasquetti, R., Popov, B.: Entropy viscosity method for nonlinear conservation laws. Journal of Computational Physics **230**(11, SI), 4248–4267 (2011). DOI {10.1016/j.jcp.2010.11.043}
183. Guezey, S., Cockburn, B., Stolarski, H.K.: The embedded discontinuous Galerkin method: application to linear shell problems. International Journal for Numerical Methods in Engineering **70**(7), 757–790 (2007). DOI {10.1002/nme.1893}
184. Gui, W., Babuska, I.: The h-version, p-version and h-p-version of the finite-element method in 1-dimension. Part 1. The error analysis of the p-version. Numerische Mathematik **49**(6), 577–612 (1986). DOI {10.1007/BF01389733}
185. Gui, W., Babuska, I.: The h-version, p-version and h-p-version of the finite-element method in 1-dimension. Part 2. The error analysis of the h- and h-p version. Numerische Mathematik **49**(6), 659–683 (1986). DOI {10.1007/BF01389733}
186. Gui, W., Babuska, I.: The h-version, p-version and h-p-version of the finite-element method in 1-dimension. Part 3. The adaptive h-p version. Numerische Mathematik **49**(6), 577–612 (1986). DOI {10.1007/BF01389733}
187. Gunther, M., Kvaerno, A., Rentrop, P.: Multirate partitioned Runge-Kutta methods. BIT Numerical Mathematics **41**(3), 504–514 (2001). DOI {10.1023/A:1021967112503}
188. Hadamard, J.: Sur les problèmes aux dérivées partielles et leur signification physique. Princeton University Bulletin (), 49–52 (1902)
189. Hartmann, R., Houston, P.: Adaptive discontinuous Galerkin finite element methods for the compressible Euler equations. Journal of Computational Physics **183**(2), 508–532 (2002). DOI {10.1006/jcph.2002.7206}
190. Heinze, T., Hense, A.: The shallow water equations on the sphere and their Lagrange-Galerkin-solution. Meteorology and Atmospheric Physics **81**(1-2), 129–137 (2002). DOI {10.1007/s007030200034}
191. Helenbrook, B.T.: On the existence of explicit hp-finite element methods using Gauss-Lobatto integration on the triangle. SIAM Journal on Numerical Analysis **47**(2), 1304–1318 (2009). DOI {10.1137/070685439}

192. Henderson, R.: Dynamic refinement algorithms for spectral element methods. Computer Methods in Applied Mechanics and Engineering **175**(3-4), 395–411 (1999). DOI {10.1016/S0045-7825(98)00363-6}

193. Hendricks, E.A., Kopera, M.A., Giraldo, F.X., Peng, M.S., Doyle James D. .and Jiang, Q.: Evaluation of the utility of static and adaptive mesh refinement for idealized tropical cyclone problems in a spectral element shallow water model. Monthly Weather Review **144**, 3697–3724 (2016)

194. Henson, V., Yang, U.: BoomerAMG: A parallel algebraic multigrid solver and preconditioner. Applied Numerical Mathematics **41**(1), 155–177 (2002). DOI {10.1016/S0168-9274(01)00115-5}. 16th IMACS World Congress, Lausanne, Switzerland, 2000

195. Heroux, M., Bartlett, R., Howle, V., Hoekstra, R., Hu, J., Kolda, T., Lehoucq, R., Long, K., Pawlowski, R., Phipps, E., Salinger, A., Thornquist, H., Tuminaro, R., Willenbring, J., Williams, A., Stanley, K.: An overview of the Trilinos Project. ACM Transactions on Mathematical Software **31**(3), 397–423 (2005). DOI {10.1145/1089014.1089021}

196. Hesthaven, J.: A stable penalty method for the compressible Navier-Stokes equations. II. One-dimensional domain decomposition schemes. SIAM Journal on Scientific Computing **18**(3), 658–685 (1997). DOI {10.1137/S1064827594276540}

197. Hesthaven, J.: A stable penalty method for the compressible Navier-Stokes equations: III. Multidimensional domain decomposition schemes. SIAM Journal on Scientific Computing **20**(1), 62–93 (1998). DOI {10.1137/S1064827596299470}

198. Hesthaven, J., Gottlieb, D.: A stable penalty method for the compressible Navier-Stokes equations. I. Open boundary conditions. SIAM Journal on Scientific Computing **17**(3), 579–612 (1996). DOI {10.1137/S1064827594268488}

199. Hesthaven, J., Warburton, T.: Nodal high-order methods on unstructured grids - I. Time-domain solution of Maxwell's equations. Journal of Computational Physics **181**(1), 186–221 (2002). DOI {10.1006/jcph.2002.7118}

200. Hesthaven, J., Warburton, T.: Nodal Discontinuous Galerkin Methods. Springer, New York (2008)

201. Hesthaven, J.S.: From electrostatics to almost optimal nodal sets for polynomial interpolation in a simplex. SIAM Journal on Numerical Analysis **35**, 655–676 (1998)

202. Higueras, I., Roldan, T.: New third order low-storage SSP explicit Runge-Kutta methods. Journal of Scientific Computing **79**(3), 1882–1906 (2019). DOI {10.1007/s10915-019-00916-3}

203. Hodur, R.: The Naval Research Laboratory's coupled ocean/atmosphere mesoscale prediction system (COAMPS). Monthly Weather Review **125**(7), 1414–1430 (1997)

204. Holton, J.: An Introduction to Dynamic Meteorology. Elsevier (2004)

205. Hou, T., Lefloch, P.: Why nonconservative schemes converge to wrong solutions - error analysis. Mathematics of Computation **62**(206), 497–530 (1994). DOI {10.2307/2153520}

206. Houston, P., Suli, E.: Adaptive Lagrange-Galerkin methods for unsteady convection-diffusion problems. Mathematics of Computation **70**(233), 77–106 (2001)

207. Hubbard, M., Nikiforakis, N.: A three-dimensional, adaptive, Godunov-type model for global atmospheric flows. Monthly Weather Review **131**(8), 1848–1864 (2003)

208. Huebner, K., Thornton, E.: The Finite Element Method for Engineers, 3rd edn. Wiley (1995)

209. Hughes, T., Feijoo, G., Mazzei, L., Quincy, J.: The variational multiscale method - a paradigm for computational mechanics. Computer Methods in Applied Mechanics and Engineering **166**(1-2), 3–24 (1998). DOI {10.1016/S0045-7825(98)00079-6}

210. Huynh, H.: A flux reconstruction approach to high-order schemes including discontinuous Galerkin methods. AIAA Paper **2007-4079** (2007)

211. Jablonowski, C., Herzog, M., Penner, J.E., Oehmke, R.C., Stout, Q.F., Van Leer, B., Powell, K.G.: Block-structured adaptive grids on the sphere: advection experiments. Monthly Weather Review **134**(12), 3691–3713 (2006)

212. Janjic, Z.: A nonhydrostatic model based on a new approach. Meteorology and Atmospheric Physics **82**(1-4), 271–285 (2003). DOI {10.1007/s00703-001-0587-6}

213. John, F.: Partial Differential Equations. Springer-Verlag (1986)

214. Johnson, A.A., Tezduyar, T.E.: Mesh update strategies in parallel finite-element computations of flow problems with moving boundaries and interfaces. Computer Methods in Applied Mechanics and Engineering **119**(1-2), 73–94 (1994). DOI {10.1016/0045-7825(94)00077-8}

215. Kamenetskiy, D.S.: On the relation of the embedded discontinuous Galerkin method to the stabilized residual-based finite element methods. Applied Numerical Mathematics **108**, 271–285 (2016). DOI {10.1016/j.apnum.2016.01.004}

216. Kanevsky, A., Carpenter, M.H., Hesthaven, J.S.: Idempotent filtering in spectral and spectral element methods. Journal of Computational Physics **220**(1), 41–58 (2006). DOI {10.1016/j.jcp.2006.05.014}

217. Kang, S., Giraldo, F.X., Bui-Thanh, T.: IMEX HDG-DG: A coupled implicit hybridized discontinuous Galerkin and explicit discontinuous Galerkin approach for shallow water systems. Journal of Computational Physics **401** (2020). DOI {10.1016/j.jcp.2019.109010}

218. Karniadakis, G., Israeli, M., Orszag, S.: High–order splitting methods for the incompressible Navier–Stokes equations. Journal of Computational Physics **97**(2), 414–443 (1991). URL http://www.sciencedirect.com/science/article/B6WHY-4DD1SM0-DY/2/a893da10fb545d7222cb0f670dc0d21c

219. Karniadakis, G., Sherwin, S.: Spectral/hp Element Methods for Computational Fluid Dynamics. Oxford University Press, New York (2005)

220. Kelly, J., Emmert, J., Eckermann, S., Giraldo, F.X., Reinecke, P., Viner, K.: Development of a ground-to-thermosphere general circulation model based on NEPTUNE: idealized test cases. In: Proceedings of AGU 2019, Abstract SA43B-3211. AGU, San Francisco, California USA (2019)

221. Kelly, J.F., Giraldo, F.X.: Continuous and discontinuous Galerkin methods for a scalable three-dimensional nonhydrostatic atmospheric model: limited-area mode. Journal of Computational Physics **231**(24), 7988–8008 (2012). DOI {10.1016/j.jcp.2012.04.042}

222. Kennedy, C., Carpenter, M.: Additive Runge-Kutta schemes for convection-diffusion-reaction equations. Applied Numerical Mathematics **44**(3), 139–181 (2003)

223. Kennedy, C., Carpenter, M., Lewis, R.: Low-storage, explicit Runge–Kutta schemes for the compressible Navier–Stokes equations. Applied Numerical Mathematics **35**(3), 177–219 (2000)

224. Kim, Y.J., Giraldo, F.X., Flatau, M., Liou, C.S., Peng, M.S.: A sensitivity study of the Kelvin wave and the Madden-Julian Oscillation in aquaplanet simulations by the Naval Research Laboratory Spectral Element Atmospheric Model. Journal of Geophysical Research-Atmospheres **113**(D20) (2008). DOI {10.1029/2008JD009887}

225. Klemp, J., Skamarock, W., Dudhia, J.: Conservative split-explicit time integration methods for the compressible nonhydrostatic equations. Monthly Weather Review **135**(8), 2897–2913 (2007)

226. Kloeckner, A., Warburton, T., Hesthaven, J.S.: Viscous shock capturing in a time-explicit discontinuous Galerkin method. Mathematical Modelling of Natural Phenomena **6**(3), 57–83 (2011). DOI {10.1051/mmnp/20116303}

227. Knoll, D., Keyes, D.: Jacobian-free Newton-Krylov methods: a survey of approaches and applications. Journal of Computational Physics **193**(2), 357–397 (2004). DOI {10.1016/j.jcp.2003.08.010}

228. Knoth, O., Wolke, R.: Implicit-explicit Runge-Kutta methods for computing atmospheric reactive flows. Applied numerical mathematics **28**(2-4), 327–341 (1998)

229. Komatitsch, D., Martin, R., Tromp, J., Taylor, M., Wingate, B.: Wave propagation in 2-D elastic media using a spectral element method with triangles and quadrangles. Journal of Computational Acoustics **9**(2), 703–718 (2001)

230. Kopera, M.A., Giraldo, F.X.: Analysis of adaptive mesh refinement for IMEX discontinuous Galerkin solutions of the compressible Euler equations with application to atmospheric simulations. Journal of Computational Physics **275**, 92–117 (2014). DOI {10.1016/j.jcp.2014.06.026}

231. Kopera, M.A., Giraldo, F.X.: Mass conservation of the unified continuous and discontinuous element-based Galerkin methods on dynamically adaptive grids with application to

atmospheric simulations. Journal of Computational Physics **297**, 90–103 (2015). DOI {10.1016/j.jcp.2015.05.010}

232. Kopriva, D.: A spectral multidomain method for the solution of hyperbolic systems. Applied Numerical Mathematics **2**(3-5), 221–241 (1986). DOI {10.1016/0168-9274(86)90030-9}

233. Kopriva, D.: Multidomain spectral solution of compressible viscous flows. Journal of Computational Physics **115**(1), 184–199 (1994). DOI {10.1006/jcph.1994.1186}

234. Kopriva, D.: Spectral solution of the viscous blunt-body problem. 2. Multidomain approximation. AIAA Journal **34**(3), 560–564 (1996). DOI {10.2514/3.13104}

235. Kopriva, D.: Metric identities and the discontinuous spectral element method on curvilinear meshes. Journal of Scientific Computing **26**(3), 301–327 (2006). DOI {10.1007/s10915-005-9070-8}

236. Kopriva, D.: Implementing Spectral Methods for Partial Differential Equations. Springer, New York (2009)

237. Kopriva, D., Woodruff, S., Hussaini, M.: Discontinuous spectral element approximation of Maxwell's equations. In: Cockburn, B and Karniadakis, GE and Shu, CW (ed.) Discontinuous Galerkin Methods: Theory, Computation and Applications, *Lecture Notes in Computational Science and Engineering*, vol. 11, pp. 355–361. Natl Sci Fdn; US DOE; USA, Army Res Off, Springer-Verlag Berlin, Heidelberger Platz 3, D-14197 Berlin, Germany (2000). 1st International Symposium on Discontinuous Galerkin Methods, Newport, RI, May 24-26, 1999

238. Kopriva, D.A.: A conservatice staggered-grid Chebyshev multidomain method for compressible flows. II: A semi-stuctured method. Tech. Rep. 2, NASA Contractor Report (1996). DOI 10.1006/jcph.1996.0225. URL http://linkinghub.elsevier.com/retrieve/pii/S0021999196902259

239. Kopriva, D.A.: A staggered-grid multidomain spectral method for the compressible Navier-Stokes equations. Journal of Computational Physics **143**(1), 125 – 158 (1998). DOI http://dx.doi.org/10.1006/jcph.1998.5956. URL http://www.sciencedirect.com/science/article/pii/S0021999198959563

240. Kopriva, D.A., Gassner, G.J.: An energy stable discontinuous Galerkin spectral element discretization for variable coefficient advection problems. SIAM Journal on Scientific Computing **36**(4), A2076–A2099 (2014). DOI {10.1137/130928650}

241. Kopriva, D.A., Gassner, G.J.: Geometry effects in nodal discontinuous Galerkin methods on curved elements that are provably stable. applied Mathematics and Computation **272**(2), 274–290 (2016). DOI {10.1016/j.amc.2015.08.047}

242. Kopriva, D.A., Winters, A.R., Bohm, M., Gassner, G.J.: A provably stable discontinuous Galerkin spectral element approximation for moving hexahedral meshes. Computers & Fluids **139**(SI), 148–160 (2016). DOI {10.1016/j.compfluid.2016.05.023}

243. Kopriva, D.A., Woodruff, S.L., Hussaini, M.Y.: Computation of electromagnetic scattering with a non-conforming discontinuous spectral element method. International Journal for Numerical Methods in Engineering **53**(1), 105–122 (2002). DOI {10.1002/nme.394}. 1st International Conference on p and hp Finite Element Methods: Mathematics and Engineering Practice (p-FEM2000), Washington Univ, St Louis, Missouri, May 31-Jun 02, 2000

244. Kovenya, V.M., Yanenko, N.N.: Numerical-method for solving the viscous-gas equations on moving grids. Computers & Fluids **8**(1), 59–70 (1980). DOI {10.1016/0045-7930(80)90033-X}

245. Kozdon, J.E., Wilcox, L.C.: An energy stable approach for discretizing hyperbolic equations with nonconforming discontinuous Galerkin methods. Journal of Scientific Computing **76**(3), 1742–1784 (2018). DOI 10.1007/s10915-018-0682-1

246. Kubatko, E.J., Bunya, S., Dawson, C., Westerink, J.J.: Dynamic p-adaptive Runge-Kutta discontinuous Galerkin methods for the shallow water equations. Computer Methods in Applied Mechanics and Engineering **198**(21), 1766–1774 (2009)

247. Läuter, M., Giraldo, F.X., Handorf, D., Dethloff, K.: A discontinuous Galerkin method for the shallow water equations in spherical triangular coordinates. Journal of Computational Physics **227**(24), 10226–10242 (2008). DOI {10.1016/j.jcp.2008.08.019}

248. Lax, P., Richtmyer, R.: Survey of the stability of linear finite difference equations. Communications on Pure and Applied Mathematics **9**, 267–293 (1956)
249. Lefloch, P.: Entropy weak solutions to non-linear hyperbolic systems in non-conservative form. Comptes Rendus de L'Academie des Sciences Serie I-Mathematique **306**(4), 181–186 (1988)
250. Lefloch, P., Liu, T.: Existence theory for nonlinear hyperbolic systems in nonconservative form. Forum Mathematicum **5**(3), 261–280 (1993)
251. LeVeque, R.: Finite Volume Methods for Hyperbolic Problems, 1st edn. Cambridge University Press (2002)
252. Lilly, D.K.: On the numerical simulation of buoyant convection. Tellus **14**, 148–172 (1962)
253. Lin, S., Rood, R.: Multidimensional flux-form semi-Lagrangian transport schemes. Monthly Weather Review **124**(9), 2046–2070 (1996). DOI {10.1175/1520-0493(1996)124<2046: MFFSLT>2.0.CO;2}
254. Liu, Y., Vinokur, M., Wang, Z.: Spectral difference method for unstructured grids I: Basic formulation. Journal of Computational Physics **216**(2), 780–801 (2006). DOI {10.1016/j. jcp.2006.01.024}
255. Lohner, R.: An adaptive finite-element solver for transient problems with moving bodies. Computers & Structures **30**(1-2), 303–317 (1988). DOI {10.1016/0045-7949(88)90236-2}
256. Lohner, R., Baum, J.: Adaptive h-refinement on 3d unstructured grids for transient problems. International Journal for Numerical Methods in Fluids **14**(12), 1407–1419 (1992). DOI {10.1002/fld.1650141204}
257. Ludwig, R.A., Flaherty, J.E., Guerinoni, F., Baehmann, P.L., Shephard, M.S.: Adaptive solutions of the Euler equations using finite quadtree and octree grids. Computers & Structures **30**(1-2), 327–336 (1988). DOI {10.1016/0045-7949(88)90238-6}
258. Lunghino, B., Santiago Tate, A.F., Mazereeuw, M., Muhari, A., Giraldo, F.X., Marras, S., Suckale, J.: The protective benefits of tsunami mitigation parks and ramifications for their strategic design. Proceedings of the National Academy of Sciences **117**(20), 10740–10745 (2020). DOI 10.1073/pnas.1911857117. URL https://www.pnas.org/content/117/20/10740
259. Lyness, J., Cools, R.: A survey of numerical cubature over triangles. Proceedings of Symposia in Applied Mathematics **48**, 127–150 (1994)
260. Lyness, J., Jespersen, D.: Moderate degree symmetric quadrature rules for triangle. Journal of the Institute of Mathematics and its Applications **15**(1), 19–32 (1975)
261. Ma, X., Zhao, G.: An automated approach to quadrilateral mesh generation with complex geometric feature constraints. Engineering with Computers **31**(2), 325–345 (2015). DOI {10.1007/s00366-014-0353-2}
262. Maday, Y., Mavriplis, C., Patera, A.T.: Nonconforming mortar element methods: Application to spectral discretizations. Institute for Computer Applications in Science and Engineering, NASA Langley Research Center (1988)
263. Marras, S.: Variational multiscale stabilization of finite and spectral elements for dry and moist atmospheric problems. Ph.D. thesis, Universidad Politecnica de Cataluyna (2012)
264. Marras, S., Kelly, J.F., Giraldo, F.X., Vazquez, M.: Variational multiscale stabilization of high-order spectral elements for the advection-diffusion equation. Journal of Computational Physics **231**(21), 7187–7213 (2012). DOI {10.1016/j.jcp.2012.06.028}
265. Marras, S., Kelly, J.F., Moragues, M., Müller, A., Kopera, M.A., Vazquez, M., Giraldo, F.X., Houzeaux, G., Jorba, O.: A review of element-based galerkin methods for numerical weather prediction: finite elements, spectral elements, and discontinuous Galerkin. Archives of Computational Methods in Engineering pp. 1–50 (2015). DOI 10.1007/s11831-015-9152-1. URL http://dx.doi.org/10.1007/s11831-015-9152-1
266. Marras, S., Kopera, M.A., Constantinescu, E.M., Suckale, J., Giraldo, F.X.: A residual-based shock capturing scheme for the continuous/discontinuous spectral element solution of the 2d shallow water equations. Advances in Water Resources **114**, 45–63 (2018). DOI {10.1016/j.advwatres.2018.02.003}
267. Marras, S., Kopera, M.A., Giraldo, F.X.: Simulation of shallow-water jets with a unified element-based continuous/discontinuous Galerkin model with grid flexibility on the sphere.

Quarterly Journal of the Royal Meteorological Society **141**(690, A), 1727–1739 (2015). DOI {10.1002/qj.2474}

268. Marras, S., Nazarov, M., Giraldo, F.X.: Stabilized high-order Galerkin methods based on a parameter-free dynamic SGS model for LES. Journal of Computational Physics **301**, 77–101 (2015). DOI {10.1016/j.jcp.2015.07.034}

269. Mavriplis, C.: Nonconforming discretizations and a posteriori error estimators for adaptive spectral element techniques. Ph.D. thesis, Massachusetts Institute of Technology (1989)

270. Miller, K.: Moving finite elements II. SIAM Journal on Numerical Analysis **18**(6), 1033–1057 (1981)

271. Miller, K., Miller, R.N.: Moving finite elements I. SIAM Journal on Numerical Analysis **18**(6), 1019–1032 (1981)

272. Moeng, C.H.: A Large-Eddy simulation model for the study of planetary boundary-layer turbulence. Journal of the Atmospheric Sciences **41**, 2052–2062 (1984)

273. Moro, D., Nguyen, N., Peraire, J.: Navier-Stokes solution using hybridizable discontinuous Galerkin methods. In: 20th AIAA Computational Fluid Dynamics Conference, p. 3407 (2011)

274. Morton, K., Priestley, A., Suli, E.: Stability of the Lagrange-Galerkin method with non-exact integration. RAIRO-Mathematical Modelling and Numerical Analysis-Modelisation Mathematique et Analyse Numerique **22**(4), 625–653 (1988)

275. Mugg, P.R.: Construction and analysis of multi-rate partitioned Runge-Kutta methods. Master's thesis, Naval Postgraduate School, Monterey, CA (2012)

276. Mulder, W.: Higher-order mass-lumped finite elements for the wave equation. Journal of Computational Acoustics **9**(2), 671–680 (2001)

277. Mulder, W.A.: New triangular mass-lumped finite elements of degree six for wave propagation. Progress in Electromagnetics Research-Pier **141**, 671–692 (2013). DOI {10.2528/PIER13051308}

278. Müller, A., Behrens, J., Giraldo, F.X., Wirth, V.: Comparison between adaptive and uniform discontinuous Galerkin simulations in dry 2d bubble experiments. Journal of Computational Physics **235**, 371–393 (2013). DOI {10.1016/j.jcp.2012.10.038}

279. Müller, A., Giraldo, F.X.: Application of an adaptive discontinuous Galerkin method to the modeling of cloud convection. In: Proceedings of ECCOMAS 2010. ECCOMAS 2010, Lisbon, Portugal (2010)

280. Müller, A., Kopera, M., Marras, S., Wilcox, L., Isaac, T., Giraldo, F.X.: Strong scaling for numerical weather prediction at petascale with the atmospheric model NUMA. The International Journal of High Performance Computing Applications **0**(0), 1094342018763966 (2018). DOI 10.1177/1094342018763966. URL https://doi.org/10.1177/1094342018763966

281. Nazarov, M.: Convergence of a residual based artificial viscosity finite element method. Computers & Mathematics with Applications **65**(4), 616–626 (2013)

282. Nazarov, M., Hoffman, J.: Residual-based artificial viscosity for simulation of turbulent compressible flow using adaptive finite element methods. International Journal for Numerical Methods in Fluids **71**, 339–357 (2013)

283. Nguyen, N.C., Peraire, J., Cockburn, B.: An implicit high-order hybridizable discontinuous Galerkin method for nonlinear convection–diffusion equations. Journal of Computational Physics **228**(23), 8841–8855 (2009)

284. Nguyen, N.C., Peraire, J., Cockburn, B.: An implicit high-order hybridizable discontinuous Galerkin method for the incompressible Navier–Stokes equations. Journal of Computational Physics **230**(4), 1147–1170 (2011)

285. Nguyen, N.C., Peraire, J., Cockburn, B.: A class of embedded discontinuous Galerkin methods for computational fluid dynamics. Journal of Computational Physics **302**, 674–692 (2015). DOI {10.1016/j.jcp.2015.09.024}

286. Niegemann, J., Diehl, R., Busch, K.: Efficient low-storage Runge-Kutta schemes with optimized stability regions. Journal of Computational Physics **231**(2), 364–372 (2012). DOI {10.1016/j.jcp.2011.09.003}

287. Nikiforakis, N.: AMR for global atmospheric modelling. In: Adaptive Mesh Refinement-Theory and Applications, pp. 505–526. Springer (2005)

288. Nordström, J., Carpenter, M.H.: Boundary and interface conditions for high-order finite-difference methods applied to the Euler and Navier-Stokes equations. Journal of Computational Physics **148**(2), 621 – 645 (1999). DOI http://dx.doi.org/10.1006/jcph.1998.6133. URL http://www.sciencedirect.com/science/article/pii/S0021999198961332

289. Oden, J., Becker, E., Carey, G.: The Finite Element Method: An Introduction. Prentice Hall (1983)

290. Oden, J.T., Strouboulis, T., Devloo, P.: Adaptive finite-element methods for the analysis of inviscid compressible flow. 1. Fast refinement unrefinement and moving mesh methods for unstructured meshes. Computer Methods in Applied Mechanics and Engineering **59**(3), 327–362 (1986). DOI {10.1016/0045-7825(86)90004-6}

291. Orszag, S.: Spectral methods for problems in complex geometries. Journal of Computational Physics **37**(1), 70–92 (1980). DOI {10.1016/0021-9991(80)90005-4}

292. Pareschi, L., Russo, G.: Implicit-explicit Runge-Kutta schemes for stiff systems of differential equations. Nova Science Publishers (2000)

293. Pareschi, L., Russo, G.: Implicit-explicit Runge-Kutta schemes and applications to hyperbolic systems with relaxation. Journal of Scientific Computing **25**(1), 129–155 (2005)

294. Parsani, M., Carpenter, M.H., Fisher, T.C., Nielsen, E.J.: Entropy stable staggered grid discontinuous spectral collocation methods of any order for the compressible Navier-Stokes equations. SIAM Journal on Scientific Computing **38**(5), A3129–A3162 (2016). DOI {10.1137/15M1043510}

295. Patankar, S.: Numerical Heat Transfer and Fluid Flow. McGraw Hill (1980)

296. Patera, A.T.: A spectral method for fluid dynamics: Laminar flow in a channel expansion. Journal of Computational Physics **54**, 468–488 (1984)

297. Pavarino, L., Widlund, O.: Iterative substructuring methods for spectral element discretizations of elliptic systems - I: Compressible linear elasticity. SIAM Journal on Numerical Analysis **37**(2), 353–374 (2000)

298. Pazner, W., Persson, P.O.: Stage-parallel fully implicit Runge-Kutta solvers for discontinuous Galerkin fluid simulations. Journal of Computational Physics **335**, 700–717 (2017). DOI {10.1016/j.jcp.2017.01.050}

299. Pazner, W., Persson, P.O.: Approximate tensor-product preconditioners for very high order discontinuous Galerkin methods. Journal of Computational Physics **354**, 344–369 (2018). DOI {10.1016/j.jcp.2017.10.030}

300. Pazner, W., Persson, P.O.: On the convergence of iterative solvers for polygonal discontinuous Galerkin discretizations. Communications in Applied Mathematics and Computational Science **13**(1), 27–51 (2018). DOI {10.2140/camcos.2018.13.27}

301. Peraire, J., Nguyen, N., Cockburn, B.: A hybridizable discontinuous Galerkin method for the compressible Euler and Navier-Stokes equations. In: 48th AIAA Aerospace Sciences Meeting Including the New Horizons Forum and Aerospace Exposition, p. 363 (2010)

302. Persson, P.O., Peraire, J.: Sub-cell shock capturing for discontinuous Galerkin methods. Proc. of the 44th AIAA Aerospace Sciences Meeting and Exhibit **AIAA-2006-112** (2006)

303. Pinty, J., Benoit, R., Richard, E., Laprise, R.: Simple tests of a semi-implicit semi-Lagrangian model on 2d mountain wave problems. Monthly Weather Review **123**(10), 3042–3058 (1995)

304. Pochet, A., Celes, W., Lopes, H., Gattass, M.: A new quadtree-based approach for automatic quadrilateral mesh generation. Engineering with Computers **33**(2), 275–292 (2017). DOI {10.1007/s00366-016-0471-0}

305. Popinet, S.: Gerris: a tree-based adaptive solver for the incompressible Euler equations in complex geometries. Journal of Computational Physics **190**(2), 572–600 (2003). DOI {10.1016/S0021-9991(03)00298-5}

306. Pozrikidis, C.: Introduction to Finite and Spectral Element Methods using Matlab. CRC, New York (2005)

307. Priestley, A.: Exact projections and the Lagrange-Galerkin method - a realistic alternative to quadrature. Journal of Computational Physics **112**(2), 316–333 (1994). DOI {10.1006/jcph.1994.1104}

308. Rachowicz, W., Oden, J.T., Demkowicz, L.: Toward a universal h-p adaptive finite element strategy part 3. design of h-p meshes. Computer Methods in Applied Mechanics and Engineering **77**(1-2), 181–212 (1989)
309. Ramakrishnan, R., Bey, K.S., Thornton, E.A.: Adaptive quadrilateral and triangular finite-element scheme for compressible flows. AIAA Journal **28**(1), 51–59 (1990). DOI {10.2514/3.10352}
310. Reed, W.H., Hill, T.R.: Triangular mesh methods for the neutron transport equation. Tech. Rep. 73, Los Alamos Scientific Laboratory - LA-UR-73-479 (1973)
311. Reisner, J., Wyszogrodzki, A., Mousseau, V., Knoll, D.: An efficient physics-based preconditioner for the fully implicit solution of small-scale thermally driven atmospheric flows. Journal of Computational Physics **189**(1), 30–44 (2003). DOI {10.1016/S0021-9991(03)00198-0}
312. Remacle, J., Flaherty, J., Shephard, M.: An adaptive discontinuous Galerkin technique with an orthogonal basis applied to compressible flow problems. SIAM Review **45**(1), 53–72 (2003). DOI {10.1137/S00361445023830}
313. Restelli, M.: Semi–Lagrangian and semi–implicit discontinuous Galerkin methods for atmospheric modeling applications. Ph.D. thesis, Politecnico di Milano (2007)
314. Restelli, M., Bonaventura, L., Sacco, R.: A semi-Lagrangian discontinuous Galerkin method for scalar advection by incompressible flows. Journal of Computational Physics **216**(1), 195–215 (2006). URL http://www.sciencedirect.com/science/article/B6WHY-4J2M0SD-1/2/43d7c88a6c6e804ad6c274caabeb29df
315. Restelli, M., Giraldo, F.X.: A conservative discontinuous Galerkin semi-implicit formulation for the Navier-Stokes equations in nonhydrostatic mesoscale modeling. SIAM Journal on Scientific Computing (2009). URL http://edoc.mpg.de/433248
316. Richardson, L.: Weather Prediction by Numerical Process, 1st edn. Cambridge University Press (1922)
317. Ritchie, H., Temperton, C., Simmons, A., Hortal, M., Davies, T., Dent, D., Hamrud, M.: Implementation of the semi-Lagrangian method in a high-resolution version of the ECMWF forecast model. Monthly Weather Review **123**(2), 489–514 (1995)
318. Riviere, B.: Discontinuous Galerkin Methods For Solving Elliptic And Parabolic Equations: Theory and Implementation. Society for Industrial and Applied Mathematics, Philadelphia, PA, USA (2008)
319. Robert, A.: Bubble convection experiments with a semi-implicit formulation of the Euler equations. Journal of the Atmospheric Sciences **50**(13), 1865–1873 (1993)
320. Roe, P.: Characteristic-based schemes for the Euler equations. Annual Review of Fluid Mechanics **18**, 337–365 (1986)
321. Rosenberg, D., Fournier, A., Fischer, P., Pouquet, A.: Geophysical-astrophysical spectral-element adaptive refinement (GASpAR): Object-oriented h-adaptive fluid dynamics simulation. Journal of Computational Physics **215**, 59–80 (2006)
322. Ruuth, S.: Implicit-explicit methods for reaction-diffusion. Journal of Mathematical Biology **34**, 148–176 (1995)
323. Ruuth, S.: Global optimization of explicit strong-stability-preserving Runge-Kutta methods. Mathematics of Computation **75**(253), 183–207 (2006)
324. Ruuth, S., Hundsdorfer, W.: High-order linear multistep methods with general monotonicity and boundedness properties. Journal of Computational Physics **209**(1), 226–248 (2005). DOI http://dx.doi.org/10.1016/j.jcp.2005.02.029
325. Ruuth, S., Spiteri, R.: High-order strong-stability-preserving Runge–Kutta methods with downwind-biased spatial discretizations. SIAM Journal on Numerical Analysis **42**(3), 974–996 (2004). DOI 10.1137/S0036142902419284. URL http://link.aip.org/link/?SNA/42/974/1
326. Saad, Y.: Iterative Methods for Sparse Linear Systems. PWS Publishing, Boston (1996)
327. Sacco, R.: Upwind mixed finite element methods with hybridization for diffusion–advection–reaction problems. In: Proceedings of World Congress on Computational Mechanics 2006. Lecture at WCCM7, Los Angeles, USA (2006)

328. Saito, K., Ishida, J.I., Aranami, K., Hara, T., Segawa, T., Narita, M., Honda, Y.: Nonhydrostatic atmospheric models and operational development at JMA. Journal of the Meteorological Society of Japan **85B**, 271–304 (2007)

329. Sandu, A., Constantinescu, E.: Multirate explicit Adams methods for time integration of conservation laws. Journal of Scientific Computing **38**(2), 229–249 (2009). DOI 10.1007/s10915-008-9235-3

330. Sarrate, J., Huerta, A.: Efficient unstructured quadrilateral mesh generation. International Journal for Numerical Methods in Engineering **49**(10), 1327–1350 (2000). DOI {10.1002/1097-0207(20001210)49:10<1327::AID-NME996>3.0.CO;2-L}

331. Sarshar, A., Roberts, S., Sandu, A.: Design of high-order decoupled multirate GARK schemes. SIAM Journal on Scientific Computing **41**(2), A816–A847 (2019). DOI {10.1137/18M1182875}

332. Savcenco, V., Hundsdorfer, W., Verwer, J.G.: A multirate time stepping strategy for stiff ordinary differential equations. BIT Numerical Mathematics **47**(1), 137–155 (2007). DOI {10.1007/s10543-006-0095-7}

333. Schlegel, M., Knoth, O., Arnold, M., Wolke, R.: Multirate Runge-Kutta schemes for advection equations. Journal of Computational and Applied Mathematics **226**(2), 345–357 (2009). DOI {10.1016/j.cam.2008.08.009}

334. Schlegel, M., Knoth, O., Arnold, M., Wolke, R.: Multirate implicit-explicit time-integration schemes in atmospheric modelling. In: Psihoyios, G and Tsitouras, C (ed.) Numerical Analysis and Applied Mathematics, Vols I-III, *AIP Conference Proceedings*, vol. 1281, pp. 1831+. European Soc Comp Methods Sci & Engn, Amer Inst Physics, 2 Huntington Quadrangle, Ste 1NO1, Melville, NY 11747-4501 USA (2010). DOI {10.1063/1.3498252}. International Conference on Numerical Analysis and Applied Mathematics, Rhodes, Greece, SEP 19-25, 2010

335. Schlegel, M., Knoth, O., Arnold, M., Wolke, R.: Implementation of multirate time integration methods for air pollution modelling. Geoscientific Model Development **5**(6), 1395–1405 (2012). DOI {10.5194/gmd-5-1395-2012}

336. Schneiders, R.: Octree-based hexahedral mesh generation. International Journal of Computational Geometry & Applications **10**(04), 383–398 (2000). DOI 10.1142/S021819590000022X. URL http://www.worldscientific.com/doi/abs/10.1142/S021819590000022X

337. Seny, B., Lambrechts, J., Comblen, R., Legat, V., Remacle, J.F.: Multirate time stepping for accelerating explicit discontinuous Galerkin computations with application to geophysical flows. International Journal for Numerical Methods in Fluids **71**(1), 41–64 (2013). DOI {10.1002/fld.3646}

338. Seny, B., Lambrechts, J., Toulorge, T., Legat, V., Remacle, J.F.: An efficient parallel implementation of explicit multirate Runge-Kutta schemes for discontinuous Galerkin computations. Journal of Computational Physics **256**, 135–160 (2014). DOI {10.1016/j.jcp.2013.07.041}

339. Sert, C., Beskok, A.: Spectral element formulations on non-conforming grids: A comparative study of pointwise matching and integral projection methods. Journal of Computational Physics **211**(1), 300–325 (2006)

340. Shahbazi, K.: An explicit expression for the penalty parameter of the interior penalty method. Journal of Computational Physics **205**(2), 401–407 (2005). DOI {10.1016/j.jcp.2004.11.017}

341. Shu, C.W.: Total-variation-diminishing time discretizations. SIAM Journal on Scientific and Statistical Computing **9**(6), 1073–1084 (1988)

342. Shu, C.W.: A survey of strong stability preserving high order time discretizations. In: D. Estep, T. Tavener (eds.) Collected Lectures On The Preservation Of Stability Under Discretization, pp. 51–65. SIAM (2002)

343. Shu, C.W., Osher, S.: Efficient implementation of essentially non-oscillatory shock-capturing schemes. Journal of Computational Physics **77**(2), 439–471 (1988)

344. Skamarock, W., Doyle, J., Clark, P., Wood, N.: A standard test set for nonhydrostatic dynamical cores of nwp models. **Poster P2.17** (2004). URL http://www.mmm.ucar.edu/projects/srnwp_tests/

345. Skamarock, W., Klemp, J.: Adaptive grid refinement for 2-dimensional and 3-dimensional nonhydrostatic atmospheric flow. Monthly Weather Review **121**(3), 788–804 (1993). DOI {10.1175/1520-0493(1993)121<0788:AGRFTD>2.0.CO;2}

346. Skamarock, W., Oliger, J., Street, R.L.: Adaptive grid refinement for numerical weather prediction. Journal of Computational Physics **80**(1), 27–60 (1989)

347. Smagorinsky, J.: General circulation experiments with the primitive equations: I. the basic experiment. Monthly Weather Review **91**, 99–164 (1963)

348. Smolarkiewicz, P., Pudykiewicz, J.: A class of semi-Lagrangian approximations for fluids. Journal of the Atmospheric Sciences **49**(22), 2082–2096 (1992). DOI {10.1175/1520-0469(1992)049<2082:ACOSLA>2.0.CO;2}

349. Solin, P., Segeth, K., Dolezel, I.: Higher-Order Finite Element Methods. Chapman and Hall, New York (2004)

350. Spijker, M.: Stepsize conditions for general monotonicity in numerical initial value problems. SIAM Journal on Numerical Analysis **45**(3), 1226–1245 (2007). DOI 10.1137/060661739

351. Spiteri, R., Ruuth, S.: A new class of optimal high-order strong-stability-preserving time discretization methods. SIAM Journal on Numerical Analysis **40**(2), 469–491 (2002). DOI http://dx.doi.org/10.1137/S0036142901389025

352. St.-Cyr, A., Jablonowski, C., Dennis, J., Tufo, H., Thomas, S.: A comparison of two shallow-water models with nonconforming adaptive grids. Monthly Weather Review **136**, 1898–1922 (2008)

353. Staniforth, A., Côté, J.: Semi-Lagrangian integration schemes for atmospheric models - a review. Monthly Weather Review **119**, 2206–2223 (1991)

354. Steensland, J., Chandra, S., Parashar, M.: An application-centric characterization of domain-based SFC partitioners for parallel SAMR. IEEE Transactions on Parallel and Distributed Systems **13**(12), 1275–1289 (2002). DOI {10.1109/TPDS.2002.1158265}

355. Strand, B.: Summation by parts for finite-difference approximations for d/dx. Journal of Computational Physics **110**(1), 47–67 (1994). DOI {10.1006/jcph.1994.1005}

356. Stroud, A.H.: Approximate Calculation of Multiple Integrals. Prentice-Hall, London (1971)

357. Sun, S., Wheeler, M.F.: Symmetric and nonsymmetric discontinuous Galerkin methods for reactive transport in porous media. SIAM Journal on Numerical Analysis **43**, 195–219 (2005)

358. Sundar, H., Sampath, R.S., Biros, G.: Bottom-up construction and 2:1 balance refinement of linear octrees in parallel. SIAM Journal on Scientific Computing **30**(5), 2675–2708 (2008). DOI 10.1137/070681727. URL http://epubs.siam.org/doi/abs/10.1137/070681727

359. Svard, M., Nordström, J.: Review of summation-by-parts schemes for initial-boundary-value problems. Journal of Computational Physics **268**, 17–38 (2014). DOI {10.1016/j.jcp.2014.02.031}

360. Sweby, P.: High resolution schemes using flux-limiters for hyperbolic conservation laws. SIAM Journal on Numerical Analysis **21**, 995–1011 (1984)

361. Szabo, B., Babuska, I.: Finite Element Analysis. Wiley and Sons, Inc., New York (1991)

362. Tadmor, E.: Numerical viscosity and the entropy condition for conservative difference-schemes. Mathematics of Computation **43**(168), 369–381 (1984). DOI {10.2307/2008282}

363. Tadmor, E.: The numerical viscosity of entropy stable schemes for systems of conservation-laws. 1. Mathematics of Computation **49**(179), 91–103 (1987). DOI {10.2307/2008251}

364. Tadmor, E.: Shock capturing by the spectral viscosity method. Computer Methods in Applied Mechanics and Engineering **80**(1-3), 197–208 (1990). DOI {10.1016/0045-7825(90)90023-F}. 1st International Conf on Spectral and High Order Methods for Partial Differential Equations (ICOSAHOM 89), Como, Italy, Jun 26-29, 1989

365. Tadmor, E., Zhong, W.: Entropy stable approximations of Navier-Stokes equations with no artificial numerical viscosity. Journal of Hyperbolic Differential Equations **3**(3), 529–559 (2006). DOI {10.1142/S0219891606000896}

366. Tang, H.Z., Tang, T.: Adaptive mesh methods for one- and two-dimensional hyperbolic conservation laws. SIAM Journal on Numerical Analysis **41**(10), 487–515 (2003)

367. Tanguay, M., Robert, A., Laprise, R.: A semi-implicit semi-Lagrangian fully compressible regional forecast model. Monthly Weather Review **118**(10), 1970–1980 (1990)

368. Taylor, M., Wingate, B.: A generalized diagonal mass matrix spectral element method for non-quadrilateral elements. Applied Numerical Mathematics **33**(1-4), 259–265 (2000)

369. Taylor, M., Wingate, B., Vincent, R.: An algorithm for computing Fekete points in the triangle. SIAM Journal on Numerical Analysis **38**(5), 1707–1720 (2000)

370. Taylor, M.A., Fournier, A.: A compatible and conservative spectral element method on unstructured grids. Journal of Computational Physics **229**(17), 5879–5895 (2010)

371. Taylor, M.A., Wingate, B.A., Bos, L.P.: A cardinal function algorithm for computing multivariate quadrature points. SIAM Journal on Numerical Analysis **45**(1), 193–205 (2007). DOI {10.1137/050625801}

372. Tezduyar, T.E.: Finite element methods for flow problems with moving boundaries and interfaces. Archives of Computational Methods in Engineering **8**(2), 83–130 (2001). DOI {10.1007/BF02897870}

373. Thomas, P., Neier, K.: Navier-Stokes simulation of 3-dimensional hypersonic equilibrium flows with ablation. Journal of Spacecraft and Rockets **27**(2), 143–149 (1990). DOI {10.2514/3.26118}. 24th Thermophysics Conf of the American Inst of Aeronautics and Astronautics, Buffalo, NY, Jun 12-14, 1989

374. Thomas, S., Girard, C., Doms, G., Schattler, U.: Semi-implicit scheme for the DWD Lokal-Modell. Meteorology and Atmospheric Physics **73**(1-2), 105–125 (2000)

375. Thomas, S.J., Hacker, J.P., Smolarkiewicz, P.K.: Spectral preconditioners for nonhydrostatic atmospheric models. Monthly Weather Review **131**, 2464–2478 (2003)

376. Thompson, J.F., Soni, B.K., Weatherill, N.P. (eds.): Handbook of Grid Generation. CRC-Press (1998)

377. Thompson, J.F., Warsi, Z.U.A., Mastin, C.W.: Numerical Grid Generation: Foundations and Applications. North-Holland (1985)

378. Tomita, H., Satoh, M.: A new dynamical framework of nonhydrostatic global model using the icosahedral grid. Fluid Dynamics Research **34**, 357–400 (2004)

379. Toro, E.: Riemann Solvers and Numerical Methods for Fluid Dynamics. Springer–Verlag, Berlin, Heidelberg (1997)

380. Trefethen, L., Bau III, D.: Numerical Linear Algebra. SIAM, Philadelphia (1997)

381. Tumolo, G., Bonaventura, L.: A semi-implicit, semi-Lagrangian discontinuous Galerkin framework for adaptive numerical weather prediction. Quarterly Journal of the Royal Meteorological Society **141**(692, A), 2582–2601 (2015). DOI {10.1002/qj.2544}

382. Tumolo, G., Bonaventura, L., Restelli, M.: A semi-implicit, semi-Lagrangian, p-adaptive discontinuous Galerkin method for the shallow water equations. Journal of Computational Physics **232**(1), 46–67 (2013). DOI {10.1016/j.jcp.2012.06.006}

383. Van Albada, G., Van Leer, B., Roberts, W.: A comparative study of computational methods in cosmic gas dynamics. Astronomy and Astrophysics **108**, 76–84 (1982)

384. Van Leer, B.: Towards the ultimate conservative difference scheme II. monotonicity and conservation combined in a second order scheme. Journal of Computational Physics **14**, 361–370 (1974)

385. Van Leer, B.: Towards the ultimate conservative difference scheme III. upstream-centered finite-difference schemes for ideal compressible flow. Journal of Computational Physics **23**, 263–275 (1977)

386. Van Leer, B.: Towards the ultimate conservative difference scheme V. a second order sequel to Godunov's method. Journal of Computational Physics **32**, 101–136 (1979)

387. Vandeven, H.: Family of spectral filters for discontinuous problems. Journal of Scientific Computing **6**, 159–192 (1991)

388. Vassilevski, P.: Multi-level Block Factorization Preconditioners: Matrix-based Analysis and Algorithms for Solving Finite Element Equations, vol. 0. Springer (2008)

389. Vater, S., Klein, R., Knio, O.: A scale-selective multilevel method for long-wave linear acoustics. Acta Geophysica **59**(6), 1076–1108 (2011). DOI 10.2478/s11600-011-0037-x. URL http://dx.doi.org/10.2478/s11600-011-0037-x

390. Vincent, P.E., Castonguay, P., Jameson, A.: A new class of high-order energy stable flux reconstruction schemes. Journal of Scientific Computing **47**(1), 50–72 (2011). DOI {10. 1007/s10915-010-9420-z}

391. Vreman, A.: An eddy-viscosity subgrid-scale model for turbulent shear flow: algebraic theory and applications. Physics of Fluids **16**(10), 3670–3681 (2004). DOI {10.1063/1.1785131}
392. Wandzura, S., Xiao, H.: Symmetric quadrature rules on a triangle. Computers & Mathematics with Applications **45**(12), 1829–1840 (2003)
393. Wang, Z.: Spectral (finite) volume method for conservation laws on unstructured grids - basic formulation. Journal of Computational Physics **178**(1), 210–251 (2002). DOI {10.1006/jcph.2002.7041}
394. Wang, Z., Liu, Y.: Spectral (finite) volume method for conservation laws on unstructured grids II. Extension to two-dimensional scalar equation. Journal of Computational Physics **179**(2), 665–697 (2002). DOI {10.1006/jcph.2002.7082}
395. Wang, Z.J., Liu, Y., May, G., Jameson, A.: Spectral difference method for unstructured grids II: extension to the Euler equations. Journal of Scientific Computing **32**(1), 45–71 (2007). DOI {10.1007/s10915-006-9113-9}
396. Warburton, T.: Application of the discontinuous Galerkin method to Maxwell's equations using unstructured polymorphic hp-finite elements. In: Cockburn, B and Karniadakis, GE and Shu, CW (ed.) Discontinuous Galerkin Methods: Theory, Computation and Applications, *Lecture Notes in Computational Science and Engineering*, vol. 11, pp. 451–458. Natl Sci Fdn; US DOE; USA, Army Res Off, Springer-Verlag Berlin, Heidelberger Platz 3, D-14197 Berlin, Germany (2000). 1st International Symposium on Discontinuous Galerkin Methods, Newport, RI, May 24-26, 1999
397. Warburton, T.: An explicit construction of interpolation nodes on the simplex. Journal of Engineering Mathematics **56**(3), 247–262 (2006)
398. Warburton, T., Lomtev, I., Du, Y., Sherwin, S., Karniadakis, G.: Galerkin and discontinuous Galerkin spectral/hp methods. Computer Methods in Applied Mechanics and Engineering **175**(3-4), 343–359 (1999). DOI {10.1016/S0045-7825(98)00360-0}
399. Warburton, T., Pavarino, L., Hesthaven, J.: A pseudo-spectral scheme for the incompressible Navier-Stokes equations using unstructured nodal elements. Journal of Computational Physics **164**(1), 1–21 (2000). DOI {10.1006/jcph.2000.6587}
400. Warburton, T., Sherwin, S., Karniadakis, G.: Basis functions for triangular and quadrilateral high-order elements. SIAM Journal on Scientific Computing **20**(5), 1671–1695 (1999). DOI {10.1137/S1064827597315716}
401. Wathen, A.J., Baines, M.J.: On the structure of the moving finite-element equations. IMA Journal of Numerical Analysis **5**(2), 161–182 (1985). DOI {10.1093/imanum/5.2.161}
402. Weinzierl, T., Mehl, M.: Peano-A traversal and storage scheme for octree-like adaptive Cartesian multiscale grids. SIAM Journal on Scientific Computing **33**(5), 2732–2760 (2011)
403. Wensch, J., Knoth, O., Galant, A.: Multirate infinitesimal step methods for atmospheric flow simulation. BIT Numerical Mathematics **49**(2), 449–473 (2009). DOI {10.1007/s10543-009-0222-3}
404. Wheeler, M.F.: An elliptic collocation-finite element method with interior penalties. SIAM Journal on Numerical Analysis **15**, 152–161 (1978)
405. Williamson, D., Rasch, P.: Two-dimensional semi-Lagrangian transport with shape-preserving interpolation. Monthly Weather Review **117**(1), 102–129 (1989). DOI {10.1175/1520-0493(1989)117<0102:TDSLTW>2.0.CO;2}
406. Wintermeyer, N., Winters, A.R., Gassner, G.J., Kopriva, D.A.: An entropy stable nodal discontinuous Galerkin method for the two dimensional shallow water equations on unstructured curvilinear meshes with discontinuous bathymetry. Journal of Computational Physics **340**, 200–242 (2017). DOI {10.1016/j.jcp.2017.03.036}
407. Wintermeyer, N., Winters, A.R., Gassner, G.J., Warburton, T.: An entropy stable discontinuous Galerkin method for the shallow water equations on curvilinear meshes with wet/dry fronts accelerated by GPUs. Journal of Computational Physics **375**, 447–480 (2018). DOI {10.1016/j.jcp.2018.08.038}
408. Xing, Y., Zhang, X., Shu, C.W.: Positivity-preserving high order well-balanced discontinuous Galerkin methods for the shallow water equations. Advances in Water Resources **33**(12), 1476–1493 (2010). DOI {10.1016/j.advwatres.2010.08.005}

409. Xiu, D., Kerniadakis, G.: A semi-Lagrangian high-order method for the Navier-Stokes equations. Journal of Computational Physics **172**, 658–684 (2001)
410. Xue, M., Droegemeier, K., Wong, V.: The advanced regional prediction system (ARPS) - a multi-scale nonhydrostatic atmospheric simulation and prediction model. Part I: model dynamics and verification. Meteorology and Atmospheric Physics **75**(3-4), 161–193 (2000)
411. Yakovlev, S., Moxey, D., Kirby, R.M., Sherwin, S.J.: To CG or to HDG: a comparative study in 3d. Journal of Scientific Computing **67**(1), 192–220 (2016)
412. Yanenko, N.N., Kroshko, E.A., Liseikin, V.V., Fomin, V.M., Shapeev, V.P., Shitov, Y.A.: Methods for the construction of moving grids for problems of fluid dynamics with big deformations. Lecture Notes in Physics **59**(1) (1976)
413. Yeh, K., Cote, J., Gravel, S., Methot, A., Patoine, A., Roch, M., Staniforth, A.: The CMC-MRB global environmental multiscale (GEM) model. Part III: nonhydrostatic formulation. Monthly Weather Review **130**(2), 339–356 (2002)
414. Yelash, L., Müller, A., Lukacova-Medvid'ova, M., Giraldo, F.X., Wirth, V.: Adaptive discontinuous evolution Galerkin method for dry atmospheric flow. Journal of Computational Physics **268**, 106–133 (2014). DOI {10.1016/j.jcp.2014.02.034}
415. Yerry, M.A., Shephard, M.S.: A modified quadtree approach to finite-element mesh generation. IEEE Computer Graphics and Applications **3**(1), 39–46 (1983). DOI {10.1109/MCG.1983.262997}
416. Yi, T.H., Giraldo, F.X.: Vertical discretization for a nonhydrostatic atmospheric model based on high-order spectral elements. Monthly Weather Review **148**(1), 415–436 (2020). DOI {10.1175/MWR-D-18-0283.1}
417. Yu, M.L., Giraldo, F.X., Peng, M., Wang, Z.J.: Localized artificial viscosity stabilization of discontinuous Galerkin methods for nonhydrostatic mesoscale atmospheric modeling. Monthly Weather Review **143**(12), 4823–4845 (2015). DOI {10.1175/MWR-D-15-0134.1}
418. Zachmanoglou, E., Thou, D.: Introduction to Partial Differential Equations with Applications. Dover (1986)
419. Zauderer, E.: Partial Differential Equations of Applied Mathematics. Wiley and Sons (1986)
420. Zhang, X., Shu, C.W.: On maximum-principle-satisfying high order schemes for scalar conservation laws. Journal of Computational Physics **229**(9), 3091–3120 (2010). DOI {10.1016/j.jcp.2009.12.030}
421. Zhang, X., Shu, C.W.: On positivity-preserving high order discontinuous Galerkin schemes for compressible Euler equations on rectangular meshes. Journal of Computational Physics **229**(23), 8918–8934 (2010). DOI {10.1016/j.jcp.2010.08.016}
422. Zhu, J.Z., Zienkiewicz, O.C., Hinton, E., Wu, J.: A new approach to the development of automatic quadrilateral mesh generation. International Journal for Numerical Methods in Engineering **32**(4), 849–866 (1991). DOI {10.1002/nme.1620320411}
423. Zienkiewicz, O., Taylor, R.: The Finite Element Method: Vol.1, The Basis, 5th edn. Butterworth-Heinemann, Oxford (2000)
424. Zienkiewicz, O., Taylor, R., Nithiarasu, P.: The Finite Element Method for Fluid Dynamics, 6th edn. Elsevier (2005)

Index

© The Editor(s) (if applicable) and The Author(s), under exclusive license
to Springer Nature Switzerland AG 2020
F. X. Giraldo, *An Introduction to Element-Based Galerkin Methods on
Tensor-Product Bases*, Texts in Computational Science and Engineering 24,
https://doi.org/10.1007/978-3-030-55069-1

Editorial Policy

1. Textbooks on topics in the field of computational science and engineering will be considered. They should be written for courses in CSE education. Both graduate and undergraduate textbooks will be published in TCSE. Multidisciplinary topics and multidisciplinary teams of authors are especially welcome.

2. Format: Only works in English will be considered. For evaluation purposes, manuscripts may be submitted in print or electronic form, in the latter case, preferably as pdf- or zipped ps-files. Authors are requested to use the LaTeX style files available from Springer at: http.//www.springer.com/gp/authors-editors/book-authors-editors/resources-guidelines/ rights-permissions-licensing/manuscript-preparation/5636#c3324 (Layout & templates – LaTeX template – contributed books).
Electronic material can be included if appropriate. Please contact the publisher.

3. Those considering a book which might be suitable for the series are strongly advised to contact the publisher or the series editors at an early stage.

General Remarks

Careful preparation of manuscripts will help keep production time short and ensure a satisfactory appearance of the finished book.

The following terms and conditions hold:

Regarding free copies and royalties, the standard terms for Springer mathematics textbooks hold. Please write to martin.peters@springer.com for details.

Authors are entitled to purchase further copies of their book and other Springer books for their personal use, at a discount of 33.3% directly from Springer-Verlag.

Series Editors

Timothy J. Barth
NASA Ames Research Center
NAS Division
Moffett Field, CA 94035, USA
barth@nas.nasa.gov

Michael Griebel
Institut für Numerische Simulation
der Universität Bonn
Wegelerstr. 6
53115 Bonn, Germany
griebel@ins.uni-bonn.de

David E. Keyes
Mathematical and Computer Sciences
and Engineering
King Abdullah University of Science
and Technology
P.O. Box 55455
Jeddah 21534, Saudi Arabia
david.keyes@kaust.edu.sa

and

Department of Applied Physics
and Applied Mathematics
Columbia University
500 W. 120 th Street
New York, NY 10027, USA
kd2112@columbia.edu

Risto M. Nieminen
Department of Applied Physics
Aalto University School of Science
and Technology
00076 Aalto, Finland
risto.nieminen@tkk.fi

Dirk Roose
Department of Computer Science
Katholieke Universiteit Leuven
Celestijnenlaan 200A
3001 Leuven-Heverlee, Belgium
dirk.roose@cs.kuleuven.be

Tamar Schlick
Department of Chemistry
and Courant Institute
of Mathematical Sciences
New York University
251 Mercer Street
New York, NY 10012, USA
schlick@nyu.edu

Editor for Computational Science
and Engineering at Springer:
Martin Peters
Springer-Verlag
Mathematics Editorial
Tiergartenstrasse 17
69121 Heidelberg, Germany
martin.peters@springer.com

Texts in Computational Science and Engineering

For further information on these books please have a look at our mathematics catalogue at the following URL: www.springer.com/series/5151

Monographs in Computational Science and Engineering

1. J. Sundnes, G.T. Lines, X. Cai, B.F. Nielsen, K.-A. Mardal, A. Tveito, *Computing the Electrical Activity in the Heart.*

For further information on this book, please have a look at our mathematics catalogue at the following URL: www.springer.com/series/7417

Lecture Notes in Computational Science and Engineering

1. D. Funaro, *Spectral Elements for Transport-Dominated Equations.*

2. H.P. Langtangen, *Computational Partial Differential Equations.* Numerical Methods and Diffpack Programming.

3. W. Hackbusch, G. Wittum (eds.), *Multigrid Methods V.*

4. P. Deuflhard, J. Hermans, B. Leimkuhler, A.E. Mark, S. Reich, R.D. Skeel (eds.), *Computational Molecular Dynamics: Challenges, Methods, Ideas.*

5. D. Kröner, M. Ohlberger, C. Rohde (eds.), *An Introduction to Recent Developments in Theory and Numerics for Conservation Laws.*

6. S. Turek, *Efficient Solvers for Incompressible Flow Problems.* An Algorithmic and Computational Approach.

7. R. von Schwerin, *Multi Body System SIMulation.* Numerical Methods, Algorithms, and Software.

8. H.-J. Bungartz, F. Durst, C. Zenger (eds.), *High Performance Scientific and Engineering Computing.*

9. T.J. Barth, H. Deconinck (eds.), *High-Order Methods for Computational Physics.*

10. H.P. Langtangen, A.M. Bruaset, E. Quak (eds.), *Advances in Software Tools for Scientific Computing.*

11. B. Cockburn, G.E. Karniadakis, C.-W. Shu (eds.), *Discontinuous Galerkin Methods.* Theory, Computation and Applications.

12. U. van Rienen, *Numerical Methods in Computational Electrodynamics.* Linear Systems in Practical Applications.

13. B. Engquist, L. Johnsson, M. Hammill, F. Short (eds.), *Simulation and Visualization on the Grid.*

14. E. Dick, K. Riemslagh, J. Vierendeels (eds.), *Multigrid Methods VI.*

15. A. Frommer, T. Lippert, B. Medeke, K. Schilling (eds.), *Numerical Challenges in Lattice Quantum Chromodynamics.*

16. J. Lang, *Adaptive Multilevel Solution of Nonlinear Parabolic PDE Systems.* Theory, Algorithm, and Applications.

17. B.I. Wohlmuth, *Discretization Methods and Iterative Solvers Based on Domain Decomposition.*

18. U. van Rienen, M. Günther, D. Hecht (eds.), *Scientific Computing in Electrical Engineering.*

19. I. Babuška, P.G. Ciarlet, T. Miyoshi (eds.), *Mathematical Modeling and Numerical Simulation in Continuum Mechanics.*

20. T.J. Barth, T. Chan, R. Haimes (eds.), *Multiscale and Multiresolution Methods.* Theory and Applications.

21. M. Breuer, F. Durst, C. Zenger (eds.), *High Performance Scientific and Engineering Computing.*

22. K. Urban, *Wavelets in Numerical Simulation.* Problem Adapted Construction and Applications.

23. L.F. Pavarino, A. Toselli (eds.), *Recent Developments in Domain Decomposition Methods.*

24. T. Schlick, H.H. Gan (eds.), *Computational Methods for Macromolecules: Challenges and Applications.*

25. T.J. Barth, H. Deconinck (eds.), *Error Estimation and Adaptive Discretization Methods in Computational Fluid Dynamics.*

26. M. Griebel, M.A. Schweitzer (eds.), *Meshfree Methods for Partial Differential Equations.*

27. S. Müller, *Adaptive Multiscale Schemes for Conservation Laws.*

28. C. Carstensen, S. Funken, W. Hackbusch, R.H.W. Hoppe, P. Monk (eds.), *Computational Electromagnetics.*

29. M.A. Schweitzer, *A Parallel Multilevel Partition of Unity Method for Elliptic Partial Differential Equations.*

30. T. Biegler, O. Ghattas, M. Heinkenschloss, B. van Bloemen Waanders (eds.), *Large-Scale PDE-Constrained Optimization.*

31. M. Ainsworth, P. Davies, D. Duncan, P. Martin, B. Rynne (eds.), *Topics in Computational Wave Propagation.* Direct and Inverse Problems.

32. H. Emmerich, B. Nestler, M. Schreckenberg (eds.), *Interface and Transport Dynamics.* Computational Modelling.

33. H.P. Langtangen, A. Tveito (eds.), *Advanced Topics in Computational Partial Differential Equations.* Numerical Methods and Diffpack Programming.

34. V. John, *Large Eddy Simulation of Turbulent Incompressible Flows.* Analytical and Numerical Results for a Class of LES Models.

35. E. Bänsch (ed.), *Challenges in Scientific Computing - CISC 2002.*

36. B.N. Khoromskij, G. Wittum, *Numerical Solution of Elliptic Differential Equations by Reduction to the Interface.*

37. A. Iske, *Multiresolution Methods in Scattered Data Modelling.*

38. S.-I. Niculescu, K. Gu (eds.), *Advances in Time-Delay Systems.*

39. S. Attinger, P. Koumoutsakos (eds.), *Multiscale Modelling and Simulation.*

40. R. Kornhuber, R. Hoppe, J. Périaux, O. Pironneau, O. Wildlund, J. Xu (eds.), *Domain Decomposition Methods in Science and Engineering.*

41. T. Plewa, T. Linde, V.G. Weirs (eds.), *Adaptive Mesh Refinement – Theory and Applications.*

42. A. Schmidt, K.G. Siebert, *Design of Adaptive Finite Element Software.* The Finite Element Toolbox ALBERTA.

43. M. Griebel, M.A. Schweitzer (eds.), *Meshfree Methods for Partial Differential Equations II.*

44. B. Engquist, P. Lötstedt, O. Runborg (eds.), *Multiscale Methods in Science and Engineering.*

45. P. Benner, V. Mehrmann, D.C. Sorensen (eds.), *Dimension Reduction of Large-Scale Systems.*

46. D. Kressner, *Numerical Methods for General and Structured Eigenvalue Problems.*

47. A. Boriçi, A. Frommer, B. Joó, A. Kennedy, B. Pendleton (eds.), *QCD and Numerical Analysis III.*

48. F. Graziani (ed.), *Computational Methods in Transport.*

49. B. Leimkuhler, C. Chipot, R. Elber, A. Laaksonen, A. Mark, T. Schlick, C. Schütte, R. Skeel (eds.), *New Algorithms for Macromolecular Simulation.*

50. M. Bücker, G. Corliss, P. Hovland, U. Naumann, B. Norris (eds.), *Automatic Differentiation: Applications, Theory, and Implementations.*

51. A.M. Bruaset, A. Tveito (eds.), *Numerical Solution of Partial Differential Equations on Parallel Computers.*

52. K.H. Hoffmann, A. Meyer (eds.), *Parallel Algorithms and Cluster Computing.*

53. H.-J. Bungartz, M. Schäfer (eds.), *Fluid-Structure Interaction.*

54. J. Behrens, *Adaptive Atmospheric Modeling.*

55. O. Widlund, D. Keyes (eds.), *Domain Decomposition Methods in Science and Engineering XVI.*

56. S. Kassinos, C. Langer, G. Iaccarino, P. Moin (eds.), *Complex Effects in Large Eddy Simulations.*

57. M. Griebel, M.A Schweitzer (eds.), *Meshfree Methods for Partial Differential Equations III.*

58. A.N. Gorban, B. Kégl, D.C. Wunsch, A. Zinovyev (eds.), *Principal Manifolds for Data Visualization and Dimension Reduction.*

59. H. Ammari (ed.), *Modeling and Computations in Electromagnetics: A Volume Dedicated to Jean-Claude Nédélec.*

60. U. Langer, M. Discacciati, D. Keyes, O. Widlund, W. Zulehner (eds.), *Domain Decomposition Methods in Science and Engineering XVII.*

61. T. Mathew, *Domain Decomposition Methods for the Numerical Solution of Partial Differential Equations.*

62. F. Graziani (ed.), *Computational Methods in Transport: Verification and Validation.*

63. M. Bebendorf, *Hierarchical Matrices.* A Means to Efficiently Solve Elliptic Boundary Value Problems.

64. C.H. Bischof, H.M. Bücker, P. Hovland, U. Naumann, J. Utke (eds.), *Advances in Automatic Differentiation.*

65. M. Griebel, M.A. Schweitzer (eds.), *Meshfree Methods for Partial Differential Equations IV.*

66. B. Engquist, P. Lötstedt, O. Runborg (eds.), *Multiscale Modeling and Simulation in Science.*

67. I.H. Tuncer, Ü. Gülcat, D.R. Emerson, K. Matsuno (eds.), *Parallel Computational Fluid Dynamics 2007.*

68. S. Yip, T. Diaz de la Rubia (eds.), *Scientific Modeling and Simulations.*

69. A. Hegarty, N. Kopteva, E. O'Riordan, M. Stynes (eds.), *BAIL 2008 – Boundary and Interior Layers.*

70. M. Bercovier, M.J. Gander, R. Kornhuber, O. Widlund (eds.), *Domain Decomposition Methods in Science and Engineering XVIII.*

71. B. Koren, C. Vuik (eds.), *Advanced Computational Methods in Science and Engineering.*

72. M. Peters (ed.), *Computational Fluid Dynamics for Sport Simulation.*

73. H.-J. Bungartz, M. Mehl, M. Schäfer (eds.), *Fluid Structure Interaction II - Modelling, Simulation, Optimization.*

74. D. Tromeur-Dervout, G. Brenner, D.R. Emerson, J. Erhel (eds.), *Parallel Computational Fluid Dynamics 2008.*

75. A.N. Gorban, D. Roose (eds.), *Coping with Complexity: Model Reduction and Data Analysis.*

76. J.S. Hesthaven, E.M. Rønquist (eds.), *Spectral and High Order Methods for Partial Differential Equations.*

77. M. Holtz, *Sparse Grid Quadrature in High Dimensions with Applications in Finance and Insurance.*

78. Y. Huang, R. Kornhuber, O. Widlund, J. Xu (eds.), *Domain Decomposition Methods in Science and Engineering XIX.*

79. M. Griebel, M.A. Schweitzer (eds.), *Meshfree Methods for Partial Differential Equations V.*

80. P.H. Lauritzen, C. Jablonowski, M.A. Taylor, R.D. Nair (eds.), *Numerical Techniques for Global Atmospheric Models.*

81. C. Clavero, J.L. Gracia, F.J. Lisbona (eds.), *BAIL 2010 – Boundary and Interior Layers, Computational and Asymptotic Methods.*

82. B. Engquist, O. Runborg, Y.R. Tsai (eds.), *Numerical Analysis and Multiscale Computations.*

83. I.G. Graham, T.Y. Hou, O. Lakkis, R. Scheichl (eds.), *Numerical Analysis of Multiscale Problems.*

84. A. Logg, K.-A. Mardal, G. Wells (eds.), *Automated Solution of Differential Equations by the Finite Element Method.*

85. J. Blowey, M. Jensen (eds.), *Frontiers in Numerical Analysis - Durham 2010.*

86. O. Kolditz, U.-J. Gorke, H. Shao, W. Wang (eds.), *Thermo-Hydro-Mechanical-Chemical Processes in Fractured Porous Media - Benchmarks and Examples.*

87. S. Forth, P. Hovland, E. Phipps, J. Utke, A. Walther (eds.), *Recent Advances in Algorithmic Differentiation.*

88. J. Garcke, M. Griebel (eds.), *Sparse Grids and Applications.*

89. M. Griebel, M.A. Schweitzer (eds.), *Meshfree Methods for Partial Differential Equations VI.*

90. C. Pechstein, *Finite and Boundary Element Tearing and Interconnecting Solvers for Multiscale Problems.*

91. R. Bank, M. Holst, O. Widlund, J. Xu (eds.), *Domain Decomposition Methods in Science and Engineering XX.*

92. H. Bijl, D. Lucor, S. Mishra, C. Schwab (eds.), *Uncertainty Quantification in Computational Fluid Dynamics.*

93. M. Bader, H.-J. Bungartz, T. Weinzierl (eds.), *Advanced Computing.*

94. M. Ehrhardt, T. Koprucki (eds.), *Advanced Mathematical Models and Numerical Techniques for Multi-Band Effective Mass Approximations.*

95. M. Azaïez, H. El Fekih, J.S. Hesthaven (eds.), *Spectral and High Order Methods for Partial Differential Equations ICOSAHOM 2012.*

96. F. Graziani, M.P. Desjarlais, R. Redmer, S.B. Trickey (eds.), *Frontiers and Challenges in Warm Dense Matter.*

97. J. Garcke, D. Pflüger (eds.), *Sparse Grids and Applications – Munich 2012*.

98. J. Erhel, M. Gander, L. Halpern, G. Pichot, T. Sassi, O. Widlund (eds.), *Domain Decomposition Methods in Science and Engineering XXI*.

99. R. Abgrall, H. Beaugendre, P.M. Congedo, C. Dobrzynski, V. Perrier, M. Ricchiuto (eds.), *High Order Nonlinear Numerical Methods for Evolutionary PDEs - HONOM 2013*.

100. M. Griebel, M.A. Schweitzer (eds.), *Meshfree Methods for Partial Differential Equations VII*.

101. R. Hoppe (ed.), *Optimization with PDE Constraints - OPTPDE 2014*.

102. S. Dahlke, W. Dahmen, M. Griebel, W. Hackbusch, K. Ritter, R. Schneider, C. Schwab, H. Yserentant (eds.), *Extraction of Quantifiable Information from Complex Systems*.

103. A. Abdulle, S. Deparis, D. Kressner, F. Nobile, M. Picasso (eds.), *Numerical Mathematics and Advanced Applications - ENUMATH 2013*.

104. T. Dickopf, M.J. Gander, L. Halpern, R. Krause, L.F. Pavarino (eds.), *Domain Decomposition Methods in Science and Engineering XXII*.

105. M. Mehl, M. Bischoff, M. Schäfer (eds.), *Recent Trends in Computational Engineering - CE2014. Optimization, Uncertainty, Parallel Algorithms, Coupled and Complex Problems*.

106. R.M. Kirby, M. Berzins, J.S. Hesthaven (eds.), *Spectral and High Order Methods for Partial Differential Equations - ICOSAHOM'14*.

107. B. Jüttler, B. Simeon (eds.), *Isogeometric Analysis and Applications 2014*.

108. P. Knobloch (ed.), *Boundary and Interior Layers, Computational and Asymptotic Methods – BAIL 2014*.

109. J. Garcke, D. Pflüger (eds.), *Sparse Grids and Applications – Stuttgart 2014*.

110. H. P. Langtangen, *Finite Difference Computing with Exponential Decay Models*.

111. A. Tveito, G.T. Lines, *Computing Characterizations of Drugs for Ion Channels and Receptors Using Markov Models*.

112. B. Karazösen, M. Manguoğlu, M. Tezer-Sezgin, S. Göktepe, Ö. Uğur (eds.), *Numerical Mathematics and Advanced Applications - ENUMATH 2015*.

113. H.-J. Bungartz, P. Neumann, W.E. Nagel (eds.), *Software for Exascale Computing - SPPEXA 2013-2015*.

114. G.R. Barrenechea, F. Brezzi, A. Cangiani, E.H. Georgoulis (eds.), *Building Bridges: Connections and Challenges in Modern Approaches to Numerical Partial Differential Equations*.

115. M. Griebel, M.A. Schweitzer (eds.), *Meshfree Methods for Partial Differential Equations VIII*.

116. C.-O. Lee, X.-C. Cai, D.E. Keyes, H.H. Kim, A. Klawonn, E.-J. Park, O.B. Widlund (eds.), *Domain Decomposition Methods in Science and Engineering XXIII*.

117. T. Sakurai, S. Zhang, T. Imamura, Y. Yusaku, K. Yoshinobu, H. Takeo (eds.), *Eigenvalue Problems: Algorithms, Software and Applications, in Petascale Computing*. EPASA 2015, Tsukuba, Japan, September 2015.

118. T. Richter (ed.), *Fluid-structure Interactions*. Models, Analysis and Finite Elements.

119. M.L. Bittencourt, N.A. Dumont, J.S. Hesthaven (eds.), *Spectral and High Order Methods for Partial Differential Equations ICOSAHOM 2016*.

120. Z. Huang, M. Stynes, Z. Zhang (eds.), *Boundary and Interior Layers, Computational and Asymptotic Methods BAIL 2016*.

121. S.P.A. Bordas, E.N. Burman, M.G. Larson, M.A. Olshanskii (eds.), *Geometrically Unfitted Finite Element Methods and Applications*. Proceedings of the UCL Workshop 2016.

122. A. Gerisch, R. Penta, J. Lang (eds.), *Multiscale Models in Mechano and Tumor Biology*. Modeling, Homogenization, and Applications.

123. J. Garcke, D. Pflüger, C.G. Webster, G. Zhang (eds.), *Sparse Grids and Applications - Miami 2016*.

124. M. Schäfer, M. Behr, M. Mehl, B. Wohlmuth (eds.), *Recent Advances in Computational Engineering*. Proceedings of the 4th International Conference on Computational Engineering (ICCE 2017) in Darmstadt.

125. P.E. Bjørstad, S.C. Brenner, L. Halpern, R. Kornhuber, H.H. Kim, T. Rahman, O.B. Widlund (eds.), *Domain Decomposition Methods in Science and Engineering XXIV*. 24th International Conference on Domain Decomposition Methods, Svalbard, Norway, February 6–10, 2017.

126. F.A. Radu, K. Kumar, I. Berre, J.M. Nordbotten, I.S. Pop (eds.), *Numerical Mathematics and Advanced Applications – ENUMATH 2017*.

127. X. Roca, A. Loseille (eds.), *27th International Meshing Roundtable*.

128. Th. Apel, U. Langer, A. Meyer, O. Steinbach (eds.), *Advanced Finite Element Methods with Applications*. Selected Papers from the 30th Chemnitz Finite Element Symposium 2017.

129. M. Griebel, M. A. Schweitzer (eds.), *Meshfree Methods for Partial Differencial Equations IX*.

130. S. Weißer, BEM-based Finite Element *Approaches on Polytopal Meshes*.

131. V. A. Garanzha, L. Kamenski, H. Si (eds.), *Numerical Geometry, Grid Generation and Scientific Computing*. Proceedings of the 9th International Conference, NUMGRID 2018/Voronoi 150, Celebrating the 150th Anniversary of G. F. Voronoi, Moscow, Russia, December 2018.

132. E. H. van Brummelen, A. Corsini, S. Perotto, G. Rozza (eds.), *Numerical Methods for Flows*.

133. E. H. van Brummelen et al. (Eds.), *Isogeometric Analysis and Applications*.

134. Sherwin, S. J., Moxey, D., Peiro, J., Vincent, P. E., Schwab, C. (Eds.) *Spectral and High Order Methods for Partial Differential Equations ICOSAHOM 2018 - Selected Papers from the ICOSAHOM Conference, London, UK, July 9-13, 2018*.

135. Barrenechea, Gabriel R., Mackenzie, John (Eds.), *Boundary and Interior Layers, Computational and Asymptotic Methods BAIL 2018*.

136. Bungartz, H.-J., Reiz, S., Uekermann, B., Neumann, P., Nagel, W.E. (Eds.), *Software for Exascale Computing - SPPEXA 2016-2019*.

137. D'Elia, Marta, Gunzburger, Max, Rozza, Gianluigi (Eds.), *Quantification of Uncertainty: Improving Efficiency and Technology - QUIET selected contributions*.

138. Haynes, R. (et al.), *Domain Decomposition Methods in Science and Engineering XXV*.

For further information on these books please have a look at our mathematics catalogue at the following URL: www.springer.com/series/3527

Printed in the United States
by Baker & Taylor Publisher Services